Peter Böger · Ko Wakabayashi   Peroxidizing Herbicides

Springer-Verlag Berlin Heidelberg GmbH

Peter Böger · Ko Wakabayashi (Eds.)

Peroxidizing Herbicides

With 135 Figures and 42 Tables

Springer

Professor Dr. Peter Böger
University of Konstanz
Department of Plant Physiology and Biochemistry
D-78457 Konstanz
Germany

Professor Dr. Ko Wakabayashi
Tamagawa University
Department of Physiology and Biochemistry
Machida-shi, Tokyo 194-8610
Japan

Cover illustration: Structures of major Peroxidizing Herbicides

ISBN 978-3-642-63674-5

Library of Congress Cataloging-in-Publication Data
Peroxidizing herbicides / P. Böger, K. Wakabayashi, (eds.).
p. cm.
Includes bibliographical references and index.
ISBN 978-3-642-63674-5 ISBN 978-3-642-58633-0 (eBook)
DOI 10.1007/978-3-642-58633-0
1. Herbicides – Peroxidation. I. Böger, Peter. II. Wakabayashi, K. (Ko), 1938- .
SB951.4.P47 1999
632'. 95 – dc21 98-40482

Originally published by Springer-Verlag Berlin Heidelberg New York in 1999
Softcover reprint of the hardcover 1st edition 1999

Cover design: Design & Production, Heidelberg
Typesetting: Best-set Typesetter Ltd., Hong Kong
SPIN 10560531 31/3137–5 4 3 2 1 0–Printed on acid-free paper

Preface

Cereal and rice production of the world – now about 2 billion tons per year – has to be increased by 2.4% annually to cope with the minimum need of a growing population. This food demand has to be achieved by higher yields from the arable land already in use. Global land resources are about 1.4 billion ha, of which 1.2 billion ha are cultivated with major crops. Experts agree that a future substantial addition of new productive areas is unlikely. Those with a high yield potential are in use; new fields with lower output may possibly be obtained by cultivation of arid or cold areas.

Raising crop productivity is not a matter of pure speculation. In the last 20 years the yield of the major crops has roughly doubled in Western agriculture. There is still the potential for further achievements, particularly in the developing countries. In 1980 four billion people had an acreage of 3000 m^2 per capita; in 2020 it will be reduced to 1600 m^2 with a total population estimated at 8 billion. This is a fascinating challenge for all involved in plant production.

There is no alternative but to optimize global agriculture by continuous development and improvement through integrated plant production systems. This has to be based on advanced technology, appropriate education of farmers and efficient and economic food distribution. Technology includes mechanization of tillage and harvest, advanced seed breeding stocks, fertilizer, and, last but not least, chemical plant protection by herbicides, fungicides and insecticides. About 50% of the market share herbicides. Without chemical control of weeds, fungal pathogens and insects, major crops would suffer losses and yield penalties up to 40% and more. Weeds compete with the crop for space, fertilizer, water and light. Ten wild oats/m^2 in a wheat stand will reduce the yield by 10% and 50 wild oats/m^2 by 24%. Such reductions will raise prices. The luxury of supposedly healthier food by "bio-farming" without chemicals is something prosperous people of the rich nations can afford; it can be no basis for the global need.

Substantial contribution of herbicides to farming technology started in the 1950s with auxin-type compounds, nitrophenol herbicides and photosynthesis inhibitors like triazines or ureas. The use rate in those days was 2–3 kg active ingredient (a.i.)/ha and higher, which had a massive chemical impact on the environment together with leaching problems resulting in contamination of ground (and tap) water. Today the agrochemical industry, the public and legislators agree that reduction of the chemical load is a major objective of

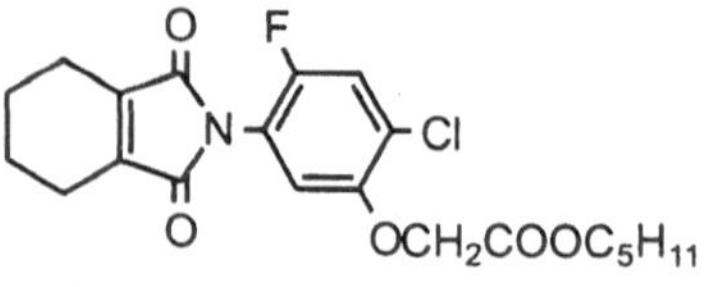

S-23031
Flumiclorac pentyl
soybean, corn; post
30-60g/ha

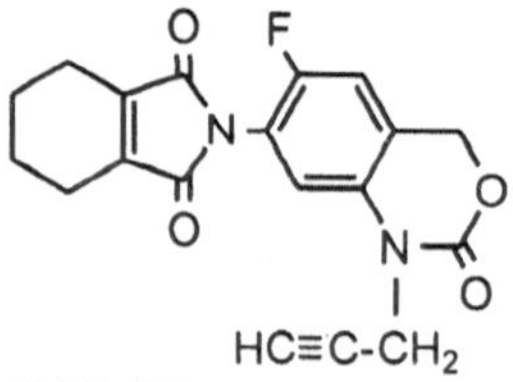

S-53482
Flumioxazin
soybean, peanuts; pre, post
50-100g/ha

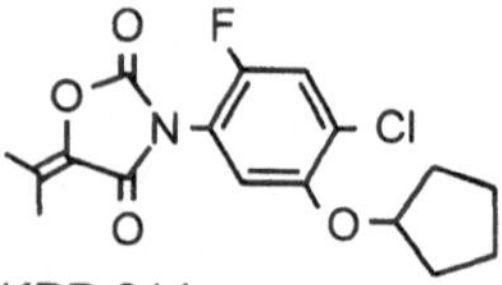

KPP-314
Pentoxazone
rice; pre
150-450g/ha

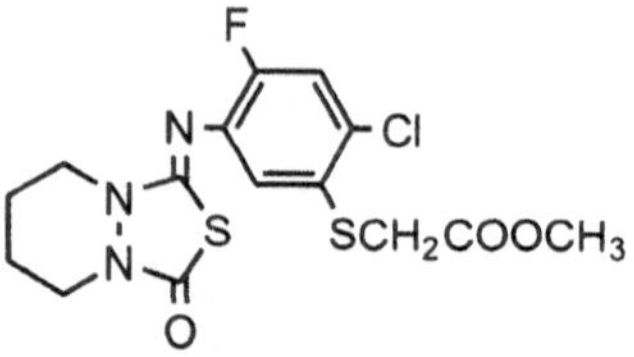

KIH-9201 or CGA-248757
Fluthiacet methyl
corn, soybean; post
3-15g/ha

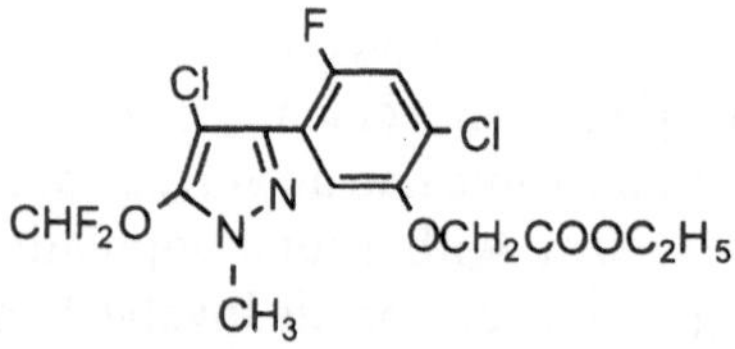

ET-751
Pyraflufen ethyl
winter wheat, barley; post
6-12g/ha

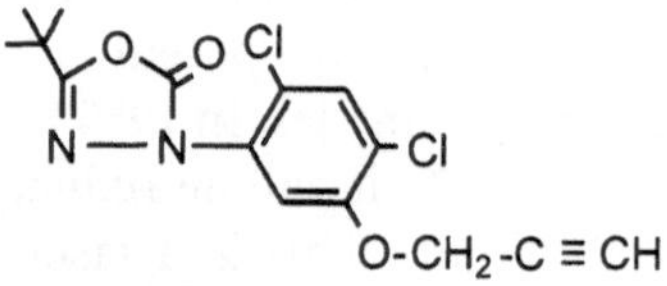

RP-020630
Oxadiargyl
rice, sugarcane, sunflower; pre
0.5kg/ha

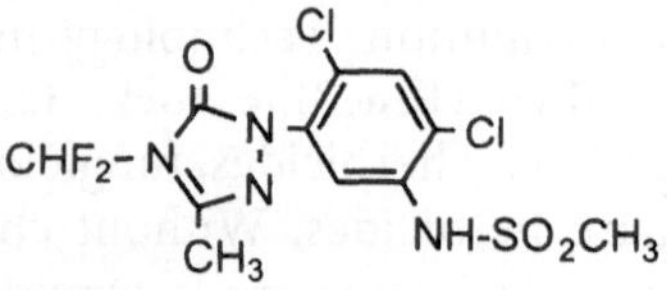

F-6285
Sulfentrazone
soybean, peanuts; pre
0.4g/ha

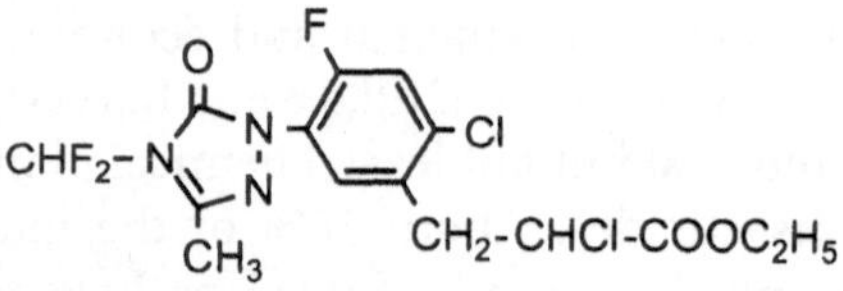

F-8426
Carfentrazone ethyl
cereals, rice; post
20-30g/ha

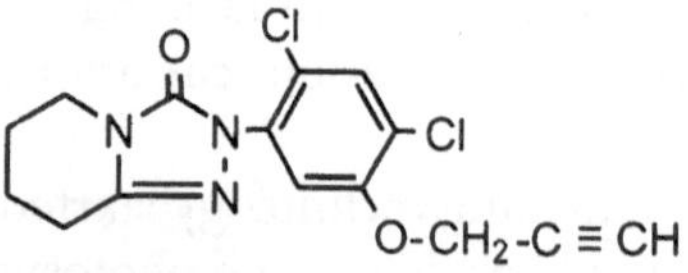

DPX-R6447
Azafenidin
vine, perennial crops, non-crops;
pre, post
100-200g/ha

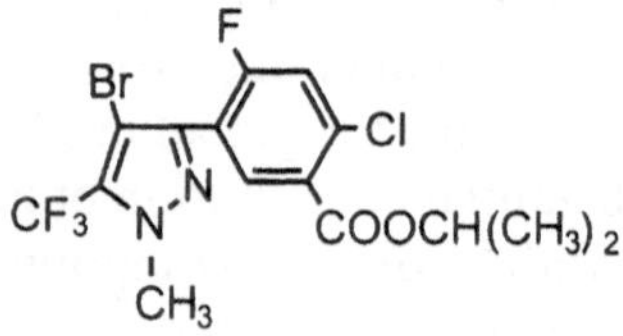

JV-485
Fluazolate
wheat; pre
175g/ha

Fig. 1. Peroxidizing herbicides in development

the development of new herbicides. At present, the use rate of modern herbicides is in the range of 100–300 g a.i./ha, with a declining tendency. Phenoxypropanoates are in the range of 100–150 g a.i./ha; acetolactate inhibitors (like sulfonylureas or imidazolinones) require an even lower amount down to 10 g a.i./ha for some commercially active ingredients. Obviously, soil overloading with chemicals or leaching problems are not an issue with such low application figures. This story of success is going on with the "peroxidizing herbicides" (peroxidizers) which are in focus of this book. Peroxidizing herbicides are inhibitors of protoporphyrinogen oxidase, a key enzyme in chlorophyll biosynthesis. They show a high affinity to their target enzyme and their radical-producing action by light activation leads to use rates down to 3 g a.i./ha, depending of course on application mode, crop and peroxidizer type.

Figure 1 demonstrates peroxidizing compounds in development as herbicides in major crops. An aryl-substituted cyclic imide, as shown in the center of Fig. 2, can be considered as the starting lead. Indeed, one of the first peroxidizers was chlorophthalim, a cyclic imide with a *p*-Cl aryl moiety. This was improved later by a 2,4,5-substitution pattern, as shown on the upper right side of Fig. 2. A fluorine in 2-position is present in most compounds. Undoubtedly the plethora of peroxidizing chemicals in the patent literature of the last 10–15 years until today was nourished by the many chemical variations possible, all yielding active substances. A few possibilities are indicated in Fig. 2. The phthalimide moiety may be modified by ring opening or by changing the imide heterocycle. Its oxygens can be replaced by S or Cl, or the number of N atoms increased. Although a heterocycle – aryl linkage via the N atom is

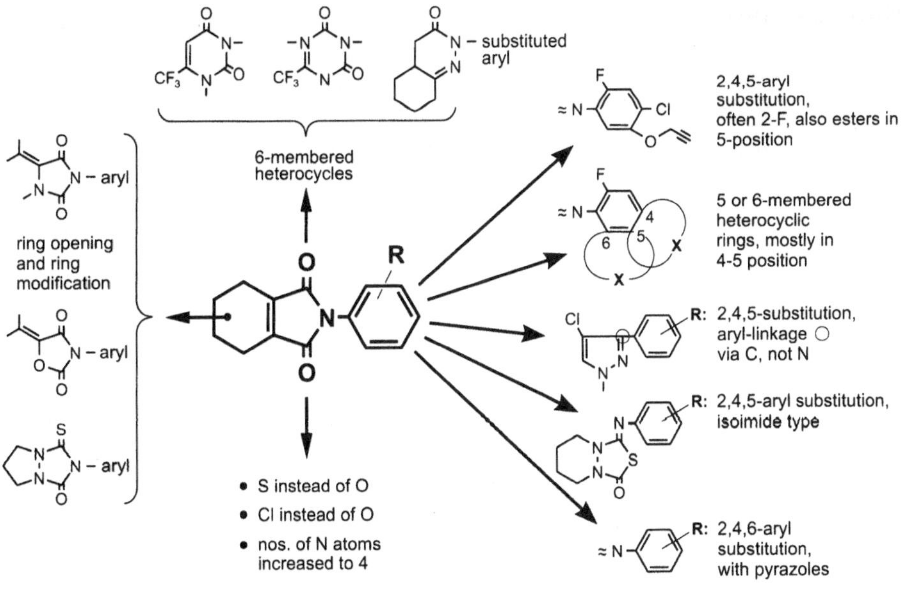

Fig. 2. Chemical modification of aryl-substituted cyclic imides

preferred, more recently a C-C bridge from aryl to heterocycle has also proved successful (see ET-751 in Fig. 1).

Peroxidizing herbicides have also raised interest in academia. Such compounds are valuable tools to study regulation of tetrapyrrole synthesis, radical production, and quenching responses of the antioxidative system of the plant cell. Basic research is performed to characterize the target enzyme and to produce it by genetic engineering. Herbicide research requires close cooperation between industry and universities at an international level. Therefore, competent authors from both agrochemical companies and basic research organizations were asked for contribution to this book. Since Japanese companies and scientists have taken the leadership in this field, the editors emphasize the Japanese input to the book.

The book covers all relevant aspects relating to preparative chemistry, structure-activity relationships, the physiological and biochemical background, mode of action, antagonizing effects to herbicidal activity and toxicological implications. The contributions should be a valuable resource for established colleagues working on plant protection and for advanced students of organic and agricultural chemistry as well as of plant biochemistry.

The editors thank the authors for their outstanding contributions and for making their expertise available. They also are grateful for the help of graduate students and technicians: Atsushi Uchida, MSc, Sagami Chemical Research Center; Ryuta Ohno, MSc, Sagami Chemical Research Center; Shinpei Ohki, MSc; Tetsuji Iida, MSc; Aiko Ohki, MSc; Yuuta Kasahara, MSc, all at Tamagawa University; Roswitha Miller, University of Konstanz.

Konstanz, Germany and Tokyo, Japan
November 1998

Peter Böger
Ko Wakabayashi

Contents

Addresses of Contributors

Aizawa, H. DuPont Agricultural Products, CPC Research Manager, 1-8-1 Shimomeguro, Meguro-ku, Tokyo, 153-0064, Japan, Tel.: +84-3-5434-6113, Fax: +81-3-5434-6187, E-mail: hiroyasu.aizawa@jpn.dupont.com

Asami, T. Plant Function Laboratory, The Institute of Physical and Chemical Research (RIKEN), 2-1 Hirosawa, Wako-shi, Saitama 351-0198, Japan, Tel.: +81-48-467-9524, Fax: +81-48-462-4674, E-mail: yshigeo@postman.riken.go.jp

Arnould, S. Laboratoire de Biochimie des Porphyrines, Départment de Microbiologie, Institut Jacques-Monod, UMR 7592 CNRS-Université Paris, 7-Université Paris 6, 2 Place Jussieu, F – Paris, Cedex 05, France, Tel.: +33-1-44 27 63 56, Fax: +33-1-44 27 57 16, E-mail: arnould@ijm.jussieu.fr

Böger, P. Faculty of Biology, University of Konstanz, D-78457 Konstanz, Germany, Tel.: +49-7531-88-2101, Fax: +49-7531-88-3042, E-mail: Peter.Boeger@uni-konstanz.de

Brown, H.M. DuPont Agricultural Products, Stine-Haskell Research Center, Newark, Delaware, USA, Tel.: +1-302-366-5554, Fax: +1-302-366-6120, E-mail: hugh.m.brown@usa.dupont.com

Camadro, J.-M. Laboratoire de Biochimie des Porphyrines, Départment de Microbiologie, Institut Jacques-Monod, UMR 7592 CNRS-Université Paris, 7-Université Paris 6, 2 Place Jussieu, F – Paris, Cedex 05, France, Tel.: +33-1-44 27 63 56, Fax: +33-1-44 27 57 16, E-mail: camadro@ijm.jussieu.fr

Dayan, F.E. USDA-ARS, NPURU, National Center for the Development of Natural Products, University of Mississippi, P.O.Box 8048, Mississippi 38677, USA, Tel.: +1-601-232-1039, Fax: +1-601-232-1035, E-mail: fdayan@ag.gov

Duke, S.O. USDA-ARS, NPURU, National Center for the Development of Natural Products, University of Mississippi, P.O.Box 8048, Mississippi 38677, USA, Tel.: +1-601-232-1039, Fax: +1-601-232-1035, E-mail: sduke@ag.gov

Fujita, T. Emil Project, 305 Heights Kyogosho, 492 Umeyacho, Kyoto 604-8057, Japan, Tel/Fax: +81-75-2563040, E-mail: PED01545@niftyserve.or.jp

Grimm, B. Institut für Pflanzengenetik und Kulturpflanzenforschung Gatersleben (IPK), Corrensstr. 3, D-06466 Gatersleben, Germany, Tel.: +49-39482 5364, Fax: +49-39482 5139, E-mail: grimm@ipk-gatersleben.de

Hirai, K. Sagami Chemical Research Center, Nishi-Ohnuma 4-1-1, Sagamihara, Kanagawa 229-0012, Japan, Tel.: 81-427-42-4791, Fax: +81-427-49-7631, E-mail: scrc12gp@alles.or.jp

Iwataki, I. Odawara Research Center, Nippon Soda Co. Ltd., 345 Takada, Odawara, Kanagawa 250-0216, Japan, Tel.: +81-465-42-4224, Fax: +81-465-42-4377, E-mail: nisso@ppp.bekkoame.or.jp

Knörzer, O. Faculty of Biology, University of Konstanz, D-78457 Konstanz, Germany, Tel.: +49-7531-88 4289, Fax: +49-7531 3042

Krijt, J. Institute of Pathophysiology, First Faculty of Medicine, Charles University, U nemocnice 5, 128 53 Prague, Czech Republic, Fax: +42-2-249-12834, E-mail: jkri@lf1.cuni.cz

Le Guen, L. Laboratoire de Biochimie des Porphyrines, Département de Microbiologie, Institut Jacques-Monod, UMR 7592 CNRS-Université Paris, 7-Université Paris 6, 2 Place Jussieu, F – Paris Cedex 05, France, Tel.: +33-1-44 27 63 56, Fax: +33-1-44 27 57 16, E-mail: leguen@ijm.jussieu.fr

Matringe, M. Laboratoire Mixte, UMR 41 CNRS-Rhône-Poulenc, BP 9163, F – 69263 Lyon Cedex 09, France, Tel.: +33-4-72 85 28 47, Fax: +33-4-72 85 22 97, E-mail: michel.matringe@rp.fr

Matsunaka, S. Kansai University, 3-3-35 Yamate-cho, Suita, Osaka 564-8680, Japan, Tel.: +81-78-982-5946, Fax: +81-78-982-5946, E-mail: matsusho@iris.dti.ne.jp

Mornet, R. Laboratoire d'Ingénierie Moléculaire et Matériaux Organiques, UMR 6501 CNRS-Université d'Angers, Faculté des Sciences, 2 Boulevard Lavoisier, F – 49045 Angers Cedex, France, Tel.: +33-2-41 73 53 77, Fax: +33-2-41 73 54 05, E-mail: Rene.Mornet@univ.angers.fr

Nagano, E. Agricultural Chemicals Research Laboratory, Sumitomo Chemical Co. Ltd., 4-2-1 Takatsukasa, Takarazuka, Hygo 665-0051, Japan, Tel.: +81-797-74-2032, Fax: +81-797-74-2129, E-mail: naganoe@sc.sumitomo-chem.co.jp

Nakayama, A. Odawara Research Center, Nippon Soda Co. Ltd., 345 Takada, Odawara, 250-0216, Japan, Tel.: +81-465-42-3511, Fax: +81-465-42-2180, E-mail: nakira@mxy.meshnet.or.jp

Reddy, K.N. USDA-ARS, Southern Weed Science Laboratory, P.O.Box 350, Stoneville, Mississippi 38776, USA, Tel.: +1-601-686-5236, Fax: +1-601-686-5422, E-mail: kreddy@ag.gov.

Santos, Renata Laboratoire de Génétique Moléculaire des Réponses Adaptatives, Département de Microbiologie, Institut Jacques-Monod, UMR 7592 CNRS-Université Paris, 7-Université Paris 6, 2 Place Jussieu, F - 75251 Paris, Cedex 05, France, Tel.: +33-1-44 27 47 19, Fax: +33-1-44 27 51 43, E-mail: santos@ijm.jussieu.fr

Sato, Y. Graduate School of Agricultural Science, Physiology and Biochemistry Division, Tamagawa University, 6-1-1 Tamagawa-gakuen, Machida-shi, Tokyo, 194-8610, Japan, Tel.: +81-427-39-8286, Fax: +81-427-39-8854, E-mail: ymsato@agr.tamagawa.ac.jp

Shimizu, T. Life Science Research Institute, Kumiai Chemical Industry Co., Ltd., 3360 Kamo, Kikukawa-cho, Ogasa-gun, Shizuoka 439-0031, Japan, Tel.: +81-537-35-3156, Fax: +81-537-36-3718, E-mail: tsutom.shimizu@nifty.ne.jp

Wakabayashi, K. Chair of Physiology & Biochemistry, Faculty of Agriculture, 6-1-1 Tamagawa University, Machida-shi, Tokyo, 194, Japan, Tel.: +81-427-39-8274, Fax: +81-427-39-8854, E-mail: kwaka@agr.tamagawa.ac.jp

Watanabe, H. Yokohama Research Center, Mitsubishi Chemical Co., 1000 Kamoshida-cho, Aoba-ku, Yokohama, Kanagawa 227-8502, Japan, Tel.: +81-45-963-3513, Fax: +81-45-963-3958, E-mail: hirohiro@rc.m-kagaku.co.jp

Yoshida, S. Plant Function Laboratory, The Institute of Physical and Chemical Research (RIKEN), 2-1 Hirosawa, Wako-shi, Saitama 351-0198, Japan, Tel.: +81-48-467-9524, Fax: +81-48-462-4674, E-mail: yshigeo@postman.riken.go.jp

CHAPTER 1

History and Early Investigation of Mode of Action of Peroxidizing Herbicides

SHOOICHI MATSUNAKA[1]

Contents

1.1 Introduction

In this chapter, the history of diphenyl ether (DPE) herbicides, which became the starting structures of present peroxidizing herbicides, and the early investigations of this herbicide group in Japan will be reviewed. First, chemical weeding in rice cultivation in the 1960s in Japan will be discussed, when pentachlorophenol (PCP) followed by phenoxy herbicides was commonly used. However, after severe fish damage by PCP, nitrofen and chloronitrofen were selected as low fish-toxicity herbicides. This was the beginning of the practical usage of DPE herbicides. There are two types of DPE herbicides. The special herbicidal activities of *ortho*-substituted DPE herbicides, especially the light requirement for the activity, will be explained in detail. Although the chemical structures are quite different from diphenyl ethers, oxadiazon, phenopylate and cyclic imides, they exhibit the same mode of action, which led to many kinds of peroxidizing herbicides. As an appendix, the larvicidal activity of DPEs on mosquitoes which seemed to contribute the control of Japanese sleeping sickness (Japanese encephalitis) mediated by mosquitoes in Japan will be discussed.

[1] Department of Biotechnology, Faculty of Engineering, Kansai University, Suita, Osaka 564-0073, Present Address: 3-10-8, Higashi-arinodai, Kita-ku, Kobe 651-1322, Japan

Peter Böger, Ko Wakabayashi
Peroxidizing Herbicides

1.2 Weeding in Rice Cultivation in the 1960s in Japan

In the latter part of the 1940s, when PCP was tested for control of a shellfish, *Katayama nosphora*, a carrier of a human pathogenic *Schistosoma japonicum* in paddy fields, by help of the Allied Occupation Forces, this compound was found to be effective for barnyardgrass control (*Echinochloa oryzicola* Vasing.), which could not be achieved by 2,4-D. The improvement of the formulation of PCP into a granular one made it safe for transplanted rice. Then, after 1962, PCP mixed with allyl ester of MCPA and MCPB, was widely used, and the former was applied to 0.5 million ha in northern Japan.

However, in 1962 after heavy rain, soil particles absorbed with PCP flooded out from paddy fields to shallow lakes or seas, causing very severe damage to fish and shellfish. This was due to the mode of action of PCP, namely the inhibition of oxidative phosphorylation which is similar in weeds and fish. Then the usage of PCP decreased suddenly. After 1977, PCP use in aquatic areas was forbidden. This situation made us recognize the importance of the fundamental research on herbicides, particularly on mode of action. Also, this accident accelerated the development of new herbicides with lower fish toxicity, because the farmers were so accustomed to using herbicides instead of the severely laborious hand weeding in the hot and muddy paddy fields.

As alternatives to PCP herbicides, in 1963 MCPCA and nitrofen were registered, and then chloronitrofen (CNP), its combination with MCPA, chlomethoxynil (chlomethoxyfen), and chloronitrofen-dymron combination followed. Nitrofen not only became the pioneer of low fish-toxicity herbicides but also opened up a new area of mode of action of herbicidal action.

The prosperity and decay of DPEs as rice herbicides in Japan are shown in Fig. 1 (Matsunaka 1987). Chloronitrofen granule application in 1974 reached 1.65 million ha, which was the maximum hectarage covered by a single herbicide formulation, but was gradually replaced by newly developed herbicides. In 1973, chlomethoxynil, whose *meta*-position at the nitro ring was substituted by a methoxy radical, was introduced into the market, and its share reached up to 0.6 million ha. Chloronitrofen-dymron combined granule application is effective for sedge, with a 250 000 ha share. Later, bifenox, another DPE, was registered in 1982 and was still applied to 54 000 ha in 1996. Registration of nitrofen was cancelled in Japan in 1982. Also, fluoronitrofen (CFNP, MO500), fluorodifen (Preforan) and DNCDE (KK60) were registered, but cancelled in 1973, 1973 and 1974, respectively. As described later (see Sect. 1.7), oxadiazon and phenopylate have no diphenyl ether structure, but the author found that their modes of action are similar to *ortho*-substituted diphenyl ethers (Matsunaka 1970; Yanai and Matsunaka 1972). Registration of phenopylate, however, was cancelled in 1977.

On the other hand, although oxadiazon granule showed some toxicity to transplanted rice, the improvement of the formulation as an emulsifiable con-

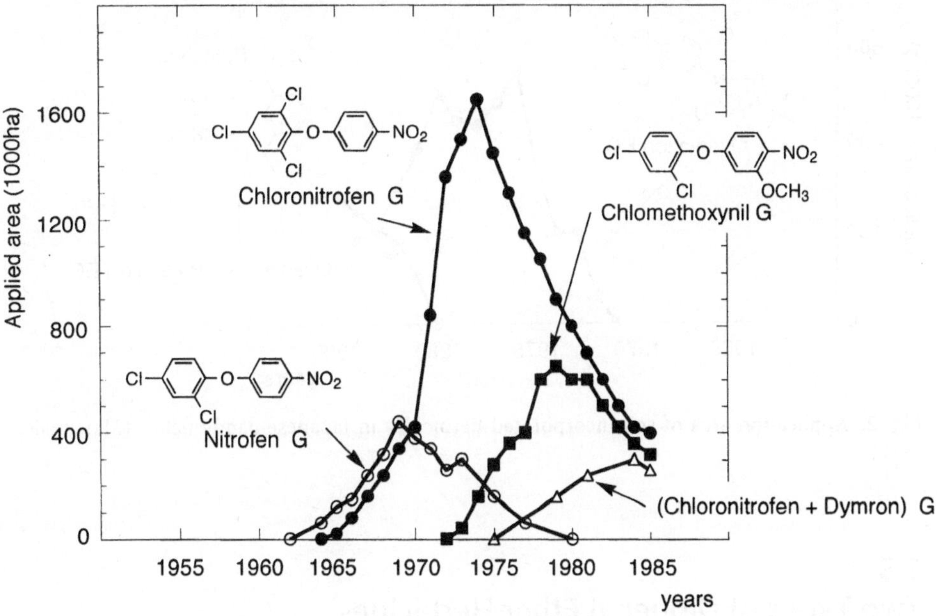

Fig. 1. Application area of DPE herbicides in Japanese paddy fields. (Matsunaka 1987)

centrate (EC) and the change of the application method made it very effective to weeds and safe to rice by application usually before transplanting, and this herbicide was registered in 1972 in Japan. Oxadiazon EC can be applied directly to the muddy water just after puddlings through three small pin holes in a fizzy drink-type bottle by hand or with a transplanting machine. The herbicide may be absorbed by suspending soil particles which will sink down to the soil surface producing a thin herbicide-soil layer. The herbicide in the layer attacks just emerged weeds very effectively, but showed almost no damage to rice plants. The share of oxadiazon was about 0.5 million ha during the latter half of the 1970s (Fig. 2; Matsunaka 1987). Since 1981 the use of oxadiazon-butachlor mixture increased, which effective also for sedges, and replaced the share of oxadiazon alone.

In the middle of the 1980s, butachor alone, pyrazolate plus butachlor, and pretilachlor alone became the big issue. Thiocarbamate herbicides, such as thiobencarb or molinate, combined with simetryn or with simetryn plus MCPB attained a larger share in rice chemical weeding. In the 1990s, sulfonylureas and other chemicals, being inhibitors of acetolactate synthesizing enzyme (ALS) of weeds, combined with other herbicides, became dominant.

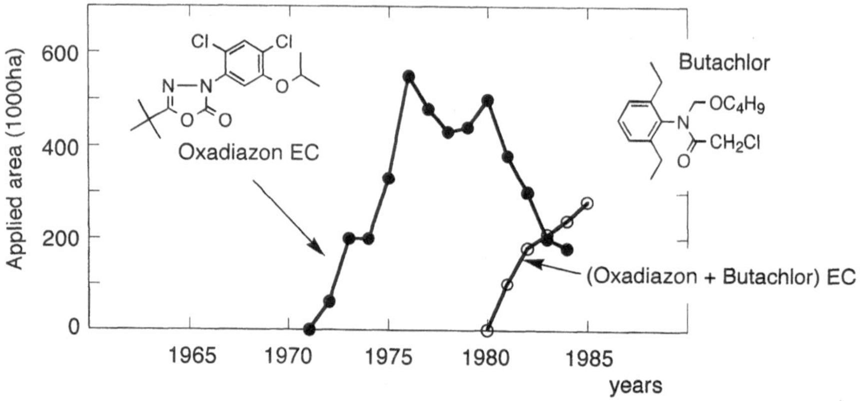

Fig. 2. Application area of soil-incorporated herbicides in Japanese paddy fields. (Matsunaka 1987)

1.3 Two Types of Diphenyl Ether Herbicides

Although the light requirement will be explained and discussed in the following section, DPEs may be classified into two groups as shown in Table 1 (Matsunaka 1976). The compounds having substituents of 2,4- or 2,4,6-position at the benzene ring are light-dependent for their herbicidal action, while 3- or 3,5-substituted ones are active even in the dark. Not only the light requirement but also the location of herbicide in the soil or mutual relationships between the locations of weed seeds and herbicides allow us to classify the diphenyl ethers into two groups, as shown in Table 1.

Nitrofen had almost no effect on buds of barnyardgrass when the herbicide was applied to the soil below the weed seeds. Low translocation of herbicide and lack of light in such soil may be the reasons for this phenomenon. Nitrofen and other *ortho*-substituted diphenyl ethers are phytotoxic only when they are applied to the soil surface above the weed seeds. Weeds absorb the chemicals during their germination and elongation through the treated soils, and after exposure to sunlight they will die.

However, three *meta*-substituted diphenyl ethers, shown in the lower part of Table 1, such as HE314, have a severe effect on barnyardgrass through the roots. When the soil beneath the seeds was treated, a 30 g/acre application of nitrofen had a slight stimulatory effect on the emergence of roots and the fresh weight of shoots of rice and barnyardgrass, while the same dosage of HE314 decreased both by 40–50% (Hayasaka 1967). Hayasaka also described that in the case of HE314 light seems not to have any influence on the action. Hisada (1969) reported similar results using DMNP (HW-40187, HE306) instead of HE314. These relationships are clearly illustrated in Fig. 3 and Table 2 (Hisada 1969).

Table 1. Light requirement for herbicidal activity in diphenyl ether herbicides and their oral acute toxicity

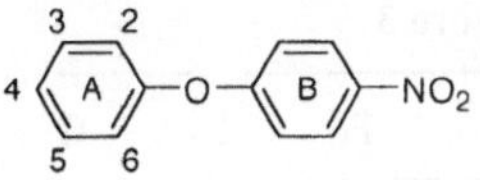

Common name or code number	2	3	4	5	6	Light requirement	Oral acute toxicity LD_{50} (mg/kg)
Nitrofen	Cl		Cl			+	3050 (R)
Chlomethoxynil[a]	Cl		Cl			+	10500 (R)
DNCDE	NO_2		Cl			+	27750 (M)
Fluorodifen	NO_2		CF_3			+	>10000 (R)
NH8902	Cl		Br			+	–
Chloronitrofen	Cl		Cl		Cl	+	10800 (R)
CFNP	Cl		Cl		F	+	2500 (M)
MO 263	Cl		Cl		CH_3	+	–
TOPE (HE314)		CH_3				–	1700 (M)
DMNP (HW-40187)		CH_3		CH_3		–	3400 (M)
MO600		Cl	F			–	–

R, rat; M, mouse.
[a] Chlomethoxynil is substituted with the methoxy group at 3′-position.

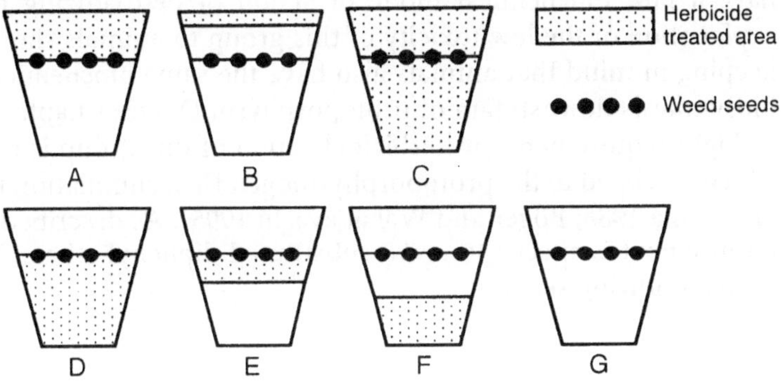

Fig. 3A–G. Location of herbicide nitrofen relative to weed seeds. **A** 3 cm below soil surface. **B** 1 cm below soil surface. **C** Whole soil treated. **D** Below sowing level. **E** Below sowing level, but bottom parts not treated. **F** Up to 3 cm above bottom. **G** Not treated. (Hisada 1969)

1.4 Properties of Herbicidal Action of *ortho*-Substituted Diphenyl Ethers

As described in the previous section, the *ortho*-substituted diphenyl ethers showed herbicidal activity only by treatment of the soil surface above the weed (Fig. 3). When a high dosage application of *ortho*-substituted DPE herbicide

Table 2. Effect of location in soil of HW-40187 (HE306,DMNP) and nitrofen on growth of millet (Hisada 1969)

	Localization shown in Figure 3						
	A	B	C	D	E	F	G
HW-40187							
Shoot	4	3	5	0	0	0	0
Root	0	0	5	5	5	2	0
Nitrofen							
Shoot	3	3	3	0	0	0	0
Root	0	0	0	0	0	0	0

Damage is graded; 5, killed or inhibited perfectly; 4, severe; 3, moderate; 2, slight; 1, weak; 0, no damage.

is administered to transplanted rice, we easily find typical brown spots on the leaf sheath. Soon the plant apparently recovers by developing newly leaf sheath.

As demonstrated in Table 1, generally, DPE herbicides show very low toxicity to animals. Their LD_{50} values of oral acute toxicity are always more than 1000 mg/kg. The low toxicity to fish is also due to this general property. Now we have a new biochemical mode of action of peroxidizing herbicides. Clear explanation of the low toxicity of this group to animals should be examined, keeping in mind that animals also have the same biochemical route of heme biosynthesis downstream of protoporphyrin IX (see Chapter 8).

Light requirement for herbicidal action of this group is a special property which is related to the protoporphyrinogen IX accumulation theory (Matringe and Scalla 1988; Böger and Wakabayashi 1995). As described above, this light requirement is special to *ortho*-substituted diphenyl ethers. Details are given in the following section.

1.5 Phenomenon of the Light-Dependent Mechanism of Herbicidal Action

Such a light requirement phenomenon was found by Arai et al. (1966). The crucial part of this paper is translated from Japanese as follows: "In the previous experiments (600 g a.i. nitrofen/10 a) we observed that young shoots elonged normally during night time and were killed after several hours under light illumination. In order to confirm it, after soaking in water the seeds of barnyardgrass were treated by 250 ppm solution of nitrofen EC at 30 °C for three days, then washed for two hours and emerged in the dark. These seeds were incubated at three conditions, under dark, under scattered light in the room, and under outdoor sunlight, respectively. After eight days, no damage was found under dark condition, some brown spots in the room light and perfect damage under sunlight. It was concluded that nitrofen showed the herbicidal activity only in light. In Petri dish with glass cover, the results were the same as those without the cover, then ultraviolet light has no effect on this activity."

As shown in Table 1, such light-dependent phytotoxic activity was found only with *ortho*-substituted diphenyl ethers, not in *meta*-substituted ones on the left benzene ring.

Matsunaka (1969) reported that with a low concentration of nitrofen (1 ppm), the herbicidal activity was effected by the intensity of illumination. On the other hand, at 5 ppm the effect of illumination was clear even at 1000 lx. Experimental data showed that the photoactivation of nitrofen is not a simple conversion to a toxic compound by light. Given together with simazine, prometryne, monuron or propanil which are well-known inhibitors of the Hill reaction in chloroplasts, nitrofen activity was not affected. This means that the activation of nitrofen by light differs from the action of bipyridilium herbicides like paraquat. Consequently, another photobiochemical activation mechanism after absorption of nitrofen into plant tissues was assumed.

In the case of photoactivation, the role of light acceptors or absorbing pigments should be mentioned. When testing the phytotoxic action of nitrofen under light conditions using normal rice seeds, the author found that a natural albino mutant of rice seedling was tolerant to nitrofen even in light. Fortunately the author could use white and yellow mutants of rice plants created by chemical treatment or radioisotopic radiation of variety Norin No 8 in the Third Laboratory of Genetics, National Institute of Agricultural Sciences, Hiratsuka at the same institute of which the author was a member at that time. In general, these mutants segregated in a ratio of three to one: three were green and one was white or yellow. White mutants may be termed albina and yellow mutants xantha. The pale green viridis mutants were also used for the experiments. Tentative line names of the mutants were as follows: viridis (pale green), CM-46 and CM-75; xantha (yellow), CM-123 and CM-213: and albina (pale yellow or true white), CM-9, CM-33, CM-37, CM-39 and CM-53.

Usually 20 slightly germinated rice seeds were set in 6-cm Petri dishes with 15 ml test solution, and incubated in a glass chamber kept at 30 °C, and illuminated by fluorescent lamps. The illumination intensity at the level of the Petri dishes was 3000 lx; after incubation for 5 days, the fresh weight of buds and roots removed from the seeds was measured. Usually the final concentration of nitrofen and ethanol was 5 ppm and 0.1%, respectively.

As shown in Figure 4, the effect of concentration of nitrofen on the growth of rice seedlings of both mutants CM-39 (white) and CM-123 (yellow) was examined. In green seedlings of both mutants, the growth in the light was almost completely inhibited even at a concentration of 2.5 ppm nitrofen. On the other hand, white mutants, CM-39, showed about 50% growth of the control even at a concentration of 20 ppm. The yellow mutants, CM-123,

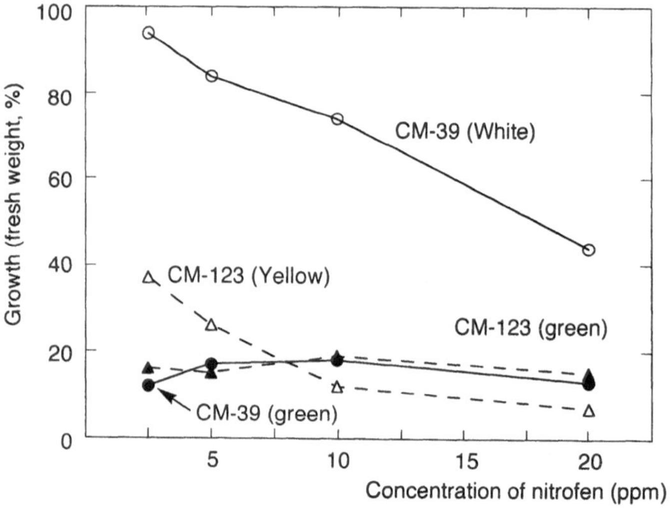

Fig. 4. Susceptibility to nitrofen of white mutant (CM-39) and yellow mutant (CM-123) of rice plants. (Matsunaka 1969)

appeared to be somewhat tolerant to these herbicides, but the difference was very low. At 10 ppm and higher, CM-123 exhibited the same susceptibility as the green mutants.

The *meta*-substituted DPE herbicides do not require light for their phytotoxic activity. In the same series of experiments as those shown in Fig. 4 the white mutant, CM-39, was found to be susceptible to HW-40187 [5-(4-nitrophenoxy)-*m*-xylene, DMNP; Matsunaka 1969]. The inhibitory rate of growth with this white mutant was over 80% and almost the same as that of its green counterpart at a concentration of 2.5 ppm HW-40187.

All other white mutants, such as CM-9, CM-33, CM-35, CM-37 and CM-53 were also tolerant to nitrofen, but their green counterparts were susceptible, as shown in Table 3. Two yellow mutants, CM-123 and CM-213, were susceptible to this herbicide regardless of having yellow or green colors. The pale green mutant, termed "viridis", was hardly distinguishable from the green plant at seedling stage. The effect of nitrofen on this mutant was examined using mixed emerged seeds in the same Petri dish. The inhibitory rate of CM-46 and CM-75 by nitrofen was the same as in the original variety, Norin No 8.

1.6 Understanding from Recent Theory of Mode of Action

DPEs or oxadiazon will inhibit protoporphyrinogen IX oxidase and chlorophyll biosynthesis (Matringe and Scalla 1988; Duke et al. 1989; Böger and Wakabayashi 1995). Yamato et al. (1990) and Yamato (1992) found that even white mutants of rice can show damage under different conditions, such as

Table 3. Susceptibility of chlorophyll mutant of rice plants to nitrofen (Matsunaka 1969) Numbers are percent reduction of growth upon addition of 5 ppm nitrofen. (cf. Fig. 5)

		Green	White			Green	Yellow
Albina	CM-9	68.3	−0.4[a]	Xantha	CM-123	41.2	37.6
	CM-33	61.0	7.8		CM-213	61.2	62.8
	CM-35	65.5	−6.0				
	CM-37	46.8	2.2	Viridis	CM-46	57.0	–
	CM-53	62.0	17.3		CM-75	64.4	–
			Original Norin No 8			63.9	

[a] Minus indicates percent increase of growth.

strong illumination, from that described in the previous section, where DPEs did not show any effect on white mutants but killed both yellow mutants and normal green rice plants. Here the contribution of a yellow pigment such as xanthophyll was assumed.

Now the discrepancy between "yellow pigment theory" (Matsunaka 1969) and new theory based on the inhibition of protoporphyrinogen IX oxidase may be explained as follows: Yamato (1992) intended to elucidate the effects of DPEs on chlorophyll mutants of rice from the new aspect of DPE mode of action. He reported that a white mutant (CM-35) showed tolerance to acifluorfen-methyl (AFM) and oxyfluorfen (OXY), which are stronger herbicide than nitrofen, under low light intensity conditions (20 $\mu E/m^2$ per s PAR) as reported by the author in 1969. However, the white mutant was damaged under high light intensity (205 $\mu E/m^2$ per s PAR) conditions. In addition, the chlorophyll mutants were examined for the accumulation of protophorphyrin IX by the treatment with DPEs under high light intensity conditions. The content of protoporphyrin IX in the white mutant was found to be much less than that in the other chlorophyll mutants (yellow and yellow-green) and normal one. This result did not correlate well with the results of the growth inhibition of the chlorophyll mutants treated with DPEs. Also, the chlorophyll mutants were treated with AFM under dark conditions in order to remove an unknown factor which results in light decomposition of protoporphyrin IX. The content of the accumulated protoporphyrin IX in the white mutant was much less than that in the other mutants under the dark conditions.

Also, the leakage of amino acids which was one of the effects caused by DPE treatment was examined. The leakage was observed with the chlorophyll mutants treated with DPEs, particularly with the white mutant. This finding was in good accordance with the result of the growth inhibition of the chlorophyll mutant with DPEs. 5-Aminolevulinic acid (ALA) synthesis, which is a rate-determining step in the chlorophyll biosynthetic pathway, was also low in the white mutant compared with the other mutants. The ALA synthesis in the white mutants was not stimulated by AFM treatment, while it was stimulated in the other mutants.

These results make it unlikely that the yellow pigments are directly related to the herbicidal activity of DPEs, and that the content of accumulated protoporphyrin IX will kill the white mutant under higher light intensity. Lower activity of biosynthesis of chlorophylls, especially that of tetrapyroles in the white mutant may be one of the reasons why the white mutant is not susceptible to DPEs.

These conclusions agree with an electronmicroscopic observation by Mayeda (1985) inferring that these chlorophyll mutants were produced by inhibition of development, with an imperfect biosynthesis pathway for chlorophylls (Fig. 5). Since the white rice mutants have small plastids, as shown by Mayeda, it is assumed that the porphyrin synthesizing system is also poorly developed. Therefore the accumulation of protoporphyrin IX in protoplasts is also small and subsequently the production of active oxygen species is low. This causes the tolerance of the white mutants to DPEs under lower intensity of light.

1.7 Similar Activities of Oxadiazon, Phenopylate and Cyclic Imides to *ortho*-Substituted Diphenyl Ether Herbicides

As described above (see Sect. 1.2), oxadiazon was registered as a rice herbicide in Japan in 1972. The author performed pot or field trials before its registration (the code number was G-315), and found that the herbicidal activities were similar to DPEs such as nitrofen although the chemical structure had not been disclosed. Then the author examined several DPE-like properties using oxadiazon, such as light requirement, brown spots on rice leaf sheath and behavior of the chlorophyll mutants of rice. In these three trials, Matsunaka (1970) found that a very similar mode of action was observed with oxadiazon as with nitrofen, although their chemical structures were very different.

Oxadiazon showed almost no activity on just emerged rice seeds in the dark even at 8 ppm, but over 50% inhibition in the light, the same as nitrofen. Oxadiazon seemed to have higher phytotoxic activity than nitrofen. In the same experiment another compound, DCPE [3,5-dichloro-1-(4-nitrophenoxy)benzene], an isomer of nitrofen, showed a somewhat mild phytotoxic activity under both light and dark conditions, as described in Section 1.3. even DCPE has a diphenyl ether structure. The brown spots on buds or leaf sheath of rice plants were observed in the case of oxadiazon as with nitrofen.

Behavior of the chlorophyll mutants of rice to oxadiazon was the same as in the case of nitrofen, where CM-35 and CM-123 were used as white and yellow mutants, respectively. Semi-quantitative experiments on the area of brown spots on rice leaf sheath also gave the same results with oxadiazon and nitrofen. From these results the author concluded that oxadiazon has a similar

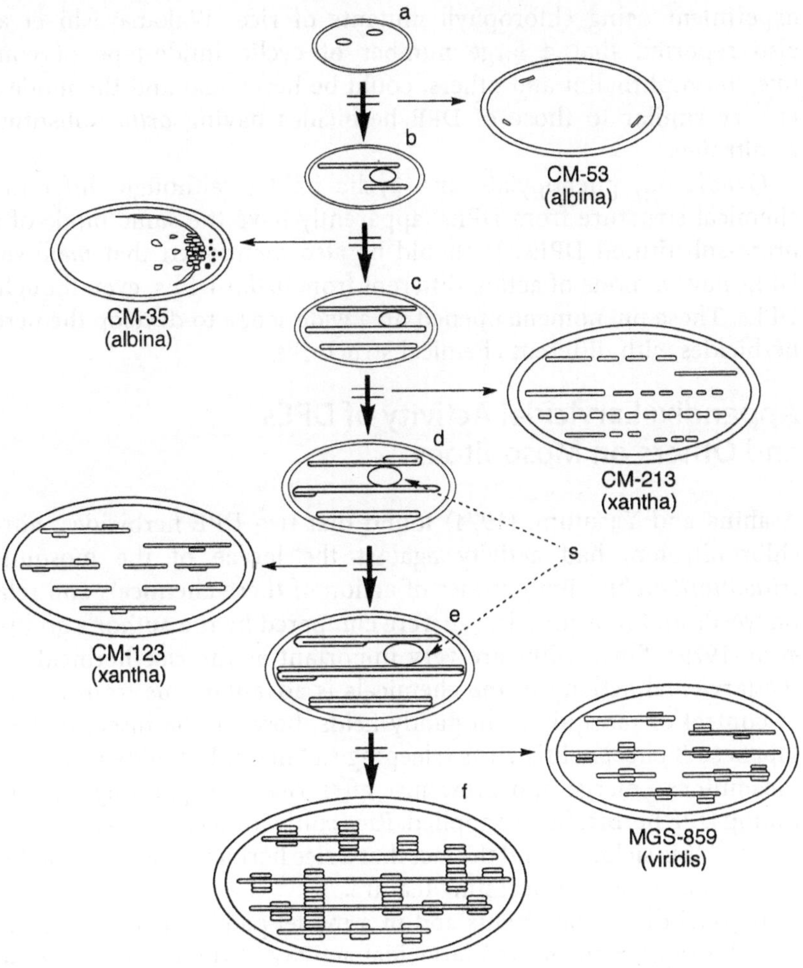

Fig. 5. Appearance of chloroplasts in chlorophyll mutant varieties (CM-53, CM-35, CM-213, CM-123 and MGS-859) based on chloroplast development in rice (Mayeda 1985). **a–f** Developing stages of normal chloroplast. *s* Starch granule

mode of action as a DPE, like nitrofen. About 20 years later, Duke et al. (1989) reported oxadiazon activity as similar to that of *p*-nitrodiphenyl ether herbicides using methods such as ultrastructure by electron microscopy, photosynthetic and porphyrin synthesis inhibitors or measuring protoporphyrin IX accumulation.

Phenopylate was registered in 1970 in Japan as a rice herbicide. Although it was cancelled in 1977, Yanai and Matsunaka (1972) reported the similarity of mode of action of this compound to *ortho*-substituted DPEs. They also tested the effect on respiration, amylase synthesis and the Hill reaction, and mutual activities with other herbicides. They concluded that the mode of action of phenopylate also resembles that of DPEs resulting from light-requirement experiment using chlorophyll mutants of rice. Wakabayashi et al. (1978) also reported that a large number of cyclic imide-type of compounds, like chlorophthalim and others, could be herbicidal and the mode of action is very similar to those of DPE herbicides having *ortho*-substitutent such as nitrofen.

Oxadiazon, phenopylate and cyclic imides, although different in their chemical structure from DPEs, apparently have the same mode of action as *ortho*-substituted DPEs. It should be also mentioned that *meta*-substituted DPEs have a mode of action different from *ortho*-DPEs, even though they are DPEs. These phenomena opened up a wide scope to develop the peroxidizing herbicides with different chemical structures.

Appendix: Larvicidal Activity of DPEs and Others on Mosquitoes

Asahina and Yasutomi (1974) found that two DPE herbicides, nitrofen and chloronitrofen, had activity against the larvae of the mosquito, *Culex tritaeniorhynchus*. Both modes of action of these chemicals and related ones on weeds and mosquito larvae were compared by the author's group (Ikeuchi et al. 1979). Since DPEs are very important as the rice herbicides in Japan, the larvicidal activity of the chemicals is advantageous from the standpoint of control of mosquitoes in paddy fields, because the insect is the vector of Japanese B encephalitis virus (sleeping sickness). In paddy fields, such vector mosquitoes emerge and grow just after rice transplanting, and with good timing DPE herbicides are applied. Rice cultivation has a bearing on the occurrence of this sickness, and the fact that a rice herbicide can control the mosquitoes seems to be an interesting feature.

In Petri dish experiments and in a model paddy field study both nitrofen and chloronitrofen showed larvicidal activity at the practical use rate in rice weed control. From this experiment, the mode of action of DPE herbicides as a larvicide is concluded not to be the same as that of herbicidal activity for the following reasons: (1) nitrofen and chloronitrofen showed larvicidal activity even in darkness, although they have almost no herbicidal activity in darkness, and (2) many chemicals were found to have high herbicidal activity but no larvicidal activity.

The fact that the DPEs with a substituent next to the *p*-nitro group (e.g. chlomethoxynil, bifenox and oxyfluorfen) showed almost no larvicidal activity is an interesting finding to elucidate the difference between larvicidal and herbicidal activities. Nowadays, DPE herbicides with larvicidal activity are not used any more in Japan for uncertain reasons. We should be on the lookout for the possibly increasing occurrence of sleeping sickness in Japan.

References

Arai M, Miyahara M, Kataoka T (1966) Persistence in the soil and movement into soil of fish-harmless herbicides in paddy rice. Weed Res 5:90–95 (in Japanese)

Asahina S, Yasutomi K (1974) Studies on colony maintenance and the development of insecticide-resistance in Japanese insects of medical importance. Environ Res Jpn 1973:16-1–16-7 (in Japanese)

Böger P, Wakabayashi K (1995) Peroxidizing herbicides (I): mechanism of action, Z. Naturforsch 50c:159–166

Duke SO, Lydon J, Paul RN (1989) Oxadiazon activity is similar to that of *p*-nitrodiphenyl ether herbicides. Weed Sci 37:152–160

Hayasaka T (1967) Studies on a new diphenylether herbicide, HE-314, part I, mode of action and selective herbicidal effect on genus of Gramineae. Weed Res (Jpn) 6:50–57 (in Japanese)

Hisada T (1969) The mode of action of a new type of diphenylether herbicides. Proc 2nd Asian-Pacific Weed Control Interchange, pp 216–234

Ikeuchi M, Yasuda S, Matsunaka S (1979) Mode of action of diphenylethers and related herbicides on mosquito larvae. Advances in pesticide science.: symposia papers, 4th Int. Congr Pesticide Chem (Zurich). Pergamon Press, Oxford, pp 470–483

Matringe M, Scalla R (1988) Studies on the mode of action of acifluorfen-methyl in non-chlorophyllous soybean cells. Plant Physiol 86:619–662

Matsunaka S (1969) Acceptor of light energy in photoactivation of diphenylether herbicides. J Agric Food Chem 17:171–175

Matsunaka S (1970) Similar mode of action in both herbicides, G-315(17623-RP) and *ortho*-substituted diphenylether compounds. Weed Res (Jpn) 10:40–43 (in Japanese)

Matsunaka S (1976) Diphenyl ethers chap 14. In: Kearney PC, Kaufman DD (eds) Herbicides: chemistry, degradation and mode of action, Dekker, New York, pp 709–739

Matsunaka S (1987) A historical review of the development of rice herbicides in Japan. Proc 11th Conf. Asian-Pacific Weed Sci Soc, pp 357–371

Matsunaka S (1991) Rice herbicide use in Japan: history, mode of action and economics. GIFAP 1991, spring meet., Kyoto, pp 1–16

Mayeda E (1985) Structure and function in organogenesis and tissue differentiation of crop. Jpn J Crop Sci 54:89–100

Wakabayashi K, Matsuya K, Jikihara T, Suzuki S, Matsunaka S (1978) Mechanism of action of cyclic imide type herbicides. Abstract 4th IUPAC Int Congr Pesticide Chem (Zürich) IV:655

Yamato S (1992) A study on the mode of action of diphenyl ether herbicides. Master thesis, Kobe University, pp 1–50

Yamato S, Onoe M, Matsunaka S (1990) The effect of light-dependent herbicide on chlorophyll mutant rice. A significant difference in tetrapyrrole accumulation, Abstr 7th IUPAC Int Congr Pesticide Chem (Hamburg), p 323

Yanai I, Matsunaka S (1972) Mode of action of phenopylate, similarity to *ortho*-substituted diphenyl ether herbicides. Proc Annu Meet Weed Sci Soc Japan, pp 75–77 (in Japanese)

Abbreviations and Chemical Names

a:	are (100m^2)
a.i.:	active ingredient
EC:	emulsifiable concentrate
G:	granule
PAR:	potosynthetically active radiation
Acifluorfen-methyl (AFM):	methyl 5-(2-chloro-4-trifluoromethylphenoxy)-2-nitrobenzoate
Bifenox:	methyl 5-(2,4-dichlorophenoxy)-2-nitrobenzoate

Butachlor:	*N*-(2-butoxymethyl)-2-chloro-2′,6′-diethylacetanilide
CFNP:	2,4-dichloro-6-fluoro-1-(4-nitrophenoxy)benzene, fluoronitrofen, MO-500
Chlo(r)methoxynil:	chlomethoxyfen 2,4-dichloro-1-(3-methoxy-4-nitrophenoxy)benzene
Chlor(o)nitrofen:	CNP; 2,4,6-trichloro-1-(4-nitrophenoxy)benzene
Chlorophthalim:	*N*-(4-chlorophenyl)-1-cyclohexene-1,2-dicarboximide
2,4-D:	2,4-dichlorophenoxyacetic acid
DPE:	diphenyl ether
DCPE:	3,5-dichloro-1-(4-nitrophenoxy)benzene
DMNP:	5-(4-nitrophenoxy) *m*-xylene, HE 306, HW-40187
DNCDE:	4-chloro-2-nitro-1-(4-nitrophenoxy)benzene
Dymron:	1-(α,α′-dimethylbenzyl)-3-(*p*-tolyl)urea, daimuron
Fluorodifen:	2-nitro-1-(4-nitrophenoxy)-4-trifluoromethylbenzene
Fluoronitrofen:	CFNP; MO 500
HE314:	3-(4-nitrophenoxy)toluene, TOPE
HW40187:	DMNP, HE 306
MCPA:	(4-chloro-*o*-tolyloxy)acetic acid
MCPB:	4-(4-chloro-*o*-tolyloxy)butyric acid
MCPCA:	2′-chloro-2-(4-chloro-*o*-tolyloxy)acetanilide
MO263:	4,6-dichloro-1-(4-nitrophenoxy)-*o*-toluene
MO600:	3-chloro-4-fluoro-(4-nitrophenoxy)benzene
Molinate:	*S*-ethyl perhydroazepine-1-carbothioate
Monuron:	3-(4-chlorophenyl)-1,1-dimethylurea
NH-8902:	4-bromo-2-chloro-1-(4-nitrophenoxy)benzene
Nitrofen:	2,4-dichloro-1-(4-nitrophenoxy)benzene
Oxadiazon:	G-315; 5-*tert*-butyl-3-(2,4-dichloro-5-isopropoxyphenyl)-1,3,4-oxadiazol-2(3H)-one
Oxyfluorfen:	OXY; 2-chloro-1-(3-ethoxy-4-nitrophenoxy)-4-trifluoromethyl-benzene
Paraquat:	1,1′-dimethyl-4,4′-bipyridinium ion
PCP:	pentachlorophenol
Phenopylate:	2,4-dichlorophenyl pyrrolidine-1-carboxylate
Prometryne:	2,4-bis(isopropylamino)-6-methylthio-1,3,5-triazine
Propanil:	3′,4′-dichloropropionanilide
Pyrazolate:	4-(2,4-dichlorobenzoyl)-(*H*-1,3-dimethylpyrazol-5-yl)*p*-toluenesulfonate
Simetryn:	2,4-bis(ethylamino)-6-methylthio-1,3,5-triazine
Thiobencarb:	benthiocarb; *S*-(4-chlorobenzoyl)diethylthiocarbamate
TOPE:	HE314

CHAPTER 2

Structural Evolution and Synthesis of Diphenyl Ethers, Cyclic Imides and Related Compounds

KENJI HIRAI[1]

Contents

2.1 Introduction

A large number of herbicides act by interfering with chlorophyll biosynthesis. The diphenyl ether (DPE) and cyclic imide classes of herbicides have been found to inhibit protoporphyrinogen IX oxidase (Protox) although their chemical structures are completely different from each other; therefore, both types of herbicides could be classified as Protox inhibitors. It is noteworthy that most recent cyclic imides consisting of a highly functionalized phenyl group and nitrogen-containing heterocycle moiety exhibit superior efficacy against weeds at low rate of application of less than 10 g/ha and some representatives also have good selectivity for crops. A few 5-membered cyclic imides have been already developed as practical herbicides and some of them are under development. Additionally, as an extension of 5-membered heterocycle

[1] Sagami Chemical Research Center, Nishi-Ohnuma 4-4-1, Sagamihara, Kanagawa 229-0012, Japan

Peter Böger, Ko Wakabayashi
Peroxidizing Herbicides

chemistry, 6-membered cyclic imides such as pyrimidine and triazine heterocycles have been actively investigated during the last decade. Particularly, 6-trifluoromethyluracils have been reported to show much more potent activity than that of 5-membered cyclic imides.

In this chapter, structural evolution of each class is chronologically reviewed from a viewpoint of molecular design in illustration of more than 700 kinds of DPEs, 5- and 6-membered cyclic imides disclosed since 1980, and general synthetic approaches to typical examples are also outlined. Cyclic imide classes cited include 5- and 6-membered nitrogen-containing heterocycles, for example, tetrahydrophthalimide, maleimide, oxazolidinedione, oxazolinone, imidazolidinedione, oxadiazolinone, pyrazole, thiadiazolidinone, triazolinone, tetrazolinone, pyrimidinedione, triazinedione and related compounds. In Figs. 1 to 23, chemical structures representing each compound are depicted together with herbicidal data cited in Chemical Abstracts.

2.2 Structural Evolution and Synthesis of Diphenyl Ethers (DPEs)

Various kinds of diphenyl ether herbicides have been developed and some of them used as practical herbicides. In particular, nitrofen, chlornitrofen and chlomethoxynil have played an important role as rice herbicides in increase the yield of rice in Japan. Moreover, DPEs such as bifenox, acifluorfen and fomesafen have been mainly applied as post-emergent contact herbicides active on broadleaf weeds in soybean areas. While DPEs such as nitrofen, which are classified in the first-generation DPE, contain two chlorine atoms at 2- and 4-positions on the phenyl ring, DPEs such as oxyfluorfen substituted with a trifluoromethyl group at 4-position belong to the second-generation DPE. Figure 1 shows representative DPEs developed as practical herbicides up until now.

DPE consists of two benzene rings (A and B) bridging with an oxygen atom, in which each benzene ring is regiospecifically functionalized to elicit potent herbicidal activity. A-ring of almost all DPEs is generally substituted with electron-withdrawing groups such as a chlorine atom and a trifluoromethyl group at 2- and 4-positions, whereas the substituent at 4′-position of B-ring is substituted with a nitro group. Introduction of a variety of substituents at 3′-position has been actively investigated because the substituents considerably affect herbicidal activity and crop selectivity. Modifications at 3′-position itself is structural evolution of DPEs.

Structural evolution of the first-generation DPEs having two chlorine atoms at 2- and 4-position of A-ring have declined in the early 1980s as shown in Fig. 1. For all that, for example, bifenox [3], functionalized with a methoxycarbonyl group at 3′-position, has been developed as a rice herbicide, which shows good efficacy against barnyard grass, annual lowland weeds, arrowhead, etc. Before 1980, 4-CF_3-DPEs bearing a variety of alkoxy groups at

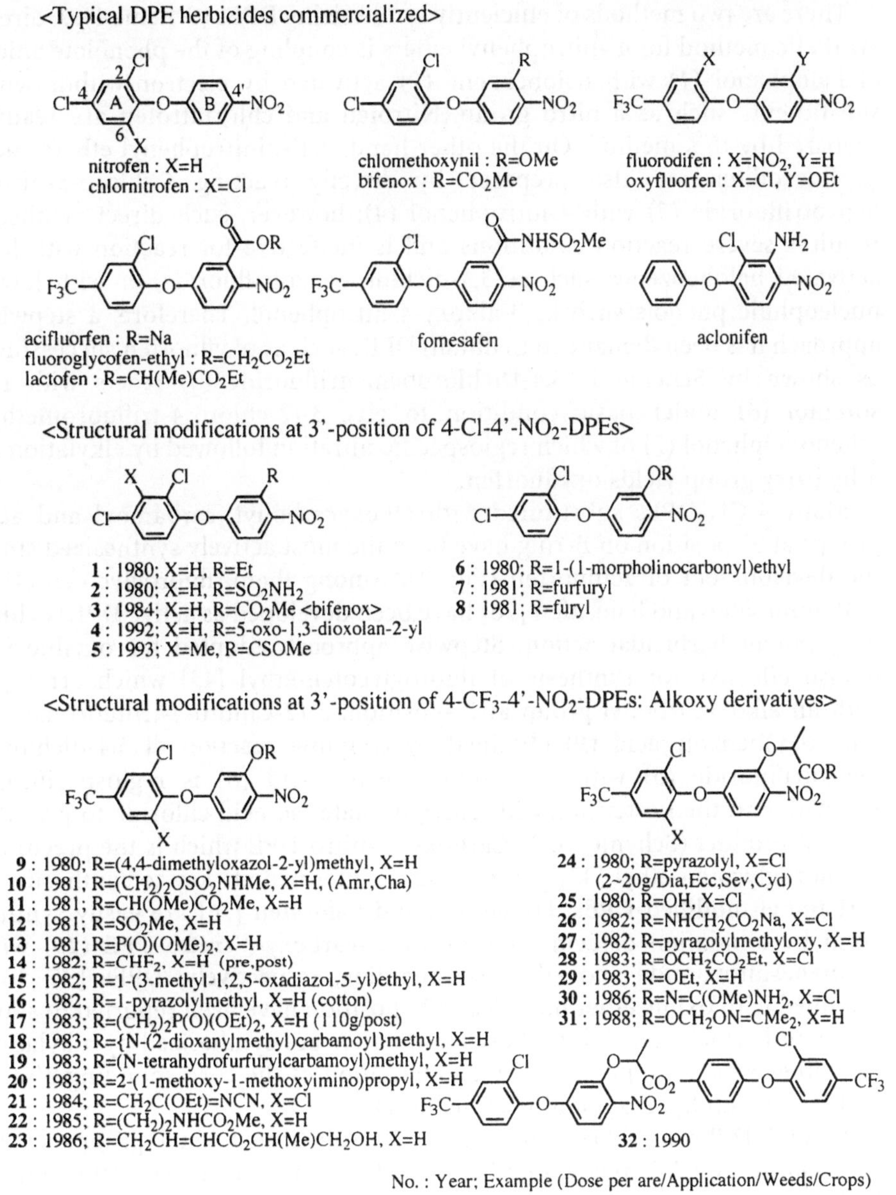

Fig. 1. Commercialized DPEs and structural modifications at 3′-position

3′-position have been aggressively synthesized since the development of oxyfluorfen belonging to the second generation. Among them, some DPEs such as CGA-84446 and MT-124 are selected as herbicidally more active compounds. Structural modifications have continued through the middle of the 1980s; however, potential analogs have still to be discovered (Fig. 1).

There are two methods of efficiently synthesizing DPEs (Scheme 1). A direct synthetic method for 4′-nitrophenyl ethers is coupling of the phenolate anion of halophenol (*1*) with halobenzene (*2*) activated by electron-withdrawing substituents such as a nitro group. Nitrofen and chlornitrofen are readily prepared by this method. On the other hand, 2,4′-dinitrophenyl ethers such as fluorodifen are also prepared by directly reacting 4-chloro-3-nitrobenzotrifluoride (*3*) with 4-nitrophenol (*4*); however, such direct synthesis requires severe reaction conditions and is ineffective for reaction with less activated halobenzenes such as 3,4-dichlorobenzotrifluoride, or with lower nucleophilic phenols such as 3-alkoxy-4-nitrophenol. Therefore, a stepwise approach has been demanded to obtain DPEs such as oxyfluorfen successfully as shown by Scheme 1. 3,4-Dichlorobenzotrifluoride (*5*) reacts with resorcinol (*6*) under basic condition to give 3-(2-chloro-4-trifluoromethylphenoxy)phenol (*7*) of which regiospecific nitration followed by alkylation of 3-hydroxy group yields oxyfluorfen.

Many 4-CF_3-DPEs substituted with alkoxycarbonyl, carbamoyl and acyl groups at 3′-position on B-ring have been the most actively synthesized since the development of acifluorfen (Fig. 2). Among them, fluoroglycofen-ethyl [43], fomesafen and halosafen [76] have been developed as active DPEs exhibiting potent herbicidal action. Stepwise approach adopted for oxyfluorfen is also effective for synthesis of fluoroglycofen-ethyl [43] which arranged with an alkoxycarbonyl group at 3′-position. 3-(2-Chloro-4-trifluoromethylphenoxy)benzoic acid (*9*) obtained by coupling reaction of 3,4-dichlorobenzotrifluoride (*5*) with 3-hydroxy benzoic acid (*8*) is regiospecifically nitrated, and then esterified with ethylglycolate *via* acid chloride to give the desired product (Scheme 1). 3′-Carboxy-4′-nitro-DPE which is the precursor of fluoroglycofen-ethyl [43] is the versatile intermediate key to synthesize 3-carbamoyl-modified DPEs. Fomesafen and halosafen [76] are easily synthesized by amidation of acid chloride of 3′-carboxy-4′-nitro-DPE (*10*) with methanesulfonamide and ethanesulfonamide, respectively. DPEs [95, 101] modified with oxime ether moieties at 3′-position have been reported to show excellent herbicidal activity. These compounds are synthesized by alkylation of oximes which are prepared by condensation of the corresponding 3′-acyl DPEs (*11*) with hydroxyl amine (Scheme 1).

Several DPEs modified with alkyl and substituted vinyl groups at 3′-position, some DPEs possessing heterocycles, and 4-CF_3-4′-NO_2-DPEs modified with nitrogen-, sulfur- and phosphorus-containing components at 3′-positions are illustrated in Fig. 3. Many DPEs of which B-rings are fused with 5- or 6-membered heterocycles have been synthesized; however, these analogs seems to be less active owing to the lack of 4′-nitro group (Fig. 4). Synthetic methodology of heterocycle-fused DPEs is entirely different from the synthetic strategy aforementioned. Details are provided in each patent literature. Miscellaneous DPEs which cannot be divided into each class in Figs. 1 to 4 and pyridyl phenyl ethers are shown in Fig. 5. These DPEs are mainly substituted

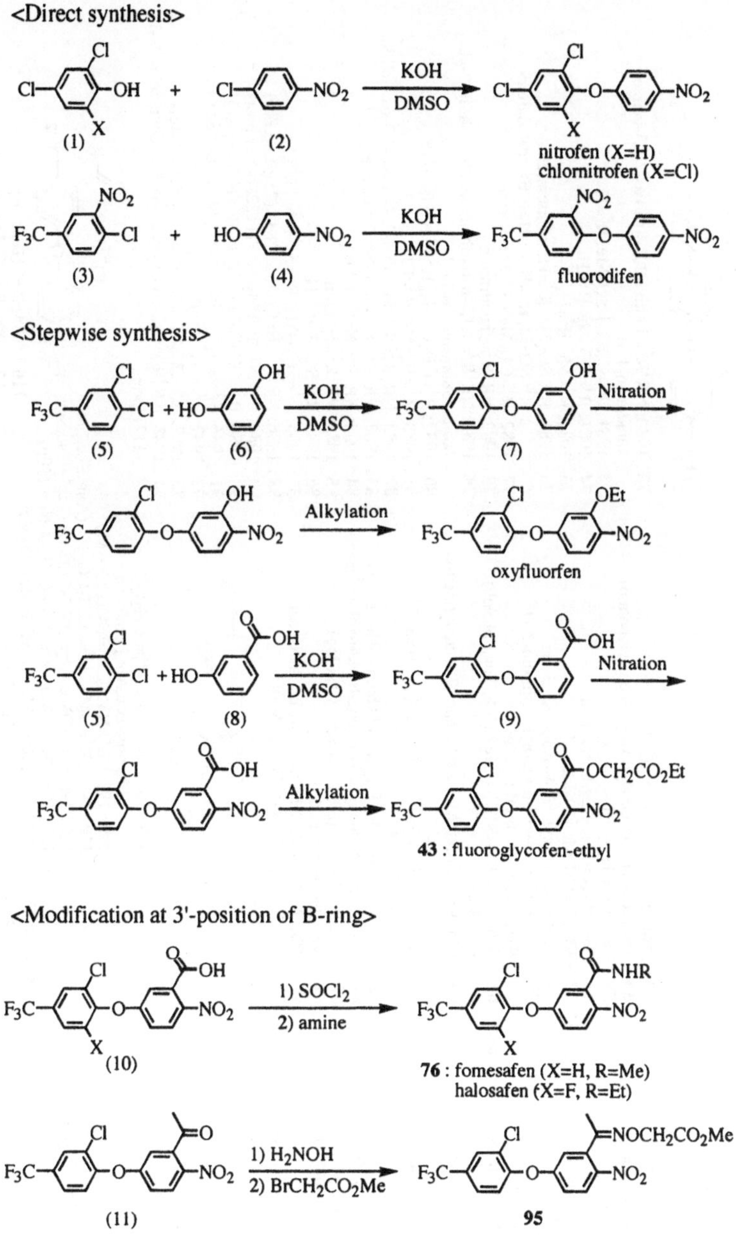

Scheme 1. Synthetic approaches to DPEs

with electron-withdrawing groups such as halogen atoms, cyano, phenylsulfonyl and formyl groups at 4′-positions instead of the nitro group. Optically active 4′-chloro-DPE [213], of which 3′-carboxy group is modified with ethyl (*S*)-lactate, is reported to show highly herbicidal activity.

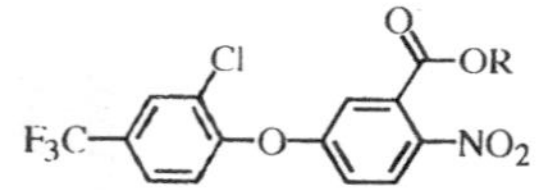

33 : 1980; R=CH_2CONH-i-Pr
34 : 1980; R=$(CH_2)_2NMe_2$ (5g)
35 : 1980; R=$(CH_2)_2OMe$ (22.7g/pre,post/Ecc)
36 : 1980; R=3-(2,2-dimethyldioxolanyl)methyl
37 : 1980; R=neopentyl
38 : 1981; R=$(CH_2)_2Cl$
39 : 1981; R=CH(Me)CO_2Et
40 : 1981; R=2-methylpiperidinocabonylmethyl, (6-H=6-Cl)
41 : 1981; R=phthalimido, X=H
42 : 1981; R=C_6H_2-2-iBu-4,6-$(NO_2)_2$
43 : 1981; R=CH_2CO_2Et <fluoroglycofen-ethyl>
44 : 1982; R=CH_2COMe
45 : 1982; R=$(CH_2)_2NO_2$
46 : 1982; R=CH_2COMe (6-H=6-Cl)
47 : 1982; R=pyrazolylmethyl (6-H=6-Cl)
48 : 1983; R=3-oxobutyliden-2-ylamino
49 : 1983; R=3-furyl
50 : 1983; R=CH(CO_2Et)CH_2CO_2Et, X=H (2.8g)
51 : 1983; R=3-(2-oxotetrahydrofuryl)
52 : 1984; R=$(CH_2)_2COCH_2O$-C_6H_3-2,4-Cl_2, (113g)

53 : 1984; R=CH(Me)CO_2N=CMe_2
54 : 1984; R=$CH_2CH_2OSO_2NMe_2$ (rice,soya,potato)
55 : 1984; R=$(CH_2)_2OH$
56 : 1984; R=N=CMe_2 (2g/Ecc/cotton)
57 : 1984; R=CH(Me)CO_2Me
58 : 1985; R=3-methyl-1,2,4-oxadiazol-5-yl
59 : 1985; R=5-Cl-4-OH-6-methylpyrimidin-2-yl
60 : 1985; R=$CH_2P(O)(OEt)_2$ (5g)
61 : 1986; R=$(CH_2)_2NHCO_2Me$ (2.84g/Sev,Sol,Soh)
62 : 1986; R=SEt
63 : 1986; R=2-Cl-3,5-Me_2-cyclopenta-2,4-dienylmethyl, (corn,wheat,rice)
64 : 1987; R=CH(CH_3)$CO_2CH_2SiMe_3$ (6-H=6-Cl) (post/Gaa,Ipl,Sis)
65 : 1987; R=$CH_2P(O)(OEt)_2$ (post/Aba,Ipl/wheat,soya)
66 : 1987; R=2-{(4-ethoxycarbonylmethyl)dioxolanyl}methyl
67 : 1987; R=CH(CCl_3)OEt, (6-H=6-Cl) (25g)
68 : 1988; R=$CH_2P(S)(OEt)_2$
69 : 1988; R=$(CH_2)_2OCH_2Ph$ (post/Cha,Sis/wheat)
70 : 1988; R=CH(4-C_5H_4N)P(O)$(OMe)_2$
71 : 1989; R=MeS^+I^-

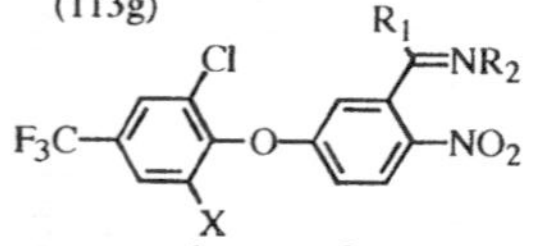

95 : 1980; R^1=Me, R^2=OCH_2CO_2Me, X=H
96 : 1983; R^1=Cl, R^2=OSO_2Me, X=F
97 : 1984; R^1=Me, R^2=OMe, X=F
98 : 1985; R^1=Me, R^2=$OCH_2CO_2CH_2$-C_6H_4-2-Cl, X=H (2g/Amr)
99 : 1986; R^1=Me, R^2=$OCH_2CO_2CH_2CO_2CH_2$-C_6H_4-2-Me, X=H
100 : 1986; R^1=Me, R^2=OCH_2CO_2N=CMe_2, X=H
101 : 1986; R^1=CH_2OMe, R^2=OCH_2CO_2Me, X=H (0.25g/pre,post)
102 : 1987; R^1=Me, R^2=$OCH_2CO_2CH(Me)CO_2Et$, X=H
103 : 1987; R^1=CH_2SMe, R^2=OCH_2COMe, X=H (2.5g/Cha)
104 : 1987; R^1=H, R^2=OCH_2CO_2Me, X=F (pre,post)
105 : 1988; R^1=H, R^2=octyl, X=H (11.4g/post)
106 : 1988; R^1=CH_2OMe, R^2=OCH_2CO_2H, X=H
107 : 1990; R^1=H, R^2=OCH_2CH=$CHCO_2Et$, X=H (113g)

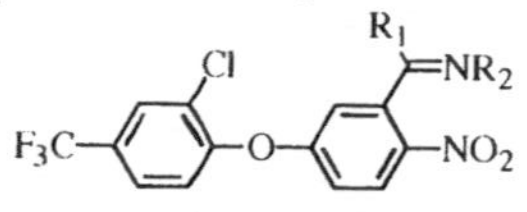

108 : 1985; R^1=OMe, R^2=$(CH_2)_2Cl$ (20g/post/Ecc,Abt,Lom/soya)
109 : 1986; R^1=NMe_2, R^2=OCH_2CO_2Me
110 : 1987; R^1=OMe, R^2=OCH_2CO_2Me (Amr,Ecc,Caf)
111 : 1987; R^1=OMe, R^2=OCH_2COSMe (1.25g/Dic,Sev)
112 : 1988; R^1=OMe, R^2=OCO-C_6H_4-4-NO_2 (10g)
113 : 1988; R^1=OMe, R^2=CH(Me)CO_2Me (1g/post/Xas/wheat)
114 : 1989; R^1=OMe, R^2=OH

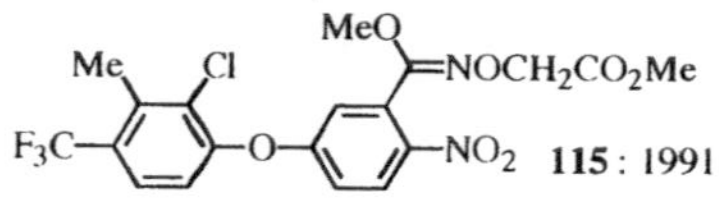

115 : 1991

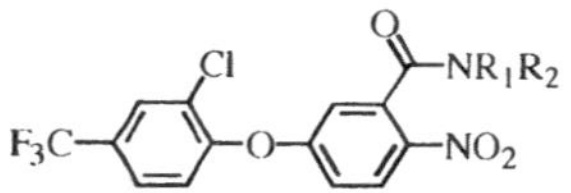

72 : 1980; NR^1R^2=3-ethyl-2,2,4-trioxo-2,1,3-benzothiadiazin-1-yl
73 : 1981; R^1=C_6H_4-4-SO_2Me, R^2=H
74 : 1981; R^1=CO_2Me, R^2=H
75 : 1981; R^1=Me, R^2=H
76 : 1981; R^1=SO_2Me, R^2=H, <fomesafen>
R^1=SO_2Et, R^2=H, (6-H=6-F) <halosafen>
77 : 1982; R^1=CH(Me)CO_2Me, R^2=H
78 : 1982; NR^1R^2=phthalimido
79 : 1982; NR^1R^2=1,1,3-trioxobenzoisothiazolidin-2-yl, (corn,cotton,soya,barlay)
80 : 1982; R^1=CH_2CO_2H, R^2=$CH_2P(O)(OH)_2$
81 : 1983; R^1=SO_2Me, R^2=Cl
82 : 1983; NR^1R^2=5,5-dimethyl-2,4-dioxo-1,3-oxazolidin-3-yl
83 : 1983; NR^1R^2=1,3-dithiolan-2-ylidene
84 : 1984; NR^1R^2=N=S(O)Me_2
85 : 1985; NR^1R^2=pyridinio
86 : 1985; R^1=SO_2F, R^2=H
87 : 1985; R^1=SCO_2Me, R^2=SO_2Me (5.7g)
88 : 1986; NR^1R^2=N=C(OMe)NH_2 (25g/Lap,Ecc)
89 : 1986; R^1=P(O)$(OPh)_2$, R^2=H (22.7g/pre,post)
90 : 1986; R^1=$CH_2P(O)(O$-i$Pr)_2$, R^2=H (pre,post)
91 : 1986; R^1=CH_2CO_2Et, R^2=H (22.7g/post/Sev,Brr)
92 : 1987; R^1=CN, R^2=Pr
93 : 1987; R^1=CH_2CO_2Me, R^2=$CH_2P(O)(OMe)_2$
94 : 1988; NR^1R^2=4-(2-ethoxyphenyl)-5-oxotetrazolin-1-yl, (112g/pre)

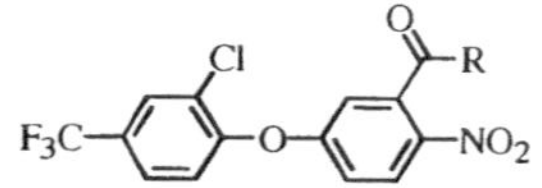

116 : 1984; R=CH=NOEt
117 : 1984; R=CO_2Me
118 : 1984; R=C(Me)$(CO_2Et)_2$ (113g/pre,post)
119 : 1986; R=CH$(CO_2Et)_2$ (10g/Cha,Sev)
120 : 1987; R=CH_2SEt (2.5g/Cha)
121 : 1989; R=CH_2CN (2g/Amr,Cha/corn)

No. :Year; Example (Dose per are/Application/Weeds/Crops)

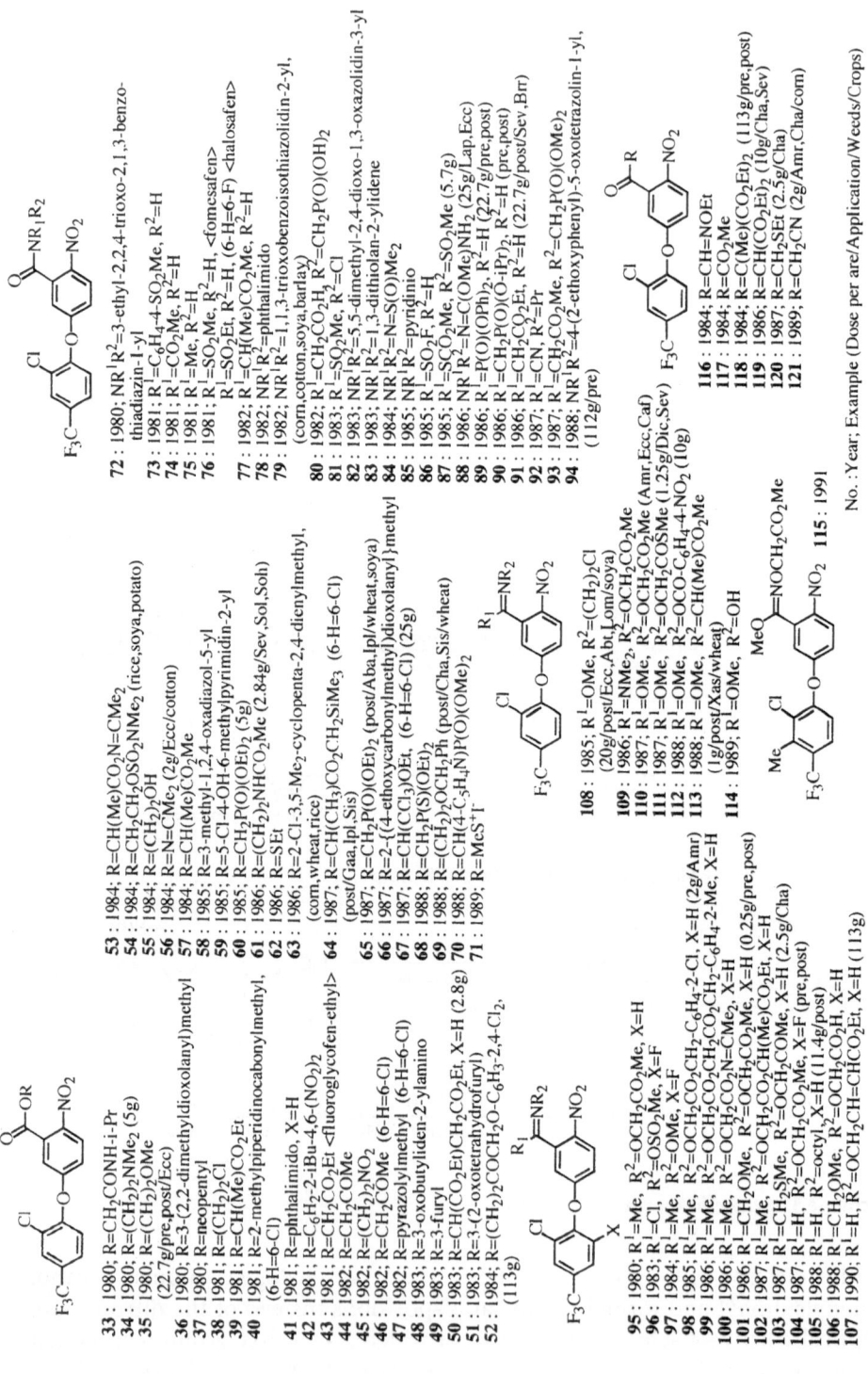

Fig. 2. Structural modifications of 4-CF_3-4′-NO_2-DPEs: 3′-alkoxycarbonyl, 3′-carbamoyl, 3′-aryl derivatives and related compounds

<Structural modifications of 4-CF_3-4'-NO_2-DPEs bearing C-C bond>

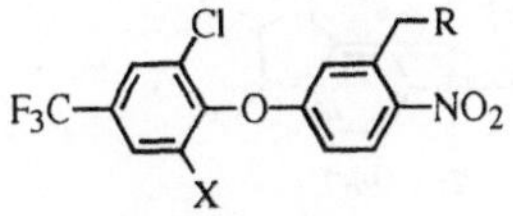

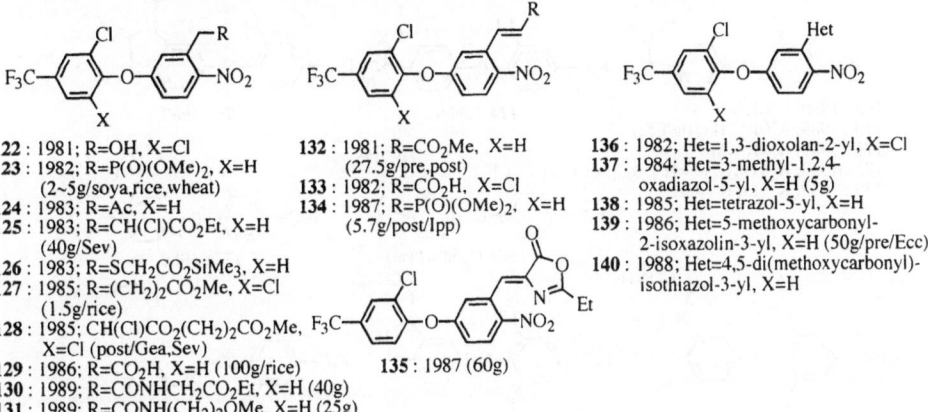

122 : 1981; R=OH, X=Cl
123 : 1982; R=$P(O)(OMe)_2$, X=H (2~5g/soya,rice,wheat)
124 : 1983; R=Ac, X=H
125 : 1983; R=$CH(Cl)CO_2Et$, X=H (40g/Sev)
126 : 1983; R=$SCH_2CO_2SiMe_3$, X=H
127 : 1985; R=$(CH_2)_2CO_2Me$, X=Cl (1.5g/rice)
128 : 1985; $CH(Cl)CO_2(CH_2)_2CO_2Me$, X=Cl (post/Gea,Sev)
129 : 1986; R=CO_2H, X=H (100g/rice)
130 : 1989; R=$CONHCH_2CO_2Et$, X=H (40g)
131 : 1989; R=$CONH(CH_2)_2OMe$, X=H (25g)

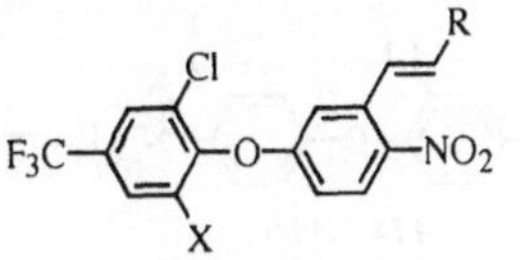

132 : 1981; R=CO_2Me, X=H (27.5g/pre,post)
133 : 1982; R=CO_2H, X=Cl
134 : 1987; R=$P(O)(OMe)_2$, X=H (5.7g/post/Ipp)

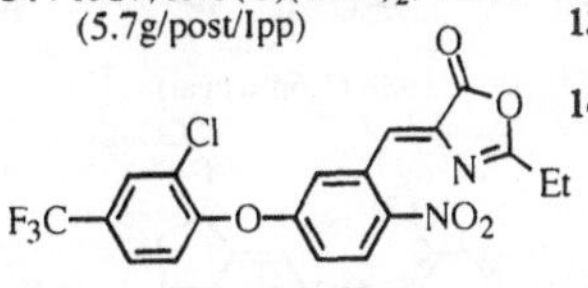

135 : 1987 (60g)

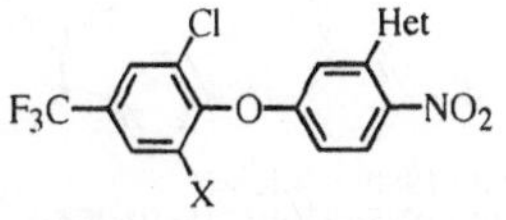

136 : 1982; Het=1,3-dioxolan-2-yl, X=Cl
137 : 1984; Het=3-methyl-1,2,4-oxadiazol-5-yl, X=H (5g)
138 : 1985; Het=tetrazol-5-yl, X=H
139 : 1986; Het=5-methoxycarbonyl-2-isoxazolin-3-yl, X=H (50g/pre/Ecc)
140 : 1988; Het=4,5-di(methoxycarbonyl)-isothiazol-3-yl, X=H

<Structural modifications of 4-CF_3-4'-NO_2-DPEs substituted by N-, S- and P-containing components>

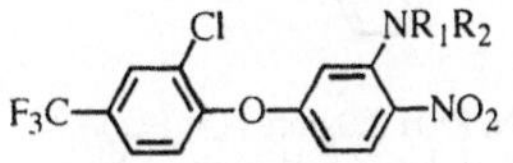

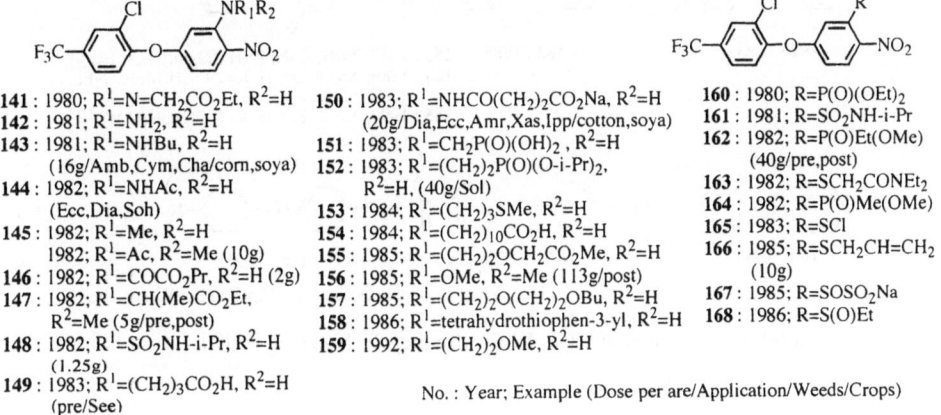

141 : 1980; R^1=$N=CH_2CO_2Et$, R^2=H
142 : 1981; R^1=NH_2, R^2=H
143 : 1981; R^1=NHBu, R^2=H (16g/Amb,Cym,Cha/corn,soya)
144 : 1982; R^1=NHAc, R^2=H (Ecc,Dia,Soh)
145 : 1982; R^1=Me, R^2=H
1982; R^1=Ac, R^2=Me (10g)
146 : 1982; R^1=$COCO_2Pr$, R^2=H (2g)
147 : 1982; R^1=$CH(Me)CO_2Et$, R^2=Me (5g/pre,post)
148 : 1982; R^1=SO_2NH-i-Pr, R^2=H (1.25g)
149 : 1983; R^1=$(CH_2)_3CO_2H$, R^2=H (pre/See)
150 : 1983; R^1=$NHCO(CH_2)_2CO_2Na$, R^2=H (20g/Dia,Ecc,Amr,Xas,Ipp/cotton,soya)
151 : 1983; R^1=$CH_2P(O)(OH)_2$, R^2=H
152 : 1983; R^1=$(CH_2)_2P(O)(O\text{-i-Pr})_2$, R^2=H, (40g/Sol)
153 : 1984; R^1=$(CH_2)_3SMe$, R^2=H
154 : 1984; R^1=$(CH_2)_{10}CO_2H$, R^2=H
155 : 1985; R^1=$(CH_2)_2OCH_2CO_2Me$, R^2=H
156 : 1985; R^1=OMe, R^2=Me (113g/post)
157 : 1985; R^1=$(CH_2)_2O(CH_2)_2OBu$, R^2=H
158 : 1986; R^1=tetrahydrothiophen-3-yl, R^2=H
159 : 1992; R^1=$(CH_2)_2OMe$, R^2=H

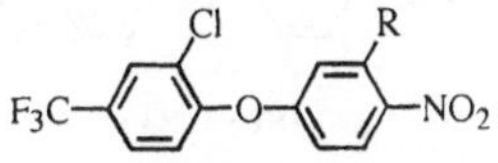

160 : 1980; R=$P(O)(OEt)_2$
161 : 1981; R=SO_2NH-i-Pr
162 : 1982; R=P(O)Et(OMe) (40g/pre,post)
163 : 1982; R=SCH_2CONEt_2
164 : 1982; R=P(O)Me(OMe)
165 : 1983; R=SCl
166 : 1985; R=$SCH_2CH=CH_2$ (10g)
167 : 1985; R=$SOSO_2Na$
168 : 1986; R=S(O)Et

No. : Year; Example (Dose per are/Application/Weeds/Crops)

Fig. 3. Structural modifications of 4-CF_3-4'-NO_2-DPEs at 3'-position

Furthermore, 3'-(*N*-ethylcarbamoyl)-modified DPE [217] is more active without electron-withdrawing substituents such as a nitro group at the 4'-position.

2.3 Structural Evolution and Synthesis of 5-Membered Cyclic Imides

Sufficient discussions on structure-activity relationships of 5-membered cyclic imides suggest that the hetero ring consisting of sp^2 carbons and heteroatoms has aromatic character and planarity which are the most significant qualities for potent herbicidal activity, although bicyclic hydantoin derivatives with sp^3

<Fused with 5-membered cycloalkyls and heterocycles>

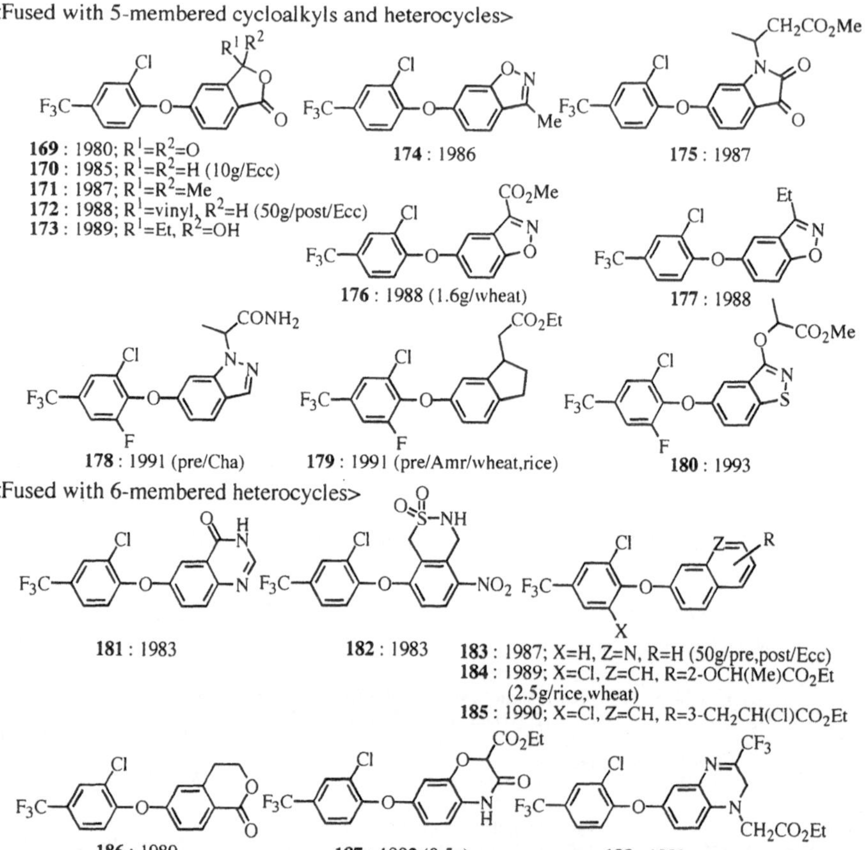

169 : 1980; R^1=R^2=O
170 : 1985; R^1=R^2=H (10g/Ecc)
171 : 1987; R^1=R^2=Me
172 : 1988; R^1=vinyl, R^2=H (50g/post/Ecc)
173 : 1989; R^1=Et, R^2=OH

174 : 1986

175 : 1987

176 : 1988 (1.6g/wheat)

177 : 1988

178 : 1991 (pre/Cha)

179 : 1991 (pre/Amr/wheat,rice)

180 : 1993

<Fused with 6-membered heterocycles>

181 : 1983

182 : 1983

183 : 1987; X=H, Z=N, R=H (50g/pre,post/Ecc)
184 : 1989; X=Cl, Z=CH, R=2-OCH(Me)CO_2Et (2.5g/rice,wheat)
185 : 1990; X=Cl, Z=CH, R=3-CH_2CH(Cl)CO_2Et

186 : 1989

187 : 1992 (0.5g)

188 : 1993

No. : Year; Example (Dose per are/Application/Weeds/Crops)

Fig. 4. Structural modifications of 4-CF_3-DPEs with bicyclic benzolog system

carbon at 5-position show strong activity, exceptionally. As other general characteristics of chemical structure, there is at least one of amide or its mimic moiety in the hetero ring on which a highly functionalized phenyl group is attached. Usually, the phenyl ring is regiospecifically substituted at 2- and 4-positions with electron-withdrawing groups such as halogen atoms, cyano and nitro groups, and also substituted at 5-position with an electron-donating group such as alkoxy, alkylthio, substituted amino and substituted alkyl groups. The most appropriate combination for these substituents is in existence for each heterocycle to elicit full herbicidal activity. Particularly, many cyclic imides with a variety of substituents at 5-position have been synthesized and evaluated, because of considerable influence of the substituents on both herbicidal activity and selectivity for crops. It is not too much to say that structural modifications on the phenyl ring are, as it is, a history of exploration for new cyclic imide class of herbicides.

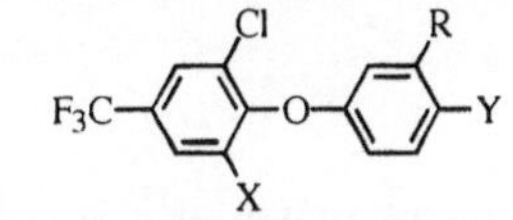

189 : 1981; R=$CONHSO_2Me$, X=H, Y=NHAc
190 : 1982; R=H, X=Cl, Y=CN
191 : 1983; R=OH, X=H, Y=SO_2-C_6H_4-4-Cl
192 : 1984; R=OMe, X=Cl, Y=SO_2-C_6H_4-4-OMe
193 : 1984; R=C_6H_3-2-Cl-4-CF_3, X=H, Y=SO_2-C_6H_4-4-OMe
194 : 1985; R=OCH(Me)CO_2Me, X=Cl, Y=SO_2-C_6H_4-4-Me
195 : 1988; R=NO_2, X=Cl, Y=1,2,3,6-tetrahydro-1,2-thiazin-1-yl
196 : 1988; R=OCH(Me)CO_2Et, X=H, Y=CO_2Me (5g/Ipp)
197 : 1990; R=$CO_2CH(Me)CO_2Et$, X=H, Y=Br (4.5g)
198 : 1990; R=OEt, X=H, Y=CH=NOH (5g/Abt)
199 : 1990; R=OCH(Me)CO_2Et, X=H, Y=CH=NOH (5g/Abt,Sol,Xas)
200 : 1990; R=NH_2, X=Cl, Y=CN
201 : 1990; R=$CH_2CH(CO_2Et)_2$, X=H, Y=SCF_3
202 : 1990; R=3-hydroxytetrahydrofuran-2-yloxycarbonyl, X=H, Y=Br
203 : 1992; R=OCH(Me)CO_2Me, X=H, Y=CHO

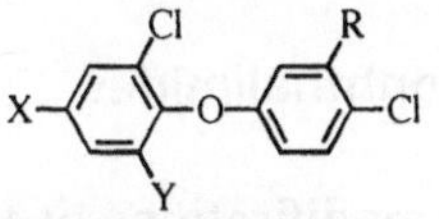

204 : 1980; R=CH_2Ac, X=CF_3, Y=Cl (20g/Dia,Ecc,Sev,Cyd)
205 : 1981; R=CH_2Ac, X=Y=Cl
206 : 1981; R=C(CN)=NOCH(Me)CO_2Me, X=CF_3, Y=H
207 : 1981; R=OSO_2Me, X=CF_3, Y=H
208 : 1981; R=OCH(Me)$(CH_2)_2CO_2Et$, X=CF_3, Y=H
209 : 1981; R=$CO_2(CH_2)_2OMe$, X=CF_3, Y=H
R=$CO_2(CH_2)_2SMe$, X=CF_3, Y=H
210 : 1983; R=OCH(Me)CO_2Me, X=Cl, Y=H (150g/Amr/wheat,corn,rice)
211 : 1983; R=OCH(Me)$CSNH_2$, X=CF_3, Y=H
212 : 1989; R=$CH_2SCH_2CO_2Et$, X=Y=Cl
213 : 1990; R=$CO_2CH(Me)CO_2Et$, X=CF_3, Y=H
214 : 1990; R=OEt, X=CF_3, Y=F (10g/pre/Sev,Abt)
215 : 1990; R=CH=NOCH(Me)CO_2Me, X=CF_3, Y=F

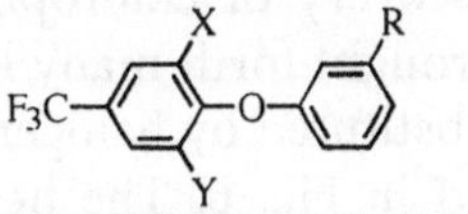

216 : 1980; R=OSO_2Me, X=Cl, Y=H
217 : 1982; R=CONHEt, X=Cl, Y=H
218 : 1983; R=NHEt, X=Cl, Y=H
219 : 1984; R=OCH(Me)CH=$CHCO_2Et$, X=Cl, Y=H
220 : 1984; R=OCH(Me)CO_2Et, X=Y=Cl
221 : 1986; R=OCH(Me)CO_2Et, X=Y=H
222 : 1987; R=C(Me)=$NOCH_2CO_2Me$, X=Cl, Y=H (12g/post)
223 : 1988; R=$C(CH_2OMe)$=$NOCH_2CO_2H$, X=Cl, Y=H
224 : 1988; R=$CONHSO_2Me$, X=Y=H
225 : 1989; R=CH=C(NHAc)CO_2Me, X=Cl, Y=H (90g/post/Abt)
226 : 1993; R=OCH(Me)CO_2Et, X=F, Y=H

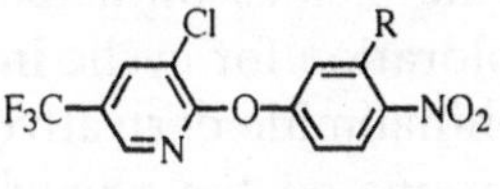

227 : 1980; R=OCH(Me)CO_2Me
228 : 1983; R=$CONHSO_2C_6H_4$-3-CF_3 (10g)
229 : 1983; R=$CONHSO_2CH$=CHC_6H_4-3-CF_3 (5g)
230 : 1986; R=C(Me)=NOEt
231 : 1989; R=C(OMe)=$NOCH_2COSMe$ (2.5g/Ecc,Dia)
232 : 1989; R=C(OMe)=$NOCH_2CO_2CH_2CH$=CH_2
233 : 1989; R=C(OMe)=$NOCH_2CO_2Me$

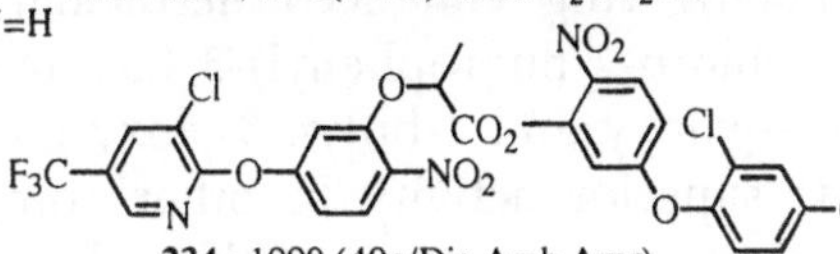

234 : 1990 (40g/Dia,Amb,Amr)

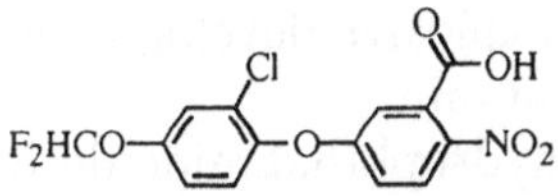

235 : 1982 (100g/pre)

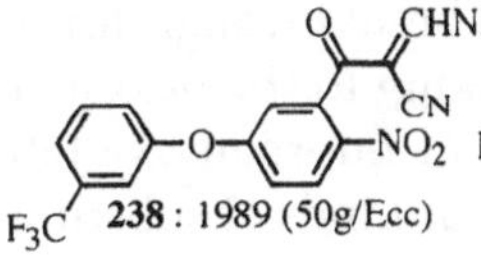

238 : 1989 (50g/Ecc)

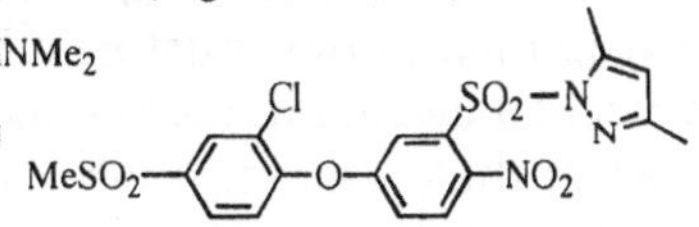

241 : 1991

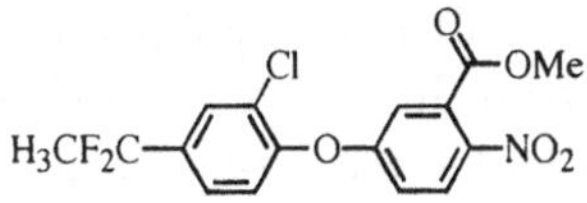

236 : 1988 (wheat)

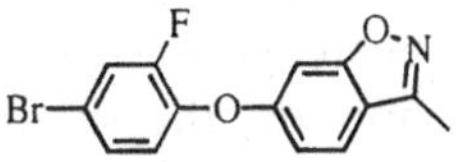

239 : 1989 (2.5g/rice,wheat)

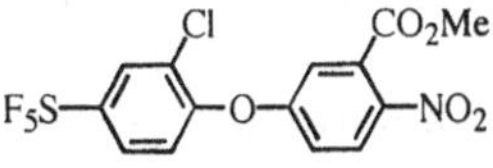

242 : 1994

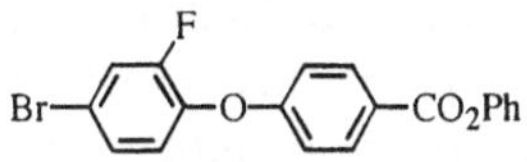

237 : 1989

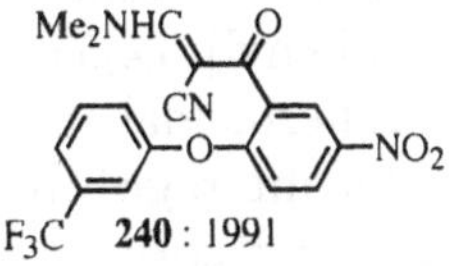

240 : 1991

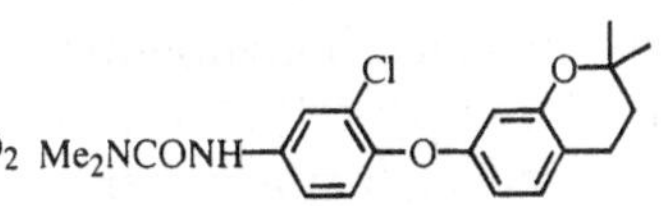

243 : 1994

No. : Year; Example (Dose per are/Application/Weeds/Crops)

Fig. 5. Miscellaneous DPEs with various substituents and pyridylphenyl ethers

2.3.1 Tetrahydrophthalimides

Structural modifications of tetrahydrophthalimide class have been most aggressively investigated as the tetrahydrophthalimide ring is a highly active component among many cyclic imide herbicides. In the first place, the origin of "cyclic imide" is *N*-(4-chlorophenyl)-3,4,5,6-tetrahydrophthalimide [245: chlorophthalim] which was discovered in the course of developments for new fungicidal tetrahydrophthalimide derivative by the Mitsubishi Chemical group in 1973. A few years before, 3-(2,4-dichloro-5-isopropoxyphenyl)-5-*tert*-butyl-1,3,4-oxadiazolin-2-one [244: oxadiazon] was disclosed by the Rhône-Poulenc group as a pioneering compound in the development of cyclic imide class of herbicides. These two compounds, which were not recognized as Protox inhibitors at that time, exhibit powerful herbicidal activity, and have been used as practical herbicides until today. Since the discovery of chlorophthalim, primary exploration for cyclic imide herbicides brought forth many kinds of tetrahydrophthalimide derivatives [247 to 253] substituted by halogen atoms or alkoxy groups on the phenyl ring, as depicted in Fig. 6. The herbicidal activity of these incipient tetrahydrophthalimides suggests that a combination of two halogen atoms at 2- and 4-positions and an alkoxy group at 5-position of phenyl ring is of importance for increasing herbicidal activity. In the early 1980s, Nagano et al. found that introduction of a fluorine atom into 2-position of phenyl ring enhanced herbicidal activity dramatically. In particular, *N*-(4-chloro-2-fluorophenyl)-3,4,5,6-tetrahydrophthalimides [254] bearing propargyloxy and 1-butyn-3-yloxy groups at 5-position on the phenyl ring show superior activity to other compounds synthesized previously. It is no exaggeration to say that the discovery of fluoro-modified tetrahydrophthalimides must be a breakthrough for an explosive development of herbicidal cyclic imides belonging to the next generation.

After that, a large number of fluoroalkoxy tetrahydrophthalimide derivatives [255 to 273] bearing various kinds of substituents on 5-oxygen atom, such as alkenyl, haloalkyl, alkoxyalkyl, alkoxycarbonylalkyl, dialkoxyphosphoryl and alkylsulfonyl groups, have been synthesized. From the results of agricultural evaluations of these compounds, the substituent on the oxygen atom performs an important role in the selectivity between weeds and crops as well as in the herbicidal activity. Among them, *N*-[4-chloro-2-fluoro-5-(pentyloxycarbonylmethyloxy)phenyl]-3,4,5,6-tetrahydrophthalimide [267: flumiclorac-pentyl] shows potent herbicidal activity against broadleaf weeds, e.g. velvetleaf, lambsquaters, cocklebur, ragweeds and morning glory, and also excellent selectivity for soybeans and corn crops at the low rates of 30 to 60 g/ha. Particularly, the most effective activity is observed in controlling velvetleaf and the post-emergence application at the rates of 60 g/ha enables enough prevention of 6- to 10-leaf stage of velvetleaf. Flumiclorac-pentyl was commercialized as the first example of new cyclic imide herbicide in Europe as early as 1993.

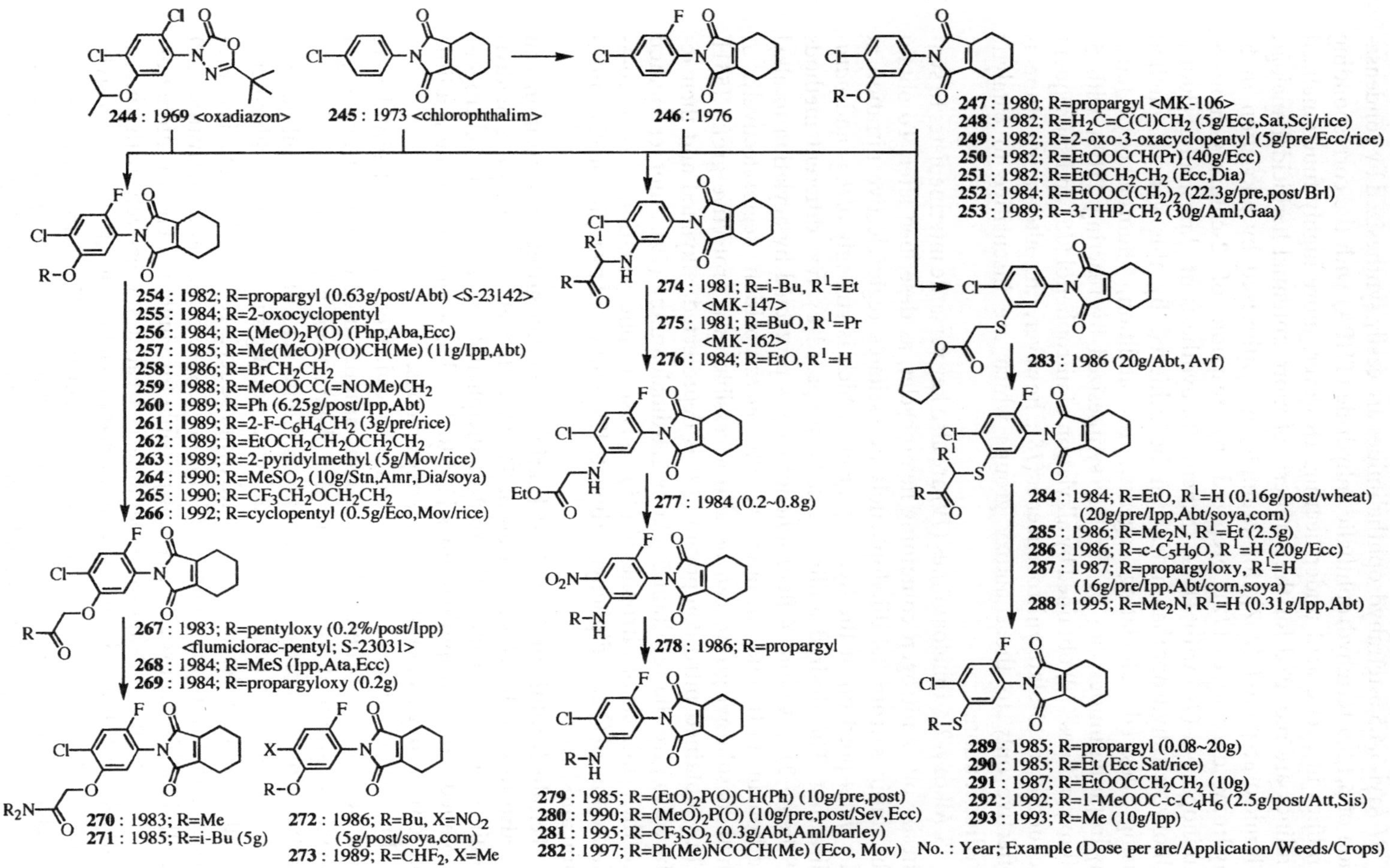

Fig. 6. Structural modifications at 5-position of the phenyl ring in tetrahydrophthalimides: alkoxy, amino and alkylthio moieties

N-Aryl-3,4,5,6-tetrahydrophthalimides are easily synthesized by condensation of 3,4,5,6-tetrahydrophthalic anhydride (THPA) with the corresponding anilines in acetic acid under heating. However, more highly functionalized anilines are required for the synthesis of more potential herbicidal analogs. Flumiclorac-pentyl [267], for example, is synthesized by alkylation of 5-hydroxy derivative (*17*) prepared by reaction of 2-chloro-4-fluoro-5-aminophenol (*16*) with THPA (Scheme 2). Synthesis of almost all of fluoroalkoxy tetrahydrophthalimides can be achieved by this general method using desired alkylating agents. In addition, fluoroalkoxy tetrahydrophthalimides can be directly synthesized by reacting THPA with 5-alkoxyanilines (*19*) obtained by alkylation of nitrophenol (*15*) and subsequent reduction of 5-alkoxynitrobenzene (*18*); however, the alternative route is effective only to those alkoxy groups which are insensitive to catalytic hydrogenation.

5-Alkoxy-2,4-dihaloanilines (*19*) are also key versatile intermediates to synthesize other nitrogen-containing heterocycles, as described later. To obtain the anilines from 2,4-dihalophenols necessitates regioselective nitration at the *meta*-position of the hydroxy group which is strongly *ortho*- and *para*-directing. The desired anilines can be prepared by two different methods (Scheme 2). 2-Chloro-4-fluorophenol (*12*) is protected by acylation to alkyl carbonate (*13*) and nitrated at the *meta*-position regioselectively. A methanesulfonyl group can be also accessible as a protective group [531]. Thus, obtained nitro compound (*14*) is deprotected, alkylated and then reduced to give the resultant 5-alkoxy-2,4-dihaloaniline (*19*). Another method for the synthesis of nitrophenol (*15*) consists of phosgenation of 2-chloro-4-fluorophenol (*12*) to diphenyl carbonate [294] followed by regioselective nitration and hydrolysis.

Tetrahydrophthalimides [273] bearing a methyl group at 4-position can be also formed by condensation of THPA with 5-alkoxy-2-fluoro-4-methylanilines derived from 4-fluoro-2-methylphenol. A series of 5-nitrogen-modified tetrahydrophthalimides [274 to 282] has a tendency to show a little bit lower activity than that of 5-alkoxy analogs.

Improved synthetic routes to 5-sulfur-modified tetrahydrophthalimides are also outlined in Scheme 2. As a conventional method for introducing a mercapto group at 5-position of the phenyl ring, a pyridinium salt of 5-acetylamino-2-chloro-4-fluorobenzenesulfonic acid obtained by sulfonation of 4-chloro-2-fluoroacetoanilide (*20*) with fuming sulfuric acid is chlorinated to chlorosulfonyl derivative (*21*), and the product is reduced by metallic zinc in acetic acid or red phosphorus with iodine to give 5-mercapto derivative (*22*) [289]. Modified approach makes it possible for single pot chlorosulfonation, in which 4-chloro-2-fluoroacetoanilide is treated successively with chlorosulfonic acid, fuming sulfuric acid and thionylchloride affording 5-acetylamino-2-chloro-4-fluorobenzenesulfonyl chloride (*21*). Direct reaction of thioglycolic acid with diazonium salt of aniline (*23*) is a more convenient method for the synthesis of 5-alkoxycarbonylmethylthio analogs [284]. As an

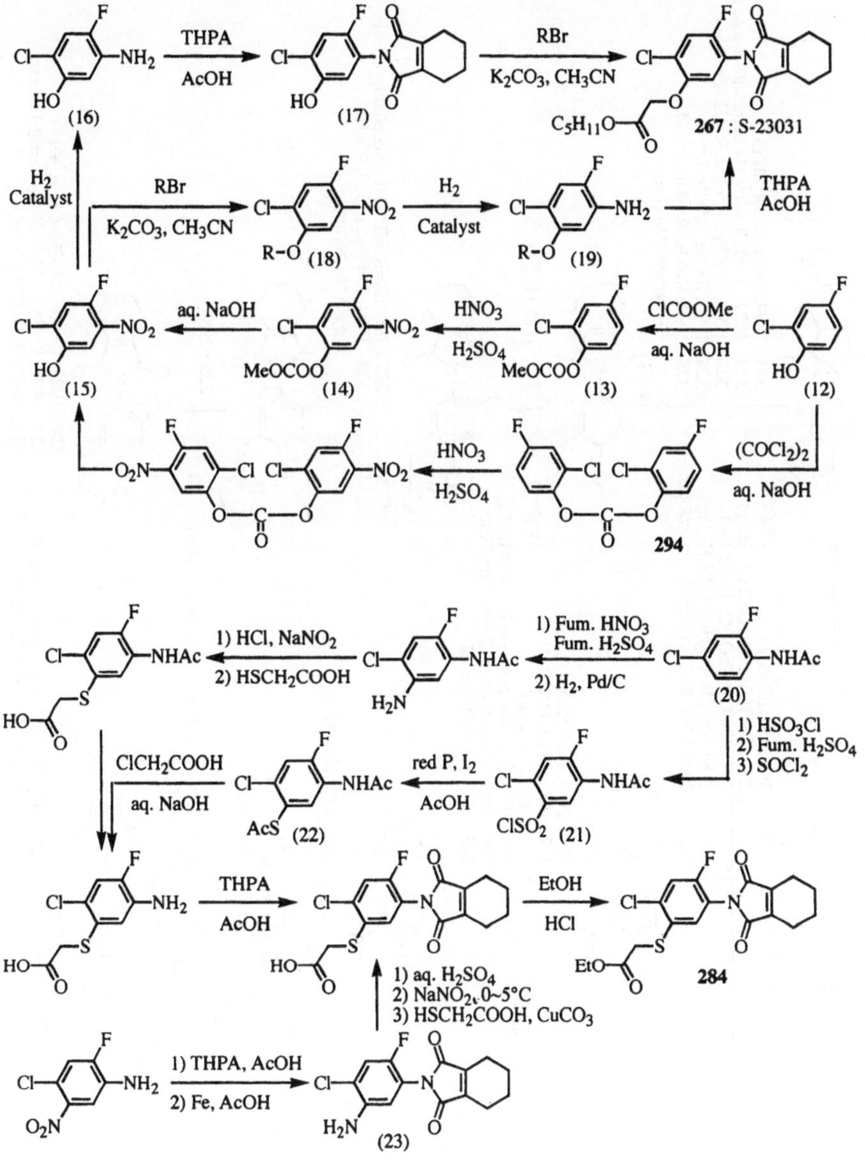

Scheme 2. Synthetic routes for 5*O*- and 5*S*-modified *N*-(4-chloro-2-fluorophenyl)-3,4,5,6-tetrahydrophthalimides

alternative approach, Ihara Chemical group has established a more efficient method for the synthesis of fluthiacet-methyl [575: KIH-9201] *via* bis(2-chloro-4-fluorophenyl)disulfide, as described later in detail.

Tetrahydrophthalimides substituted with alkoxycarbonyl, acyl, protected formyl and substituted vinyl groups at 5-position of each phenyl ring are summarized in Fig. 7. These compounds are generally synthesized *via*

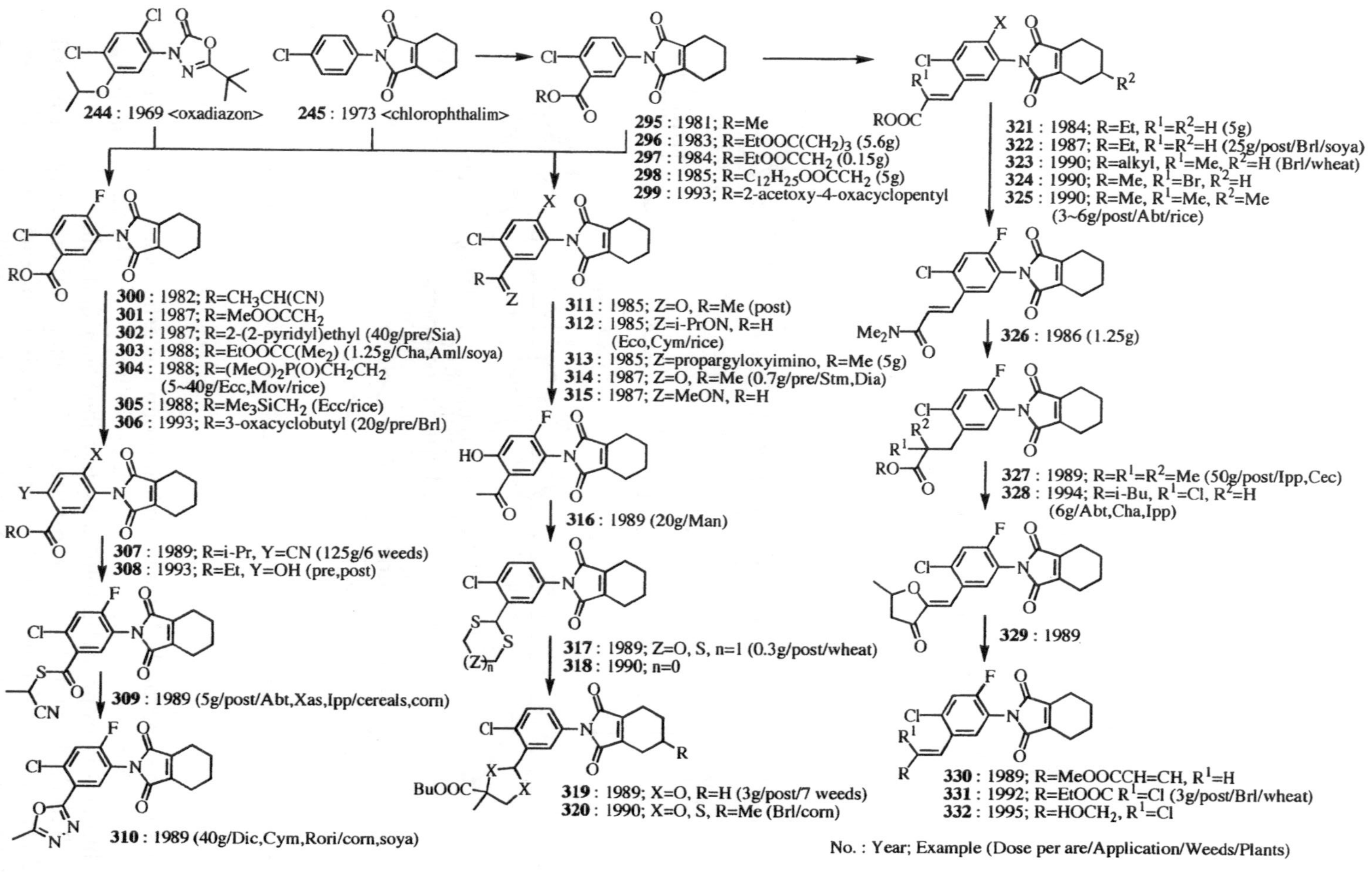

Fig. 7. Structural modifications at 5-position of phenyl ring in tetrahydrophthalimides: Alkyloxycarbonyl, acyl and vinyl moieties

halogenated benzoic acid or benzaldehyde prepared by oxidation of the corresponding halotoluene. As shown in Scheme 3, introduction of a cyano group at 4-position is carried out by nucleophilic substitution of 4-iodized analog (*24*) with cuprous cyanide [307].

Modifications with an alkoxycarbonylvinyl group at 5-position have been attempted taking account of side pendants on pyrrole rings of protoporphyrinogen IX as the substrate of Protox. Several methods for introduction of the alkoxycarbonylvinyl group have been reported, as shown in Scheme 3. For example, ethyl (2-chloro-5-nitrophenyl)glycidate (*25*) prepared by Darzens condensation is treated dropwise with thionylchloride in DMF at 100 °C affording ethyl dichloronitrocinnamate (*26*), which is reduced and condensed with THPA to give resultant tetrahydrophthalimides [331]. Wittig reaction is the most direct route for transformation of a formyl group (*27*) to alkoxycarbonylvinyl moiety. The C—C double bond can be readily hydrogenated to an alkoxycarbonylethyl group. Carfentrazone-ethyl [631: F-8426] consists of 1,2,4-triazolinone heterocycle and the phenyl ring decorated with the ethoxycarbonylethyl group at 5-position; however, the synthetic methodology for F-8426 is strategically different from these approaches, as shown in Scheme 3.

Figure 8 shows structural modifications of tetrahydrophthalimide derivatives bearing bicyclic benzologs fused with heterocycles at 4- and 5-positions of

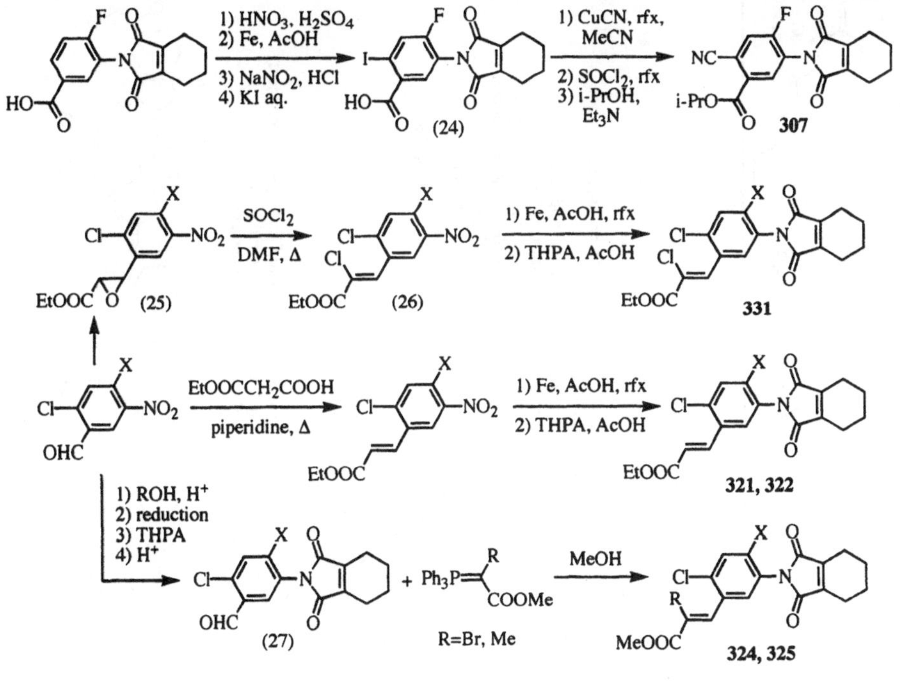

Scheme 3. Synthetic routes for *N*-aryl-3,4,5,6-tetrahydrophthalimides modified with ester moiety at 5-position

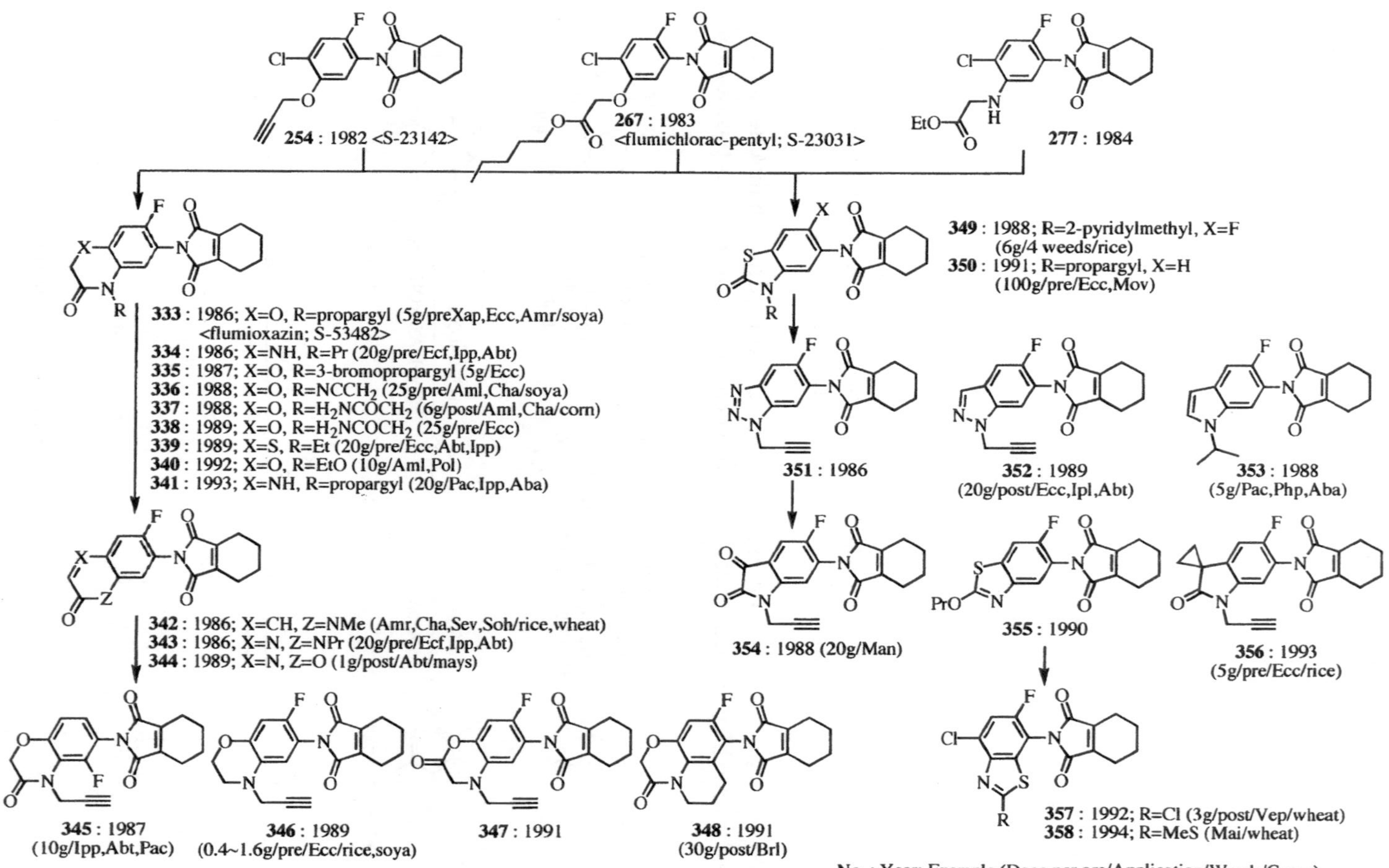

No. : Year; Example (Dose per are/Application/Weeds/Crops)

Fig. 8. Conversion of phenyl ring into bicyclic benzologs fused with 5- and 6-membered heterocycles

the phenyl ring instead of the 2,4,5-trisubstituted phenyl moiety. A variety of benzologs such as benzoxazinone, benzothiazinone, quinolinone, quinoxalinone, indolinone, benzothiazolinone, indazole and benzotriazole are introduced on the nitrogen atom of tetrahydrophthalimide ring. Among them, benzoxazinone analogs seem to be more active. Herbicidal activity of the series is considerably influenced by the substituents on the nitrogen atom at 3-position of benzoxazine ring; in particular, the herbicidally most active analog, flumioxazin [333: S-53482], has a propargyl group at that position. Flumioxazin has been developed by the Sumitomo Chemical group as a pre-emergent herbicide active on annual broadleaf weeds such as velvetleaf, lambsquaters, amaranths, morning glory and prickly sida at the rates of 50 to 100 g/ha.

While many methods exist for the preparation of benzoxazinone-substituted tetrahydrophthalimides, efficient synthetic approaches to S-53482 are summarized in Scheme 4. Benzoxazinone skeleton is cleanly constructed by reduction and intramolecular amidation of 4-methoxycarbonylmethyloxy-5-nitrobenzene derivatives (*28–30*) prepared by alkylation of nitrophenol with haloacetic acid or its ester, or by nucleophilic substitution of a fluorine atom activated by the neighbor nitro group with glycolic acid or its ester.

Recently, FMC group reported that an indolinone analog [356] having spirocyclopropane component. 3-Fluorophenylacetic acid (*31*) is bis-nitrated in middle yield because of its low regioselectivity. Reduction of dinitro compound (*32*) over palladium on carbon gives the diamino compound which is induced to cyclize upon treatment with acid, affording 6-aminoindolin-2-one (*33*). Condensation with THPA, *N*-protection and bis-alkylation with 1,2-dibromoethane give the spirocyclopropyl indolin-2-one (*34*), which is converted to the final product [356] by deprotection followed by *N*-alkylation, typically with propargyl bromide (Scheme 4).

Miscellaneous tetrahydrophthalimides which cannot be divided into any classes shown in Figs. 5 to 8 are represented in Fig. 9. Whereas structural modifications of the phenyl ring have been effectively spread over more than 15 years, there are not so many trials to modify tetrahydrophthalimide ring (Fig. 10). Transformation of the carbonyl group into thiocarbonyl, imino and methylidene groups maintains the aromaticity and planarity of imide ring; however, conversion to methylthio, chloro and hydroxy groups leads to disappearance of aromatic character increasing herbicidal activity. Reduction of cyclohexene ring to cyclohexane also decreases herbicidal activity by reason of the same. Several maleimide derivatives combined with the typical 2,4,5-trisubstituted phenyl groups or herbicidally more active benzolog moieties have been synthesized, too.

2.3.2
Tetrahydrophthalamides and Related Compounds

In the synthesis of tetrahydrophthalimides by condensation of THPA with anilines, the reaction is carried out below room temperature yielding *N*-aryl-

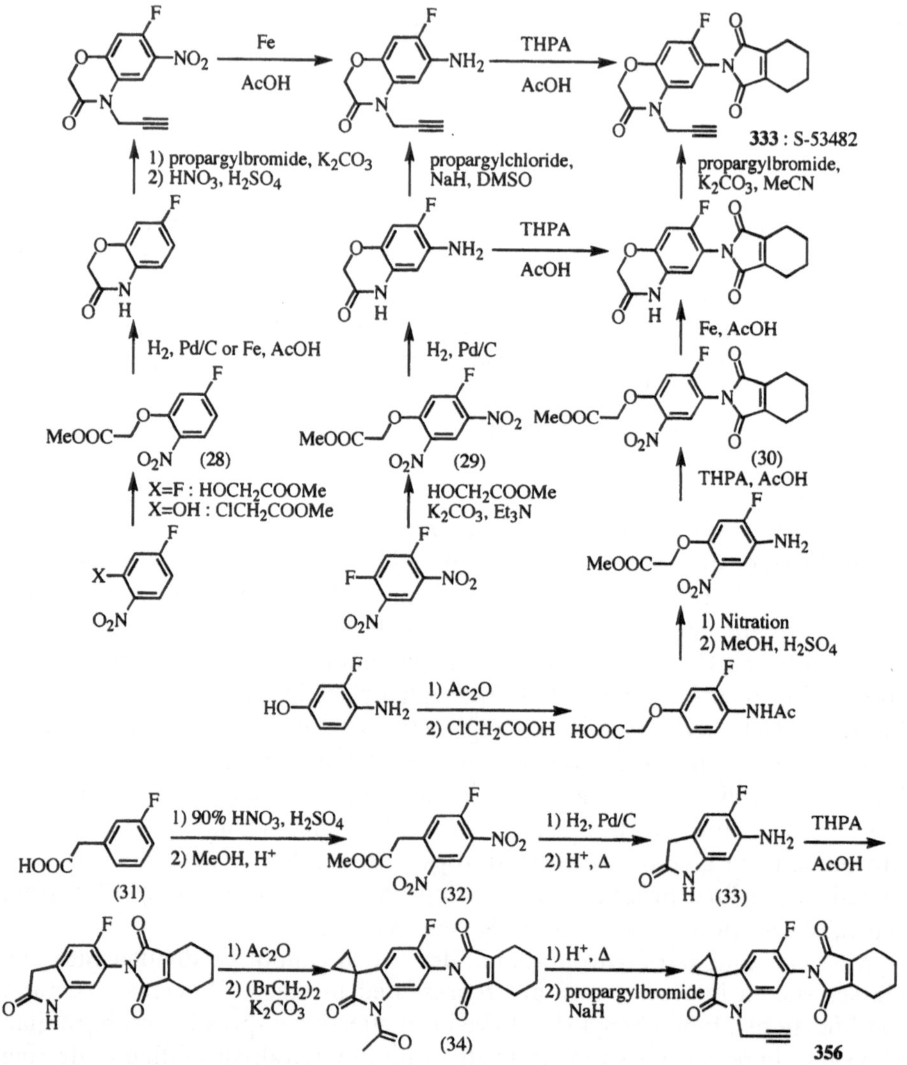

Scheme 4. Synthetic routes for 3,4,5,6-tetrahydrophthalimides substituted with bicyclic benzologs

3,4,5,6-tetrahydrophthalamic acids [407, 424, 425] in good yields, which are key precursors to synthesize *N*-aryl-3,4,5,6-tetrahydrophthalisoimides [569 to 571]. That is, thus obtained tetrahydrophthalamic acids are treated with dehydration agents such as dicyclohexylcarbodiimide (DCC) affording isoimides in good yields, while tetrahydrophthalamic acids are heated in acetic acid to give tetrahydrophthalimides (Scheme 5). In addition, *N*-aryl-3,4,5,6-tetrahydrophthalamic acid esters are prepared by treating tetrahydrophthalic acid chloride with anilines. *N*-Aryl-3,4,5,6-tetrahydrophthalamides [408 to 423] are cleanly synthesized by reacting tetrahydrophthalimides with primary

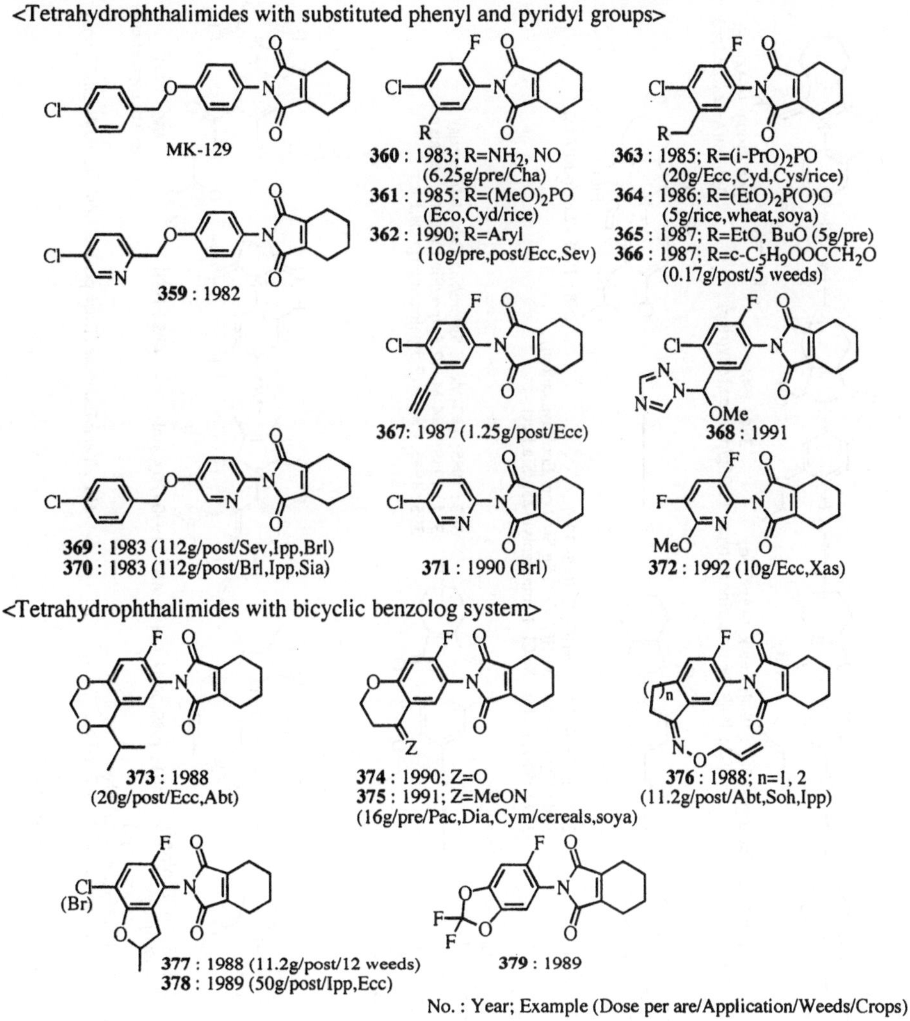

Fig. 9. Miscellaneous tetrahydrophthalimides and maleimides

or secondary amines in the presence of bases. Using bulky amines or amines with low nucleophilicity, reactions with isoimides give rather good results (Fig. 11).

N-Aryl-3,4,5,6-tetrahydrophthalamic acids, their esters, isoimides and tetrahydrophthalamides are as potent as the corresponding tetrahydrophthalimides; however, these compounds, which are easily transformed to imides biologically, are thought to be the so-called pro-drugs of tetrahydrophthalimides. Moreover, *N*-aryltetrahydrophthalamic ester analogs [431 to 437], in which the nitrogen atom neighboring the phenyl ring is decorated with acyl groups such as chloroacetyl and benzoyl groups, are stable against recyclization to imides. These compounds are prepared by treatment of

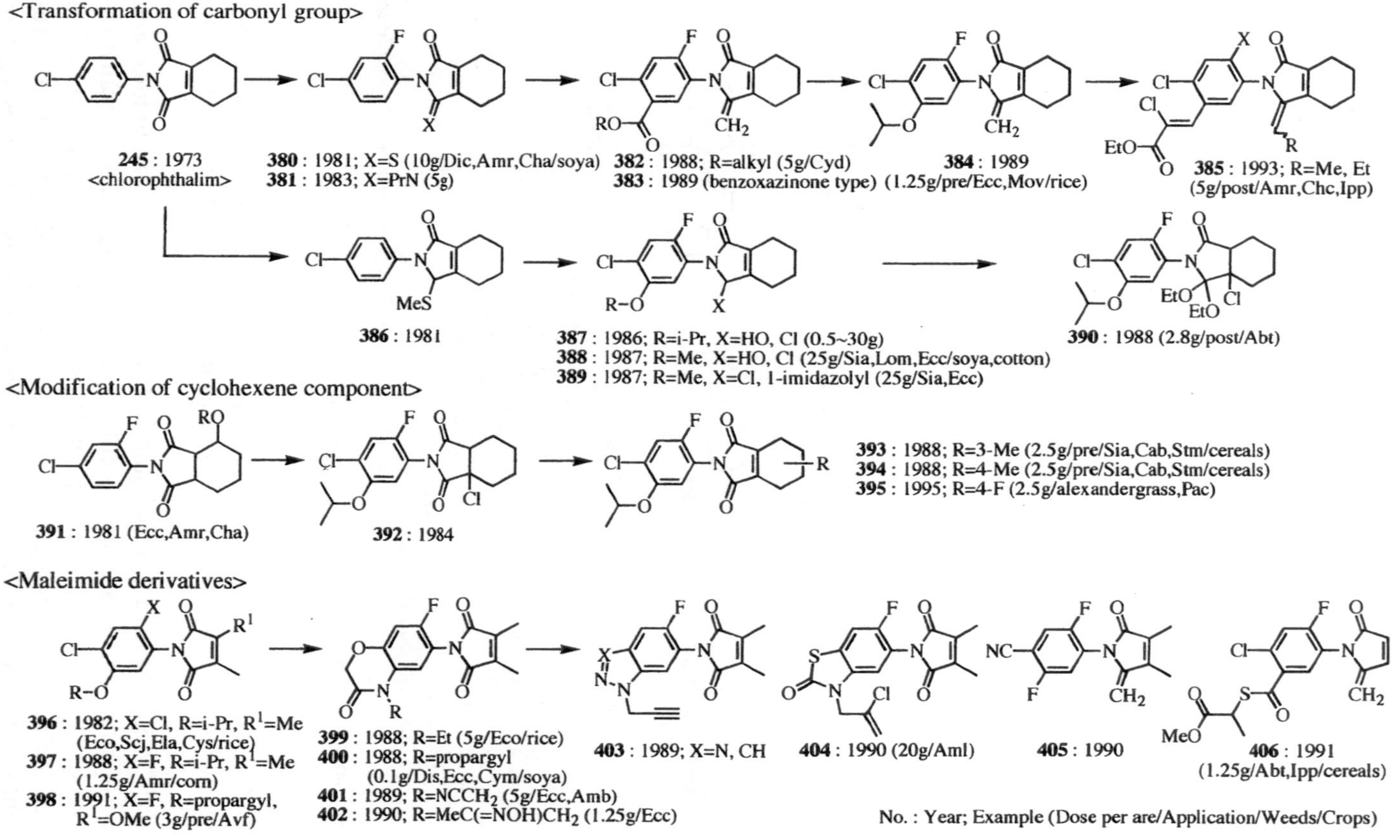

Fig. 10. Structural modifications of the tetrahydrophthalimide ring

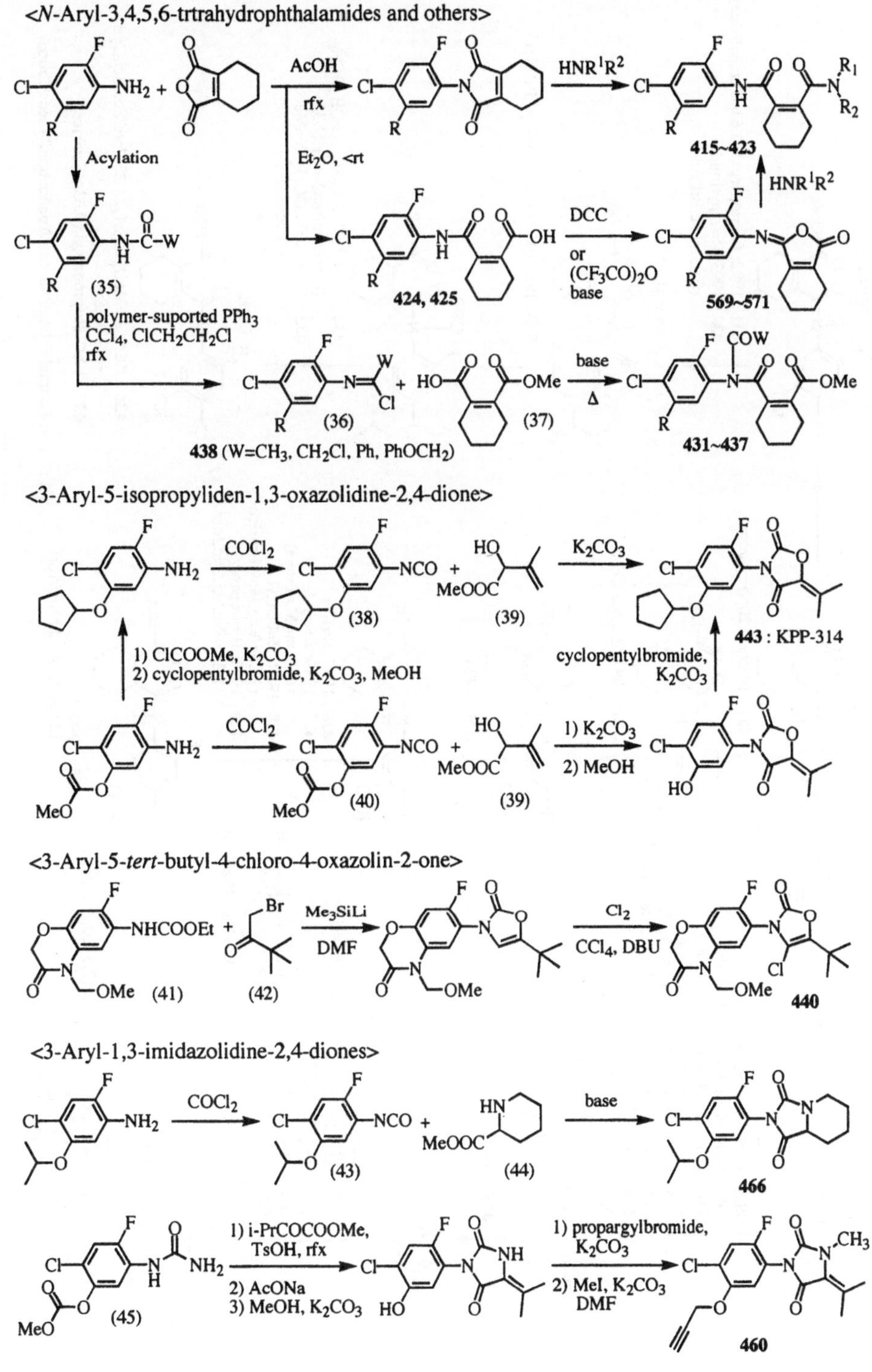

Scheme 5. Synthetic routes for tetrahydrophthalamides, oxazolidinedione and imidazolidinediones

<Tetrahydrophthalamide> <Tetrahydrophthalamic acid and ester>

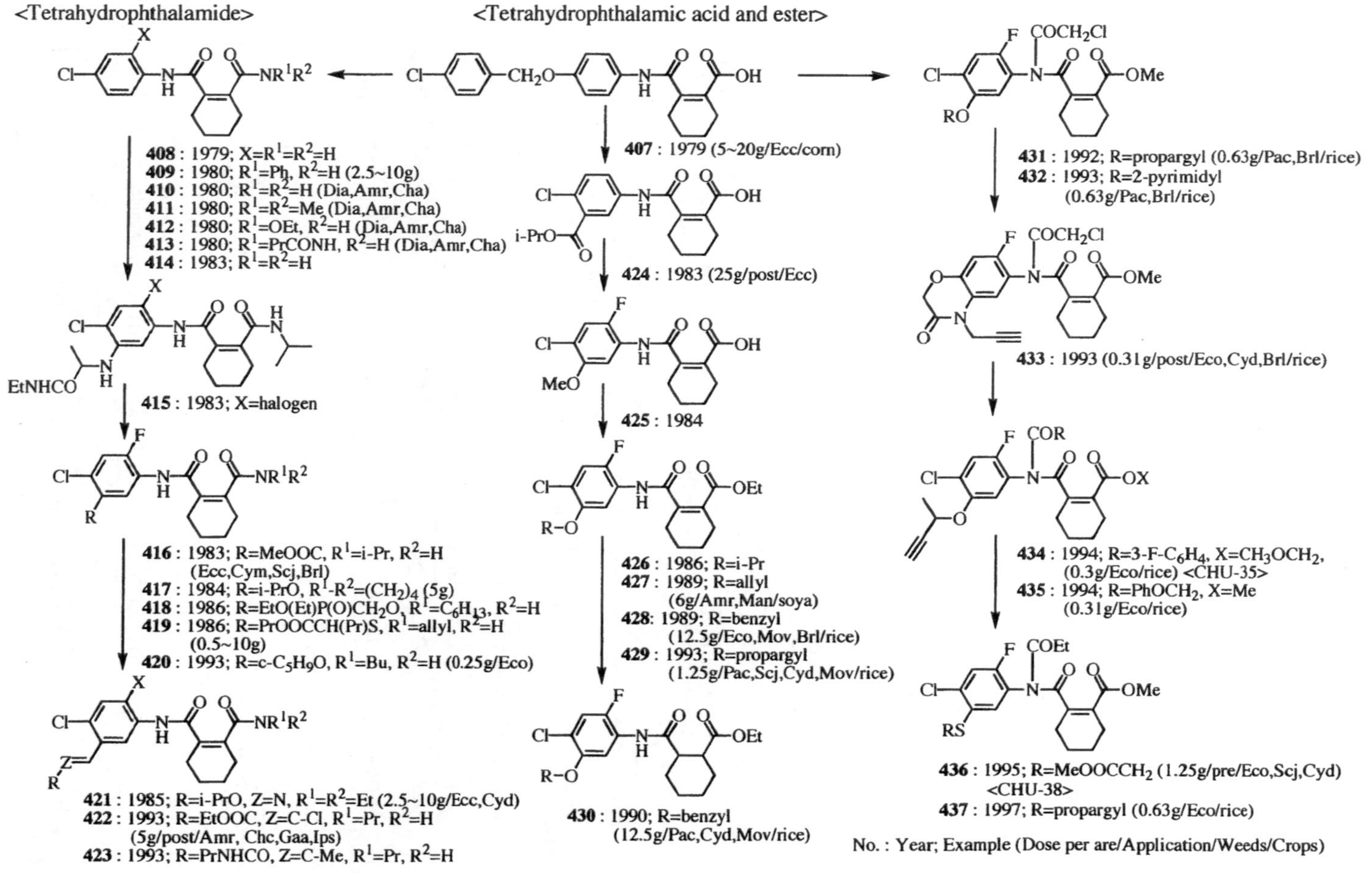

408 : 1979; X=R¹=R²=H
409 : 1980; R¹=Ph, R²=H (2.5~10g)
410 : 1980; R¹=R²=H (Dia,Amr,Cha)
411 : 1980; R¹=R²=Me (Dia,Amr,Cha)
412 : 1980; R¹=OEt, R²=H (Dia,Amr,Cha)
413 : 1980; R¹=PrCONH, R²=H (Dia,Amr,Cha)
414 : 1983; R¹=R²=H

415 : 1983; X=halogen

416 : 1983; R=MeOOC, R¹=i-Pr, R²=H (Ecc,Cym,Scj,Brl)
417 : 1984; R=i-PrO, R¹-R²=$(CH_2)_4$ (5g)
418 : 1986; R=EtO(Et)P(O)CH_2O, R¹=C_6H_{13}, R²=H
419 : 1986; R=PrOOCCH(Pr)S, R¹=allyl, R²=H (0.5~10g)
420 : 1993; R=c-C_5H_9O, R¹=Bu, R²=H (0.25g/Eco)

421 : 1985; R=i-PrO, Z=N, R¹=R²=Et (2.5~10g/Ecc,Cyd)
422 : 1993; R=EtOOC, Z=C-Cl, R¹=Pr, R²=H (5g/post/Amr, Chc,Gaa,Ips)
423 : 1993; R=PrNHCO, Z=C-Me, R¹=Pr, R²=H

407 : 1979 (5~20g/Ecc/corn)

424 : 1983 (25g/post/Ecc)

425 : 1984

426 : 1986; R=i-Pr
427 : 1989; R=allyl (6g/Amr,Man/soya)
428: 1989; R=benzyl (12.5g/Eco,Mov,Brl/rice)
429 : 1993; R=propargyl (1.25g/Pac,Scj,Cyd,Mov/rice)

430 : 1990; R=benzyl (12.5g/Pac,Cyd,Mov/rice)

431 : 1992; R=propargyl (0.63g/Pac,Brl/rice)
432 : 1993; R=2-pyrimidyl (0.63g/Pac,Brl/rice)

433 : 1993 (0.31g/post/Eco,Cyd,Brl/rice)

434 : 1994; R=3-F-C_6H_4, X=CH_3OCH_2, (0.3g/Eco/rice) <CHU-35>
435 : 1994; R=PhOCH_2, X=Me (0.31g/Eco/rice)

436 : 1995; R=Me$OOCCH_2$ (1.25g/pre/Eco,Scj,Cyd) <CHU-38>
437 : 1997; R=propargyl (0.63g/Eco/rice)

No. : Year; Example (Dose per are/Application/Weeds/Crops)

Fig. 11. Structural modifications of tetrahydrophthalimides and related compounds

tetrahydrophthalic acid half ester (*37*) with *N*-arylimidochloride (*36*) obtained quantitatively by refluxing *N*-acyl anilines (*35*) with polymer-bound triphenylphosphine in carbontetrachloride. Among them, *N*-acyl substituted *N*-aryltetrahydrophthalates, CHU-35, [434] shows good selectivity between soybeans and velvetleaf with pre-emergence application, and CHU-38 [436] shows potent growth control of barnyard grass with no damage to transplanting rice.

2.3.3
Teraconimides, Oxazolidinediones and Oxazolinones

There are not so many 5-membered cyclic imide herbicides containing one nitrogen atom in the hetero ring except for tetrahydrophthalimides and related compounds (Fig. 12). Teraconimides [441] have been developed in the forefront of this series, however, herbicidal activity is slightly low because of lack of aromaticity. The methylene unit in teraconimide ring is replaced with an oxygen atom to produce oxazolidinedione derivative [443: KPP-314], which is the first example of herbicidally active oxazolidinediones, although a few derivatives such as dichlozoline and vinchlozoline are well known to fungicides. Major improvement of the chemical structure is achieved by introducing an isopropylidene group at 5-position of the oxazolidine ring in order to install the aromaticity. In fact, X-ray crystallographic analysis of KPP-314 reveals that the oxazolidine ring is approximately planar and twists by 76 degrees about the axis of the C—N bond joining it to the phenyl ring.

5-Isopropyliden-2,4-oxazolidinedione skeleton is easily formed by addition of 2-hydroxy-3-methyl-3-butenoate (*39*) to 2,4,5-trisubstituted phenylisocyanate (*38*) followed by intramolecular cyclization and olefin isomerization in basic condition (Scheme 5). Using *O*-protected phenylisocyanate (*40*), alkylation of deprotected phenol following construction of oxazolidine ring gives efficiently various analogs with a variety of substituents on the phenyl ring. Among these compounds, pentoxazone [443: KPP-314] provides excellent growth control of barnyard grass and a wide range of annual broadleaf weeds at the rates of 145 to 450 g/ha with pre- and early post-emergence applications in paddy field, and bears good selectivity between barnyard grass and rice. Noteworthy is that pentoxazone keeps its efficacy for more than 45 days because of its good soil adsorption. Pentoxazone is now under development for rice herbicide by Kaken Pharm. and will be marketed in the near future.

Oxazolin-2-one derivatives [439, 440] are a new type of 5-membered cyclic imide herbicides containing one nitrogen atom together with three sp^2 carbons and one oxygen atom. Benzoxadinone-modified oxazolinone analog [440] is synthesized by condensation of *N*-arylcarbamate (*41*) with α-bromoketone (*42*) in the presence of hindered strong base as shown in Scheme 5.

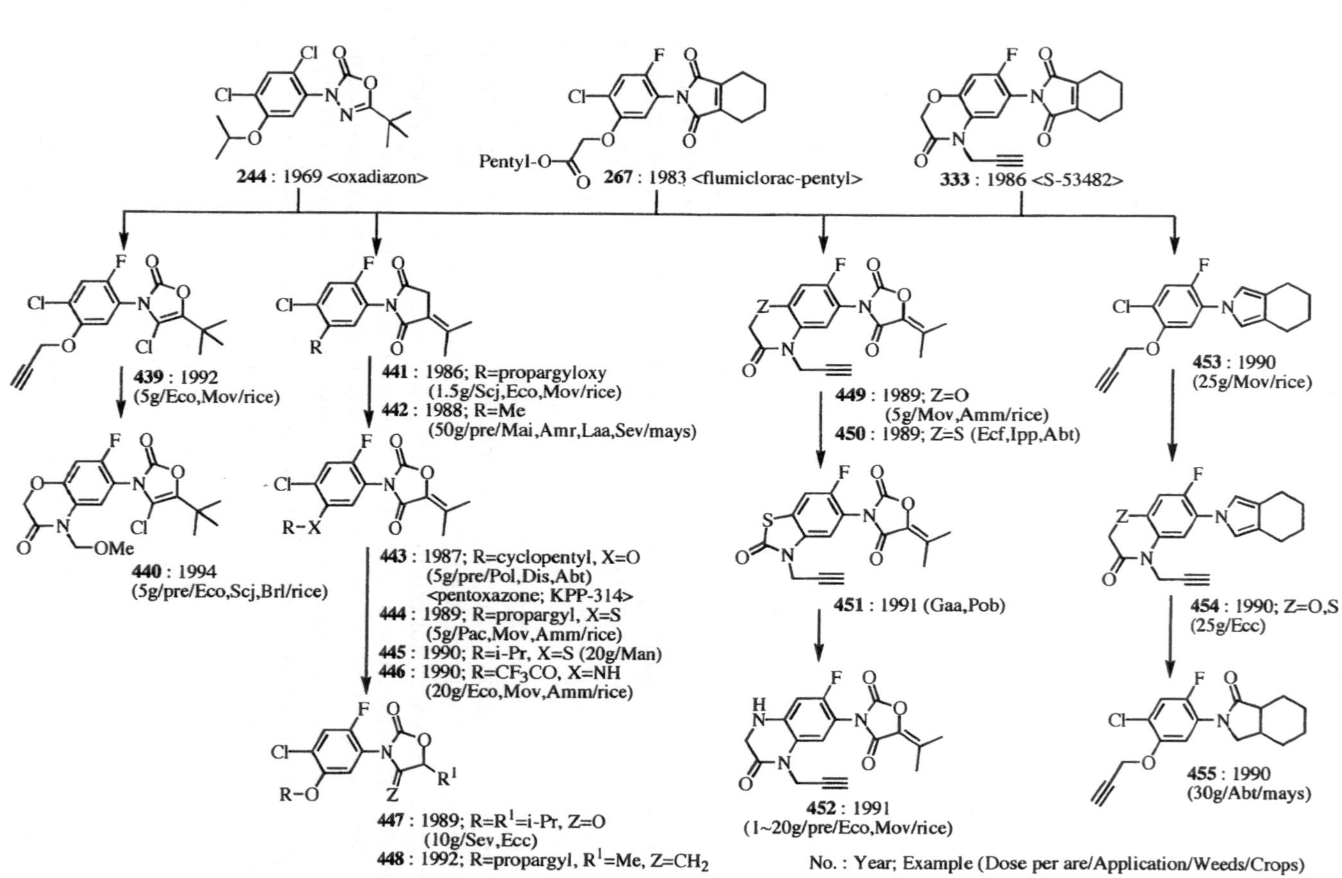

Fig. 12. Structural modifications of cyclic imides containing one nitrogen atom: oxazolinones, teraconimides, oxazolidinediones and pyrroles

2.3.4
Imidazolidinediones (Hydantoins)

A modification map of monocyclic and bicyclic imidazolidine-2,4-diones (hydantoins) is depicted in Fig. 13. Since the early 1980s, hydantoin derivatives have been actively synthesized because hydantoins themselves are useful precursors for the production of α-amino acids and therefore synthetic methodology has been already established. Monocyclic hydantoin derivatives with the regiospecifically functionalized phenyl rings were first reported in 1982 at the same time as the corresponding tetrahydrophthalimides were disclosed. However, structural modifications of bicyclic hydantoins have been more actively studied than monocyclic analogs because bicyclic hydantoins only exhibit strong herbicidal activity among non-aromatic cyclic-imide herbicides.

Bicyclic hydantoin [466] is readily prepared by reacting 4-chloro-5-isopropyloxy-2-fluorophenylisocyanate (*43*) with pipecolinate (*44*) in the presence of base (Scheme 5). Use of 4,5-dehydropipecolinate decorates piperidine component with a C—C double bond [482, 483]. Furthermore, reaction of 2,3-dehydropipecolinate with arylisocyanate gives a new type of bicycled aromatic hydantoins [486] having a double bond at the bridge head of imidazopiperidine ring. Synthesis of aromatic monocyclic-hydantoins [460, 461] is accomplished by condensation of *N*-arylureas (*45*) with α-ketoesters in the presence of acids and bases. Long chained alkyl groups on 1-nitrogen atom decrease herbicidal activity and propargyl-modified analog [460] is the most active.

2.3.5
Oxadiazolinones and Pyrazoles

Structural modifications of cyclic imide class containing two nitrogen atoms in the hetero ring such as 1,3,4-oxadiazolin-2-one and pyrazole are demonstrated in Fig. 14. Oxadiazon [244] has been used as a practical rice herbicide for more than 20 years in Japan. Note that oxadiazon shows powerful herbicidal efficacy even though the *ortho*-position of phenyl ring is substituted with a chlorine atom, while 2-chlorinated analogs of cyclic imides aforementioned are usually not so active. This characteristics is also observed in triazolinone derivatives such as sulfentrazone [629: F-6285] and azafenidin [640: DPX-R6447] mentioned later. A point in common with these compounds is that each heterocyclic ring has no imide component but an amide on which the phenyl ring is attached. In all cyclic imide classes, the chlorine atom at *ortho*-position markedly disturbs rotational flexibility by large barrier with two carbonyl groups on the hetero ring. On the other hand, in the case of cyclic amide class, the molecule has some rotational flexibility because there is just one carbonyl group which brings about large steric hindrance with *ortho*-chlorine atom, which is considered to be the primary cause for showing higher herbicidal activity.

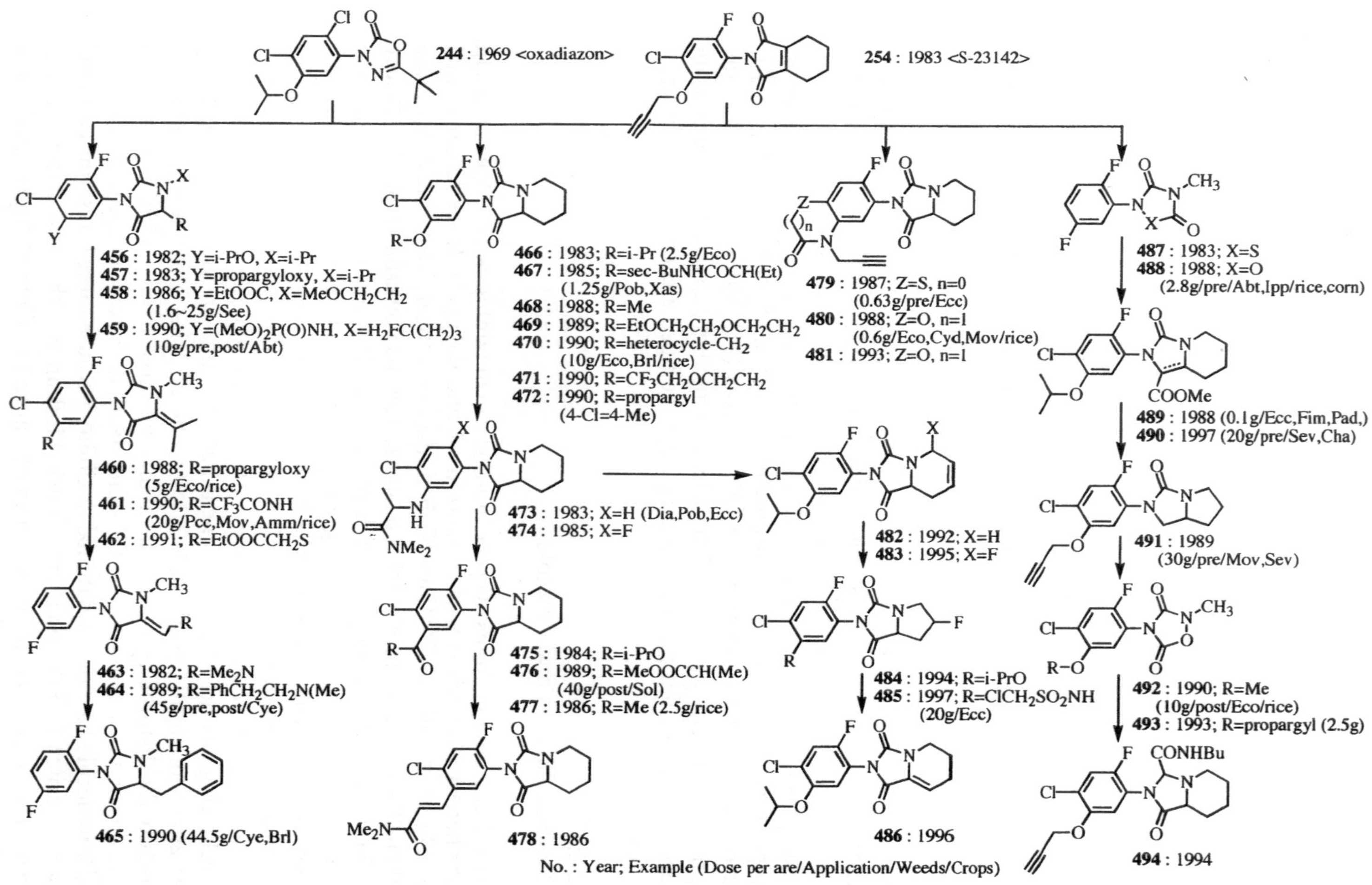

Fig. 13. Structural modifications of cyclic imides containing two nitrogen atoms: imidazolidinediones and related compounds

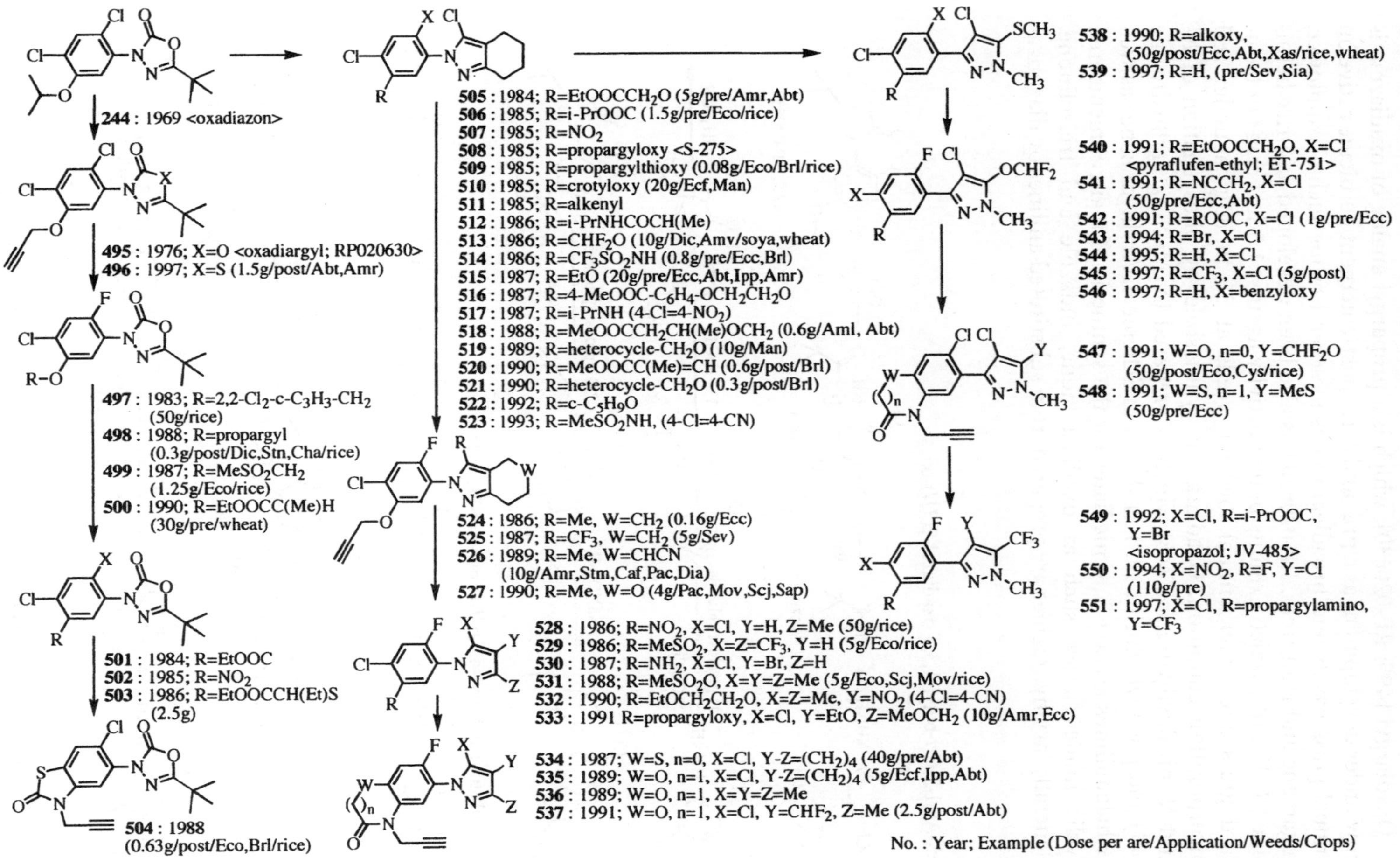

Fig. 14. Structural modifications of cyclic imides containing two nitrogen atoms: oxadiazolinones and pyrazoles

Oxadiargyl [495: RP-020630], which is a propargyl analog of oxadiazon, is now under development as a pre- and early post-emergent herbicide active on annual grass weeds and broadleaf weeds in several crops such as sunflower, sugarcane and fruit trees. RP-020630 has also been developed as a rice herbicide (RYH-118), which exhibits potent activity against 1.5-leaf stage of barnyard grass and other annual lowland weeds at 35 to 80 g/ha under soil incorporation condition. Oxadiargyl is synthesized by phosgenation of *N*-aryl-*N'*-pivaloylhydrazine (*48*) which is prepared by reacting arylhydrazine (*46*) and pivaloylchloride (*47*) as shown in Scheme 6. Arylhydrazines as well as haloanilines are key intermediates for the synthesis of many 5-membered cyclic imide classes such as oxadiazolinone, indazole and triazolinone. Generally, arylhydrazines are prepared from substituted anilines by diazotiza-

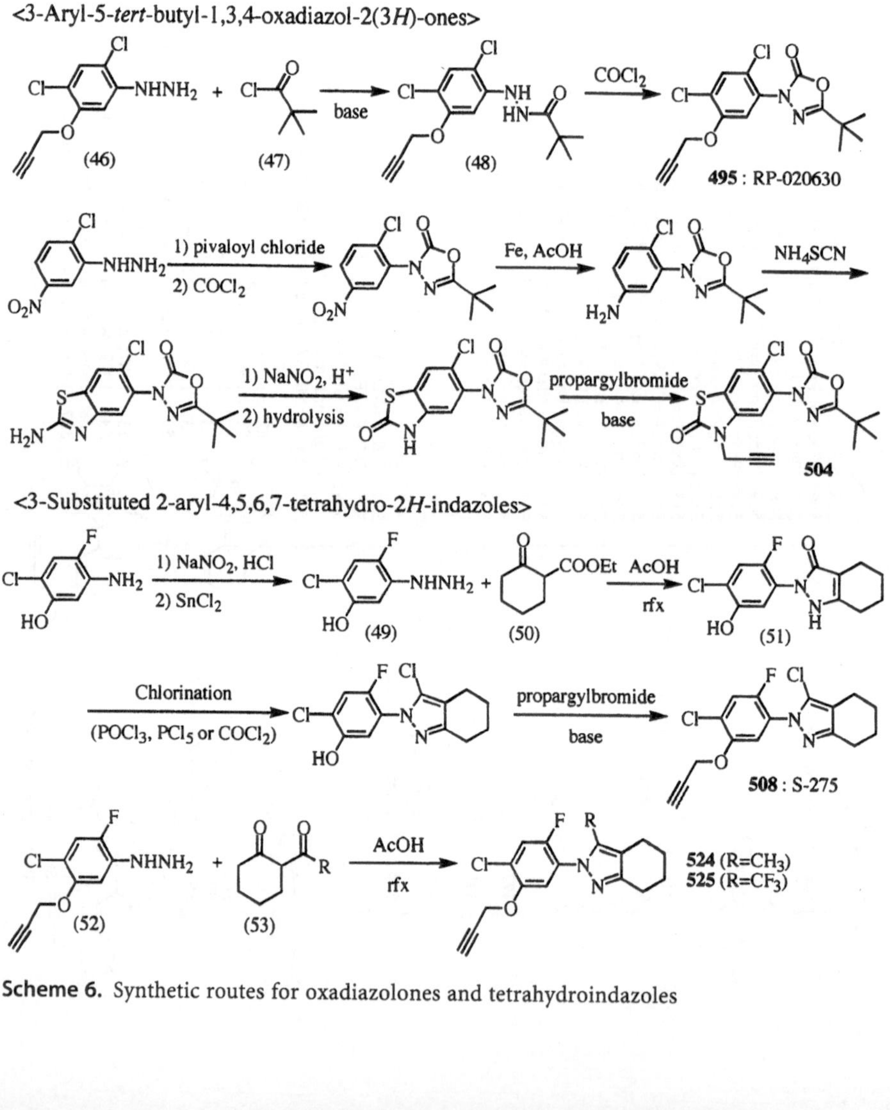

Scheme 6. Synthetic routes for oxadiazolones and tetrahydroindazoles

tion and reduction with stannous chloride. A synthetic method for benzothiazolone-modified oxadiazolinone [504] is also illustrated in Scheme 6.

Pyrazole derivatives have been much more actively investigated among cyclic imide classes containing two nitrogen atoms in the hetero ring. There are two classes of *N*-aryl- and *C*-arylpyrazoles. Structural modifications of *N*-arylpyrazole class have been focused on bicyclic 4,5,6,7-tetrahydroindazole derivative exemplified by S-275 [508]. As illustrated in Scheme 6, arylhydrazine (*49*) is cyclocondensed with 2-(alkoxycarbonyl)cyclohexanone (*50*) to give hexahydro-3*H*-indazol-3-one (*51*) which is chlorinated with phosphorus pentachloride or phosgene to give 2-aryl-3-chloro-4,5,6,7-tetrahydroindazole (S-275) after *O*-propargylation. Reaction of arylhydrazine (*52*) with 2-(trifluoroacetyl)cyclohexanone (*53*: R = trifluoromethyl group) directly gives 3-trifluoromethyl indazole [525].

Another pyrazole analog represented by pyraflufen-ethyl [540: ET-751] has an entirely different chemical structure from the previous cyclic imide class, that is, pyrazole and phenyl rings bond together with a C—C bond at 3-position of hetero ring. Therefore, synthetic strategy for C—N bridged cyclic imides outlined above is of no use for such C—C bridged analogs. For instance, synthetic routes for ET-751 are depicted in Scheme 7. 5-(Ethoxycarbonylmethoxy)benzoylacetate (*54*) is cyclocondensed with methylhydrazine yielding 3-aryl-5-hydroxypyrazole (*55*) which is converted to the final product by *O*-difluoromethylation and chlorination with phosphorus pentachloride. As an alternative preparation, Sandmeyer hydroxylation of 5-amino precursor (*56*) followed by alkylation with ethyl chloroacetate yields the target compound.

Pyraflufen-ethyl [540: ET-751] shows excellent selectivity for winter cereals at the extremely low rates of 6 to 12 g/ha, and also long-term residual activity brought out by chemical and biological stabilities. ET-751 is now under development as a post-emergent contact herbicide active on broadleaf weeds such as cleavers, henbit, chickweeds and wild chamomile; especially, ET-751 provides effective control of 5- to 6-leaf stage of cleavers at the low rate.

5-Methylthiopyrazole analogs [539] are synthesized by condensation of methylhydrazine with benzoylketene dithioacetal (*57*) instead of benzoylacetate (Scheme 7). Multistep reaction from 5-aminophenyl derivative (*58*) gives 3-(5-alkoxyphenyl)-5-methylthiopyrazole [538]. Synthesis of 5-trifluoromethyl analog, for example, isopropazol [549: JV-485], requires cyclocondensation of trifluoroacetylacetophenone (*59*) with hydrazine followed by methylation and bromination. The nitrated product [550] can be subjected to further structural modifications by electrophilic substitution and Sandmeyer chlorination following reduction of the nitro group.

Isopropazol [549: JV-485] is a new selective herbicide discovered by the Monsanto group and jointly developed with the Bayer company. With pre-emergence application to winter wheat at the rates of 125 to 175 g/ha, JV-485 provides effective control of important grass weeds such as water foxtail,

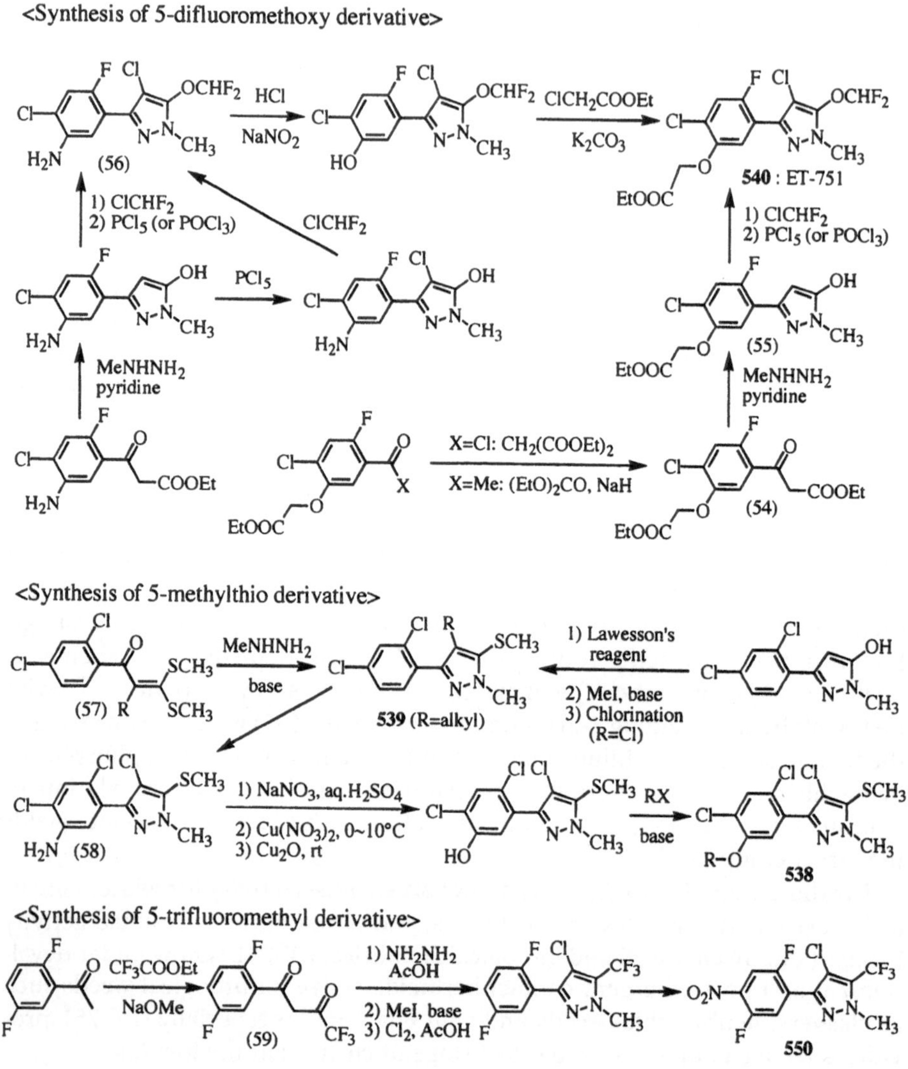

Scheme 7. Synthetic routes for 5-substituted 3-aryl-4-chloro-1-methylpyrazoles

ryegrass, bluegrass and canarygrasses as well as many dicotyledonous weeds including cleavers.

2.3.6 Triazolidinediones, Thiadiazolidinones and Triazolinones

Cyclic imides containing three nitrogen atoms include 4-aryl-1,2,4-triazolidine-3,5-diones and 2-aryl-1,2,4-triazolin-3-ones. Moreover, 2-arylimino-1,3,4-thiadiazolidin-5-ones are also included in this class as structural isomers of 1,2,4-triazolidine-3,5-diones. These heterocycles have

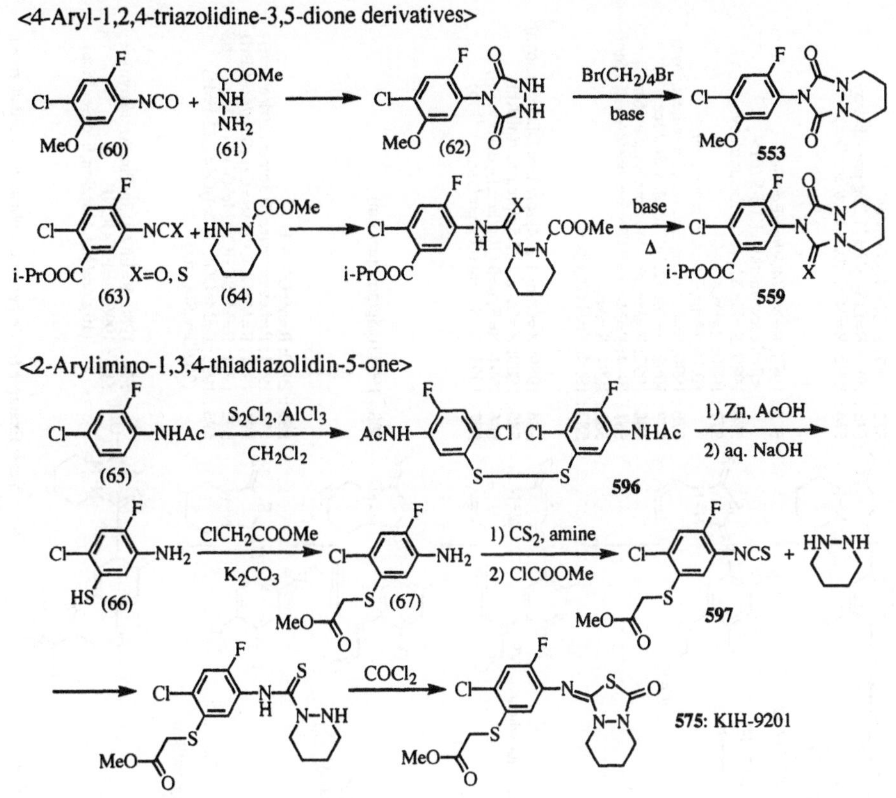

Scheme 8. Synthetic routes for triazolidinediones and thiadiazolidinone

been continuously investigated since 1980, and further structural modification stimulated by the discovery of practically active 2-arylimino-1,3,4 thiadiazolidin-5-one, fluthiacet-methyl [575: KIH-9201], is the most aggressive of all 5-membered cyclic imides since 1990 (Fig. 15).

Chemical structure of bicyclic 1,2,4-triazolidine-3,5-dione is considered to be isoelectric of tetrahydrophthalimide by replacing the double bond with a pair of nitrogen atoms, and possesses an aromatic character. Two methods are provided for the synthesis of bicyclic triazolidinediones, as shown in Scheme 8. 4-Chloro-2-fluoro-5-methoxyphenylisocyanate (*60*) condenses with hydrazinocarboxylate (*61*), and the product (*62*) is *N,N*-dialkylated with α,ω-dihaloalkane to give bicyclic 4-aryl-1,2,4-triazolidine-3,5-dione [553]. Cyclocondensation of arylisocyanate (*63*) with pyridazinecarboxylate (*64*) enables one-pot synthesis of triazolidinedione [559]. In each approach, arylisothiocyanate gives the corresponding thioxo analog.

2-Arylimino-1,3,4-thiadiazolidin-5-ones as structural isomers of 4-aryl-1,2,4-triazolidine-3,5-diones exhibit strong peroxidizing action. Herbicidally active species are thought to be triazolidinedione because 1,3,4-thiadiazolidin-5-ones are easily isomerized to triazolidinediones biologically

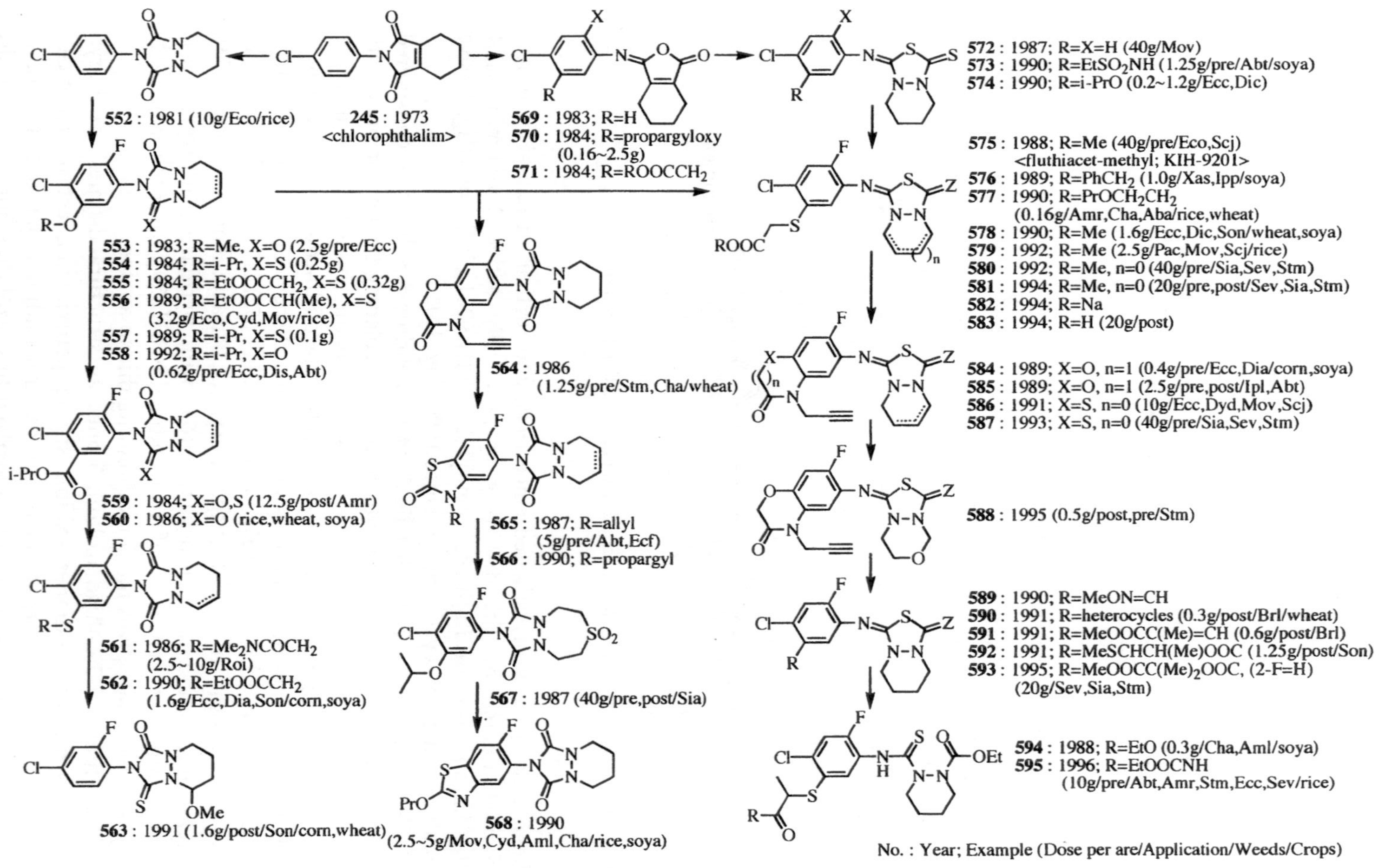

Fig. 15. Structural modifications of cyclic imides containing three nitrogen atoms: triazolidinediones and thiadiazolidinones

by glutathion-*S*-transferase (GST). Mode of isomerization is described in detail in Chapter 7.

Fluthiacet-methyl [575: KIH-9201] is typical of 2-arylimino-1,3,4-thiadiazolidin-5-one belonging to Protox inhibitors (cf. Chapter 6). In the synthesis of KIH-9201, introduction of a mercapto group at 5-position of the phenyl ring is a key reaction. Ihara Chemical group has developed a unique method grown out of conventionally available methodology consisting of sulfonation with fuming sulfuric acid and reduction by red phosphorus with iodine. That is, 4-chloro-2-fluoroacetanilide (*65*) is treated with dichlorodisulfide in the presence of aluminum trichloride to give diphenyldisulfide [596], which is reductively cleaved and deprotected, affording the desired mercaptoaniline (*66*) as depicted in Scheme 8.

Another noticeable improvement is observed in the preparation of arylisothiocyanate [597]. *S*-Alkylated aniline (*67*) reacts with carbondisulfide in the presence of bulky amines such as 1,4-diazabicyclo[2.2.2]octane and 4-pyrazolidinopyridine yielding dithiocarbamic acid ammonium salts, which are converted into arylisothiocyanates [597] in more than 95% yields by treating with methyl chloroformate. Dithiocarbamic acid salts crystallized out from the reaction system, are readily isolated by filtration and the amines are easily recovered and recycled. When triethylamine and 1,8-diazabicyclo[5.4.0]undec-7-ene are used as the base, 0 and 35% of salt is obtained, respectively. In this process, arylisothiocyanates [597], which are useful as intermediates for other herbicides, are produced in high yields without using toxic thiophosgene.

Fluthiacet-methyl [KIH-9201] is active on a wide range of broadleaf weeds, e.g. velvetleaf, lambsquaters and amaranths, at the low rates of 3 to 10 g/ha; especially, fully grown velvetleaf 60 cm in height can be well controlled at 3 g/ha application. Fluthiacet-methyl was initially commercialized as a post-emergent contact herbicide in corn and soybean crops in 1997 in North America.

Structural modifications of 2-aryl-1,2,4-triazolin-3-ones are summarized in Fig. 16. Among the many herbicidally active compounds synthesized, a few potent candidates have been discovered. Several synthetic approaches distinguished from the substituents at 4- and 5-positions of triazoline ring have been developed for these compounds, while arylhydrazines are common reagents for each method (Scheme 9).

Bicyclic 1,2,4-triazolin-3-ones [639, 640: DPX-R6447] are prepared by reacting arylhydrazine (*68, 70*) with 2-piperidinone (*69*) and its derivative (*71*). Recently, it was reported that DPX-R6447 is readily prepared by reacting *N*-chlorocarbonyl-2-piperidinone (*73*) with *N*-(2,4-dichloro-5-propargyloxyphenyl)-*N′*-formylhydrazine (*72*). Alkylation following cyclic condensation of arylhydrazine (*74*) with *N*-(ethoxycarbonyl)acetimidate (*75*) allows various substituents to be introduced at 4*N*-position of 5-methyltriazolinone [604]. Arylhydrazine (*76*) reacts with acid chloride such as pivaloylchloride to give acylhydrazine, which can then be treated with methylisocyanate to

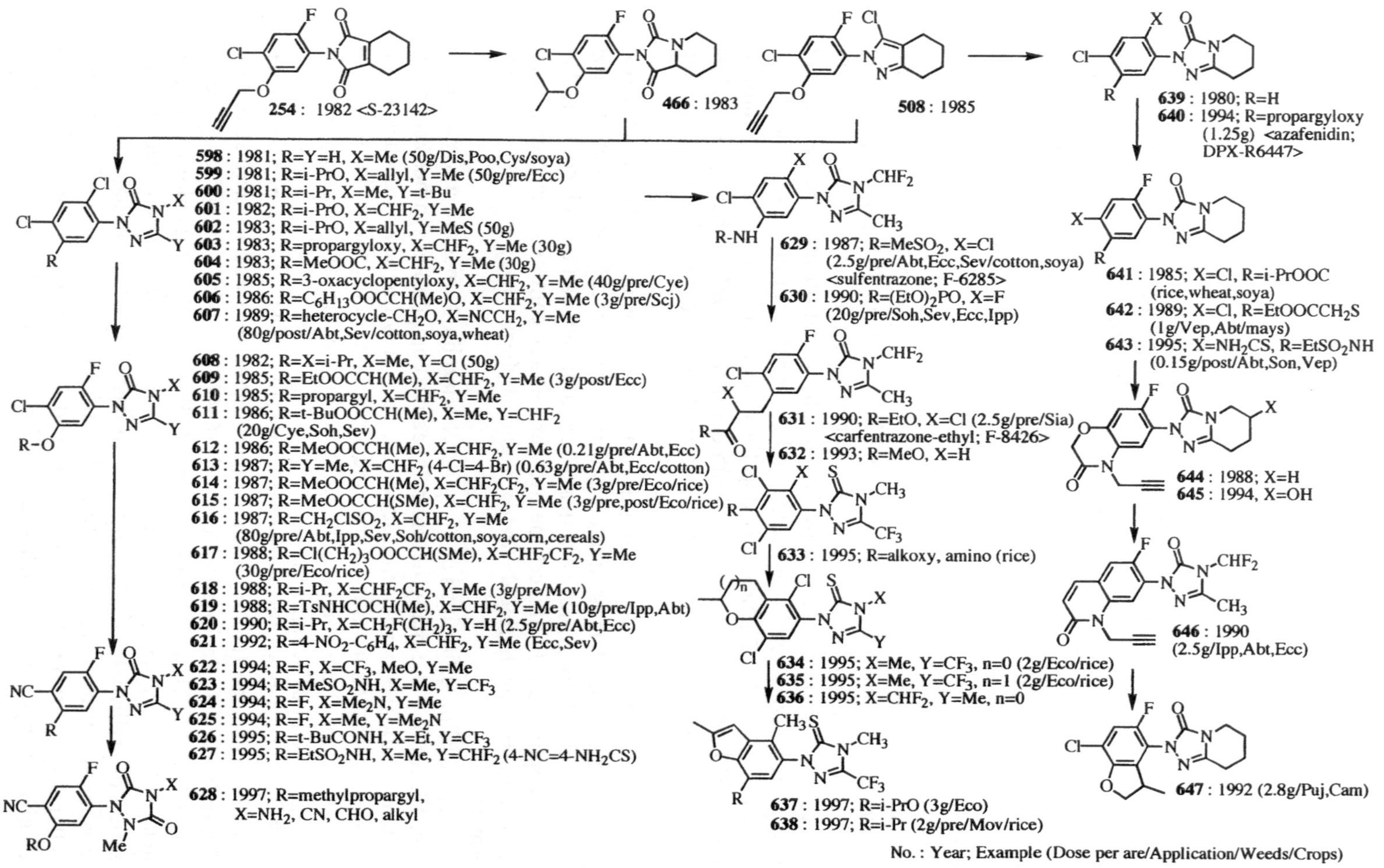

Fig. 16. Structural modifications of cyclic imides containing three nitrogen atoms: triazolinones and related compounds

<Bi- and monocyclic 2-aryl-Δ^2-1,2,4-triazolin-5-one>

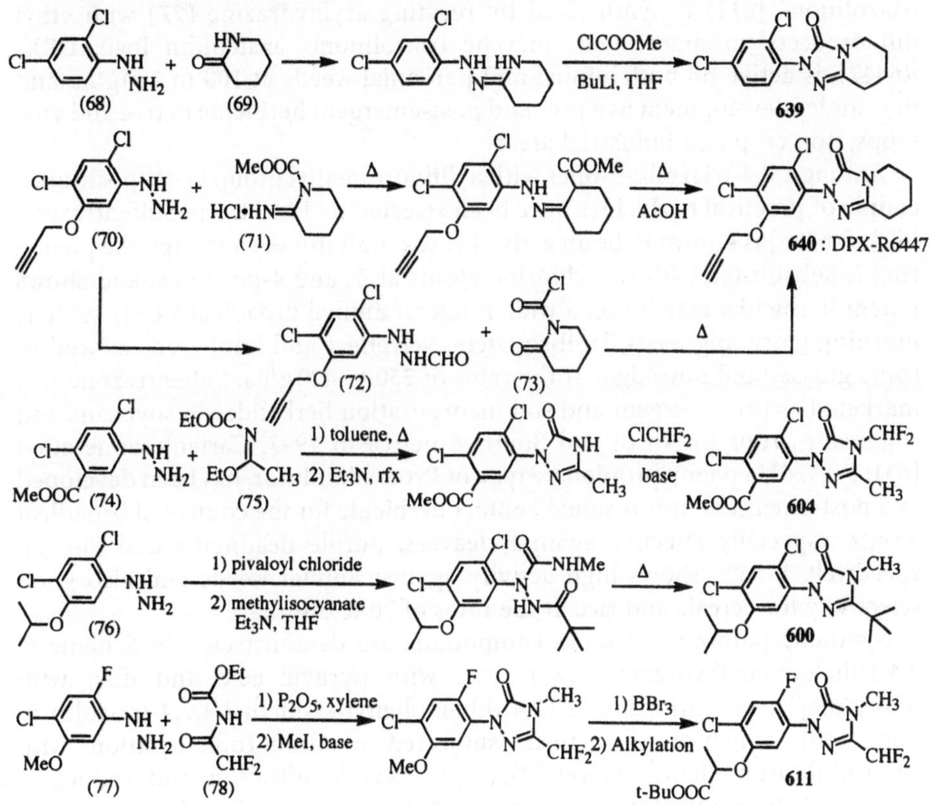

<2-Aryl-5-methyl-4-difluoromethyl-Δ^2-1,2,4-triazolin-5-ones>

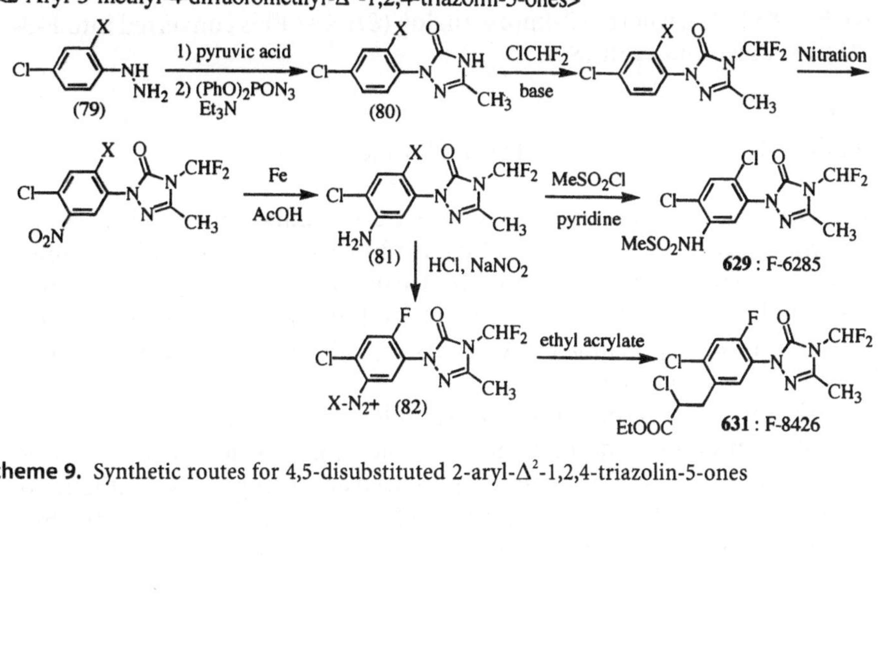

Scheme 9. Synthetic routes for 4,5-disubstituted 2-aryl-Δ^2-1,2,4-triazolin-5-ones

yield, for example, 5-*tert*-butyl-1,2,4-triazolin-3-one [600]. 5-Difluoromethyl triazolinone [611] is synthesized by reacting arylhydrazine (*77*) with ethyl difluoroacetylcarbamate (*78*). Bicyclic triazolinone, azafenidin [640: DPX-R6447], is active on both annual and perennial weeds at 100 to 200 g/ha and now under development as a pre- and post-emergent herbicide in tree and vine crops, non crop and industrial areas.

Among 1,2,4-triazolin-5-ones with a difluoromethyl group at 4*N*-position, a couple of practical herbicides have been selected by FMC Corp. Sulfentrazone [629: F-6285] is known to be an active Protox inhibitor even though the phenyl ring is substituted with two chlorine atoms at 2- and 4-positions, and shows potent herbicidal activity on a wide range of annual broadleaf weeds such as morning glory, pigweeds, lambsquaters, velvetleaf and knotweeds as well as some grasses and nutsedges at the rates of 350 to 420 g/ha. Sulfentrazone was marketed as pre-emergent and soil incorporation herbicides in soybeans and sugarcane crops in North and South America in 1997. Carfentrazone-ethyl [631: F-8426], a phenylpropionate type of Protox inhibitor, has been developed as a post-emergent non-residual contact herbicide for the control of broadleaf weeds, especially effective against cleavers, purple deadnettle and Persian speedwell. F-8426 shows high activity against annual weeds and also good selectivity for cereals and rice at the rates of 20 to 30 g/ha.

Synthetic pathways of these compounds are demonstrated in Scheme 9. 2,4-Dihalophenylhydrazines (*79*) react with pyruvic acid and then with phosphoryl azide affording 2-(2,4-dihalophenyl)-5-methyl-1,2,4-triazolin-5-ones (*80*), which are, in turn, subjected to difluoromethylation with chlorodifluoromethane (Freon-22), regioselective nitration and reduction to give 4-difluoromethyl-2-(5-amino-2,4-dihalophenyl)-5-methyl-1,2,4-triazolin-5-ones (*81*). The 2-chloro analog (*81*: X = Cl) is directly sulfonated to give F-6285 [629], whereas 2-fluoro analog (*81*: X = F) is converted into F-8426 [631] *via* diazonium salt (*82*).

2.3.7
Thiadiazolines, Triazoles and Tetrazolinones

There are a few examples of 5-arylimino-1,2,4-thiadiazoline heterocycles (Fig. 17), which are prepared by reacting isothiocyanates (*83*) with 2-amino-1-pyrroline derivatives (*84*) followed by treatment with bromine (Scheme 10). Thidiazimin [650: SN-124085] has been developed as a post-emergent contact herbicide showing potent activity against broadleaf weeds in winter cereals at the rates of 20 to 40 g/ha.

Tetrazolinone heterocycle is an example of cyclic imide class containing four nitrogen atoms. The most active analog [658] which bears a unique 3-fluoropropyl group at 4*N*-position is prepared according to the synthetic route shown in Scheme 10. Tetrazolinone skeleton (*86*) is formed from phenyl-isocyanate (*85*) and trimethylsilyl azide. Regioselective nitration, reduction of

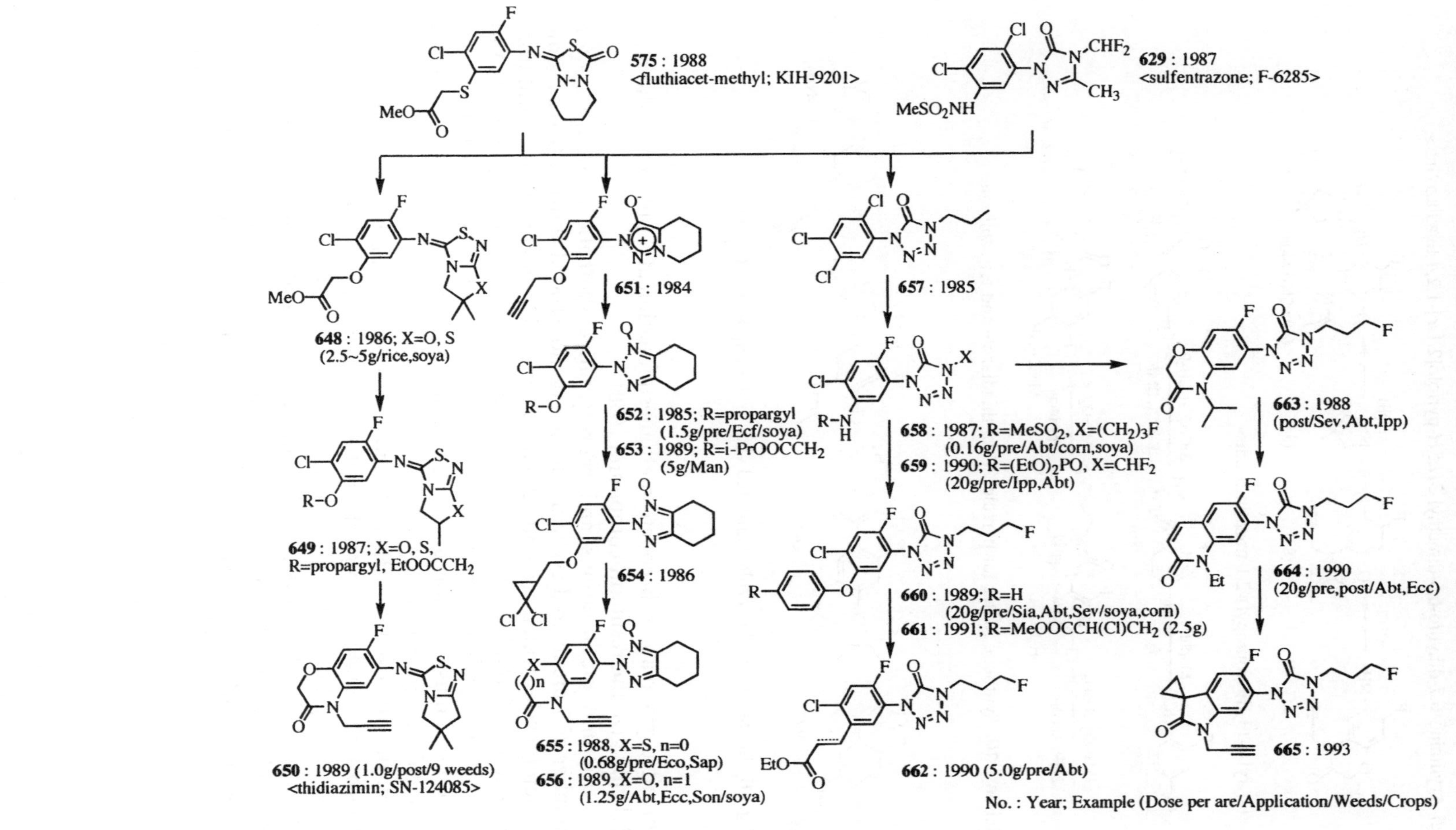

Fig. 17. Structural modifications of thiadiazolines, triazoles and tetrazolinones

<5-Arylimino-6,7-dihydro-6,6-dimethyl-3*H*,5*H*-pyrrolo[2,1-*c*]-1,2,4-thiadiazoline>

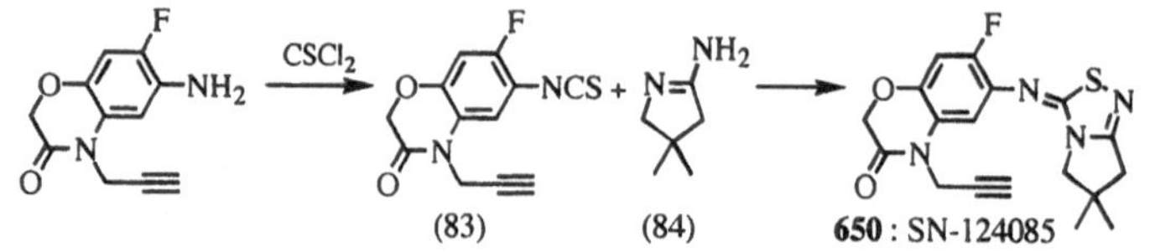

<1-Aryl-4-(3-fluoropropyl)-2-tetrazolin-5-one>

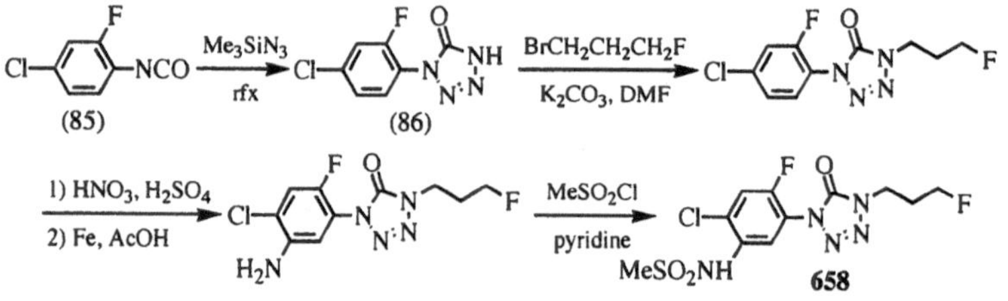

Scheme 10. Synthetic routes for pyrrolo[2,1-*c*] thiadiazole and tetrazolinone

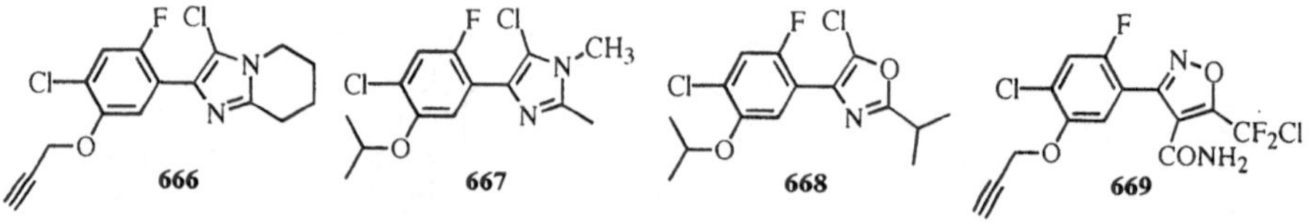

Fig. 18. Imidazoles and oxazole connected with aryl group by C—C bond

the nitro group and subsequent methanesulfonation at 5-position of the phenyl ring give the resultant compound [658].

Structural modifications and representative synthetic approaches of a number of 5-membered cyclic imides are demonstrated above, whereas some of heterocycles which are undoubtedly identified with non-Protox inhibitors have to be left out because of space limitation. However, novel imidazole and oxazole analogs [666 to 669] which are connected with the phenyl rings by a C—C bond must be noticed, taking account of ET-751 exhibiting an excellent herbicidal performance (Fig. 18).

On the other hand, DHL-1777 and LS82-556 have been known to inhibit Protox remarkably, although their chemical structures are definitely different from the cyclic imides aforementioned (Fig. 19). Furthermore, pyrazole analog, MB-39279, shows peroxidizing action even though it consists of the same phenyl ring system and similarly functionalized heterocycle to fipronil (MB-46030), which has been developed as a practical insecticide by the Rhône-Poulenc group. A review of these heterocycles will be presented, taking advantage of another opportunity, as it is doubtful and not easy to select and divide these chemicals into appropriate classes on the basis of their own chemical structures.

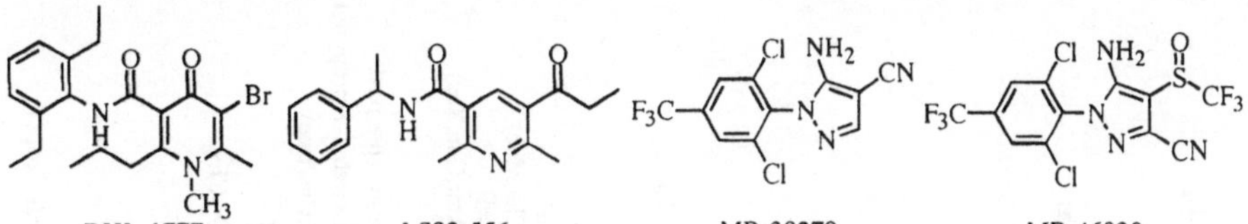

DHL-1777 LS82-556 MB-39279 MB-46030

Fig. 19. Other class of Protox inhibitors and related compounds

2.4 Structural Evolution and Synthesis of 6-Membered Cyclic Imides

It is well known that the herbicidal mode of action of 5-membered cyclic imides such as tetrahydrophthalimides, thiadiazolidinones, oxazolidinediones and so on is a peroxidizing effect, that of light-induced ethane formation and Protox inhibition, including bleaching action. On the other hand, 6-membered cyclic imides have been actively developed all over the world. Some 6-membered heterocycles as well as 5-membered cyclic imides have been reported to show potential peroxidizing action, although the inhibitory mechanism of all these compounds is not always obviously confirmed. In this section, whereas mode of action of all compounds has not been determined, structural evolution and synthesis of 6-membered cyclic imides disclosed for the last decade are summarized, taking into account the similarity of each chemical structure.

2.4.1 6-Trifluoromethylpyrimidinediones (6-Trifluoromethyluracils)

Herbicidally active pyrimidine-2,4-diones (uracils) have been continuously investigated since the early 1980s. The incipient types of uracils such as bromacil and terbacil have no phenyl group, and DCMD is substituted with 3,4-dichlorophenyl group on the amide nitrogen atom. These compounds are not recognized as Protox inhibitors. The first example of 6-membered cyclic imide is thought to be a new uracil analog [670] discovered by the Hoffman-La Roche group in 1986, which bears a highly functionalized phenyl group and a trifluoromethyl group at 6-position. The trifluoromethyl group is the most important and favorable substituent, and enhances herbicidal activity extremely; in fact, 6-methyl uracils [671, 672], 6-phenyl analogs [700] and bicyclic uracil [694] are a little less active than 6-trifluoromethyluracils. Since then, structural modifications of 6-trifluoromethyluracils have been aggressively investigated up to the present date (Figs. 20 to 22). Among them, UCC C-4243 [674] has been developed as a pre-emergent herbicide effective on annual weeds in wheat production, and it is also useful as a desiccant for potatoes.

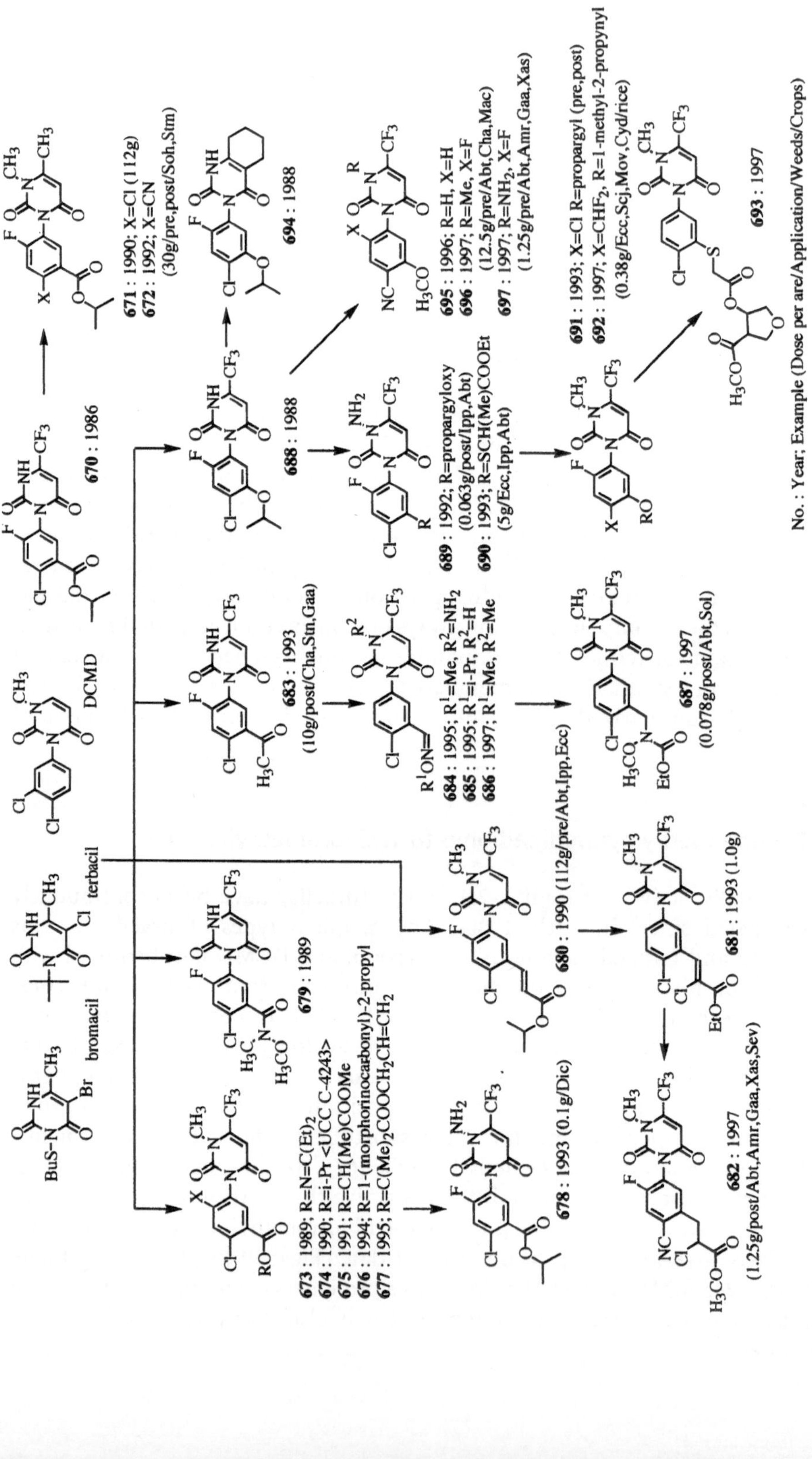

Fig. 20. Structural modifications at 5-position of phenyl ring in 6-CF_3-uracil-type cyclic imides

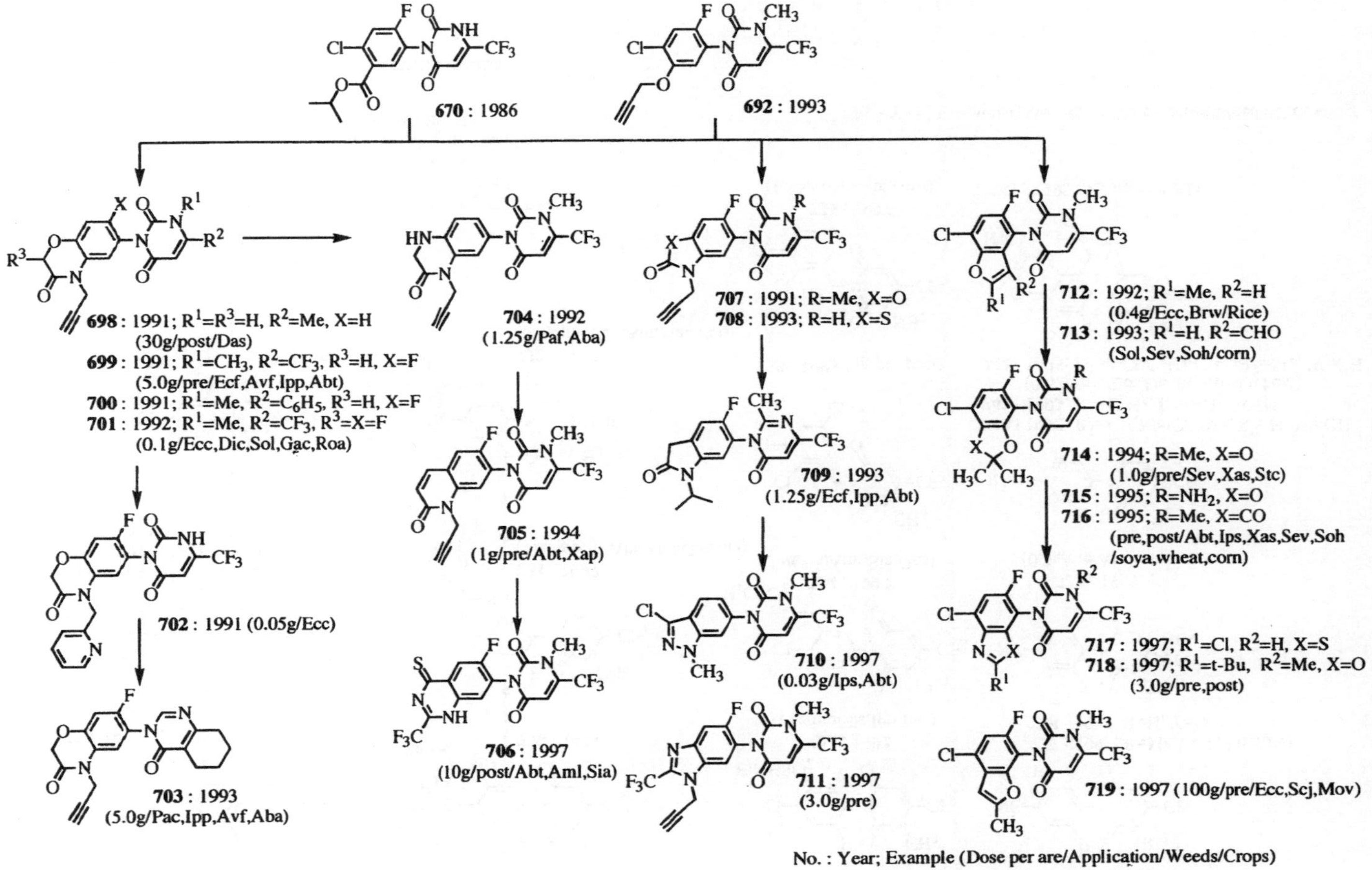

Fig. 21. Structural modifications of the phenyl ring into bicyclic benzologs fused with 5- and 6-membered heterocycles

<Uracils with N-modified phenyl ring systems>

<Other uracil-type cyclic imides>

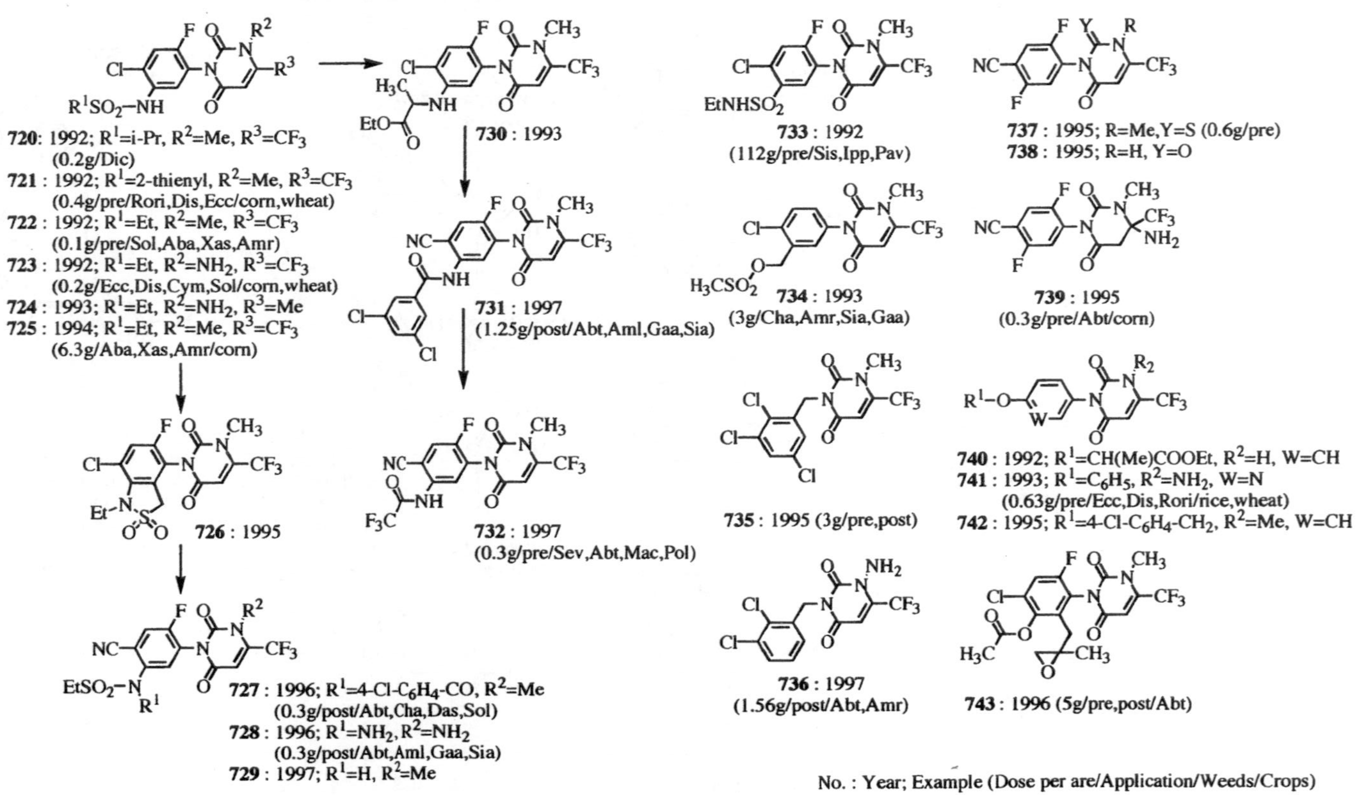

720: 1992; R^1=i-Pr, R^2=Me, R^3=CF_3 (0.2g/Dic)
721 : 1992; R^1=2-thienyl, R^2=Me, R^3=CF_3 (0.4g/pre/Rori,Dis,Ecc/corn,wheat)
722 : 1992; R^1=Et, R^2=Me, R^3=CF_3 (0.1g/pre/Sol,Aba,Xas,Amr)
723 : 1992; R^1=Et, R^2=NH_2, R^3=CF_3 (0.2g/Ecc,Dis,Cym,Sol/corn,wheat)
724 : 1993; R^1=Et, R^2=NH_2, R^3=Me
725 : 1994; R^1=Et, R^2=Me, R^3=CF_3 (6.3g/Aba,Xas,Amr/corn)

730 : 1993

731 : 1997 (1.25g/post/Abt,Aml,Gaa,Sia)

732 : 1997 (0.3g/pre/Sev,Abt,Mac,Pol)

726 : 1995

727 : 1996; R^1=4-Cl-C_6H_4-CO, R^2=Me (0.3g/post/Abt,Cha,Das,Sol)
728 : 1996; R^1=NH_2, R^2=NH_2 (0.3g/post/Abt,Aml,Gaa,Sia)
729 : 1997; R^1=H, R^2=Me

733 : 1992 (112g/pre/Sis,Ipp,Pav)

737 : 1995; R=Me,Y=S (0.6g/pre)
738 : 1995; R=H, Y=O

734 : 1993 (3g/Cha,Amr,Sia,Gaa)

739 : 1995 (0.3g/pre/Abt/corn)

735 : 1995 (3g/pre,post)

740 : 1992; R^1=CH(Me)COOEt, R^2=H, W=CH
741 : 1993; R^1=C_6H_5, R^2=NH_2, W=N (0.63g/pre/Ecc,Dis,Rori/rice,wheat)
742 : 1995; R^1=4-Cl-C_6H_4-CH_2, R^2=Me, W=CH

736 : 1997 (1.56g/post/Abt,Amr)

743 : 1996 (5g/pre,post/Abt)

No. : Year; Example (Dose per are/Application/Weeds/Crops)

Fig. 22. Miscellaneous uracil-type cyclic imides

Three types of synthetic routes to form 6-trifluoromethyluracils have been developed. General synthetic method for 6-trifluoromethyluracil skeleton consists of addition and cyclization of arylisocyanate and trifluorocrotonate. For example, isopropyl 2-chloro-4-fluoro-5-[3,6-dihydro-4-trifluoromethyl-2,6-dioxo-1(2*H*)-pyrimidinyl]benzoate [670] is readily prepared by reaction of isopropyl 2-chloro-4-fluoro-5-isocyanatobenzoate (*87*) with ethyl 3-amino-4,4,4-trifluorocrotonate (*88*) in the presence of sodium hydride in DMF (Scheme 11). The alternative pathway for 6-trifluoromethyluracils has been reported by the Nissan Chemical group, in which *N*-arylcarbamate (*89*) instead of arylisocyanate cyclocondenses with trifluorocrotonate (*88*) to yield, for example, 3-(4-chloro-5-ethylsulfonylamino-2-fluorophenyl)-6-trifluoromethylpyrimidine-2,4-dione [722], as shown in Scheme 11.

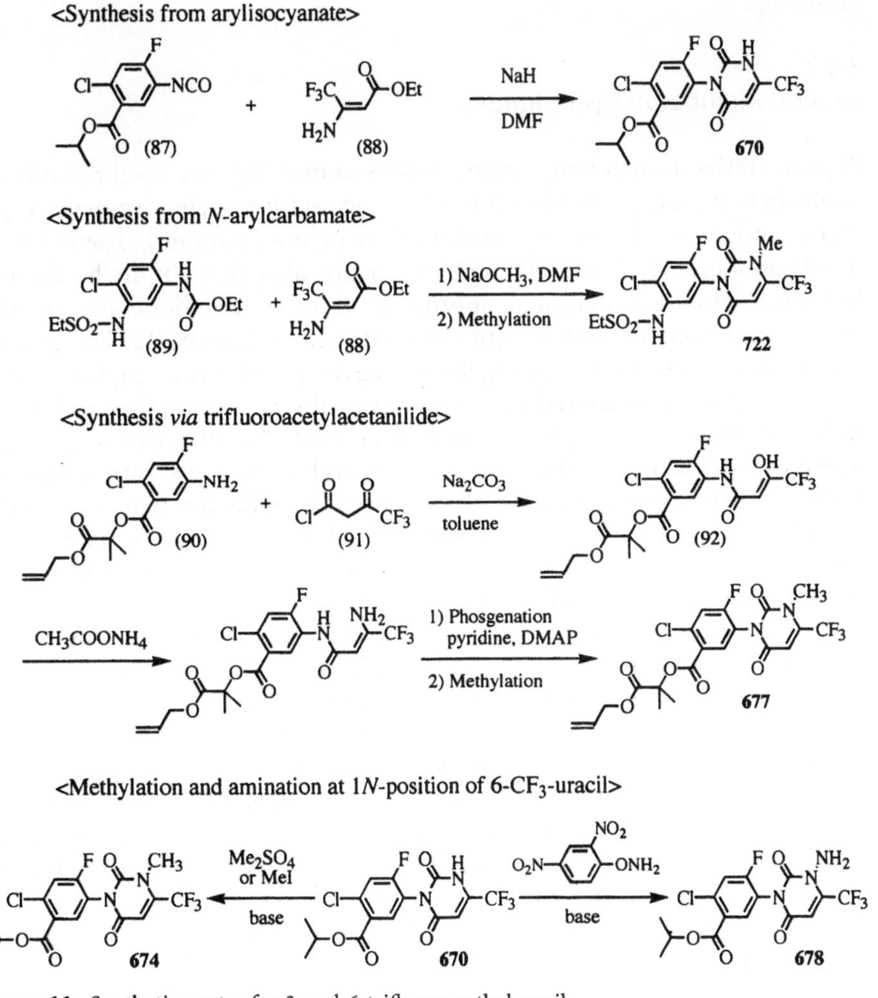

Scheme 11. Synthetic routes for 3-aryl-6-trifluoromethyluracils

Another method published begins from anilines which are precursors of arylisocyanates (Scheme 11). 5-Alkoxycarbonyl-4-chloro-2-fluoroaniline (*90*) reacts with 2-(trifluoroacetyl)acetylchloride (*91*) affording 2-(trifluoroacetyl)acetanilide (*92*), which is subjected to amination and cyclization with phosgene, and then accomplished by methylation to give 3-aryl-1-methyl-6-trifluoromethylpyrimidine-2,4-dione [677].

While herbicidally active 6-trifluoromethyluracils are usually substituted with a methyl group at 1*N*-position of uracil ring, some analogs are unsubstituted or substituted with an amino group. Amination at such position is achieved by reaction of 1*N*-unsubstituted derivative [670] with *O*-(2,4-dinitrophenyl)hydroxylamine; on the other hand, methylation is easy using dimethylsulfate or methylhalide in the presence of base (Scheme 11). 1*N*-Methyl analogs [674] have been known to show potent peroxidizing action; however, the mode of action of 1*N*-amino analogs [678] seems to be bleaching.

2.4.2
Other 6-Membered Cyclic Imides

Representative 6-membered cyclic imides containing one to three nitrogen atoms in heterocycles are shown in Fig. 23, except for 6-trifluoromethyluracils aforementioned. These compounds such as piperidinediones, oxazinediones, pyridazinones, triazinediones and so on are also thought to be included in Protox inhibitors. Synthetic approaches to piperidinedione, oxazinedione and pyridazinone derivatives are demonstrated in Scheme 12. For instance, 1-(4-chloro-2-fluoro-5-propargyloxyphenyl)-4-trifluoromethylpiperidine-2,4-dione [746] is prepared by imidation of the corresponding aniline (*93*) with 3-(trifluoromethyl)glutalic anhydride (*94*). 2,2-Dimethyl-4,5,6,7-tetrahydrocyclopenta-1,3-dioxin-4-one (*96*) is cyclocondensed with 4-chloro-2-fluoro-5-nitrophenylisocyanate (*95*) followed by reduction over iron powder to give oxazinedione [752]. Pyridazinone [754] is easily synthesized by allylation following condensation of arylhydrazine (*97*) with γ-hydroxy-γ-butyrolactone (*98*), which is derived from THPA.

1,2,4-Triazine-3,5-dione derivatives, which also belong to 6-membered cyclic imides with three nitrogen atoms, have been steadily studied in the last decade. 1,2,4-Triazine-3,5-diones fall into two structural isomers distinguished from the position of respective aryl group. In the case of 2-aryl-1,2,4-triazine-3,5-dione [767], the 2-position of triazine ring is substituted with an aryl group which is somewhat different from the phenyl ring systems of usual Protox inhibitor. Treatment of 2-aryl-1-benzylidene-3,5-dichlorophenyl-4-(2,2-diethoxyethyl)semicarbazide (*99*) with hydrogen chloride followed by oxidation with hydrogen peroxide produces 2-[3,5-dichloro-4-(4-chlorobenzyl)phenyl]-1,2,4-triazine-3,5(2*H*,4*H*)-dione [767] illustrated in Scheme 12.

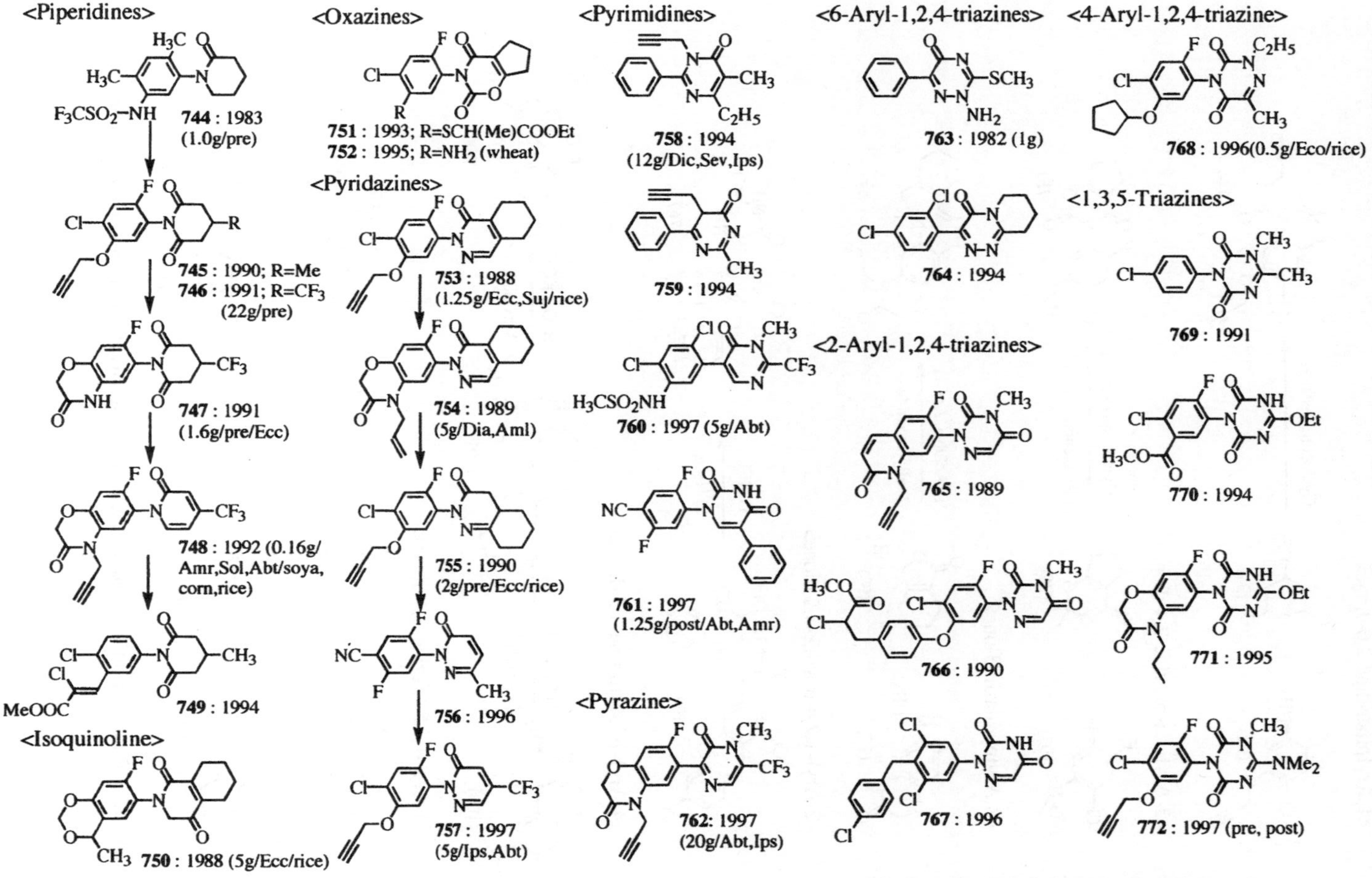

No. : Year; Example (Dose per are/Application/Weeds/Crops)

Fig. 23. Structural modifications of other 6-membered heterocycles belonging to cyclic imide class

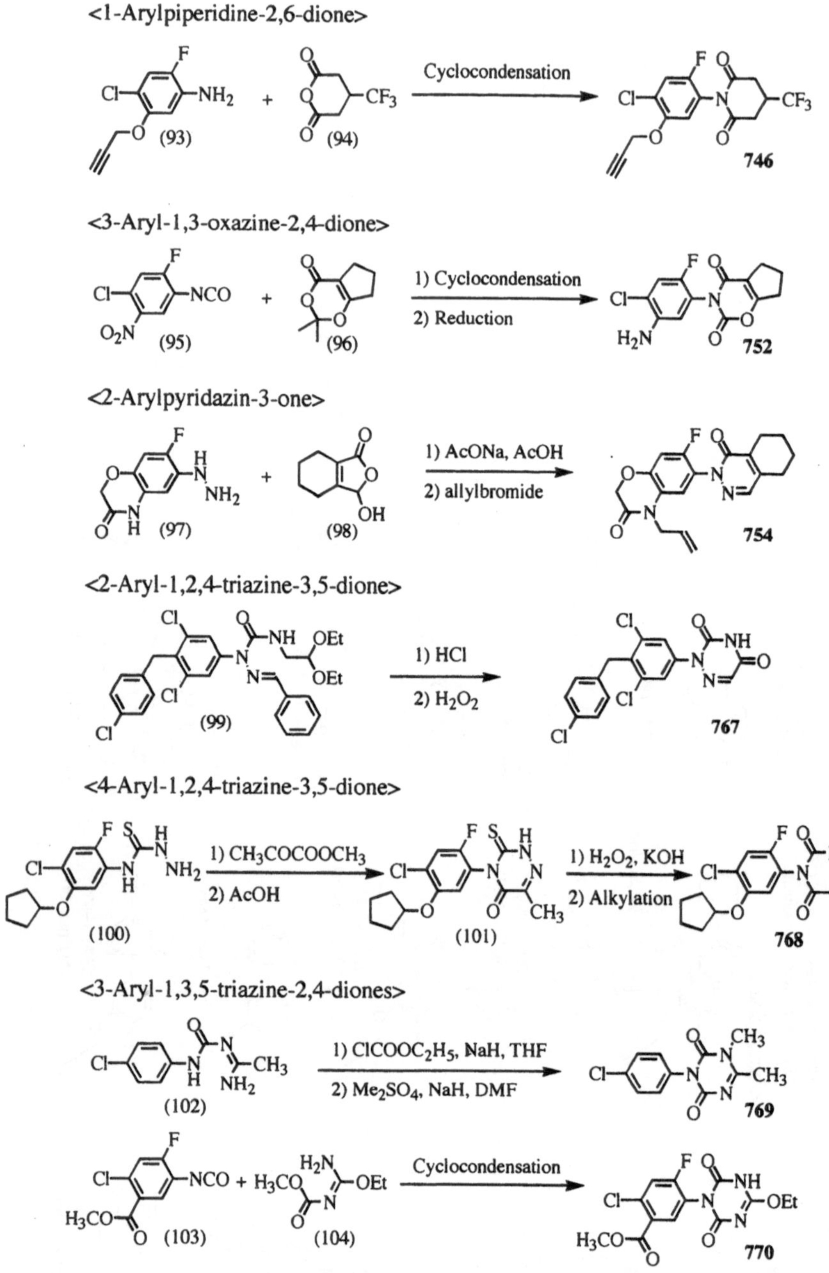

Scheme 12. Synthesis of piperidinedione, oxazinedione, pyridazinone and triazinediones

Another type of 1,2,4-triazine-3,5-dione is substituted with a phenyl group at 4*N*-position. For example, 4-(4-chloro-5-cyclopentyloxy-2-fluorophenyl)-6-methyl-2,3,4,5-tetrahydro-1,2,4-triazine-3,5-dione [768] has been recognized to show peroxidizing action. Herbicidal activity is influenced by the substituents at 2- and 6-positions, and alkyl groups with long carbon chains decrease activity. 3-Carbonyl group is essential for herbicidal activity because the corresponding thiocarbonyl analog is inactive.

Synthetic route of 4-aryl-1,2,4-triazine-3,5-dione [768] is shown in Scheme 12. 4-(4-Chloro-5-cyclopentyloxy-2-fluorophenyl)thiosemicarbazide (*100*), which is obtained by reacting arylisothiocyanate with hydrazine, is cyclized with methyl pyruvate, and the thiocarbonyl group of the product (*101*) is oxidized with hydrogen peroxide to yield 4-aryl-6-methyl-1,2,4-triazine-3,5-dione [768].

Some 1,3,5-triazine-2,4-dione derivatives as 6-membered cyclic imide class of Protox inhibitor have been reported since 1991. Two synthetic approaches to 1,3,5-triazine analogs are depicted in Scheme 12. Cyclization following acylation of *N*-(4-chlorophenyl)carbamoylacetamidine (*102*) with ethyl chloroformate gives 1*N*-unsubstituted 1,3,5-triazine-2,4-dione, which is methylated with dimethylsulfate to yield the resultant product [769]. Cyclocondensation of 3-(methoxycarbonyl)-2-ethyl-3-isourea (*104*) with methyl 2-chloro-4-fluoro-5-isocyanatobenzoate (*103*) yields 3-aryl-6-ethoxy-1,3,5-triazine-2,4-dione [770].

Recently, new types of 6-membered heterocycles such as pyrimidinones [758 to 760], pyrazinone [762] and 6-aryl-1,2,4-triazinones [763, 764] have been investigated, in which each aryl group and hetero ring are bridged together with a C—C bond. These compounds must be noticed as a new class of Protox inhibitors, because 5-membered cyclic imides such as ET-751 having the same bridging system exhibit more potent peroxidizing performance.

2.5 Epilogue

The major purpose of this chapter was to catalog every type of DPE and cyclic imide herbicides to provide a comprehensive guide to the patent literature released from 1980 through to the middle of 1997. As an ambiguous presentation in the patent literature does not always reveal the most active compound, the compounds shown in the Figures are selected as representative examples, for which synthetic methods are briefly cited in Chemical Abstracts except for the practical herbicides and the potential candidates under development. However, structural evolution of recent DPE and cyclic imide herbicides is clearly demonstrated. Hopefully, this chapter will make some of the unexplored class of Protox inhibitors more apparent, will encourage further progress of the underexplored class and will stimulate new ideas for development of next-generation herbicides. *Acknowledgments.* I wish to express my gratitude to Mr. Ryuta Ohno and Mr. Atsushi Uchida of Sagami Chemical Research Center for the collection and arrangement of enormous data and also for their generous assistance in preparing this chapter.

References and Notes

In this chapter, the agricultural data (application style, practical dosage, controlled weeds, applied crops, etc) of the practical herbicides and the potential candidates under development are mainly quoted from [i] Hopkins W L (1995–97) Ag Chem New Compound Review Vol. 13–15. Ag Chem Information Services, Indiana, [ii] Sekai no Shin-noyaku Chukantai Business (Japanese): New agrochemicals reports. CMC, Tokyo and [iii] Ikura K (1997) Development of New Agrochemicals (Japanese). CMC, Tokyo.

For easy access to references, the number assigned to each compound corresponds to a reference number. Company names used in the patent literature are abbreviated as follows: ACC:American Cyanamid Co.; A-Kanesho:Agro-Kanesho Co.; Asahi:Asahi Chemical Ind. Co.; BASF:BASF A.-G.; Bayer:Bayer A.-G.; Budapesti:Budapesti Vegyimuvek; Central Glass:Central Glass Co.; Chevron:Chevron Chemical Co.; Chisso:Chisso Corp.; Ciba-Geigy:Ciba-Geigy A.-G.; Degussa:Degussa A.-G.; DIC:Dainippon Ink & Chemicals, Inc.; Dow:Dow Chemical Co.; Dupher:Dupher Nederland B.V.; DuPont:E. I. Du Pont de Nemours and Co.; Eli Lilly:Eli Lilly and Co.; FMC:FMC Corp.; GAF:GAF Corp.; Hoechst:Hoechst A.-G.; Hokko:Hokko Chemical Ind. Co.; ICI:Imperial Chemical Ind.; ICI Americas:ICI Americas, Inc.; Ihara:Ihara Chemical Ind. Co.; Isagro:Isagro Ricerca S.R.L.; Ishihara:Ishihara Sangyo K.K.; Kaken:Kaken Pharmaceutical Co.; Kanesho:Kanesho Co.; Kayaku:Nippon Kayaku Co.; Kumiai:Kumiai Chemical Ind. Co.; Kuraray:Kuraray Co.; Mitsubishi:Mitsubishi Chemical Ind. Co.; Mitsui:Mitsui-Toatsu Chemicals Inc.; Mobil:Mobil Oil Corp.; Monsanto:Monsanto Co.; Nichino:Nihon Nohyaku Co.; Nihon Bayer:Nihon Bayer Agrochem K.K.; Nissan:Nissan Chemical Ind.; Nisso:Nippon Soda Co.; Otsuka:Otsuka Chemical Co.; PPG:PPG Ind. Inc.; R & H:Rohm and Haas Co.; Roche:Hoffman-La Roche, F. and Co. A.-G.; RP:Rhône-Poulenc Agrochimie.; Sagami:Sagami Chemical Research Center; Sandoz:Sandoz A.-G.; Sankyo:Sankyo Co.; Schering:Schering A.-G.; Shell:Shell International Research M.B.V.; Shionogi:Shionogi and Co.; Sho-Den:Showa Denko K.K.; Showa-R:Showa Rodia Kagaku.; Stauffer:Stauffer Chemical Co.; Sumitomo:Sumitomo Chemical Co.; Suntory:Suntory Ltd.; Takeda:Takeda Chemical Ind.; Teijin:Teijin Ltd.; Tokuno:Nihon Tokushu Nohyaku Seizo K.K.; Tokuso:Tokuyama Soda K.K.; Tosoh:Tosoh Corp.; Toyo Soda:Toyo Soda Mfg. Co.; Uniroyal:Uniroyal Chemical Co.; Velsicol:Velsicol Chemical Corp.; Zeneca:Zeneca Agrochemicals; Zoecon:Zoecon Corp.

Patent Literature

[1] Yokomichi I et al. (1980) JP5507403 Ishihara. [2] Kawaguchi Y et al. (1980) US4213775 Ube Ind. [3] Martinuzzi EA et al. (1984) BR PI8207374. [4] Ishizaki M et al. (1992) JP04149177 Tokuso. [5] Taniyama E et al. (1993) JP0551360 Mitsubishi. [6] Schoenowsky H et al. (1980) DE2850902 Hoechst. [7] Yoshimoto T et al. (1981) JP5602965 Mitsui. [8] Yoshimoto T et al. (1981) FR2471979 Mitsui. [9] Yoshimoto T et al. (1980) JP5515450 Mitsui. [10] Yoshimoto T et al. (1981) JP5602954 Mitsui. [11] Swithenbank C et al. (1981) EP22626 R & H. [12] Takematsu T et al. (1981) JP5679665 Mitsui. [13] Forester H et al. (1981) DE3013264 Bayer. [14] Rohr O et al. (1982) EP59167 Ciba-Geigy. [15] Umemoto M et al. (1982) JP57175178 Mitsui. [16] Forester H et al. (1982) DE3122174 Bayer. [17] Lee SF et al. (1983) EP78536 Zoecon. [18] Hokama T et al. (1983) US4402729 Velsicol. [19] Stach LJ et al. (1983) US4404018 Velsicol. [20] Forester H et al. (1983) DE3220526 Bayer. [21] Futher W et al. (1984) DE3224984 Bayer. [22] Winternitz P et al. (1985) EP138183 Roche. [23] Chan D et al. (1986) DE3521934 Chevron. [24] Forester H et al. (1980) DE2906237 Bayer. [25] Schmidt RR et al. (1980) DE2906087 Bayer. [26] Forester H et al. (1982) DE3116298 Bayer. [27] Forester H et al. (1982) DE3122175 Bayer. [28] Shenem HP et al. (1983) DE3127225 Bayer. [29]

Hokama T et al. (1983) BE895664 Velsicol. [30] Lieble R et al. (1986) DE3502933 Hoechst. [31] Winternitz P et al. (1988) US4792616 Roche. [32] Fukami J et al. (1990) EP303415 Suntory. [33] Theissen RJ et al. (1980) EP33629 Mobil. [34] Durr D et al. (1980) EP13660 Ciba-Geigy. [35] Johnson WO et al. (1980) BR PI790287 R & H. [36] Grove WS et al. (1980) US4293329 PPG. [37] Johnson WO et al. (1980) EP20052 R & H. [38] Swithenbank C et al. (1981) EP21692 R & H. [39] Grove WS et al. (1981) FR2463119 PPG. [40] Forester H et al. (1981) DE2950401 Bayer. [41] Grove WS et al. (1981) US4288243 PPG. [42] Theissen RJ et al. (1981) EP34883 Mobil. [43] Downing CR et al. (1981) EP40898 Mobil. [44] Grove WS et al. (1982) US4311515 PPG. [45] Theissen RJ et al. (1982) US4339268 RP. [46] Forester H et al. (1982) EP63741 Bayer. [47] Forester H et al. (1982) JP57212140 Bayer. [48] Ashnore JW et al. (1983) EP72649 R & H. [49] Wu F et al. (1983) BE895662 Velsicol. [50] Theissen RJ et al. (1983) BR PI8203522 RP. [51] Gough STD et al. (1983) BR PI8204355 RP. [52] Liu KC et al. (1984) US4429146 GAF. [53] Frater G et al. (1984) EP105527 Roche. [54] Kou C et al. (1984) ZA8306161 GAF. [55] Liu KC et al. (1984) US4448982 GAF. [56] Cartwright D et al. (1984) EP122037 ICI. [57] Cartwright D et al. (1984) BR PI84015870 ICI. [58] Kondo N et al. (1985) JP6061573 Sumitomo. [59] Kondo N et al. (1985) JP6061571 Sumitomo. [60] Peter JD et al. (1985) JP60237094 Ciba-Geigy. [61] Nagubandi S et al. (1986) US4602946 Stauffer. [62] Theissen RJ et al. (1986) US4606758 RP. [63] Ehrhardt H et al. (1986) DE3506878 Hoechst. [64] Forester H et al. (1987) DE3526975 Bayer. [65] Forester H et al. (1987) DE3538689 Bayer. [66] Schegel G et al. (1987) DE3531007 Bayer. [67] Schegel G et al. (1987) DE3531006 Bayer. [68] Forester H et al. (1988) DE3627615 Bayer. [69] Forester H et al. (1988) DE3628317 Bayer. [70] Pflster T et al. (1988) DE3702921 Bayer. [71] Yih RY et al. (1989) US4806682 R & H. [72] Theissen RJ et al. (1980) US4340417 RP. [73] Cartwright D et al. (1981) EP23100 ICI. [74] Barton J et al. (1987) US4263041 PPG. [75] Johnson WO et al. (1981) BR PI7908321 R & H. [76] Cartwright D et al. (1981) US4285723 ICI. [77] Parg A et al. (1982) EP53679 Bayer. [78] Wu F et al. (1982) US4427441 Velsicol. [79] Lee GH et al. (1982) DE3215991 RP. [80] Gough STD et al. (1982) US4364767 RP. [81] Gough STD et al. (1983) BE893939 RP. [82] Thessen RJ et al. (1983) US4367090 RP. [83] Gough STD et al. (1983) US4382815 RP. [84] Frater G et al. (1984) EP109575 Roche. [85] West PJ et al. (1985) EP142248 FMC. [86] Heiba E et al. (1985) US4528145. [87] Heiba E et al. (1985) US4562279. [88] Lieble R et al. (1986) DE3502629 Hoechst. [89] Nagubandi S et al. (1986) US4627870 Stauffer. [90] Die PJ et al. (1986) DE3620217 Ciba-Geigy. [91] Nagubandi S et al. (1986) US4632697 Stauffer. [92] Lange A et al. (1987) DE353664 BASF. [93] Felix A et al. (1987) US4654072 Stauffer. [94] Nowakowski MA et al. (1988) US4772712 Uniroyal. [95] Krass DK et al. (1980) DE3017795 PPG. [96] Borrod G et al. (1983) EP91387 RP. [97] Cartwright D et al. (1984) EP124985 ICI. [98] Krass DK et al. (1985) EP155613 PPG. [99] Krass DK et al. (1986) DD232915 PPG. [100] Krass DK et al. (1986) US4584403 PPG. [101] Hayashi Y et al. (1986) EP186989 Asahi. [102] Krass DK et al. (1987) DD242743 PPG. [103] Hayashi Y et al. (1987) JP6236352 Asahi. [104] Krass DK et al. (1987) US4657580 PPG [105] Bakos J et al. (1988) DE3722427 Budapesti. [106] Furuhashi S et al. (1988) JP6366156 Asahi. [107] Schwindeman JA et al. (1990) US4900862 ACC. [108] Chene A et al. (1985) DE3425537 RP. [109] Munro D et al. (1986) EP174046 Shell. [110] Go A et al. (1987) JP62201860 Mitsubishi. [111] Go A et al. (1987) EP273432 Mitsubishi. [112] Go A et al. (1988) JP63165359 Mitsubishi. [113] Go A et al. (1988) JP63313758 Mitsubishi. [114] Go A et al. (1989) JP01246250 Mitsubishi. [115] Jikihara K et al. (1991) JP03123707 Mitsubishi. [116] Omichi H et al. (1984) JP59210063 Asahi. [117] Parg A et al. (1984) DE3247669 BASF. [118] Schmideman JA et al. (1984) US4596883 PPG. [119] Komives T et al. (1986) GB2181133 Budapesti. [120] Hayashi Y et al. (1987) JP6219566 Asahi. [121] Magyar AT et al. (1989) HU47368 Budapesti. [122] Swithenbank C et al. (1981) US4306900 R & H. [123] Yoshimoto T et al. (1982) JP57154194 Mitsui. [124] Forester H et al. (1983) DE3141861 Bayer. [125] Duerr D et al. (1983) DE3239449 Ciba-Geigy. [126] Acker RD et al. (1983) DE3210336 BASF. [127] Hayashi Y et al. (1985) EP161529 Asahi. [128] Kirste R et al. (1985) DE3419937 Bayer. [129] Hayashi Y et al. (1986) JP61180740 Asahi. [130] Higuchi N et al. (1989) JP64135751 Suntory. [131] Sclegel G et al. (1989) DE3744021 Hoechst. [132] Huang IC et al. (1981) BE885594 Chevron. [133] Forester H et al. (1982) EP53321 Bayer. [134] Prisbylla MP et al. (1987) US4714491 Stauffer. [135] Bacos J et al. (1987) DE3641046 Budapesti. [136] Forester H et al. (1982) DE3118371 Bayer. [137] Kondo S et al. (1984) JP59196876 Sumitomo. [138] Bertron C et al. (1985) JP6064972 RP. [139] Munro D et al. (1986) EP174685

Shell. [140] Munro D et al. (1988) GB2201672 Shell. [141] Yoshimoto T et al. (1980) DE2944783 Mitsui. [142] Hosono A et al. (1981) ZA8002066 Mitsui. [143] Hosono A et al. (1981) JP56169603 Mitsui. [144] Hosono A et al. (1982) JP5728039 Mitsui. [145] Hosono A et al. (1982) JP5772948 Mitsui. [146] Hosono A et al. (1982) JP5772947 Mitsui. [147] Hosono A et al. (1982) JP5772946 Mitsui. [148] Buck W et al. (1982) DE3110894 Bayer. [149] Duerr D et al. (1983) EP72348 Ciba-Geigy. [150] Yoshimoto T et al. (1983) ZA8301656 Mitsui. [151] Hosono A et al. (1983) JP58110597 Mitsui. [152] Meier L et al. (1983) EP93081 BASF. [153] Parg A et al. (1984) DE3300585 BASF. [154] Paeres GB et al. (1984) EP126294 Dupher. [155] Bealieu AH et al. (1985) EP130041 R & H. [156] Lee SF et al. (1985) US4526608 Zoecon. [157] Duerr D et al. (1985) EP149427 Ciba-Geigy. [158] Munro D et al. (1986) GB2194530 Shell. [159] Ishizuka M et al. (1992) JP04342505 Tokuso. [160] Maier LD et al. (1980) EP14684 Ciba-Geigy. [161] Nagai S et al. (1981) USSR843695 Ube Ind. [162] Maier LD et al. (1982) EP43801 Ciba-Geigy. [163] Yoshimoto T et al. (1982) JP5724355 Mitsui. [164] Hirose Y et al. (1982) JP5748990 Mitsui. [165] Parg A et al. (1983) DE3212165 BASF. [166] Aly EA et al. (1985) US4548640 FMC. [167] Parg A et al. (1985) DE3436175 BASF. [168] Parg A et al. (1986) DE3434315 BASF. [169] Hiraga K et al. (1980) JP5535016 Nichino. [170] Hoddock EC et al. (1985) EP145078 Shell. [171] Michael TC et al. (1987) JP6261973 Shell. [172] Clark MT et al. (1988) EP265020 Shell. [173] Briggs SP et al. (1989) EP306096 Shell. [174] Nielser DR et al. (1986) US4571255 PPG. [175] Nielser DR et al. (1987) EP238266 PPG. [176] Munro D et al. (1988) GB2192878 Shell. [177] Munro D et al. (1988) GB2192879 Shell. [178] Barton J et al. (1991) EP448206 ICI. [179] Chrystal EJT et al. (1991) EP453066 ICI. [180] Chrystal EJT et al. (1993) GB2258233 ICI. [181] Steffen JJ et al. (1983) US4377408 RP. [182] Bahr JT et al. (1983) US4388105 RP. [183] Munro D et al. (1987) GB2189238 Shell. [184] Jackson LA et al. (1989) US4888041 Dow. [185] Andree R et al. (1990) DE3837464 Bayer. [186] Clark MT et al. (1989) GB2207425 Shell. [187] Kawaguchi N et al. (1992) EP479420 Suntory. [188] Barton J et al. (1993) GB2257970 ICI. [189] Barton J et al. (1981) EP23392 ICI. [190] Rohr L et al. (1982) CA1120497 Bayer. [191] Konishi H et al. (1983) EP94641 Sumitomo. [192] Konishi H et al. (1984) JP59116259 Sumitomo. [193] Konishi H et al. (1984) US4486223 Sumitomo. [194] Konishi H et al. (1985) JP6081162 Sumitomo. [195] Munro D et al. (1988) GB2199577 Shell. [196] Azuma S et al. (1988) EP281103 Teijin. [197] Bacos J et al. (1990) DE3943016 Budapesti. [198] Azuma S et al. (1990) WO90/02113 Teijin. [199] Azuma S et al. (1990) WO90/01874 Teijin. [200] Busse U et al. (1990) EP379915 Bayer. [201] Haug M et al. (1990) EP384230 Bayer. [202] Jikihara T et al. (1990) EP403938 Mitsubishi. [203] Azuma S et al. (1992) JP0466549 Teijin. [204] Yoshimoto T et al. (1980) JP5520712 Mitsui. [205] Yoshimoto T et al. (1981) JP5625143 Mitsui. [206] Pissotas G et al. (1981) EP23890 Ciba-Geigy. [207] Takematsu T et al. (1981) JP5646859 Mitsui. [208] Meyer W et al. (1981) EP34120 Ciba-Geigy. [209] Hiraga K et al. (1981) EP34457 Nichino. [210] Rohr O et al. (1983) CH637361 Ciba-Geigy. [211] Rohr O et al. (1983) EP86747 Ciba-Geigy. [212] Busee U et al. (1989) EP342440 Bayer. [213] Bacos J et al. (1990) DE3943015 Budapesti. [214] Cartwright D et al. (1990) GB2225014 ICI. [215] Busse U et al. (1990) DE3916527 Bayer. [216] Takematsu T et al. (1980) JP55154953 Mitsui. [217] Lee GH et al. (1982) FR2525593 RP. [218] Swithenbank C et al. (1983) US4419124 R & H. [219] Chan DCK et al. (1984) US4468247 Chevron. [220] Forester H et al. (1984) DE3319290 Bayer. [221] Santel HJ et al. (1986) DE3425702 Bayer. [222] Philips JG et al. (1987) US4710582 PPG. [223] Yamamoto M et al. (1988) JP63162666 Asahi. [224] Brand WW et al. (1988) US4754052 ICI. [225] Bacos J et al. (1989) HU47203 Budapesti. [226] Forester H et al. (1993) DE4133674 Bayer. [227] Szczepanski H et al. (1980) EP4317 Ciba-Geigy. [228] Fujikawa K et al. (1983) JP5890555 Ishihara. [229] Fujikawa K et al. (1983) JP5890556 Ishihara. [230] Pissiotas G et al. (1986) EP202195 Ciba-Geigy. [231] Go A et al. (1989) JP01268606 Mitsubishi. [232] Go A et al. (1989) JP01272505 Mitsubishi. [233] Go A et al. (1989) JP01268607 Mitsubishi. [234] Fukami J et al. (1990) JP0225454 Suntory. [235] Cartwright D et al. (1982) EP63873 ICI. [236] Swithenbank C et al. (1988) US4743703 R & H. [237] Orvik JA et al. (1989) EP322983 Dow. [238] Gilkerson T et al. (1989) EP348002 Shell. [239] Jackson LA et al. (1989) US4888041 Dow. [240] Forester H et al. (1991) GB2238789 Shell. [241] Andree R et al. (1991) EP457140 Bayer. [242] Barton J et al. (1994) GB2276379 Zeneca. [243] Takematsu T et al. (1994) JP06135806 Mitsui. [244] GB1110500 (1969) RP. [245] Matsui K et al. (1973) JP4811940 Mitsubishi. [246] Steven JG et al. (1976) JP5151521 DuPont. [247] Wakabayashi O et al. (1980) DE3013162 Mitsubishi. [248] Jikihara T et al. (1982)

JP5756403 Mitsubishi. [249] Jikihara T et al. (1982) EP49511 Mitsubishi. [250] Jikihara T et al. (1982) EP49508 Mitsubishi. [251] Jikihara T et al. (1982) JP5762256 Mitsubishi. [252] Swithenbank C et al. (1984) US4439229 R & H. [253] Rueb L et al. (1989) EP313963 BASF. [254] Nagano E et al. (1982) EP61741 Sumitomo; Takemoto K et al. (1985) JP6064961 JP6064959 EP150064 JP60158147 Sumitomo. [255] Takemoto K et al. (1984) JP5982360 Sumitomo. [256] Nagano E et al. (1984) JP5933293 Sumitomo. [257] Lee SF et al. (1985) DE3426634 Sandoz; Lee SF et al. (1985) US4560752 Zoecon. [258] Haga T et al. (1986) JP61280471 Sumitomo. [259] Stetter J et al. (1988) EP288789 Bayer. [260] Theodoridis G et al. (1989) US4816065 FMC. [261] Hanasaki Y et al. (1989) EP323271 Tosoh, A-Kanesho. [262] Schallner O et al. (1989) DE3819439 Bayer. [263] Kume T et al. (1989) EP337151 Tokuno. [264] Ito M et al. (1990) JP0273065 Tosoh, A-Kanesho. [265] Schallner O et al. (1990) DE3905006 Bayer. [266] Hirai K et al. (1992) WO92/01671 Sagami, Kaken, Chisso. [267] Nagano E et al. (1983) EP83055 Sumitomo. [268] Nagano E et al. (1984) JP5965074 Sumitomo. [269] Nagano E et al. (1984) JP59181256 Sumitomo. [270] Nagano E et al. (1983) EP95192 Sumitomo. [271] Okada I et al. (1985) JP60123469 Mitsubishi. [272] Haga T et al. (1986) JP61134373 Sumitomo. [273] Fischer R et al. (1989) DE3731516, (1990) DE3827222 Bayer; Marhold A et al. (1989) EP307777 Bayer. [274] JP56169846 (1981) Mitsubishi. [275] JP56199847 (1981) Mitsubishi. [276] Takemoto K et al. (1984) JP59108766 Sumitomo. [277] Nagano E et al. (1984) JP5980661 Sumitomo. [278] Haga T et al. (1986) JP61118364 Sumitomo. [279] Anderson RJ et al. (1985) DE3504051 Sandoz. [280] Theodoridis G et al. (1990) US4954159 FMC. [281] Drewes MW et al. (1995) DE4414568 Bayer. [282] Natsume B et al. (1997) EP786453 Mitsubishi. [283] Haga T et al. (1986) JP61161259, (1986) JP61161260, (1990) US4902832 Sumitomo; Osada S et al. (1987) JP62187449 Sumitomo. [284] Nagano E et al. (1984) EP126419, JP60152453, (1985) JP60248657, JP60248663 Sumitomo; Miyata N et al. (1987) JP6233148, JP6233149, JP6256467, JP6256468 Sumitomo; Nagano E et al. (1987) US4709049, JP06199774, (1994) JP06199775 Sumitomo; Yoshida K et al. (1994) JP06128218 Mitsubishi. [285] Jikihara T et al. (1986) JP6140261 Mitsubishi. [286] Haga T et al. (1986) JP61118363 Sumitomo. [287] Nagano E et al. (1987) US4684397 Sumitomo. [288] Hiratsuka M et al. (1995) CA2130419 Sumitomo. [289] Haga T et al. (1985) JP6067461, JP6078959 Sumitomo. [290] Jikihara T et al. (1985) JP60152464 Mitsubishi. [291] Haga T et al. (1987) JP6236356 Sumitomo. [292] Pissiotas G et al. (1992) EP468924 Ciba-Geigy. [293] Akutagawa K et al. (1993) JP05194386 Takeda. [294] Hirai K et al. (1992) EP496347, (1994) WO94/06753, (1995) JP0782224 Sagami, Kaken. [295] Yamada O et al. (1981) GB2071100 Kayaku. [296] Swithenbank C et al. (1983) EP68822 R & H. [297] Someya S et al. (1984) JP59128371 Kanesho. [298] Akahira R et al. (1985) JP60109563 Kanesho, Toyo Soda. [299] Natsume B et al. (1993) CA2084001 Mitsubishi. [300] Yanagi M et al. (1982) JP57149267 Kayaku. [301] Liebl R et al. (1987) DE3533440 Hoechst. [302] Pissiotas G et al. (1987) EP233151 Ciba-Geigy. [303] Kasahara I et al. (1988) JP6345253 Nisso. [304] Diel PJ et al. (1988) DE3737152 Ciba-Geigy. [305] Moser H et al. (1988) DE3743127 Ciba-Geigy. [306] Brunner H et al. (1993) EP530642 Ciba-Geigy. [307] Moser H et al. (1989) EP303573 Ciba-Geigy. [308] Schallner O et al. (1993) DE4139636 Bayer. [309] Pissiotas G et al. (1989) EP338987 Ciba-Geigy. [310] Sato J et al. (1989) JP01139581 Nissan. [311] Clark MT et al. (1985) GB2150929 Shell. [312] Okada I et al. (1985) JP60152465 Mitsubishi. [313] Matsumoto S et al. (1985) JP60246367 Mitsubishi. [314] Pissiotas G et al. (1987) EP207894 Ciba-Geigy. [315] Eicken K et al. (1987) DE3607300 BASF. [316] Enomoto M et al. (1989) JP0152755, (1995) JP07145112 Sumitomo. [317] Rueb L et al. (1989) DE3741273 BASF. [318] Grossmann K et al. (1990) DE3905916 BASF. [319] Rueb L et al. (1989) DE3741272 BASF. [320] Rueb L et al. (1990) EP398115 BASF. [321] Nagano E et al. (1984) JP59155358 Sumitomo. [322] Plath P et al. (1987) DE3603789 BASF. [323] Plath P et al. (1990) EP358108 BASF. [324] Freund W et al. (1990) DE3904082 BASF; Schaefer B et al. (1991) DE3941562 BASF. [325] Rueb L et al. (1990) EP400427 BASF. [326] Matsumoto S et al. (1986) JP6127962 Mitsubishi. [327] Schwalge B et al. (1989) EP300398 BASF. [328] Heistracher E et al. (1994) DE4237981 BASF. [329] Schwalge B et al. (1989) DE3819464 BASF. [330] Plath P et al. (1989) EP300387 BASF. [331] Schaefer B et al. (1992) DE4042194, (1993) DE4209497 BASF. [332] Klintz R et al. (1995) DE19517597 BASF. [333] Nagano E et al. (1986) EP170191 Sumitomo; Haga T et al. (1986) JP61140570, JP61140572, JP61140573 Sumitomo; Takemoto K et al. (1987) JP62221677 Sumitomo; Haga T et al. (1988) JP63132881 Sumitomo; Enomoto M et al. (1989)

JP01125351 Sumitomo; Tanabe A et al. (1992) JP04305556 Sumitomo; Fukushima M et al. (1993) JP0597826 Sumitomo; Matsumoto M et al. (1993) JP0597848 Sumitomo. [334] Haga T et al. (1986) EP177032 Sumitomo. [335] Haga T et al. (1987) JP6251685 Sumitomo. [336] Kume T et al. (1988) JP6368587 Tokuno. [337] Kume T et al. (1988) EP290863 Tokuno. [338] Kume T et al. (1989) US4798620 Tokuno. [339] Enomoto M et al. (1989) JP0147784 Sumitomo. [340] Kume T et al. (1992) EP477677 Nihon Bayer. [341] Haga T et al. (1993) JP0586028 Sumitomo. [342] Haga T et al. (1986) JP61165383 Sumitomo. [343] see Reference 334. [344] Ganzer M et al. (1989) DE3810706 Schering. [345] Haga T et al. (1987) JP62277383, (1993) JP05345781 Sumitomo. [346] Sakata I et al. (1989) JP0134982 Nissan. [347] Kunisch F et al. (1991) DE3936826 Bayer. [348] Ganzer M et al. (1991) EP406993 Schering. [349] Kume T et al. (1988) EP296416 Tokuno. [350] Yamaguchi M et al. (1991) JP03284678 Kumiai, Ihara. [351] Haga T et al. (1986) EP188259 Sumitomo. [352] Enomoto M et al. (1989) JP0166182 Sumitomo. [353] Enomoto M et al. (1988) JP63174970 Sumitomo. [354] Haga T et al. (1988) JP63313770 Sumitomo. [355] Kume T et al. (1990) EP373461, JP0204784 Tokuno. [356] Condon ME et al. (1993) EP549892 ACC; Karp GM; Condon ME et al. (1994) J Heterocyclic Chem **31**:1513–1520. [357] Ganzer M et al. (1992) DE4117508 Schering. [358] Ganzer M et al. (1994) DE4241658 Schering. [359] Anderson RJ et al. (1982) US4332944 Zoecon. [360] Jikihara T et al. (1983) EP77938 Mitsubishi; Nagano E et al. (1984) JP5967261 Sumitomo; Takemoto K et al. (1985) JP6067459, JP6067460 Sumitomo. [361] Okada I et al. (1985) JP60228494 Mitsubishi. [362] Sato J et al. (1990) JP02229157 Nissan. [363] Okada I et al. (1985) JP60246392 Mitsubishi. [364] Okada I et al. (1986) JP61103887 Mitsubishi. [365] Chang CH et al. (1987) WO87/04049 FMC. [366] Wheeler TN et al. (1987) WO87/07602 RP. [367] Haga T et al. (1987) JP62114921, JP62114938, JP62114941, JP62114961 Sumitomo. [368] Seele R et al. (1991) EP456125 BASF. [369] Anderson RJ et al. (1983) US4406690 Zoecon. [370] Anderson RJ et al. (1983) US4406689 Zoecon. [371] Andree R et al. (1990) DE3917469 Bayer. [372] Sano H et al. (1992) WO92/00976 Sankyo. [373] Enomoto M et al. (1988) JP63275580 Sumitomo. [374] Kunisch F et al. (1990) EP400403 Bayer. [375] Sato J et al. (1991) JP0324076 Nissan. [376] Semple JE et al. (1988) EP275131 Shell. [377] Semple JE et al. (1988) EP271170 Shell. [378] Semple JE et al. (1989) US4881967 DuPont. [379] Daum W et al. (1989) EP3000307 Bayer. [380] Ishida Y et al. (1981) EP40849 Takeda. [381] Nagase H et al. (1983) JP58174362 Takeda. [382] Saito K et al. (1988) JP6339859 Nisso. [383] Saito K et al. (1989) JP01139580 Nisso. [384] Pissiotas G et al. (1989) EP305333 Ciba-Geigy. [385] Klintz R et al. (1993) DE4217137 BASF. [386] Ishida Y et al. (1981) EP41256 Takeda. [387] Boehner B et al. (1986) DE3618501 Ciba-Geigy. [388] Liebl R et al. (1987) DE3533442 Hoechst. [389] Liebl R et al. (1987) DE3604599 Hoechst. [390] Ray JA et al. (1988) Statutory Invent Regist US 531. [391] Ishida Y et al. (1981) JP56164166 Takeda. [392] Nagano E et al. (1984) JP59108767 Sumitomo. [393] Bohner B et al. (1988) EP259265 Ciba-Geigy. [394] Bohner B et al. (1988) EP259264 Ciba-Geigy. [395] Kilama JJ et al. (1995) WO95/27698 DuPont. [396] Yamaguchi H et al. (1982) JP57144204 Sho-Den. [397] Mose H et al. (1988) EP260228 Ciba-Geigy. [398] Dorfmeister G et al. (1991) DE3927438 Schering. [399] Kume T et al. (1988) JP6314782 Tokuno. [400] Sakata I et al. (1988) JP63222167 Nissan. [401] Kume T et al. (1989) JP01299288 Tokuno. [402] Kume T et al. (1990) EP360072 Tokuno. [403] Ganzer M et al. (1989) DE3819823 Schering. [404] Kume T et al. (1990) EP351676 Tokuno; Dorfmeister G et al. (1991) DE4006555 Schering. [405] Elbe HL et al. (1990) EP403891 Bayer. [406] Pissiotas G et al. (1991) EP420810 Ciba-Geigy. [407] Wakabayashi O et al. (1979) JP54125640 Mitsubishi. [408] Wakabayashi O et al. (1979) DE2921002 Mitsubishi. [409] Nagase H et al. (1980) JP55154949 Takeda. [410] Nagase H et al. (1980) JP55157545 Takeda. [411] Nagase H et al. (1979) JP5476577, (1980) JP55157546 Takeda. [412] Nagase H et al. (1980) JP55157547 Takeda. [413] Nagase H et al. (1980) JP55157552 Takeda. [414] Nagase H et al. (1983) JP58210056 Takeda. [415] Okada I et al. (1983) JP58188848 Mitsubishi. [416] Yanagi M et al. (1983) JP58219152 Kayaku. [417] Takemoto K et al. (1984) JP5951250 Sumitomo. [418] Lee SF et al. (1986) US4613675 Sandoz. [419] Jikihara T et al. (1986) JP6143160 Mitsubishi. [420] Hirai K et al. (1993) WO93/19039 Sagami, Kaken. [421] Matsumoto S et al. (1985) JP60252457 Mitsubishi. [422] Klintz R et al. (1993) DE4213715 BASF. [423] Klintz R et al. (1993) WO93/22280 BASF. [424] Yanagi M et al. (1983) EP97056 Kayaku. [425] Takemoto K et al. (1984) JP5967255 Sumitomo. [426] Akahira R et al. (1986) JP6133154 Toyo Soda. [427] Tokunaga T et al. (1989) JP01175961 Tosoh, A-Kanesho. [428] Tokunaga T et al. (1989) EP326764

Tosoh, A-Kanesho. [429] Takematsu T et al. (1993) JP05155836 Central Glass. [430] Tokunaga T et al. (1990) JP0272145 Tosoh, A-Kanesho. [431] Takematsu T et al. (1992) WO92/03407 Central Glass. [432] Takematsu T et al. (1993) JP05230034 Central Glass. [433] Takematsu T et al. (1993) WO93/17005 Central Glass. [434] Takematsu T et al. (1994) WO94/12468 Central Glass. [435] Takematsu T et al. (1994) WO94/12469 Central Glass. [436] Takematsu T et al. (1995) WO95/19962 Central Glass. [437] Takematsu T et al. (1997) JP09337504 Central Glass. [438] Kume K et al. (1996) JP08259521, JP08259522, JP08259523, JP08259524 Central Glass. [439] Sato K et al. (1992) WO92/12139 Sankyo. [440] Sato K et al. (1994) JP0616664 Sankyo. [441] Shoji H et al. (1985) EP190755, (1986) JP61103801 Asahi. [442] Haeberle N et al. (1988) DE3642372 Dow. [443] Hirai K et al. (1987) WO87/02357 Sagami, Kaken, Chisso; Takahashi J et al. (1989) JP01316306 Sumitomo; Hirai K et al. (1993) JP0517427 Sagami, Kaken, Chisso; Fujita A et al. (1992) JP04139163 Chisso; Sato K et al. (1993) JP05230042 Chisso. [444] Hirai K et al. (1989) JP01203306 Sagami, Kaken, Chisso. [445] Takahashi J et al. (1990) JP0273058 Sumitomo. [446] Hirai K et al. (1990) JP0288565 Sagami, Chisso. [447] Hirai K et al. (1989) JP01316305 Sagami, Kaken, Chisso; Hirai K et al. (1990) JP0285269 Sagami, Chisso. [448] Ooms P et al. (1992) EP484776 Bayer. [449] Hirai K et al. (1989) EP328001 Sagami, Kaken, Chisso; Hirai K et al. (1990) WO90/10626 Sagami, Chisso. [450] Takahashi J et al. (1989) EP338533 Sumitomo. [451] Ooms P et al. (1991) DE3922107 Bayer. [452] Hirai K et al. (1991) JP0356485 Sagami, Kaken, Chisso. [453] Kume T et al. (1990) EP351641 Tokuno. [454] Kume T et al. (1990) EP369262 Tokuno. [455] Franke W et al. (1990) EP397295 Schering. [456] Nagano E et al. (1982) JP57197268 Sumitomo. [457] Nagano E et al. (1983) JP58219167 Sumitomo. [458] Shimano S et al. (1986) JP6169763 Kayaku. [459] Theodoridis G et al. (1990) US4902338 FMC. [460] Hirai K et al. (1988) EP262428 Sagami, Kaken, Chisso; Hirai K et al. (1989) JP01226875 Sagami. [461] Hirai K et al. (1990) JP0288565 Sagami, Chisso. [462] Ooms P et al. (1991) DE3929170 Bayer. [463] Thibault TD et al. (1982) US4345935 Eli Lilly. [464] Prisbylla MP et al. (1989) EP300882 ICI Americas. [465] Prisbylla MP et al. (1990) US4911748 ICI Americas. [466] Nagano E et al. (1983) EP70389, (1982) JP57209290 Sumitomo. [467] Okada I et al. (1985) JP60233075 Mitsubishi. [468] Liebl R et al. (1988) DE3643749 Hoechst. [469] see Reference 262. [470] Onodera N et al. (1990) EP384973 A-Kanesho. [471] Andree R et al. (1990) DE3917469 Bayer. [472] Fischer R et al. (1990) DE3827221 Bayer. [473] Okada I et al. (1983) JP58189178 Mitsubishi. [474] Okada I et al. (1985) JP6094980, JP6097980 Mitsubishi. [475] Shimano S et al. (1984) EP104532 Kayaku. [476] Pissiotas G et al. (1989) DE3919320 Ciba-Geigy. [477] Matsumoto S et al. (1986) JP6127985 Mitsubishi. [478] see Reference 326. [479] Haga T et al. (1987) JP62158280 Sumitomo. [480] Kume T et al. (1988) EP263299 Tokuno. [481] Ganzer M et al. (1993) DE4208778 Schering. [482] Seckinger K et al. (1992) EP468930 Sandoz. [483] Schaefer M et al. (1995) WO95/23509 DuPont, Degussa. [484] Schaefer M et al. (1994) WO94/05668 Degussa. [485] Adams EJ et al. (1997) WO97/15576 DuPont. [486] Hirai K et al. (1996) WO96/20195 Sagami, Kaken. [487] Tanabe Y et al. (1983) JP58216177 Sumitomo. [488] Krenzer J et al. (1988) US4758263 Sandoz. [489] Blume F et al. (1988) DE3701666 Schering. [490] Kilama JJ (1997) US5643855 DuPont. [491] Frank W et al. (1989) DE3809390 Schering. [492] Clark MT et al. (1990) GB2224732 Shell. [493] Kim HJ et al. (1993) Bull. Korean Chemical Soc **14**:717–722. [494] Kilama JJ (1994) WO94/14817 DuPont. [495] US3385862 (1976) RP. [496] Bettarini F et al. (1997) EP780385 Isagro. [497] Takematsu T et al. (1983) JP5890570 Sho-Den. [498] Yokoo H et al. (1988) WO88/04653 Showa-R. [499] Takematsu T et al. (1987) JP62161772 Kuraray. [500] Clark MT et al. (1990) EP354622 Shell. [501] Yamada O et al. (1984) JP59148769 Kayaku. [502] Okada I et al. (1985) JP60109578 Mitsubishi. [503] Jikihara T et al. (1986) JP6140277 Mitsubishi. [504] Haga T et al. (1988) EP258773 Sumitomo. [505] Nagano E et al. (1984) GB2127410, (1986) US4624699 Sumitomo. [506] Yanagi M et al. (1985) EP138527 Kayaku. [507] Jikihara T et al. (1985) JP60233061 Mitsubishi. [508] Nagano E et al. (1985) JP60252465 Sumitomo. [509] Haga T et al. (1985) EP152890 Sumitomo. [510] Nagano E et al. (1985) JP60255772 Sumitomo. [511] Haga T et al. (1985) JP60246372 Sumitomo. [512] Okada I et al. (1986) JP6136268 Mitsubishi. [513] Hayase Y et al. (1986) EP197495 Shionogi. [514] Yanagi M et al. (1986) EP191303 Kayaku. [515] Nagano E et al. (1987) US4670043 Sumitomo. [516] Hagiwara K et al. (1987) JP6251669 Nisso. [517] Haga T et al. (1987) JP6212761 Sumitomo. [518] Hagiwara K et al. (1988) JP63287766 Nisso. [519] Enomoto M et al. (1989) JP0109987 Sumitomo. [520] Rueb L et al. (1990) DE3901705 BASF;

Schaefer B et al. (1992) DE4042194 BASF. [521] Rueb L et al. (1990) DE3901550 BASF. [522] Hirai K et al. (1992) WO92/01671 Sagami, Kaken, Chisso. [523] Schallner O et al. (1993) EP558999 Bayer. [524] Nagano E et al. (1986) JP6178768 Sumitomo. [525] Hagiwara K et al. (1987) JP62286970 Nisso. [526] Moriyasu K et al. (1989) JP01305065 Mitsui. [527] Moriyasu K et al. (1990) JP0232080 Mitsui. [528] Yanagi M et al. (1986) JP61165373 Kayaku. [529] Oyama H et al. (1986) EP202169 Hokko. [530] Kawada S et al. (1987) JP62123173 Kayaku. [531] Oyama H et al. (1988) JP6341448, JP6341449 Hokko. [532] Jensen KU et al. (1990) DE3903799 Bayer. [533] Endo Y et al. (1991) JP0395162 Otsuka. [534] Haga T et al. (1987) EP235567, (1987) EP235567 Sumitomo. [535] Enomoto M et al. (1989) EP304935 Sumitomo. [536] Schallner O et al. (1989) DE3832348 Bayer; Lantzsch R et al. (1990) EP379894 Bayer. [537] Enomoto M et al. (1991) EP422639 Sumitomo. [538] Miura Y et al. (1990) EP361114 Nichino. [539] Brunner H-G et al. (1997) WO97/00246 Ciba-Geigy. [540] Miura Y et al. (1991) JP0372460 Nichino; Takahashi H et al. (1992) JP04225937 Nichino; Ohtani T et al. (1993) FR2692258 Nichino; Nakano I et al. (1994) JP06199804 Nichino; Kodaira T et al. (1994) JP06199806 Nichino; Bussche-H CD et al. (1995) DE4417837 BASF. [541] Miura Y et al. (1991) JP03151367 Nichino. [542] Miura Y et al. (1991) EP443059 Nichino. [543] Miura Y et al. (1994) JP06199805 Nichino. [544] Bansol HS et al. (1995) WO95/19967, (1994) WO94/26109 Zeneca. [545] Cyrill Z et al. (1997) DE19542520 BASF. [546] Theodoridis G et al. (1997) WO97/05115 FMC. [547] Machitani K et al. (1991) JP0347180 Nichino. [548] Machitani K et al. (1991) JP0381275 Nichino. [549] Dutra GA et al. (1992) WO92/06962 Monsanto; Woodard SS et al. (1996) US5489571 Monsanto; Chupp JP et al. (1996) US5587485 Monsanto. [550] Wooderd SS et al. (1994) US5281571 Monsanto; Hamper BC et al. (1997) US5668088 Monsanto. [551] Cyrill Z et al. (1997) WO97/15559 BASF. [552] Jikihara T et al. (1981) US4249934 Mitsubishi. [553] Nagano E et al. (1983) EP75267 Sumitomo. [554] Nagano E et al. (1984) JP5982372, JP5967286, JP5970672 Sumitomo. [555] Nagano E et al. (1984) JP59172491 Sumitomo. [556] Yamaguchi M et al. (1989) JP01121290 Kumiai, Ihara. [557] Dorfmeister G et al. (1989) DE3813884 Schering. [558] Amuti KS et al. (1992) WO92/13453 DuPont. [559] Kobayashi S (1984) EP104484, JP5942384 Kayaku. [560] Shimano S et al. (1986) JP6169778 Kayaku. [561] Jikihara T et al. (1986) JP6176487 Mitsubishi. [562] Sato J et al. (1990) JP02289573 Nissan. [563] Sato J et al. (1991) JP0363278 Nissan. [564] Haga T et al. (1986) EP176101 Sumitomo. [565] Haga T et al. (1987) EP230874 Sumitomo. [566] Kume T et al. (1990) EP349832 Tokuno. [567] Boehner B et al. (1987) EP210137 Ciba-Geigy. [568] see Reference 355. [569] Nagase H et al. (1983) JP58216181 Takeda; Wakabayashi O et al. (1979) JP541256652 Mitsubishi. [570] Takemoto K et al. (1984) JP5970682 Sumitomo. [571] Takemoto K et al. (1984) JP59204181 Sumitomo. [572] Yamaguchi M et al. (1987) JP6200091 Kumiai, Ihara. [573] Chang JH et al. (1990) US4906281 FMC. [574] Sato J et al. (1990) JP02188588, JP02191261 Nissan. [575] Yamaguchi M et al. (1988) EP273417 Kumiai, Ihara. [576] Yamaguchi M et al. (1989) EP312064 Kumiai, Ihara. [577] Yamaguchi M et al. (1990) JP0245488 Kumiai, Ihara. [578] Sato J et al. (1990) JP02289573 Nissan. [579] Yamaguchi M et al. (1992) JP04145087 Kumiai, Ihara. [580] Brunner HG et al. (1992) WO92/21684 Ciba-Geigy. [581] Brunner HG et al. (1994) EP6000833 Ciba-Geigy. [582] Pissiotas G et al. (1994) EP611768 Ciba-Geigy. [583] Pissiotas G et al. (1994) WO94/25467 Ciba-Geigy. [584] Sato J et al. (1989) EP304920, (1990) JP02193961 Nissan. [585] Chang JH et al. (1989) US4830659 FMC. [586] Yamaguchi M et al. (1991) JP03284685 Kumiai, Ihara. [587] Pissiotas G et al. (1993) EP528765 Ciba-Geigy. [588] Hong W et al. (1995) WO95/06643 DuPont, Degussa. [589] Chang JH et al. (1990) US4913723 FMC. [590] Rueb L et al. (1991) DE3927388 BASF. [591] Rueb L et al. (1991) EP410265 BASF. [592] Pissiotas G et al. (1991) EP457714 Ciba-Geigy. [593] Pissiotas G et al. (1995) WO95/21174 Ciba-Geigy. [594] Hagiwara K et al. (1988) JP6345268 Nisso. [595] Takasuka S et al. (1996) JP08253455 A-Kanesho. [596] Sato J et al. (1990) JP02221254 Nissan; Ohi H et al. (1992) JP0412127 Ihara. [597] Yamaguchi M et al. (1992) WO92/13835 Ihara. [598] Kajioka M et al. (1981) JP5632403 Nichino. [599] Kajioka M et al. (1981) DE3024316, JP5632468, JP5653662, JP5653663 Nichino. [600] Kajioka M et al. (1981) JP5653665 Nichino. [601] Kajioka M et al. (1982) FR2497201 Nichino. [602] Kajioka M et al. (1983) JP5823680 Nichino. [603] Kajioka M et al. (1983) JP58157771 Nichino; Matsui H et al. (1985) JP6048965, JP6048974, JP6048975, JP6048976, JP6048977, JP6051180, JP60136572, JP60136573 Nichino. [604] Kajioka M et al. (1983) JP58225070 Nichino. [605] Maravetz LL et al. (1985) WO85/01637 FMC. [606] Kajioka M et al.

(1986) JP61205264 Nichino. [607] Chang JHS et al. (1989) ZA8805308 FMC. [608] Uematsu T et al. (1982) EP55105 Sumitomo. [609] Kajioka M et al. (1985) DE3514057 Nichino; Matsui H et al. (1986) JP61197569 Nichino; Nakanishi H et al. (1986) JP61197566, JP61197567, JP61197568 Nichino; Matsui H et al. (1986) EP193660 Nichino; Kajioka M et al. (1986) JP61243070 Nichino; Machitani K et al. (1987) JP62265275 Nichino. [610] Maravetz LL et al. (1985) WO85/04307 FMC. [611] Maravetz LL et al. (1986) WO86/04481 FMC. [612] Maravetz LL et al. (1986) WO86/02642 FMC. [613] Theodoridis G et al. (1987) WO87/0030 FMC. [614] Kajioka M et al. (1987) DE3636318, JP6299364, JP6299366 Nichino; Yabutani K et al. (1987) EP220952 Nichino. [615] Kawada S et al. (1987) JP6299365 Nichino. [616] Maravetz LL et al. (1987) US4705557 FMC. [617] Kajioka M et al. (1988) JP6327483 Nichino; Ikeda M et al. (1989) JP0193577 Nichino. [618] Kajioka M et al. (1988) JP6330475 Nichino. [619] Maravetz LL et al. (1988) US4743291, (1989) US4806145 FMC. [620] Theodoridis G et al. (1990) US4906284 FMC. [621] Theodoridis G et al. (1992) US5084085 FMC. [622] Haas W et al. (1994) EP597360 Bayer; Schallner O et al. (1994) DE4239296, (1995) US5378681 Bayer. [623] Haas W et al. (1994) EP609734 Bayer. [624] Linker K-H et al. (1994) EP617026 Bayer. [625] Linker K-H et al. (1994) DE4303676 Bayer. [626] Linker K-H et al. (1995) DE4405614 Bayer. [627] Linker K-H et al. (1995) DE19500439 Bayer. [628] Linker K-H et al. (1997) WO97/26248 Bayer. [629] Theodoridis G et al. (1987) WO87/03782, (1989) US4818275, (1990) GB2230261 FMC. [630] Theodoridis G et al. (1990) US4898606, WO90/03731 FMC. [631] Poss KM et al. (1990) WO90/02120 FMC. [632] Poss KM et al. (1993) US5208212 FMC. [633] Endo Y et al. (1995) JP0789813, JP07187919 Otsuka. [634] Endo Y et al. (1995) JP07188220 Otsuka. [635] Endo Y et al. (1995) JP07188221 Otsuka. [636] Amuti KS et al. (1995) WO95/32621 DuPont, Degussa. [637] Endo Y et al. (1997) WO97/09326 Otsuka. [638] Nakagawa H et al. (1997) JP0971583 Otsuka. [639] Wolf A et al. (1980) US4213773 DuPont; Weckbecker C et al. (1997) DE19601189 Degussa. [640] Amuti KS et al. (1994) US5332718, WO94/22828 DuPont; Weckbecker C et al. (1997) EP784053 Degussa. [641] Kobayashi S et al. (1985) JP6169776 Kayaku; Ager JW et al. (1997) WO97/07107 FMC. [642] Blume F et al. (1989) EP317947; Dorfmeister G et al. (1991) DE402629 Schering. [643] Linker K-H et al. (1995) WO95/30661 Bayer. [644] Chang JH et al. (1988) US4761174 FMC. [645] Hong W et al. (1994) WO94/22860 DuPont. [646] Theodoridis G et al. (1990) US4894084 FMC. [647] Amuti KS et al. (1992) WO92/04827 DuPont. [648] Hagiwara K et al. (1986) DE3528583 Nisso. [649] Hagiwara K et al. (1987) JP62181283 Nisso. [650] Ganzer M et al. (1989) EP311135 Schering. [651] Nagano E et al. (1984) EP116928 Sumitomo. [652] Haga T et al. (1985) EP142769 Sumitomo. [653] Enomoto M et al. (1989) JP0116774 Sumitomo. [654] Nagano E et al. (1986) JP6143175 Sumitomo. [655] Enomoto M et al. (1988) JP63150281 Sumitomo. [656] Enomoto M et al. (1989) EP305923 Sumitomo. [657] Theodoridis G et al. (1985) WO85/01939 FMC. [658] Theodoridis G et al. (1987) WO87/03973 FMC. [659] see Reference 630. [660] Theodoridis G et al. (1989) US4885025 FMC. [661] Theodoridis G et al. (1991) US4985065 FMC. [662] Poss KM et al. (1990) US4913724 FMC. [663] Chang JH et al. (1988) US4734124 FMC. [664] Theodoridis G et al. (1990) US4909829 FMC. [665] see Reference 356. [666] Selby T et al. (1993) WO93/15074 DuPont. [667] Nakanishi H et al. (1994) EP590834 Nichino. [668] Nakanishi H et al. (1994) JP06340643 Nichino. [669] Dayan FE, Duke SO, Reddy KN, Hamper BC, Leschinsky KL (1997) J. Agric. Food Chem., **45**, 967–975. [670] Wenger J et al. (1986) EP195346 Roche. [671] Brouwer W et al. (1990) US4927451 Uniroyal. [672] Suchy M et al. (1992) EP473551 Ciba-Geigy. [673] Suchy M et al. (1989) WO89/03825 Roche. [674] Bell AR et al. (1990) US4943309 Uniroyal. [675] Suchy M et al. (1991) WO91/00278 Ciba-Geigy. [676] Winternitz P et al. (1994) WO94/10155 Ciba-Geigy. [677] Kunz W et al. (1995) WO95/32952 Ciba-Geigy. [678] Kawamura Y et al. (1993) JP0525143 Nissan. [679] Wenger J et al. (1989) WO89/02891 Roche. [680] Brouwer WG et al. (1990) US4979982 Uniroyal. [681] Klintz R et al. (1993) DE4131038 BASF. [682] Andree R et al. (1997) DE19528186 Bayer. [683] Wenger J et al. (1993) EP542685 Ciba-Geigy. [684] Schaefer P et al. (1995) DE4329537 BASF. [685] Brouwer WG et al. (1995) WO95/25725 Uniroyal. [686] Klintz R et al. (1997) WO97/02253 BASF. [687] Klintz R et al. (1997) DE19524617 BASF. [688] Wenger J et al. (1988) EP255047 Roche. [689] Enomoto M et al. (1992) EP517181 Sumitomo. [690] Nagano H et al. (1993) JP0504972 Sumitomo. [691] Amuti KS et al. (1993) WO93/14073 DuPont. [692] Koiso T et al. (1997) JP09241245 DIC. [693] Ota C et al. (1997) JP09188676 Mitsubishi. [694] Blume F et al. (1988) DE3712782 Schering. [695] Klintz R et al. (1996) DE19504188 BASF. [696] Andree R et al. (1997) DE19547475 Bayer. [697] Andree R et al. (1997) DE19527570 Bayer. [698] Strunk RJ et al. (1991) US4981508 Uniroyal. [699] Enomoto M et al. (1991) EP420194 Sumitomo. [700] Fukuda K et al. (1991) JP03215476 Nissan. [701] Sato J et al. (1992) US5084084 Nissan. [702] Sato J et al. (1991) EP408382 Nissan. [703] Takemura S et al. (1993) JP05140155 Sumitomo. [704] Enomoto M et al. (1992) JP04193876 Sumitomo. [705] Theodoridis G et al. (1994) US5310723 FMC. [706] Drewes MW et al. (1997) DE19530451 Bayer. [707] see Reference 699. [708] Suchy M et al. (1993) US5232898 Ciba-Geigy. [709] Nagano E et al. (1993) EP568041 Sumitomo. [710] Maravetz LL et al. (1997) WO97/12884 FMC. [711] Crawford SD et al. (1997) US5661108 FMC. [712] Enomoto M et al. (1992) EP476697 Sumitomo. [713] Takemura S et al. (1993) EP561319 Sumitomo. [714] Theodoridis G et al. (1994) US5346881 FMC. [715] Theodoridis G et al. (1995) US5441925 FMC. [716] Theodoridis G et al. (1995) WO95/05079 FMC. [717] Heistracher E et al. (1997) DE19532048 BASF. [718] Crawford SD et al. (1997) WO97/08170 FMC. [719] Miyazaki M et al. (1997) WO97/29105 Kumiai. [720] Sato J et al. (1992) US5154755 Nissan. [721] Sato J et al. (1992) WO92/11244 Nissan. [722] Kawamura Y et al. (1992) EP496595 Nissan. [723] Kawamura Y et al. (1992) JP04346981 Nissan. [724] Sato J et al. (1993) JP0501043 Nissan. [725] Kawamura Y et al. (1994) JP0692943 Nissan. [726] Kamireddy B et al. (1995) WO95/33746 DuPont. [727] Andree R et al. (1996) DE19516785 Bayer. [728] Andree R et al. (1996) DE19517732 Bayer. [729] Kawamura Y et al. (1997) JP0948761 Nissan. [730] Enomoto M et al. (1993) EP540023 Sumitomo. [731] Andree R et al. (1997) DE19523640 Bayer. [732] Andree R et al. (1997) DE19532344 Bayer. [733] Strunk RJ et al. (1992) US5169430 Uniroyal. [734] Wenger J et al. (1993) DE4237920 Ciba-Geigy. [735] Konz MJ et al. (1995) US5391541 FMC. [736] Menke O et al. (1997) DE19523372 BASF. [737] Andree R et al. (1995) DE4424401 Bayer. [738] Andree R et al. (1995) DE4412079 Bayer. [739] Drewes M-W et al. (1995) WO95/31440 Bayer. [740] Sato J et al. (1992) JP04178373 Nissan. [741] Kawamura Y et al. (1993) JP05202031 Nissan. [742] Hotzman FW et al. (1995) WO95/17096 FMC. [743] Takano M et al. (1996) EP705829 Sumitomo. [744] Tobler H et al. (1983) US4394156 Ciba-Geigy. [745] Moser H et al. (1990) EP391847 Ciba-Geigy. [746] Lange BC et al. (1991) EP415642 R & H. [747] Sato J et al. (1991) EP454444 Nissan. [748] Uekawa T et al. (1992) EP488220 Sumitomo. [749] Klintz R et al. (1994) DE4236880 BASF. [750] Enomoto M et al. (1988) JP63275580 Sumitomo. [751] Ooms P et al. (1993) DE4128031 Bayer. [752] Ooms P et al. (1995) EP638563 Bayer. [753] Saito K et al. (1988) JP63156779 Nisso. [754] Saito K et al. (1989) JP01139580 Nisso. [755] Moriyasu K et al. (1990) JP0269466 Mitsui. [756] Linker K-H et al. (1996) DE19520613 Bayer. [757] Katayama T et al. (1997) WO97/07104 Sumitomo. [758] Tice CM et al. (1994) EP579424 R & H. [759] Tice CM et al. (1994) EP579425 R & H. [760] Enomoto M et al. (1997) WO97/06150 Sumitomo. [761] Linker K-H et al. (1997) DE19528305 Bayer. [762] Shuto A et al. (1997) WO97/11060 Sumitomo. [763] Sanemitsu Y et al. (1982) JP5785378 Sumitomo. [764] Taylor ED et al. (1994) WO94/03454 DuPont. [765] Theodoridis G et al. (1989) US4878941 FMC. [766] Theodoridis G et al. (1990) US4956004 FMC. [767] Miki H et al. (1996) EP737672 Takeda. [768] Hirai K et al. (1996) WO96/22285 Sagami, Kaken. [769] Sato J et al. (1991) JP0377874 Nissan. [770] Schallner O et al. (1994) EP584655 Bayer. [771] Schallner O et al. (1995) EP640600 Bayer. [772] Amuti KS et al. (1997) US5602077 DuPont.

Notes

Weed names in parentheses in each Figure are abbreviated according to the acronyms of their botanical name as follows: Aba:*Abutilon avicennae*, Abt:*Abutilon theophrasti*, Amb:*Amaranthus blitum*, Aml:*Amaranthus lividus*, Amr:*Amaranthus retroflexus*, Amv:*Amaranthus viridis*, Ama:*Ambrosia artemisiaefolia*, Amm:*Ammannia multiflora*, Aps:*Apera spica-venti*, Avf:*Avena fatua*, Brr:*Brassica rapa*, Brl:Broadleaf weeds, Cam:*Calapogonium mucunoides*, Cab:*Capsella bursa-pastoris*, Caf:*Cardamine flexuosa*, Cec:*Centaurea cyanus*, Cha:*Chenopodium album*, Chc:*Chrysanthemum coronarium*, Cyd:*Cyperus difformis*, Cye:*Cyperus esculentus*, Cym:*Cyperus microiria*, Cys:*Cyperus serotinus*, Das:*Datura stramonium*, Dia:*Digitaria adscendens*, Dic:*Digitaria ciliaris*, Dis:*Digitaria sanguinalis*, Ecc:*Echinochloa crus-galli*, Ecf:*Echinochloa frumentacea*, Eco:*Echinochloa oryzicola*, Ela:*Eleocharis acicularis*, Fim:*Fimbristylis miliacea*,

Gaa:*Galium aparine*, Gac:*Galinsoga ciliata*, Gas:*Galium spurium*, Goh:*Gossypium hirsutum*, Ipl:*Ipomoea lacumosa*, Ipp:*Ipomoea purpurea*, Ips:*Ipomoea subspecies*, Laa:*Lamium amplexicaule*, Lap:*Lamium purpureum*, Lom:*Loium multiflorum*, Mac:*Matricaria chamomilla*, Mai:*Matricaria inodora*, Mar:*Matricaria recutita*, Mov:*Monochoria vaginalis*, Pac:*Panicum crus-galli*, Pad:*Paspalum distichum*, Pav:*Panicum virgatum*, Pel:*Persicaria longiseta*, Php:*Pharbitis purpurea*, Pob:*Polygonum blumei*, Pol:*Polygonum laphathifolium*, Poo:*Portulaca oleracea*, Puj:*Pueraria japonica*, Roa:*Rorripa austriaca*, Rori:*Rorippa indica*, Roti:*Rotala indica*, Sap:*Sagittaria pygmaea*, Sat:*Sagittaria trifolia*, Sch:*Scirpus hotarui*, Scj:*Scirpus juncoides*, See:*Sesbania exaltata*, Sev:*Setaria viridis*, Sis:*Sida spinosa*, Sia:*Sinapis arvensis*, Sol:*Solanum lycopersicum*, Son:*Solanum nigrum*, Soh:*Sorghum halepense*, Stm:*Stellaria media*, Stn:*Stellaria neglecta*, Suj:*Sucirpus juncoides*, Vep:*Veronica persica*, Xap:*Xanthium pennsylvanicum*, Xas:*Xanthium strumarium*.

CHAPTER 3

Recent Advances in Peroxidizing Herbicides (Patent Overview)

Isao Iwataki[1]

Contents

3.1 Introduction

It is well known that diphenyl ethers, 1-phenylpyrazoles, and O-phenylcarbamates are herbicides which inhibit protoporphyrinogen oxidase (Protox) (Duke et al. 1991; Nandihalli et al. 1992, Nandihali and Duke 1994 see Chapter 6). In addition to these compounds, cyclic imide type Protox inhibitors have been thoroughly investigated recently.

Oxadiazon **1** which was the first lead compound of the cyclic imide family was introduced into the market for the control of annual grasses and broadleaf weeds in pre-emergence or early post-emergence treatment by Rhone-Poulenc in 1969. Its basic patent was applied in 1965 and filed in 1968[1]. The second cyclic imide herbicide, chlorophthalim **2**, was filed by Mitsubishi Kasei in 1972[2]. Some years later, their family of compounds appeared in patents[3] which had the specific substitution pattern at the benzene ring (mostly X = F, Y = Cl; see formula **3**). Since then, various cyclic imide derivatives have been developed by many companies. Recently, a compound group appeared in patents,

[1] Odawara Research Center, Nippon Soda Co. Ltd., 345 Takada, Odawara, Kanagawa 250-0216, Japan

Peter Böger, Ko Wakabayashi
Peroxidizing Herbicides

whose basic structure is largely modified. Therefore, it became difficult to distinguish whether it is a Protox inhibitor or not, but the author boldly assumes such patent compounds are the inhibitors when they have at least 2,4-phenyl substitution, like the partial structure **3** with a claim of halogen included. This chapter summarizes the patent literature of the cyclic imide family of compounds between years the 1986 and 1995.

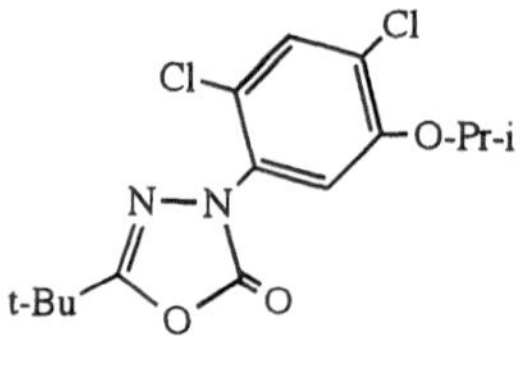

1

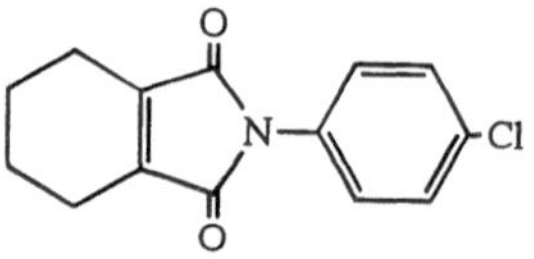

2

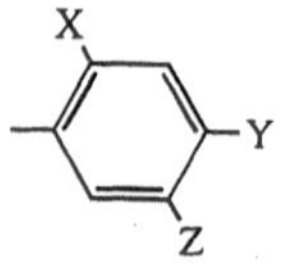

3

3.2 State of Patent Application

The state of patent filing for the 10 years 1986 and 1995 is shown in Fig. 1. Total patents of cyclic imide herbicides have amounted to 431. The development research has been made by 45 companies (7 American, 11 European, 26 Japanese, and 1 other). Detailed structure analysis will be presented later, but the imide herbicides can be roughly classified into the following eight groups (the general structures of each type (**4–10**) are shown in Fig. 2):

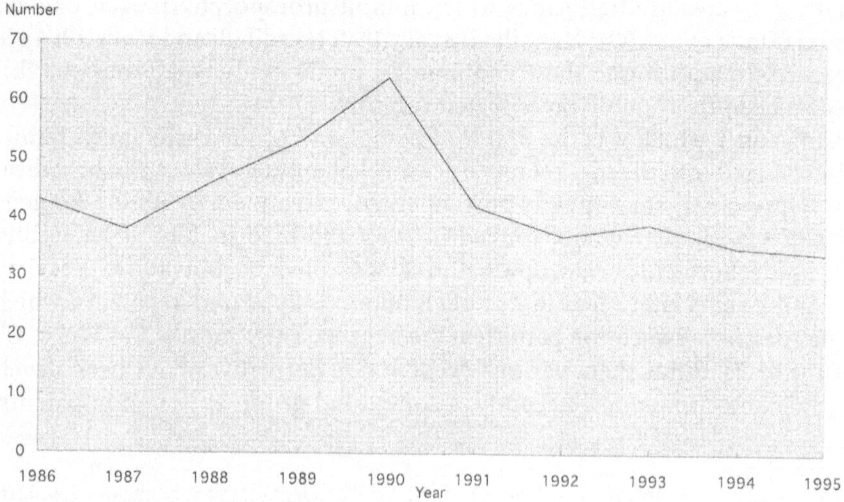

Fig. 1. Patent numbers of Protox herbicides

Type 1. Tetrahydrophthalimide derivatives **4**
Type 2. Bicyclic five-membered heterocyclic derivatives **5**
Type 3. Bicyclic six-membered heterocyclic derivatives **6**
Type 4. Five-membered heterocyclic derivatives **7**
Type 5. Six-membered heterocyclic derivatives **8**
Type 6. Heterocyclic imino derivatives **9**
Type 7. Heterocyclic C-phenyl derivatives **10**
Type 8. Others

The number of the patents in each type is shown in Table 1.

Looking at the content of patents, the initial lead structures, types 1, 2 and 4, are still being intensively investigated. However, as a recent trend, compounds with the unique core structure of types 7 and 8 have been produced. It

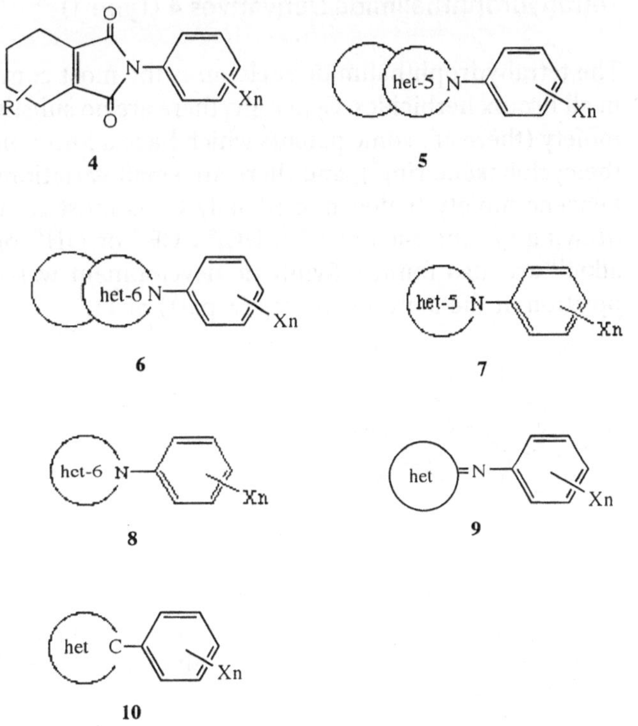

Fig. 2. General structures of imide herbicides applied in patents

Table 1. Classified herbicide patents[a]

Type 1	Type 2	Type 3	Type 4	Type 5	Type 6	Type 7	Type 8
118	86	9	143	67	45	25	29

[a] Duplication is due to contents of patents. Data are based on a patent search from Derwent Chemical Patent Index.

is noteworthy that this new compound type first developed by Nihon Noyaku in the late 1980s[4], in which the 2,4-dihalo-5-substituted phenyl group directly bonded with a carbon atom of a five-membered heterocycle (see **10**, Fig. 2). Instead of the original 2,4,5-trisubstituted phenyl moiety, Sumitomo Chemical Ind. found the 2-fluoro-4,5-heterocyclized phenyl substitution in the late 1980s[5].

It is difficult to estimate the degree of herbicidal activity of filed compounds from patent information only, but the following section will summarize how each company attacks the development of Protox herbicides.

3.3
Characteristics of the Each Type of Compounds

3.3.1
Tetrahydrophthalimide Derivatives 4 (Type 1)

The tetrahydrophthalimide skeleton is the most common and basic structure in all Protox herbicides. Generally, there are no substituents at the cyclohexene moiety (there are some patents which have a lower alkyl or haloalkyl group at the cyclohexene ring[6]), and there are small variations at the 4 position of the benzene moiety (halogen, especially Cl, is most common, but electron withdrawing groups such as CN[7], NO_2[8], CF_3[9] or OH[10] or donating ones such as alkyl[11] are also found). Synthetic development was mostly focused on the 5 position of the benzene moiety (Z part) of **11**.

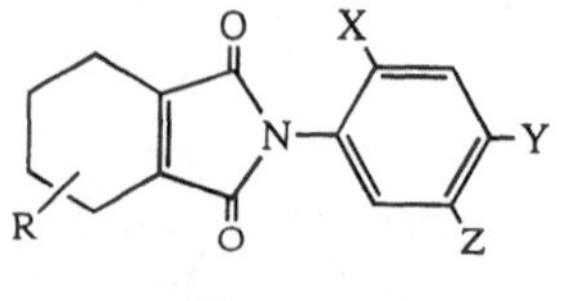

11

The typical substituents of Z are as follows:

SH, S-alkyl, S-haloalkyl, S-alkoxyalkyl, S-alkoxyalkoxyalkyl, S-cyanoalkyl, S-CH(Me)CO-alkyl, S-CH(Me)COS-alkyl, S-CH(Me)CO_2-alkyl, S-$CH_2CH_2CO_2$-alkyl, S-heterocycle, $SO_2N(alkyl)_2$

OH, O-alkyl, O-haloalkyl, O-alkoxyalkyl, O-alkoxyalkoxyalkyl, O-cyanoalkyl, O-CH(Me)CO-alkyl, O-CH(Me)COS-alkyl, O-CH(Me)CO_2-alkyl, O-$CH_2CH_2CO_2$-alkyl, O-heterocycle, O-SO_2-alkyl

N(alkyl)-P(=O)(O-alkyl)$_2$, $NHCH_2CO_2$-alkyl, $NHCH_2$COS-alkyl, N(alkyl)-SO_2-alkyl, NO_2

CO_2-alkyl, CO_2-alkoxycarbonylalkyl, CO_2-alkoxyalkoxyalkyl, CO_2-heterocycle substituted alkyl, CO_2-trialkylsilylalkyl, CO_2-dialkoxyphosphonoalkyl, CO_2-N-alkylidene, COS-substituted alkyl, CON-substituted alkyl, O-heterocycle

C≡C-H, C≡C-alkyl, C≡C-substituted phenyl

CH=C(halo)CO_2-substituted alkyl, CH=C(alkyl)-alkoxycarbonylalkyl, CH=C(alkyl)-CH=C(alkyl)-CO_2-alkyl, CH=C(cyano)CO_2-alkyl,

CH=N-O-alkyl, CH=N-O-alkenyl, CH=N-O-alkynyl, CH=N-O-alkoxycarbonylalkyl

acyl, CO-CH_2O-substituted alkyl

5,6-membered substituted heterocycles

CH_2OH, CH_2OP(=O)(O-alkyl)$_2$, CH_2O-alkyl, CH_2-N(alkyl)$_2$, CH_2S-alkyl, CH_2O-acyl, CH_2O-alkoxycarbonylalkyl, haloalkyl, alkoxycarbonylalkylene, CH_2P(=O)(O-alkyl)$_2$

P(=O)(O-alkyl)$_2$

There are some variations in the tetrahydrophthalimide moiety, as shown in Fig. 3.

As described in Section 3.2 above, instead of the original 2,4,5-trisubstituted phenyl moiety, Sumitomo Chemical Ind. found 2-fluoro-4,5-heterocyclized phenyl substitution **12** in the middle of the 1980s[5] and many such types of substituents have been found since. In the case of this type, bulky lipophilic

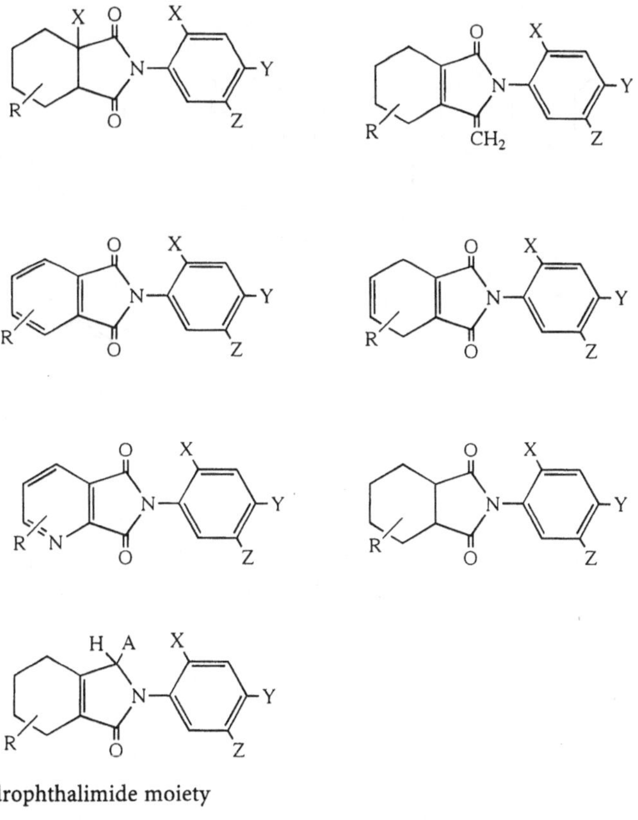

Fig. 3. Variation of tetrahydrophthalimide moiety

substituents such as 1-substituted phenyl benzotriazole seem still to maintain a potent herbicidal activity[12]. Some of the representative substituents are shown in Fig. 4.

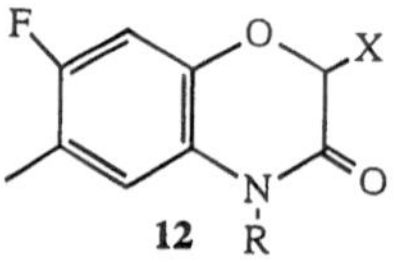

12

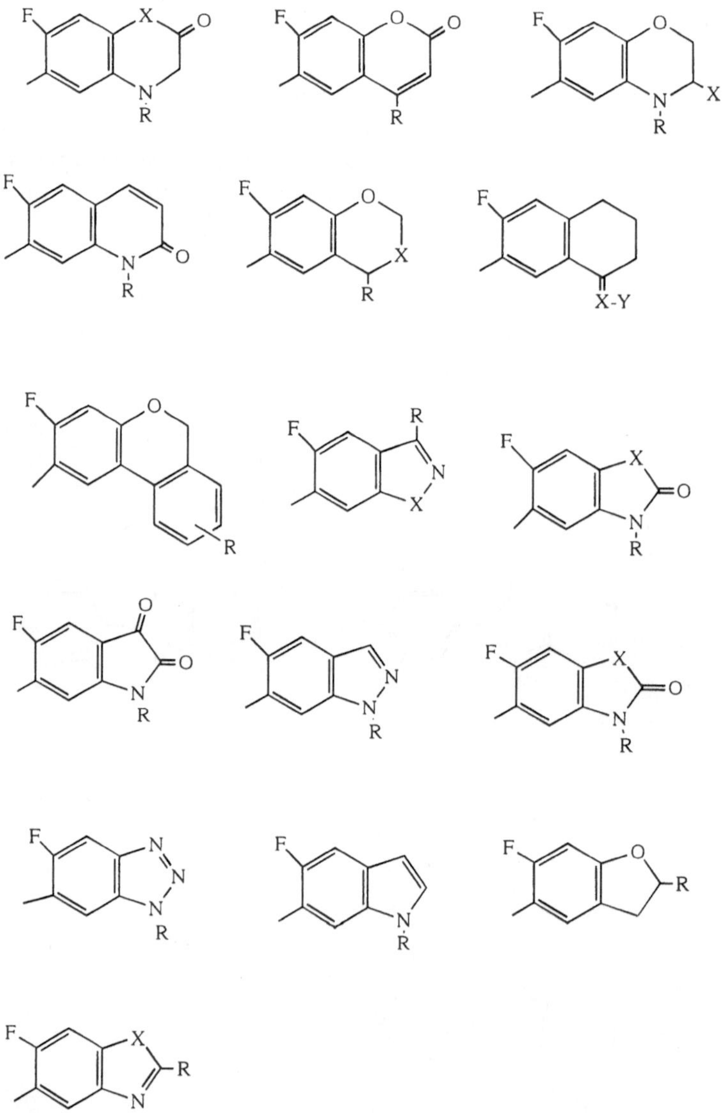

Fig. 4. 2-Fluoro-4,5-heterocyclized phenyl substitutions (Type 1)

3.3.2
Bicyclic Five-Membered Heterocyclic Derivatives 5 (Type 2)

As a one of variations of the authentic N-phenyltetrahydrophthalimide derivatives, bicyclic five-membered heterocyclic derivatives **5** appeared. The representative core skeletons are shown in Fig. 5. The phenyl substitutions are similar to those of tetrahydrophthalimide derivatives.

3.3.3
Bicyclic Six-Membered Heterocyclic Derivatives 6 (Type 3)

As shown in Table 1, bicyclic six-membered heterocyclic derivatives **6** are less developed. The representative core skeletons are shown in Fig. 6.

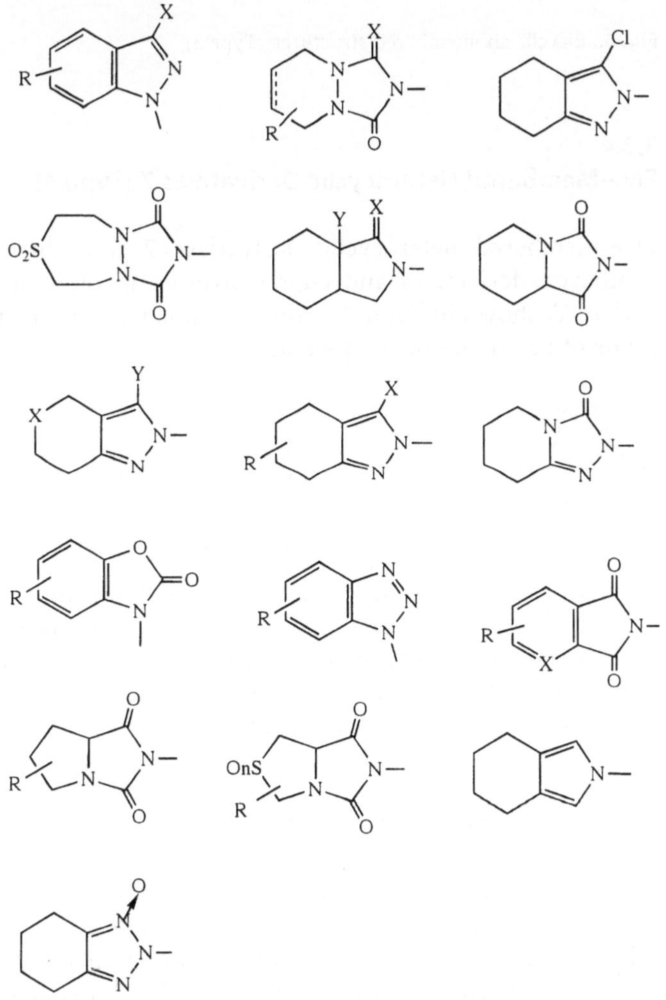

Fig. 5. Bicyclic five-membered structures (Type 2)

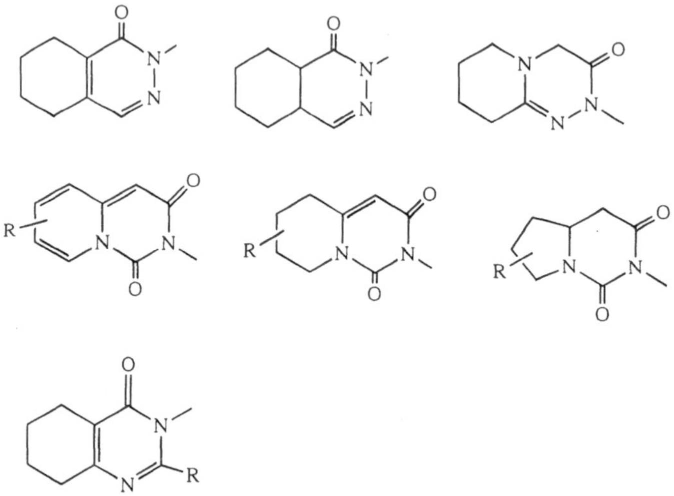

Fig. 6. Bicyclic six-membered structures (Type 3)

3.3.4
Five-Membered Heterocyclic Derivatives 7 (Type 4)

Five-membered heterocyclic derivatives 7 were first developed as basic oxadiazon derivatives and later converted to other five-membered heterocycles. As shown in Table 1, patents of this type are the most developed of all kinds of Protox herbicide patents.

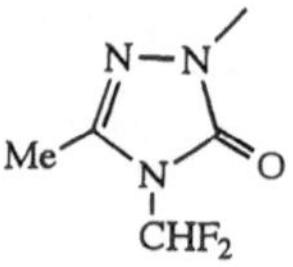

13

Nihon Noyaku developed 1,2,4-oxadiazolinone derivatives which have special substituents at the 4 and 5 positions of the 1,2,4-triazolinone ring, like **13**[13]. Substituents at the five-membered heterocyclic skeletons except for the N-phenyl substitution are mostly lower alkyls, halogens or haloalkyls. The representative skeletons are shown in Fig. 7.

3.3.5
Six-Membered Heterocyclic Derivatives 8 (Type 5)

Patents of six-membered heterocyclic derivatives 8 increased after 1990. The degree of herbicidal activity is not clear, but 11 European, American and Japanese companies have developed compounds using this type of skeleton. The representative skeletons are shown in Fig. 8. Remaining substituents at the

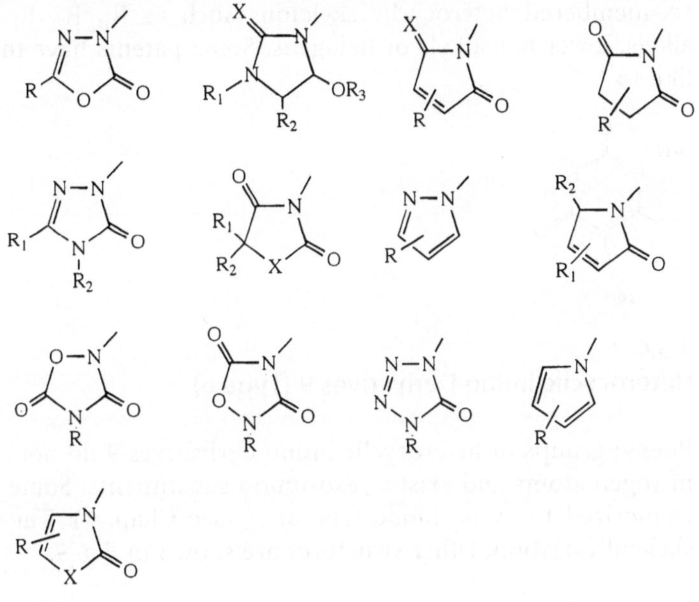

Fig. 7. Five-membered heterocyclic structures (Type 4)

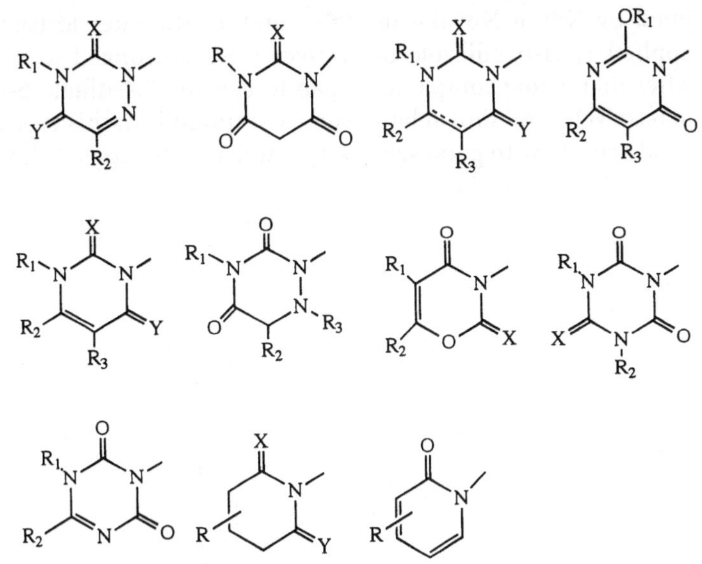

Fig. 8. Six-membered heterocyclic structures (Type 5)

six-membered heterocyclic skeletons such as R_1, R_2, R_3 are mainly lower alkyls, lower haloalkyls or halogens. Some patents have the N-amino group, like **14**[14].

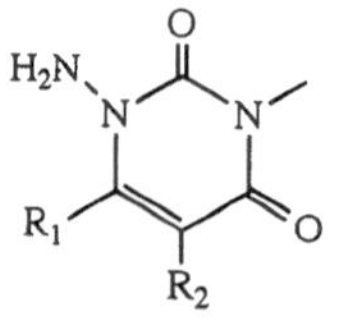

14

3.3.6 Heterocyclic Imino Derivatives 9 (Type 6)

Phenyl groups of heterocyclic imino derivatives **9** do not bond with the ring nitrogen atoms and exist as exo-imino substituents. Some compounds easily isomerized to cyclic imide type ones (see Chap. 7). There is not so much skeletal variation. Other structures are shown in Fig. 9.

3.3.7 Heterocyclic C-Phenyl Derivatives 10 (Type 7)

Heterocyclic C-phenyl derivatives **10** with a pyrazole skeleton were first applied by Nihon Noyaku in 1988[15] and a little later Mitsubishi Petro. Chem. applied 1,2,4-oxadiazole derivatives which belonged to the same category[17]. After that, some companies began to develop 2,4-dihalo-5-substituted phenyl heterocycles. As shown later, some compounds of this type have potent herbicidal activity with good selectivity toward some crops. Nihon Noyaku's deriva-

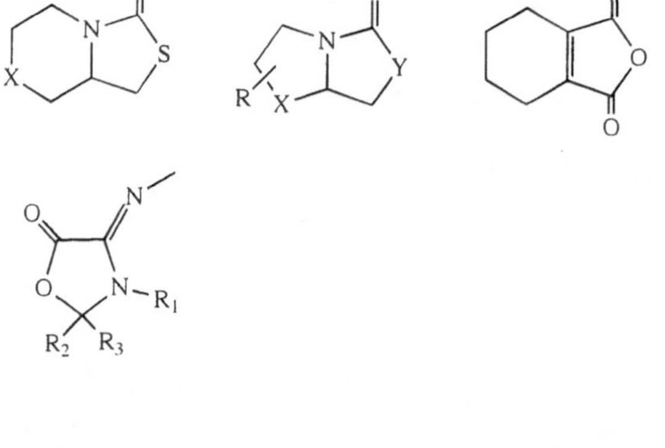

Fig. 9. Heterocyclic imino structures (Type 6)

tives show protoporphyrinogen oxidase inhibiting activity similar to type 1–6 compounds, but mode of action of others listed here is still obscure. Substituents at the benzene ring X and Y are mostly halogens, and Z has wide variation like that of type 1. The representative structures are shown in Fig. 10.

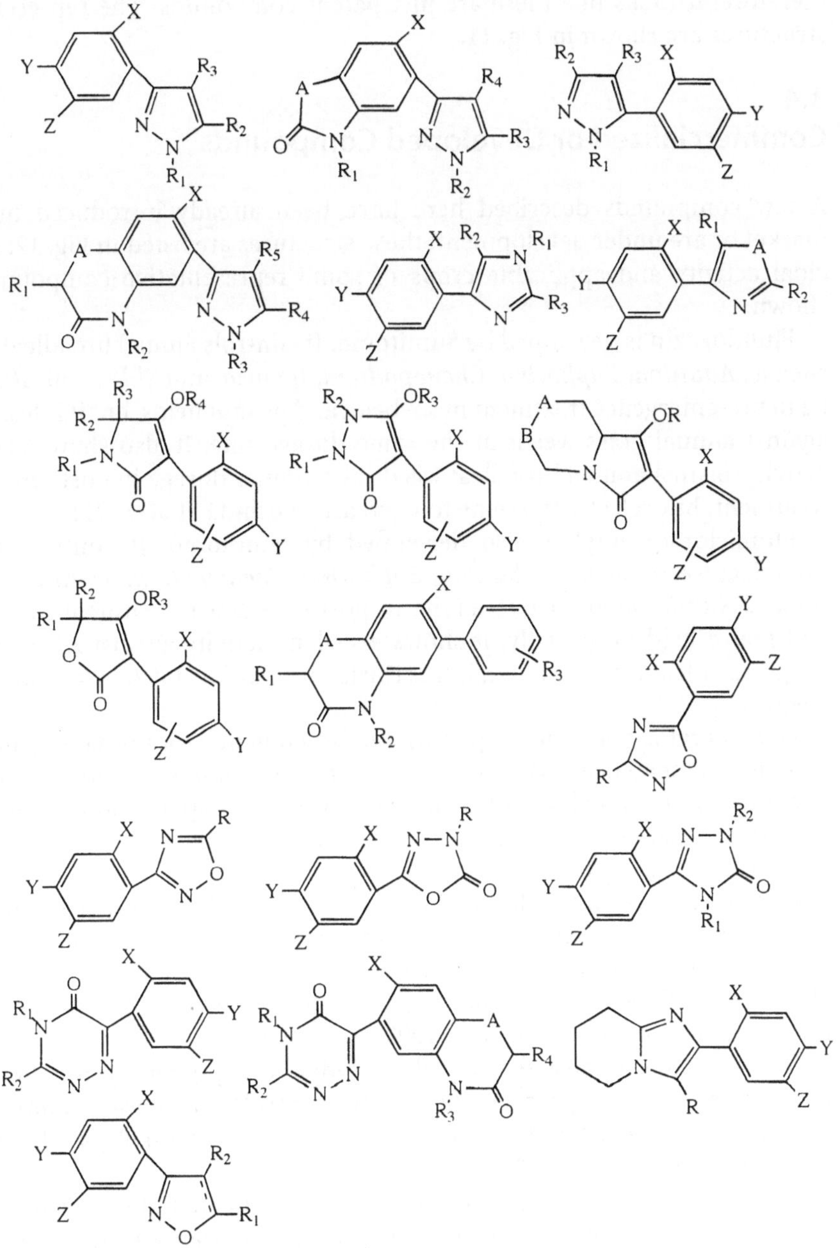

Fig. 10. C-Phenyl structures (Type 7)

3.3.8
Others (Type 8)

There are many patents which do not belong to categories type 1–7. Some of them seem intermediates of type 1–5 synthesis, like type 6. No information on this group was not given anywhere that these are developing compounds; therefore, it looks like these are just patent compounds. The representative structures are shown in Fig. 11.

3.4
Commercialized or Developed Compounds

A few compounds described here have been already introduced into the market or are under development; these structures are listed in Fig. 12. Herbicidal activity and applicable crops of some representative compounds are shown.

Flumioxazin is developed by Sumitomo. It controls annual broadleaf weeds such as *Abutilon*, *Euphorbia*, *Chenopodium*, *Ipomea*, and *Sida* at 50–100 g a.i./ha in pre-emergence treatment in soybean and peanut fields, but it is less active against annual grass weeds at the same dosage rate. It also shows excellent activity against annual broadleaf weeds at a similar dosage in post-emergence treatment, but selectivity seems to decrease (Yoshida et al. 1991).

Flumiclorac-pentyl is also developed by Sumitomo. It controls annual broadleaf weeds such as *Abutilon*, *Euphorbia*, *Chenopodium*, *Datura*, *Ambrosia* and *Xanthium* at 30–60 g a.i./ha in post-emergence treatment in soybean and maize fields; especially, it shows excellent activity against *Abutilon* at a progressed leaf stage at 60 g a.i./ha (Kurts and Pawlak 1992, 1993; Saito et al. 1993).

Oxadiargyl is being developed by Rhone-Poulenc as a pre-emergence and early post-emergence herbicide in sugarcane and sunflower fields or orchards to control both annual broadleaf and grass weeds. Its application rate is rather high at 500–2000 g a.i./ha. It is also developed as a herbicide for rice and turf in Japan (Oe et al. 1995).

Carfentrazone-ethyl is developed by FMC as a post-emergence herbicide in rice, cereals and maize fields. It shows excellent activity against annual broadleaf weeds such as *Galium*, *Lamium* and *Veronica* in wheat at 20–35 g a.i./ha. It also controls *Euphorbia*, *Polygonum*, *Abutilon*, *Ipomea*, *Kochia*, *Salsola*, etc. in foliar application (Vansaun et al. 1993; Mize 1995).

Sulfentrazone is also developed by FMC as a pre-emergence herbicide in soybean and sugarcane fields. It controls *Ipomoea*, *Amaranthus*, *Chenopodium*, *Abutilon*, *Polygonum*, *Datura*, etc. at 350–420 g a.i./ha (Oliver et al. 1996; Vidrine et al. 1996).

Fluthiacet-methyl was discovered by Kumiai Chem. and now is developed by Novartis as a post-emergence herbicide in soybean and maize fields. It shows excellent activity against annual broadleaf weeds such as *Abutilon*,

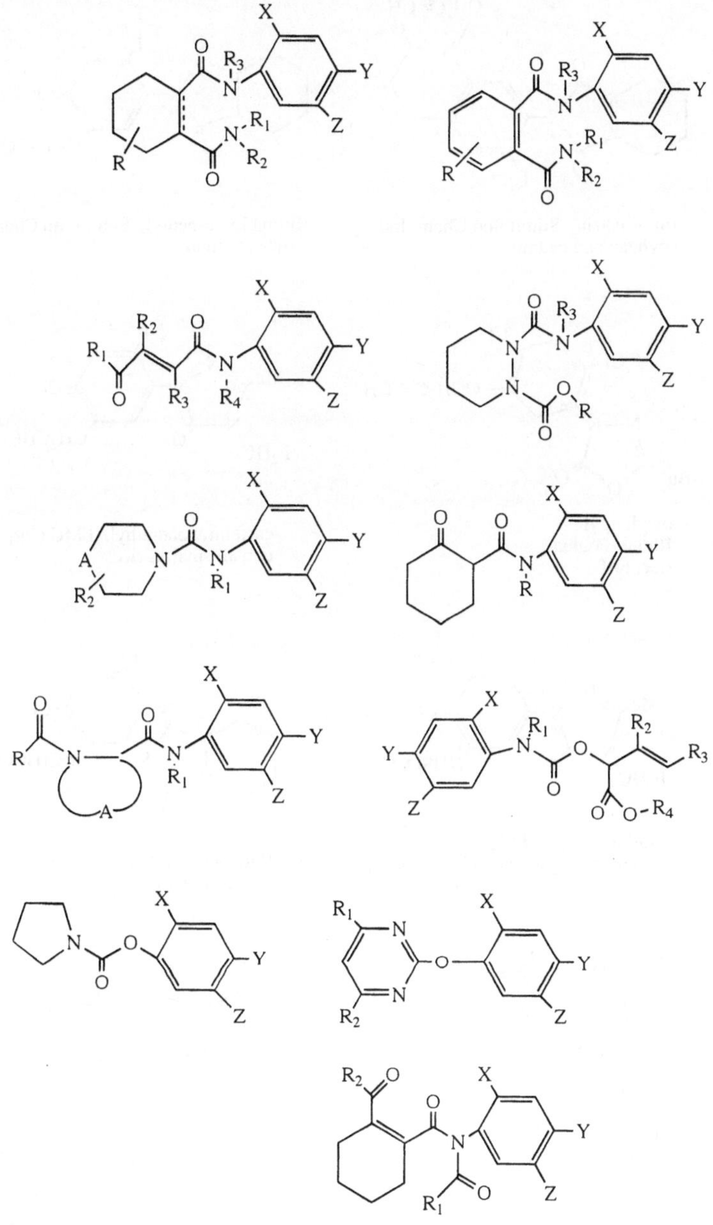

Fig. 11. Others (Type 8)

Chenopodium and *Euphorbia* at a quite low use rate of 3–10 g a.i./ha (Vitolo et al. 1995).

Pentoxazone was discovered by Sagami Chemical Research Center and Kaken Pharmaceutical as a pre-emergence and early post-emergence herbicide in rice. It shows excellent activity against both annual and perrenial weeds

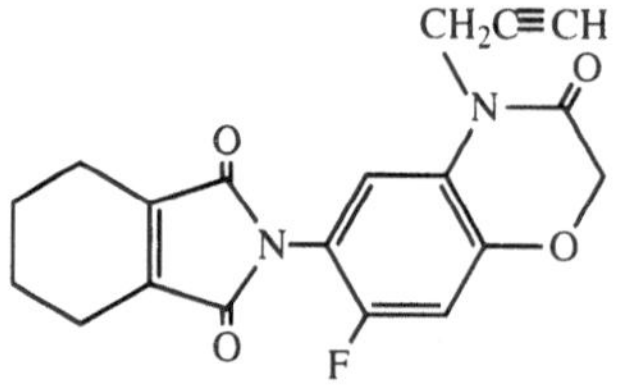

flumioxazin, Sumitomo Chem. Ind.
soybean and peanut

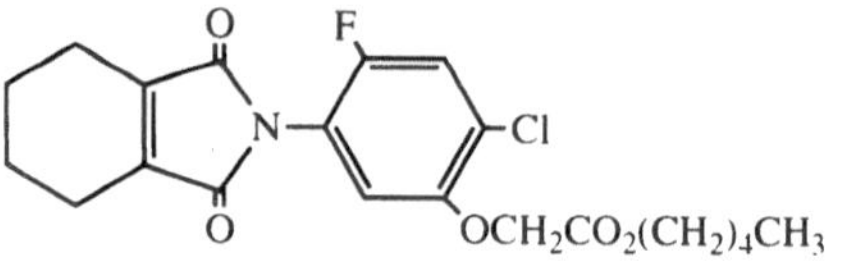

flumiclorac-pentyl, Sumitomo Chem. Ind.
soybean, maize

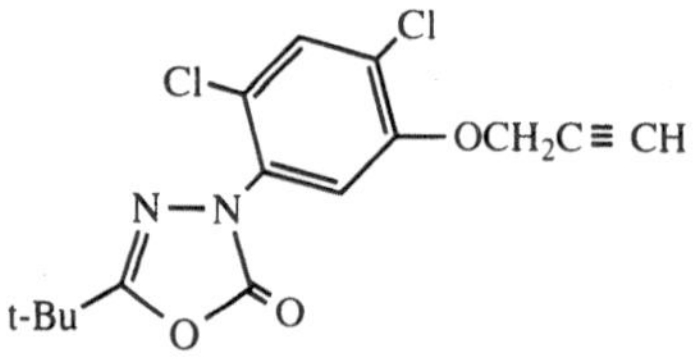

oxadiargyl
Rhone-Poulenc
rice, turf

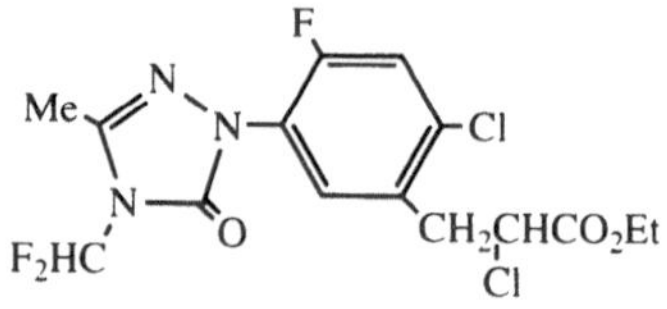

carfentrazone-ethyl, FMC Corp.
cereals, maize, rice

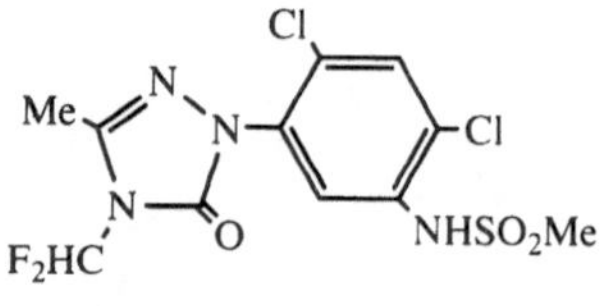

sulfentrazone, FMC Corp.
soybean, sugarcane, peanut, tobacco

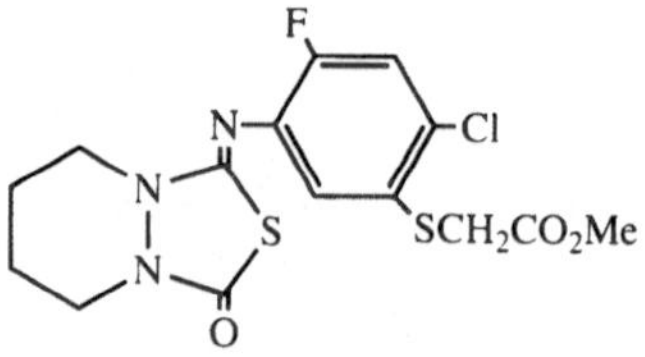

fluthiacet-methyl, Kumiai Chem. Ind.
soybean, maize

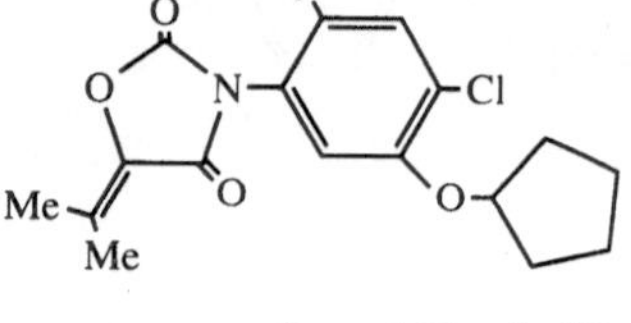

pentoxazone, Sagami Chemical Research Center,
Kaken Pharm. Co.
rice

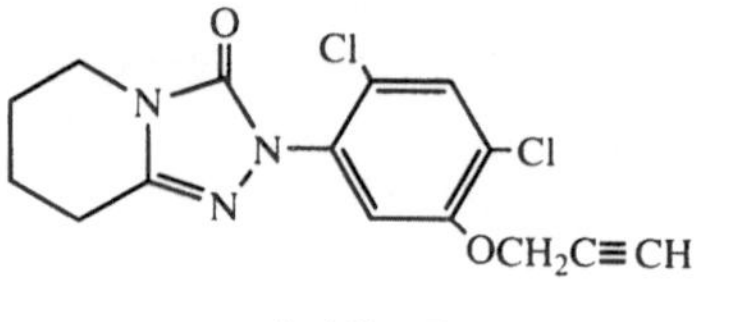

azafenidin, Du Pont
non-crop land, fruit trees

Fig. 12. Commercialized or Developed Compounds (including experimental compounds)

such as *Echinochloa, Eleocharis, Sagittaria* and *Cyperus* at 145–450 g a.i./ha (Yoshimura et al. 1992; Hirai et al. 1995).

Azafenidin is developed by Du Pont as a non-selective pre- and post-emergence herbicide for non-cropland and orchard. It has wide herbicidal

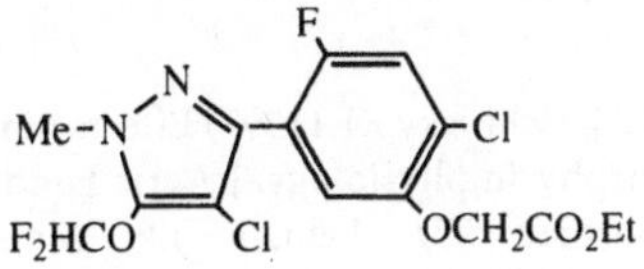

pyraflufen-ethyl, Nihon Noyaku
cereals

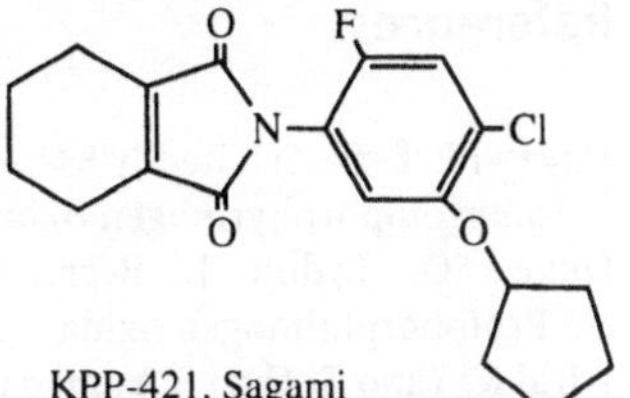

KPP-421, Sagami
Chemical Research Center,
Kaken Pharm. Co.
soybean, cereals

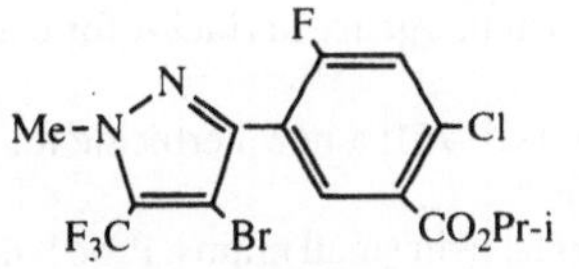

JV-485, Monsanto
cereals

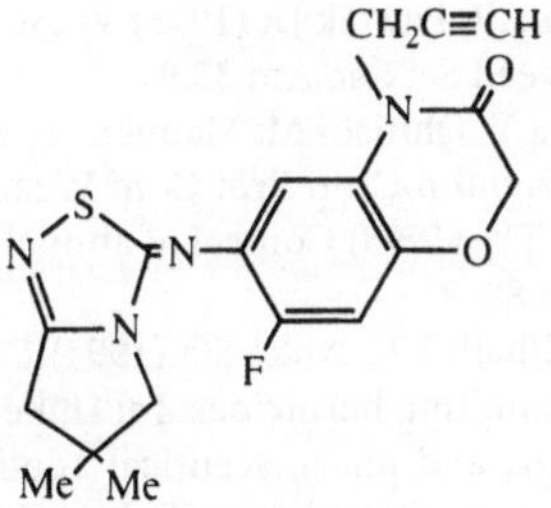

thidiazimin, Schering AG
cereals

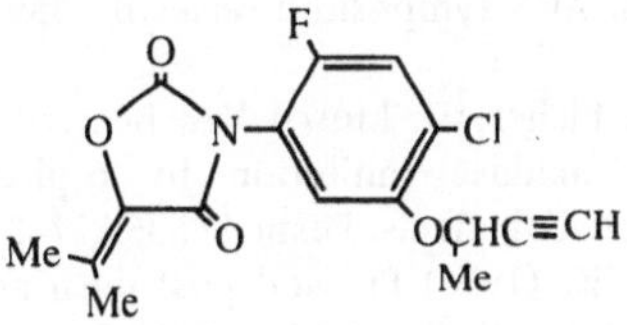

KPP-300, Sagami Chemical Research
Center, Kaken Pharm. Co.
rice

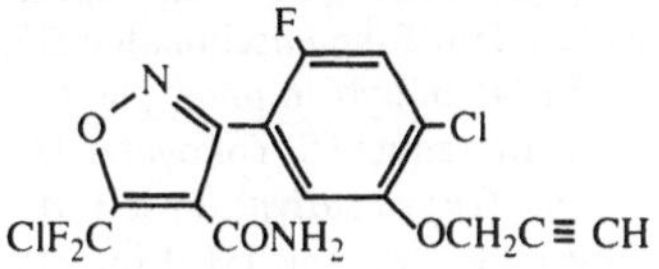

Monsanto

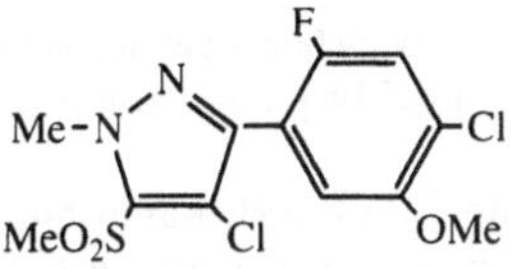

Monsanto

Fig. 12. *(Continued)*

spectra and controls both annual and perrenial weeds at 560 g a.i./ha (Netzer et al. 1996).

Herbicidal activity of other compounds in Fig. 12 are reported the following (pyraflufen-ethyl: Miura et al. 1993; KPP-421: Ukai et al. 1993; JV-485: Prosch et al. 1997; thidiazimin: Weiler et al. 1993; KPP-300: Ukai et al. 1992; Monsanto experimental compound: Dayan et al. 1997). By changing the skeleton or phenyl substitution, not only herbicidal activity but also selectivity largely differ and seem to be applicable to many crop species. To confirm how much products will be practically used, we must await future development.

References

Dayan FE, Duke SO, Reddy KN, Hamper BC, Leschinsky KL (1997) Effects of isoxazole herbicides on protoporphyrinogen oxidase and porphyrin physiology. J Agric Food Chem 45:967–975

Duke SO, Lydon J, Becerril JM, Sherman TD, Lehnen LP, Matsumoto H (1991) Protoporphilinogen oxidase inhibiting herbicides. Weed Sci 39:465–473

Hirai K, Yano T, Ugai S, Yamada O, Yoshimura T, Hori M (1995) A new low-rate pre-emergence herbicide KPP-314 for rice. Proc 15th Conf Asian Pacific Weed Sci Soc, pp 840–845

Kurtz AR, Pawlak JA (1992) Postemergence weed control in field corn with V-23031 herbicide. Proc North Central Weed Sci Soc 47:47

Kurtz AR, Pawlak JA (1993) V-23031 – a new postemergence herbicide for use in field corn. Abstr Weed Sci Soc Am 33:9

Miura Y, Ohnishi M, Mabuchi T, Yanai I (1993) ET-751: a new herbicide for use in cereals. Proc Brighton Crop Prot Conf Weeds, pp 35–40

Mize TW (1995) Control of annual weeds with F8426 in small grains. Proc Southern Weed Sci Soc 48:83–84

Nandihalli UB, Duke SO (1994) Structure-activity relationships of protoporphylinogen oxidase inhibiting herbicides. In: Duke SO, Rebeiz CA (eds) Porphyric pesticides: chemistry, toxicology, and pharmaceutical applications. ACS symposium series no 559. American Chemical Society, Washington DC, pp 133–146

Nandihalli UB, Sherman TD, Duke MV, Fisher JD, Musco VA, Becerril JM, Duke SO (1992) Correlation of protoporphylinogen oxidase inhibition by o-phenylpyrrolidino- and piperidino-carbamates with their herbicidal effects. Pestic Sci 35:227–235

Netzer DA, Riemenschneider DE, Bauer EO (1996) Pre and post flush application of DuPont R6447 in hybrid polar plantations. Proc North Central Weed Sci Soc 51:105

Oe Y, Kawaguchi S, Yokoyama M, Jikihara K (1995) RP020630 – a new herbicide for transplanted rice (*Oryza sativa*). Proc 15th Conf Asian Pacific Weed Sci Soc, pp 239–243

Oliver LR, Swantek JM, King CA (1996) Weed control in soybeans with sulfentrazone. Abstr Weed Sci Soc Am 36:2

Prosch SD, Ciha AJ, Grogna R, Hamper BC (1997) JV-485: a new herbicide for pre-emergence broad spectrum weed control in winter wheat. Proc Brighton Crop Prot Conf Weeds, pp 45–50

Saito K, Sakaki M, Sato R, Nagano E, Hashimoto S, Oshio H, Kamoshita K (1993) New post-emergence herbicide S-23031; herbicidal activity and selectivity. Abstr 18th Conf Pest Sci Soc Jpn, p 32

Ukai S, Nagato S, Yoshimura T, Taguchi T, Hirai K, Yano T, Ejiri E, Hirose H, Fujita A, Sato K (1992) New oxazolidinedione herbicides. Abstr 18th Conf Pest Sci Soc Jpn, p 47

Ukai S, Yamada O, Yoshii T, Yoshimura T, Hori M (1993) New tetrahydrohydrophthalimide herbicides. Abstr 18th Conf Pest Sci Soc Jpn, p 35

Vansaun WA, Bahr JT, Bourdouxhe LJ, Gargantiel FJ, Hotzman FW, Shires SW, Sladen NA, Tutt SF, Wilson KR (1993) F8426 – a new rapidly acting, low rate herbicide for the post-emergence selective control of broad-leaved weeds in cereals. Proc Brighton Crop Prot Conf Weeds, pp 19–22

Vidrine PR, Griffin JL, Jordan DL, Reynolds DB (1996) Broadleaf weed control in soybean (*Glycine max*) with sulfentrazone. Weed Technol 10:762–765

Vitolo DB, Schnappinger MG, Pruss SW, Porpiglia PJ (1995) CGA-248757 – postemergence broadleaf weed control in corn and soybeans. Proc North Central Weed Sci Soc 49:44

Yoshida R, Sakaki M, Sato R, Haga T, Nagano E, Oshio H, Kamoshita K (1991) S-53482 – a new N-phenyl phthalimide herbicide. Proc Brighton Crop Prot Conf Weeds, pp 69–75

Weiler R, Johann G, Ganzer M, Mach M (1993) Thidiazimin – a novel herbicide for use in cereals. Proc Brighton Crop Prot Conf Weeds, pp 29–34

Yoshimura Y, Ukai S, Nagato S, Hori M, Hirai K, Yano T, Ejiri E (1992) New oxazolidinedione herbicide; mode of action of paddy field herbicide KPP-314. Abstr 17th Conf Pest Sci Soc Jpn, p 48

Patent Literature

1 Rhone Poulenc S.A., Brit 1110500 (1968)
2 Mitsubishi Chem. Ind., Fr 2119703 (1972)
3 Sumitomo Chem. Ind., EP 69855 (1983)
4 Nihon Nouyaku Co. Ltd., EP 361114 (1990)
5 Sumitomo Chem. Ind., EP 170191 (1986)
6 Ciba-Geigy AG, EP 207894 (1987), EP 259264, 259265 (1988), Hoechst AG, DE 3533440 (1987), EP 215424 (1987), EP 288960, 289910 (1988), DE 3717828 (1988), Bayer AG, DE 3819439 (1989), etc.
7 Bayer AG, DE 3835168 (1990), Takeda Chem. Ind., JP 194386 (1993), BASF AG, DE 4217137 (1993), 4237984 (1994)
8 Sumitomo Chem. Ind., JP 1134373, 118363, 118364 (1986), EP 188259 (1986), BASF AG, DE 4217137 (1993), 4237984 (1994) Hoechst AG, EP 288960, etc.
9 BASF AG, DE 4217137 (1993), FMC Corp., US 4816065 (1989), 4846882 (1989), WO 8704049, Agro Kanesho Co., EP 323271 (1989)
10 Sumitomo Chem. Ind., JP 52755 (1989), Bayer AG, DE 3835168 (1990), DE 4139636 (1993)
11 Bayer AG, EP 308702 (1989)
12 American Cyanamid Co., US 5523277 (1996)
13 Nihon Noyaku Co., BE 902306 (1985)
14 Sumitomo Chem. Ind., EP 517181 (1992), JP 4972 (1993)
15 Nihon Noyaku Co., EP 361114 (1990)
16 Mitsubishi Petro. Chem. Ind., JP 250873 (1990)

CHAPTER 4

Structure-Activity Relationship and Molecular Design of Peroxidizing Herbicides with Cyclic Imide Structures and Their Relatives

TOSHIO FUJITA[1] and AKIRA NAKAYAMA[2]

Contents

4.1 Introduction

Since oxadiazon (1) and chlorophthalim (2: X = 4-Cl) were patented around 1970 (Boesch and Metivier 1963; Matsui et al. 1973), numerous trials have been undertaken to optimize and elaborate their structural features as being *N*-phenyl-nitrogen-heterocyclic. At present, there is a great variety of their

[1] EMIL Project, #305 Heights Kyogosho, 492 Umeyacho, Kyoto 604-8057, Japan
[2] Odawara Research Center, Nippon Soda Co. Ltd, 345 Takada, Odawara 250-0216, Japan

Peter Böger, Ko Wakabayashi
Peroxidizing Herbicides

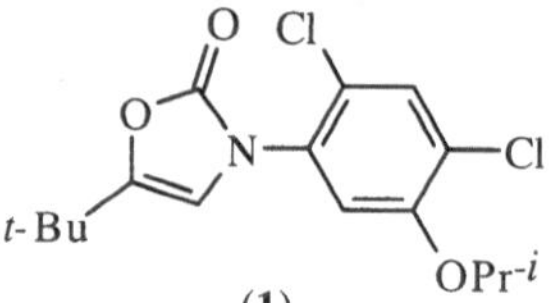

(1)

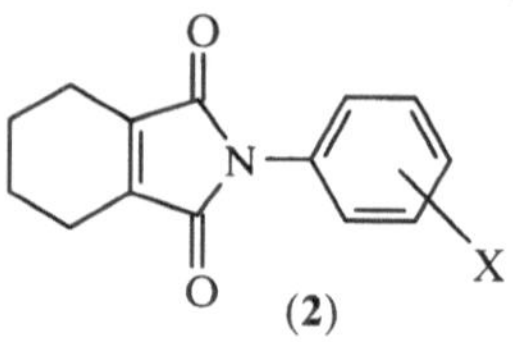

(2)

structural analogs including *C*-phenyl-nitrogen-heterocycles known as peroxidizing herbicides and candidates. As recognized from preceding Chapters (2 and 3) of this volume, there seem to be broadly two facets in the structure transformation trials. One is to disclose optimal substituents as well as to identify the best substitution patterns in the phenyl moiety. The other is to discover novel nitrogen-heterocyclic skeletons with well-matched substituents.

It is very important to analyze structure-activity data accumulated during past trials when formulating rational structure-activity relationships (SAR). The information about (sub)molecular mechanisms of biological action may be extracted from the relationships. The relationships could also be utilized as possible guiding principles for further structural transformation leading to novel peroxidizers. The aim of this chapter is to review structure-activity studies for peroxidizing herbicidal compounds of imide type. Because we believe that the rational formulation of SAR of bioactive compounds should be accompanied at least with quantitative procedures, SAR studies covered here are mainly those with analyses based on QSAR (quantitative structure-activity relationships) methodologies (Draber and Fujita 1992; Hansch and Fujita 1995; Hansch and Leo 1995). SAR's for various heterocyclic skeletal series are also discussed in terms of *bioisosterism* in a broader sense than usual, which has been proposed as bioanalogy (Floersheim et al. 1992; Fujita 1996).

4.2 QSAR Studies of *N*-Phenyltetrahydrophthalimides and Analogs

Publications of Wakabayashi, Ohta, and associates of the Mitsubishi Chemical group in the mid-1970s are probably the earliest in this area of QSAR studies of *N*-phenyltetrahydrophthalimides (Wakabayashi et al. 1978, 1979). For series of *N*-phenyltetrahydrophthalimides and related compounds, the effects of substituents in the *N*-phenyl moiety on the herbicidal potency were examined quantitatively using physicochemical substituent parameters and regression analyses. The regression results were represented as correlation equations in which contributions from various physicochemical effects are expressed by corresponding parameter terms with respective weights. Wakabayashi and

Mitsubishi coworkers suggested a specific contribution from the dipole moment of substituents besides their hydrophobicity. The dipole moment is, however, a vector and inappropriate to be used without physicochemical rationalization. In the QSAR correlation equations, the parameters to be utilizable should be additive and constitutive, preferably free-energy-related. Moreover, according to the Mitsubishi group's original analyses, higher alkoxy groups such as *p*-OEt and -OPr were outliers.

Subsequent efforts of the Mitsubishi group to formulate improved correlation equations revealed (Ohta et al. 1980) that the same data sets are clearly analyzable without outliers using STERIMOL parameters proposed by Verloop and coworkers (1976). Recently, Wakabayashi, moved to Tamagawa University, Sato and their group formulated very clear correlation equations for sets of compounds including newly synthesized and assayed analogs (Sato et al., unpubl.) using the revised STERIMOL parameters (Verloop 1983). They, jointly with Böger at Universität Konstanz, also measured the inhibition against the protoporphyrinogen oxidase preparation for an expanded set of substituted derivatives. In this section, the QSAR analyses based on the structure-activity data accumulated by the Mitsubishi, Tamagawa, and Konstanz groups and collaborated by one of the present authors (T. F.) are mainly described.

4.2.1
Herbicidal QSAR for Effects of Substituents in the *N*-Phenyl Moiety

The structure and inhibitory potencies of series 2 compounds against the root-growth of sawa millet (*Echinochloa utilis*) seedlings (Ohta et al. 1980; Sato et al., unpubl.) and the proliferation of unicellular green microalgae (*Scenedesmus acutus*) measured under light conditions (Sato et al., unpubl.) are listed in Table 1. As easily observed, the range of potency variations is broadest at the *para* and narrowest at the *meta* position among common substituents. The most potent compounds are found in *para*-, 2,4-di-, and 3,4-di-substituted derivatives. Among various combinations of substituent parameters, Eqs. (1) and (2) with electronic and position-specific steric parameters were selected as equations giving the highest quality of the correlation (Sato et al., unpubl.):

Root growth inhibition of sawa millet seedlings:

$$\begin{aligned} pI_{50}(\text{Ech.}) = & -0.909(\pm 0.374)\sum\sigma^0 + 1.144(\pm 0.769)\Delta L(o) \\ & - 0.770(\pm 0.488)[\Delta L(o)]^2 + 2.112(\pm 0.327)\Delta L(p) \\ & - 0.379(\pm 0.095)[\Delta L(p)]^2 - 1.069(\pm 0.211)\Delta B_5(p) \\ & + 4.061(\pm 0.216) \\ & n = 40, r = 0.928, s = 0.323, F = 34.19 \end{aligned} \quad (1)$$

Table 1. Biological activities and physicochemical parameters of *N*-phenyltetrahydrophthalimides

						pI_{50}(Ech.)			pI_{50}(Sce.)		
No.	X[a]	$\Sigma\sigma^0$ [b]	$\Delta L(o)$[b]	$\Delta L(p)$[b]	$\Delta B_5(p)$[b]	Obs.[c]	Cal.[d]	Dev.	Obs.[e]	Cal.[f]	Dev.
1	H	0.00	0.00	0.00	0.00	3.81	4.06	−0.25	5.25	5.56	−0.31
2	2-F	0.17	0.59	0.00	0.00	4.39	4.31	0.08	6.01	5.95	0.06
3	2-Cl	0.27	1.46	0.00	0.00	3.49	3.85	−0.36	5.22	5.40	−0.18
4	2-Me	−0.12	0.81	0.00	0.00	4.64	4.59	0.05	6.40	6.28	0.12
5	2-CF_3	0.54	1.24	0.00	0.00	3.09	3.81	−0.72	4.31	5.38	−1.07
6	2-OMe	−0.16	1.92	0.00	0.00	3.81	3.57	0.24	5.25	4.99	0.26
7	3-F	0.34	0.00	0.00	0.00	3.87	3.75	0.12	5.33	5.21	0.12
8	3-Cl	0.37	0.00	0.00	0.00	3.92	3.72	0.20	5.15	5.18	−0.03
9	3-Br	0.32	0.00	0.00	0.00	4.22	3.77	0.45	5.79	5.23	0.56
10	3-Me	−0.07	0.00	0.00	0.00	3.78	4.13	−0.35	5.40	5.64	−0.24
11	3-CF_3	0.48	0.00	0.00	0.00	3.88	3.62	0.26	5.34	5.07	0.27
12	3-OH	0.12	0.00	0.00	0.00	3.90	3.95	−0.05	5.37	5.44	−0.07
13	3-OMe	0.12	0.00	0.00	0.00	3.86	3.95	−0.09	5.32	5.44	−0.12
14	4-F	0.17	0.00	0.59	0.35	4.96	4.65	0.31	6.67	6.33	0.43
15	4-Cl	0.27	0.00	1.46	0.80	5.25	5.24	0.01	7.00	7.09	−0.09
16	4-Br	0.26	0.00	1.76	0.95	5.63	5.35	0.28	7.46	7.24	0.22
17	4-I	0.31	0.00	2.17	1.15	5.17	5.35	−0.18	7.04	7.24	−0.20
18	4-Me	−0.12	0.00	0.81	1.04	4.34	4.52	−0.18	6.00	6.13	−0.13
19	4-Et	−0.13	0.00	2.05	2.17	4.20	4.60	−0.40	5.76	6.23	−0.47
20	4-CF_3	0.54	0.00	1.24	1.61	3.43	3.89	−0.46	4.75	5.41	−0.66
21	4-OMe	−0.16	0.00	1.92	2.07	4.90	4.65	0.25	6.30	6.29	0.01
22	4-OEt	−0.20	0.00	2.74	2.36	4.42	4.66	−0.24	6.05	6.30	−0.25
23	4-OPr	−0.24	0.00	3.99	3.42	3.22	3.02	0.20	4.48	4.20	0.28
24	4-SMe	−0.02	0.00	2.24	2.26	4.32	4.49	−0.17	5.92	6.11	−0.19
25	4-NO_2	0.82	0.00	1.38	1.44	3.76	3.97	−0.21	6.00	5.55	0.45
26	4-CN	0.71	0.00	2.17	0.60	5.16	5.57	−0.41	7.02	7.57	−0.55
27	4-COOH	0.41	0.00	1.87	2.67	3.28	3.46	−0.18	4.56	4.85	−0.29
28	4-SCN	0.59	0.00	2.02	3.45	3.12	2.56	0.56	4.35	3.73	0.62
29	4-SO_2Me	0.75	0.00	2.05	2.17	3.44	3.80	−0.36	4.77	5.33	−0.56
30	4-CH_2Ph*	−0.10	0.00	2.56	5.02	<2.00	1.71	–	–[g]	2.57	–
31	4-OPh*	−0.06	0.00	2.45	4.89	<2.00	1.79	–	–[g]	2.67	–
32	4-OCH_2Ph(4-Cl)*	−0.16	0.00	7.60	2.50	6.11	−4.29	10.40	8.40	−5.12	13.52
33	2,4-Cl_2	0.54	1.46	1.46	0.80	5.33	5.02	0.31	7.25	6.93	0.32
34	2-F,4-Cl	0.44	0.59	1.46	0.80	5.93	5.49	0.44	8.03	7.48	0.55
35	2-F,4-Br	0.43	0.59	1.76	0.95	5.84	5.61	0.23	7.92	7.63	0.29
36	3,4-Cl_2	0.64	0.00	1.46	0.80	4.70	4.90	−0.20	6.42	6.71	−0.29
37	3-Cl,4-F	0.54	0.00	0.59	0.35	4.69	4.31	0.38	6.41	5.95	0.46
38	3-Cl,4-Br	0.63	0.00	1.76	0.95	4.86	5.02	−0.16	6.63	6.86	−0.23
39	3-Me,4-Cl	0.20	0.00	1.46	0.80	5.35	5.30	0.05	7.27	7.16	0.11
40	3-Me,4-Br	0.19	0.00	1.76	0.95	5.34	5.42	−0.08	7.26	7.31	−0.05
41	3-Me,4-OMe	−0.23	0.00	1.92	2.07	4.80	4.72	0.08	6.55	6.36	0.19
42	3-OMe,4-Br	0.38	0.00	1.76	0.95	5.59	5.24	0.35	7.59	7.11	0.48
43	3-NO_2,4-Cl	0.98	0.00	1.46	0.80	4.79	4.59	0.20	6.54	6.36	0.18

Obs., observed; Cal., calculated; Dev., deviation.

[a] Asterisked compounds are not included in the analysis.

[b] Values are either cited or estimated from related parameters in compilations of Charton (1981) and Hansch et al. (1995).

[c] From Ohta et al. (1980); Sato et al., unpubl.

[d] By Eq. (1).

[e] Sato et al., unpubl.

[f] By Eq. (2).

[g] Not measured.

Inhibition of growth of green microalgae:

$$\begin{aligned} pI_{50}(\text{Sce.}) = & -1.027(\pm0.469)\sum\sigma^0 + 1.554(\pm0.964)\Delta L(o) \\ & -1.010(\pm0.612)[\Delta L(o)]^2 + 2.683(\pm0.410)\Delta L(p) \\ & -0.482(\pm0.119)[\Delta L(p)]^2 - 1.357(\pm0.264)\Delta B_5(p) \\ & +5.563(\pm0.271) \\ & n = 40, r = 0.930, s = 0.404, F = 35.39 \end{aligned} \tag{2}$$

In these and the following QSAR equations, n is the number of compounds included in the analysis, r is the correlation coefficient, s is the standard deviation, F is the ratio of regression and residual variances, and figures in parentheses are the 95% confidence intervals. Unless otherwise noted, the terms insignificant at the 95% confidence level were disregarded. The pI_{50} is the log value of the reciprocal molar concentration required for the 50% growth/proliferation inhibition. L and B_5 are the revised STERIMOL parameters for substituents with the fully extended and staggered conformation at designated positions (Verloop 1983). The L parameter is that for the length (in angstrom) along the bond axis connecting the α atom of substituents with the rest of the molecule. The B_5 parameter is for the largest width (in angstrom) perpendicular to the L axis. The reference point of STERIMOL parameters is shifted to L and B_5 values of hydrogen, so Δ is placed ahead of notations. $\Sigma\sigma^0$ means that σ^0 values of substituents in multi-substituted compounds are summed up. The σ^0 is one of the Hammett-type parameters representing the electronic effect of substituents applicable when the direct conjugation of substituents with the side-chain functional group is not operative (Hansch and Leo 1995). The benzene ring and the cyclic dicarboximide moiety of this class of compounds are not coplanar, but twisted 50–60 degrees in the X-ray crystallographic analysis as well as in the MO-based most stable conformation (Kohno et al. 1993). Thus, the conjugation of the cyclic imide function with the benzene ring and its substituents is not important so that the use of σ^0 for substituents is well warranted.

It should be noted that the electronic effect of substituents is directed to the "side-chain" functional group. In the present case, the functional group is the dicarboximido-moiety. The effect is usually not specific to the position of substituents. When the effect of *para* substituents is observed, that of *meta* and *ortho* substituents should also be operated. Therefore, $\Sigma\sigma^0$ term is used for multisubstituted compounds. For *ortho* substituents, σ^0 is taken as that for the corresponding *para* substituents, and parameter terms for proximity effects, such as steric and field/inductive, on the functional group are added whenever significant (Fujita and Nishioka 1976) with use of the STERIMOL or Taft-Kutter-Hansch E_s (Hansch and Leo 1995) and σ_I or the Swain-Lupton-Hansch F parameters (Hansch et al. 1991; Hansch and Leo 1995), respectively. The steric effect of *meta* and *para* substituents usually would not be directed to the functional group, but work by their own rights at each position intermolecularly. Thus, they are position-specific. For *ortho* substituents, the intramolecular proximity steric effect on the functional group and the intermolecular steric effect exerted in processes involved up to the "receptor" binding are usually unseparable. The relevant parameter values and the pI_{50} values calculated by Eqs. (1) and (2) are listed in Table 1.

Equations (1) and (2) with negative $\Sigma\sigma^0$ term indicate that the electron donating property of substituents at any position potentiates the growth inhibitory activity. The proximity electronic effect parameter term expressed by σ_I or F does not appear significantly. The steric effect of substituents in terms of STERIMOL parameters is significant specifically at *ortho* and *para* positions. There seem to be optimal lengths at about 0.76 and 2.8 in the ΔL value for *ortho* and *para* substituents, respectively, in both Eqs. (1) and (2). This would mean that the steric restriction for the *ortho* substituents is much greater than that for the *para* substitution. The colinearity between ΔL and ΔB_5 parameters in the range of *para* substituents used here is high ($n = 40$, $r = 0.854$) including values of $\Delta L = \Delta B_5 = 0$ of *para*-H for *ortho* and *meta* substituted derivatives ($n = 13$). The exchange of $[\Delta L(p)]^2$ with $[\Delta B_5(p)]^2$ gives the poorer correlation. Moreover, the selection of quadratic $\Delta L(p)$ terms clearly predict the unmeasurably low sawa millet activity of *p*-CH_2Ph and -OPh compounds (nos. 30, 31 in Table 1) in which the ratio $\Delta B_5/\Delta L$ of substituents is much higher than others. The negative $\Delta B_5(p)$ term means that the smaller the maximum width of the *para* substituent, i.e., the smaller as well as the more symmetric the shape, the greater the potency. The colinearity between ΔL and ΔB_5 parameters of *ortho* substituents is also high ($n = 40$, $r = 0.933$). The quality of the correlation in Eqs. (1) and (2) was slightly lowered by using the quadratic $\Delta B_5(o)$ terms. Similar to those for *para* substituents, the quadratic $\Delta L(o)$ terms are selected here. To validate the selection of quadratic $\Delta L(o)$ terms, compounds with *ortho* substituents in which the ΔB_5 and ΔL values vary as widely and independently as possible should hopefully be further included, although the high level of the activity is not expected because of the steric restriction of *ortho* substituents. Without significant steric terms, *meta* substituents included here, mostly electron-withdrawing, do not participate much in the potency variations. The variations in the potency throughout the series compounds are governed most importantly by the position-specific steric effect of substituents. The most potent compounds included in Eqs. (1) and (2) generally satisfy the steric requirements of substituents.

The only outlier among compounds in Table 1 is the 4-OCH_2Ph(4-Cl) derivative (no. 32). The activities against sawa millet seedlings and green microalgae are much higher than those calculated by Eqs. (1) and (2). In this compound, the benzyloxy-benzene ring with the *p*-Cl substituent should parcitipate in an additional interaction with the target enzyme. As will be shown later in this chapter, there are similar examples. Substituents longer than a certain threshold make the activity, once lowered by increasing the length, revive enormously. In the present series, the *p*-OPh and -CH_2Ph derivatives show unmeasurably low activity. Thus, the threshold should be located between the length of OPh and OCH_2Ph(4-Cl).

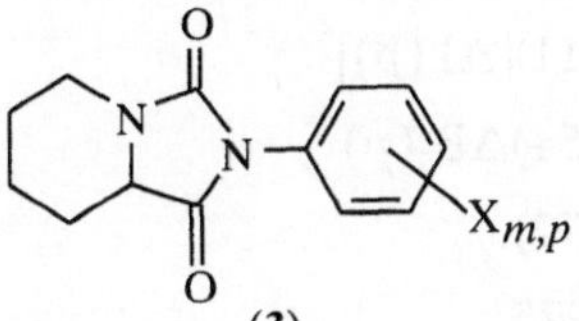

(3)

$$pI_{50}(\text{Ech.}) = -0.534(\pm 0.529)\sum\sigma^0 + 1.829(\pm 0.288)\Delta L(p) - 1.428(\pm 0.266)\Delta B_5(p) + 4.065(\pm 0.248)$$
$$n = 18, r = 0.965, s = 0.283, F = 63.15 \qquad (3)$$

$$pI_{50}(\text{Sce.}) = -0.647(\pm 0.637)\sum\sigma^0 + 2.191(\pm 0.347)\Delta L(p) - 1.714(\pm 0.321)\Delta B_5(p) + 5.652(\pm 0.299)$$
$$n = 18, r = 0.965, s = 0.341, F = 62.52 \qquad (4)$$

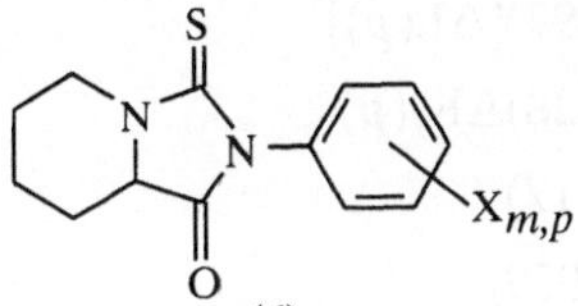

(4)

$$pI_{50}(\text{Ech.}) = 2.200(\pm 0.407)\Delta L(p) - 0.429(\pm 0.202)[\Delta L(p)]^2 - 0.886(\pm 0.157)\Delta B_5(p) + 3.207(\pm 0.144)$$
$$n = 18, r = 0.980, s = 0.167, F = 115.55 \qquad (5)$$

$$pI_{50}(\text{Sce.}) = 2.743(\pm 0.522)\Delta L(p) - 0.512(\pm 0.259)[\Delta L(p)]^2 - 1.134(\pm 0.202)\Delta B_5(p) + 4.603(\pm 0.184)$$
$$n = 18, r = 0.980, s = 0.213, F = 113.73 \qquad (6)$$

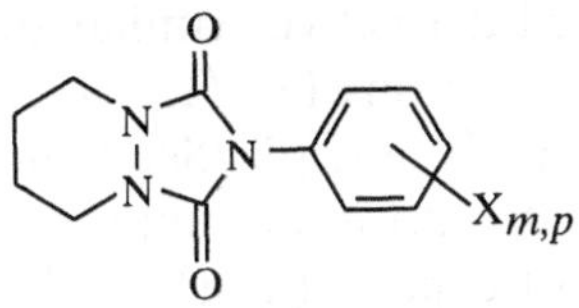

(5)

$$pI_{50}(\text{Ech.}) = 1.425(\pm 0.278)\Delta L(p) - 1.027(\pm 0.363)\Delta B_5(p) + 3.521(\pm 0.164)$$
$$n = 18, r = 0.958, s = 0.217, F = 83.35 \qquad (7)$$

$$pI_{50}(\text{Sce.}) = 1.687(\pm 0.340)\Delta L(p) - 1.209(\pm 0.444)\Delta B_5(p) + 4.990(\pm 0.200)$$
$$n = 18, r = 0.956, s = 0.265, F = 78.84 \qquad (8)$$

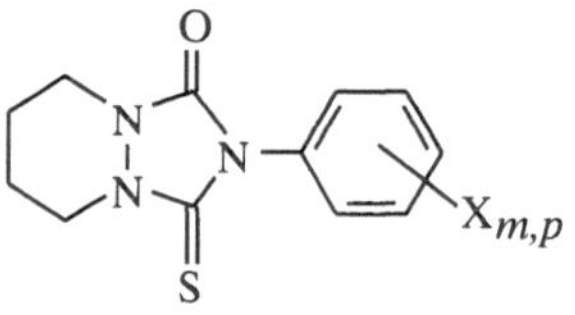

(6)

$$
\begin{aligned}
pI_{50}(\text{Ech.}) = {} & 2.025(\pm 0.392)\Delta L(p) \\
& - 0.496(\pm 0.211)[\Delta L(p)]^2 \\
& - 0.760(\pm 0.154)\Delta B_5(p) \\
& + 4.807(\pm 0.172) \\
& n = 19, r = 0.975, \\
& s = 0.202, F = 98.08
\end{aligned}
\quad (9)
$$

$$
\begin{aligned}
pI_{50}(\text{Sce.}) = {} & 2.450(\pm 0.470)\Delta L(p) - 0.608(\pm 0.253)[\Delta L(p)]^2 \\
& - 0.911(\pm 0.184)\Delta B_5(p) + 6.553(\pm 0.207) \\
& n = 19, r = 0.976, s = 0.242, F = 98.87
\end{aligned}
\quad (10)
$$

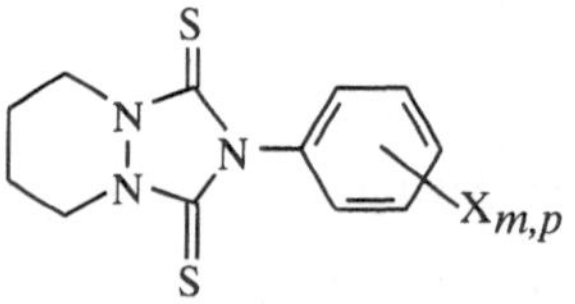

(7)

$$
\begin{aligned}
pI_{50}(\text{Ech.}) = {} & 1.571(\pm 0.521)\Delta L(p) \\
& - 0.403(\pm 0.297)[\Delta L(p)]^2 \\
& - 0.529(\pm 0.228)\Delta B_5(p) \\
& + 4.927(\pm 0.217) \\
& n = 17, r = 0.939, \\
& s = 0.250, F = 32.03
\end{aligned}
\quad (11)
$$

$$
\begin{aligned}
pI_{50}(\text{Sce.}) = {} & 1.922(\pm 0.637)\Delta L(p) - 0.487(\pm 0.363)[\Delta L(p)]^2 \\
& - 0.654(\pm 0.279)\Delta B_5(p) + 6.696(\pm 0.265) \\
& n = 17, r = 0.939, s = 0.306, F = 32.31
\end{aligned}
\quad (12)
$$

Analogous series of compounds (**3–7**) also exhibit the light-dependent herbicidal activity. For sets of *meta* and *para* substituted derivatives similar to those included in Eqs. (1) and (2), the correlation equations (3)–(12) were formulated for pI_{50}(Ech.), the sawa millet herbicidal activity, and pI_{50}(Sce.), the green microalgae growth inhibition (Ohta et al. 1980; Sato et al., unpubl.). Because mono-*ortho* substituted compounds were not expected to show the potent activity similar to compound series **2**, they were neither synthesized nor included.

No compound behaves as outlier in Eqs. (3)–(12) [the corresponding *p*-OCH_2Ph(4-Cl) derivatives were not included]. As observed in each pair of consecutive two equations including Eqs. (1) and (2) for series **2**, quantitative structure-activity patterns in terms of the effects of substituents attached to the phenyl moiety are very similar between sawa millet seedlings and green microalgae in each series **2–7** in spite of the difference in bioassay conditions. The root growth inhibition of the millet seedlings is measured 7 days after sowing on a filter paper soaked with the test sample solution/emulsion, whereas the proliferation inhibition of the green microalgae is assayed 24h after the onset of the treatment of the cell suspension. The structure of millet roots is much more complex than that of unicellular microalgae. The similarity in substituent effects on the activity between two species clearly indicates that such differences in experimental conditions and biological test systems are not critical for determining the variations in the growth-inhibiting potencies.

Quantitative substituent effects are similar among tetrahydrophthalimides (**2**) and analogs (**3–7**) in that position-specific steric effects represented by length and width are decisive. Parameter terms significant in Eqs. (3)–(12) for series **3–7** are, in fact, found as components almost as such in the most "comprehensive" Eqs. (1) and (2). Because the negative $[\Delta L(p)]^2$ term is significant at the 90% level in adding into Eqs. (3) and (4) and only at the 80% level into Eqs.(7) and (8), it is not indicated in these equations. Although it is too much to draw any definite conclusion before effects of additional substituents are analyzed together, the optimum $\Delta L(p)$ values could be around 3.3 in Eqs. (3) and (4) and around 4.0 in Eqs. (7) and (8). These values are not far from the optimum $\Delta L(p)$ values: 2.8 in Eqs. (1) and (2), 2.6 in Eqs. (5) and (6), and 2.0 in Eqs. (9)–(12).

In Eqs. (3) and (4), the negative $\Sigma\sigma^0$ term is significant as in Eqs. (1) and (2). In Eqs. (5)–(12) for compound series **4–6**, the addition of $\Sigma\sigma^0$ term is not significant even at the 60% level. The instability of the $\Sigma\sigma^0$ term in these closely related compound series will be discussed later (Sect. 4.5.1).

Lyga and coworkers of the FMC Corporation have also examined the QSAR of this class of compounds (**8**), listed in Table 2 (Lyga et al. 1991). For the effects of *para* substituents X, they formulated Eq. (13).

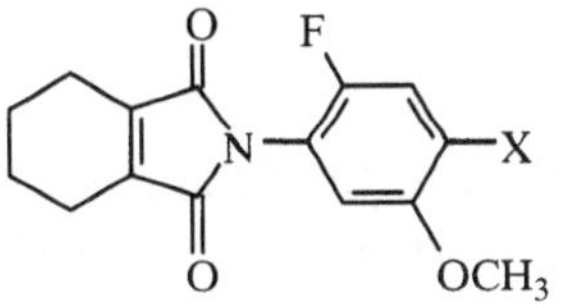

(**8**)

$$\begin{aligned} pI_{50}(\text{Cuc.}) = {} & 1.08(\pm 0.18)\pi + 2.41(\pm 0.92)F \\ & + 1.40(\pm 0.60)R \\ & - 0.072(\pm 0.019)MR + 5.25 \\ & n = 16, r = 0.917, \\ & s = 0.59, F = 14.3 \end{aligned} \quad (13)$$

In Eq. (13), I_{50} is the molar concentration to provide 50% fresh weight reduction of hydroponically grown cucumber seedlings (5-day-old; *Cucumis sativus* cv. Wisconsin) for 10 days under light conditions. F and R are field/inductive and resonance electronic parameters, respectively, so that F + R = σ (*para*) (Hansch and Leo 1995). MR used as a bulk effect parameter is the molecular refractivity assigned to substituents (Hansch and Leo 1995).

Their procedure to select substituents is based on a cluster analysis (Hansch et al. 1973) so that physicochemical parameters such as π, F and R, and MR could vary as independently as possible among substituents selected. The principle is all right, but the selection of parameters does not seem to be appropriate physicochemically. First, F and R parameters are used indepen-

Table 2. Biological activity and physicochemical parameters of *N*-(2-F-4-X-5-OMe-phenyl)-tetrahydrophthalimides

				pI_{50}		
No.	X[a]	$\sigma^{0\,b}$	π(X/imide)[c]	Obs.[d]	Cal.[e]	Dev.
1	H	0	0	4.90	4.96	−0.06
2	SMe*	−0.02	0.57	4.50	5.48	−0.98
3	NO_2	0.82	0.31	6.70	6.86	−0.16
4	NH_2	−0.38	−1.02	2.70	2.65	0.05
5	NHAc	0.03	−0.94	4.10	3.58	0.52
6	$NHSO_2Me$	0.20	−0.95	3.50	3.89	−0.39
7	I	0.27	1.30	6.30	6.40	−0.10
8	OH	−0.14	−0.41	3.70	4.15	−0.45
9	OCOMe*	0.12	−0.50	3.20	4.51	−1.31
10	OMe	−0.16	−0.14	4.80	4.49	0.31
11	O-*i*-Pr	−0.17	0.67	4.70	5.27	−0.57
12	O-*n*-Pent	−0.16	1.85	5.50	5.59	−0.09
13	OSO_2Ph*	0.33	1.24	4.30	6.50	−2.20
14	Br	0.26	1.00	6.70	6.28	0.42
15	Me	−0.12	0.53	5.60	5.26	0.34
16	Cl	0.27	0.87	6.40	6.23	0.17

[a] Asterisked compounds are not included in the analysis.
[b] See footnote [b] in Table 1.
[c] Estimated by correlation equation derived from log P values of *ortho*, *meta*, and *para* mono-substituted compound series **9**:
$\pi(X/imide) = 0.956\pi(X/PhH) + 0.728\sigma^0 + 0.227E_s(o) - 0.344F(o) + 0.371\rho_x - 2.95\delta_x(o) + 0.36\rho_x^1(o) + 0.046$. The ρ_x and δ_x parameters are for backward effects of the imido group on X. The parameters with (*o*) are for *ortho* effects. For detail see Fujita (1983).
[d] From Lyga et al. (1991).
[e] By Eq. (14).

dently instead of σ(*para*) in Eq. (13). The reason why the weight of the F term is far greater than that of the R term is not readily understood unless a novel electronic mechanism occurring in the substituted phenyl system is to be confirmed. Secondly, the substituent hydrophobicity π used here is that derived as the $\Delta\log P$ (P: 1-octanol/water partition coefficient) value between mono- and un-substituted benzenes denoted as π(X/PhH).

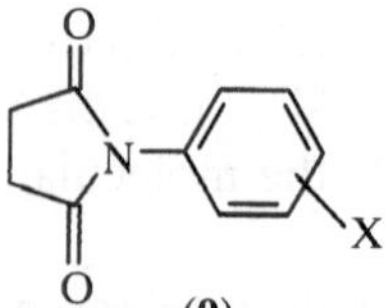

(9)

The aromatic π value of a substituent X is generally defined between XC_6H_4Y and C_6H_5Y, in which Y is a constant group (Fujita et al. 1964). Being denoted as π(X/PhY), the value, especially that of hydrogen-bonding X substituents, varies from one system to another depending on the stereoelectronic functions of the group Y (Fujita 1983; Nakagawa et al. 1992). Thus, while they adopted a value of −0.28 for the *p*-NO_2 as that of the $\pi(NO_2/PhH)$, the corresponding value, $\pi(p\text{-}NO_2)$, in a set of closely related compounds (**9**) is +0.40. Even for the Cl substituent, $\pi(Cl/PhH) = 0.71$, whereas $\pi(p\text{-}Cl \text{ in } \mathbf{9}) = 0.90$. Thus, we reexamined their analysis using π values from *N*-phenyl-succinimides **9**, π(X/imide), measured by Takayama and Fujinami (1979). The electronic effect of the succinimido group is such that its $\sigma_m = 0.34$ and $\sigma_p = 0.31$ (Hansch et al. 1995), while that of the maleimido group, which is the core skeletal structure of the tetrahydrophthalimido substituent, is represented by $\sigma_m = 0.33$ and $\sigma_p = 0.27$ (Hansch et al. 1995). The electronic features of the Y group affecting the π value of substituents on the benzene ring are well expected to be similar between the two series. For substituents the π value of which had not been measured in series **9**, the value was estimated by an empirical equation correlating π(X/imide) with π(X/PhH) and by considering proximity stereoelectronic effects of OMe (Fujita 1983; Nakagawa et al. 1992) on the π(*p*-X/imide) value in compounds **8**. Thus, Eq. (13) was revised as Eq. (14).

$$\begin{aligned} pI_{50}(\text{Cuc.}) &= 1.907(\pm 0.897)\sigma^0 + 1.185(\pm 0.353)\pi(\text{X/imide}) \\ &\quad - 0.369(\pm 0.343)[\pi(\text{X/imide})]^2 + 4.964(\pm 0.360) \\ &\quad n = 13, r = 0.964, s = 0.397, F = 39.8 \end{aligned} \tag{14}$$

In Eq. (14), SMe, OCOMe, and OSO_2Ph compounds (nos. 2, 9, and 13 in Table 2, respectively) are not included. Their activity is one tenth to one hundredth lower than predicted. This is not unreasonable considering the assay conditions. The SMe group would be oxidized to the SO_2Me and the OCOMe and OSO_2Ph groups could be hydrolyzed in the 10 day cucumber assay. Since no specific electronic mechanism has to be considered in terms of physical organic chemistry, Eq. (14) seems more reasonable than Eq. (13) even though three compounds are deleted. It is interesting to note that the QSAR features with the cucumber hydroponic assay differ from those with sawa millet seedlings and green microalgae growth inhibition. In Eq. (14), the σ^0 term is positive and the hydrophobic effect of substituents is quadratically significant, but the steric effect is not at least in the range of substituents included in the set. The optimum π value in Eq. (14) is about 1.6 which is close to $\pi(p\text{-I/imide}) = 1.30$ and $\pi(p\text{-OPent/imide}) = 1.85$, but the positive σ^0 term makes the *p*-NO_2 the most favorable one.

4.2.2 QSAR of Protoporphyrinogen Oxidase Inhibition for the Effects of Substituents in *N*-Phenyl Moiety

The inhibitory activity in terms of pI_{50}(M) against protoporphyrinogen oxidase (protox) preparation from corn etioplasts has been measured by Wakabayashi, Böger, and coworkers with compound series of various substitution patterns (Böger and Wakabayashi, unpubl.). For compounds **2** in which X is mono- and 2,4-di-substituted listed in Table 3, Eq. (15) was formulated (Sato et al., unpubl.).

$$\begin{aligned} pI_{50}(\text{protox.}) = {} & 2.756(\pm 0.811)\Delta L(p) - 0.447(\pm 0.363)[\Delta L(p)]^2 \\ & - 1.220(\pm 0.542)\Delta B_5(p) + 1.640(\pm 1.201)\Delta L(o) \\ & - 1.388(\pm 0.935)[\Delta L(o)]^2 - 1.571(\pm 0.539)I_H(p) \\ & + 5.572(\pm 0.336) \\ & n = 18, r = 0.976, s = 0.300, F = 37.34 \end{aligned} \quad (15)$$

The number of compounds in Eq. (15) is rather low, but the correlation is highly significant. The number of *meta* substituents included in the compound set is only two, Me and Cl. Their steric effect is not clearly analyzable, unless substituents are selected more systematically. The colinearity between $\Delta L(p)$

Table 3. Protox inhibition and physicochemical parameters of *N*-(mono- and 2,4-di-substituted phenyl)-tetrahydrophthalimides

						pI_{50}		
No.	X[a]	$\Delta L(o)$[b]	$\Delta L(p)$[b]	$\Delta B_5(p)$[b]	$I_H(p)$	Obs.[c]	Cal.[d]	Dev.
1	H	0	0	0	0	5.80	5.57	0.23
2	2-Me	0.81	0	0	0	5.89	5.99	−0.10
3	2-Cl	1.46	0	0	0	5.02	5.01	0.01
4	2-F	0.59	0	0	0	5.64	6.06	−0.42
5	3-Me	0	0	0	0	5.89	5.57	0.32
6	3-Cl	0	0	0	0	5.60	5.57	0.03
7	4-$CONH_2$	0	2.00	2.07	1	4.97	5.20	−0.23
8	4-NH_2	0	0.72	0.97	1	4.80	4.57	0.23
9	4-F	0	0.59	0.35	0	6.37	6.62	−0.25
10	4-NO_2	0	1.38	1.44	0	6.40	6.77	−0.37
11	4-CF_3	0	1.24	1.61	0	6.55	6.34	0.21
12	4-SMe	0	2.24	2.26	0	6.68	6.75	−0.07
13	4-OMe	0	1.92	2.07	0	7.00	6.69	0.31
14	4-Cl	0	1.46	0.80	0	7.60	7.67	−0.07
15	4-Br	0	1.76	0.95	0	7.67	7.88	−0.21
16	4-Me	0	0.81	1.04	0	6.08	6.24	−0.16
17	4-OCH_2Ph(4-Cl)*	0	7.60	2.50	0	8.30	−2.36	10.66
18	2-F, 4-Cl	0.59	1.46	0.80	0	8.43	8.15	0.28
19	2-F, 4-Br	0.59	1.76	0.95	0	8.60	8.36	0.24

[a] Compound 17 (asterisked) is not included in the analysis.
[b] From compilation of Hansch et al. (1995).
[c] Sato et al., unpubl.
[d] By Eq. (15).

and $\Delta B_5(p)$ is high (r = 0.896). The exchange of the $[\Delta L(p)]^2$ with the $[\Delta B_5(p)]^2$ parameter gave an almost equivalent result (r = 0.975). Equation (15) was selected because position-specific steric effects of substituents in this equation are similar to those appearing in Eqs. (1) and (2) for the whole body/cellular activities. The optimum $\Delta L(o)$ and $\Delta L(p)$ values in Eq. (15), 0.6 and 3.1, respectively, are very close to corresponding values in Eqs. (1) and (2). I_H is an indicator variable taking the value of unity for *para* substituents having the hydrogen-bonding donor such as $CONH_2$ (no. 7 in Table 3) and NH_2 (no. 8). Thus, at the enzyme level, there seems to be an unfavorable hydrogen-bonding effect. The *p*-OCH_2Ph (4-Cl) derivative (no. 17) is not included in the correlation, because its activity is much higher than that expected from Eq. (15). As suggested for its very high herbicidal activity, there should be an additional interaction site for this type of lengthy substituents.

For 3,4-disubstituted derivatives in which the 4-substituent is fixed as Cl and the 3-substituent is varied as listed in Table 4, Eq. (16) was formulated (Sato et al., unpubl.).

$$\begin{aligned} pI_{50}(\text{protox.}) = & -1.863(\pm 1.234)\sum\sigma^0 - 0.102(\pm 0.037)[\Delta L(m)]^2 \\ & + 1.114(\pm 0.311)\Delta L(m) - 0.357(\pm 0.109)\Delta B_5(m) \\ & + 8.029(\pm 0.573) \\ & n = 18, r = 0.929, s = 0.207, F = 20.54 \end{aligned} \tag{16}$$

Among 18 compounds included, 16 have ester and amide substituents at the *meta* position. Because structural features of these substituents are varied among double-, unsaturated alkyl-, and thiol-esters, colinearity between ΔL and ΔB_5 is not very high (r = 0.73). Equation (16) shows that the steric effects of *meta* substituents are revealed in terms of the quadratic ΔL and the negative ΔB_5 terms similar to those of *para* substituents, when lengthy and bulky substituents are examined together. The coefficient of the $[\Delta L(m)]^2$ term (−0.10) in Eq. (16) is closer to zero (i.e. the parabolic curvature is much more gentle) than that of the $[\Delta L(p)]^2$ (−0.45) in Eq. (15) and the optimum $\Delta L(m)$ value (5.6) in Eq. (16) is larger than that of $\Delta L(p)$ (3.1) in Eq. (15). Thus, the pI_{50} value is much less susceptible to the variations in the substituent length at the *meta* position than at the *para* position. In Eq. (16), the 4-Cl-3-COOH (no. 2 in Table 4) and 4-Cl-3-CO_2Oct (no. 12) compounds are not included. Their activity is, respectively, much lower and higher than that predicted. The 4-Cl-3-COOH compound is ionized under assay conditions (pH 7.3). The ionized form would be much less active than the non-ionized form at the enzyme level. On the whole body/cellular level, the 4-COOH compound (no. 27 in Table 1) was not the outlier. Possible reasonings are the differences in assay conditions as well as pK_A. The 4-Cl-3-COOH compound is more acidic than the 4-COOH compound. The 4-Cl-3-CO_2Oct compound has a too lengthy straight chain substituent with ten heavy atoms. The situation is similar to the very high activity of the *p*-OCH_2Ph(4-Cl) derivative.

Table 4. Protox inhibition and physicochemical parameters of *N*-(3-X-4-Cl-phenyl)-tetrahydrophthalimides

						pI_{50}				
							Eq. (16)		Eq. (18)	
No.	X[a]	$\Sigma\sigma^{0}$ [b]	$\Delta L(m)$[b]	$\Delta B_5(m)$[b]	I(2-F)[c]	Obs.[d]	Cal.	Dev.	Cal.	Dev.
1	H	0.27	0.00	0.00	0.00	7.60	7.53	0.07	7.92	−0.32
2	COOH*	0.63	1.85	1.66	0.00	5.73	7.97	−2.24	8.16	−2.43
3	COOMe	0.63	2.67	2.36	0.00	8.05	8.26	−0.21	8.37	−0.31
4	COOEt	0.63	3.89	3.41	0.00	8.43	8.42	0.01	8.43	0.00
5	COO-*n*-Pr	0.63	4.71	3.83	0.00	8.66	8.47	0.19	8.41	0.25
6	COO-*i*-Pr	0.63	3.91	2.43	0.00	8.90	8.78	0.12	8.73	0.17
7	COO-*n*-Bu	0.63	5.94	4.85	0.00	8.04	8.13	−0.09	8.02	0.02
8	COO-*s*-Bu	0.63	4.71	3.83	0.00	8.60	8.47	0.13	8.41	0.19
9	COO-*i*-Bu	0.63	4.71	4.71	0.00	8.32	8.15	0.17	8.14	0.17
10	COO-*t*-Bu	0.63	3.91	3.41	0.00	7.85	8.43	−0.58	8.44	−0.59
11	COO-*n*-Pent	0.63	6.75	5.30	0.00	7.70	7.82	−0.12	7.68	0.02
12	COO-*n*-Oct*	0.63	10.02	7.76	0.00	7.60	4.97	2.63	4.86	2.74
13	COSMe	0.84[e]	3.26	2.49	0.00	8.13	8.12	0.01	8.16	−0.03
14	COSEt	0.84[e]	4.54	3.56	0.00	8.26	8.14	0.12	8.10	0.16
15	$COOCH_2COOMe$	0.63	5.75	4.71	0.00	8.30	8.20	0.10	8.10	0.20
16	COOCH(Me)COOMe	0.63	5.75	4.71	0.00	8.12	8.20	−0.08	8.10	0.02
17	$COOCH_2C{\equiv}CH$	0.63	3.76	4.82	0.00	8.00	7.88	0.12	7.98	0.02
18	$COOCH_2CH{=}CH_2$	0.63	4.90	4.12	0.00	8.43	8.39	0.04	8.33	0.10
19	$CONHCH_2Ph$	0.62	3.99	6.61	0.00	7.30	7.33	−0.03	7.51	−0.21
20	$OCH_2C{\equiv}CH$	0.45[f]	4.52	2.07	0.00	9.40	9.40	0.00	9.26	0.14

[a] Asterisked compounds are not included in the analysis.
[b] See footnote [b] in Table 1. Σ means that σ^{0}_{p}(Cl) is added. Some STERIMOL parameters were calculated by the program of Verloop (1983).
[c] Indicator variable used in the combination with data in Table 5.
[d] Sato et al., unpubl.
[e] Estimated from calculated $\sigma_m(c\sigma_m)$ value in Hansch et al. (1991) as: σ^{0}_{m}(COOMe) + $c\sigma_m$(COSMe) − $c\sigma_m$(COOMe).
[f] Estimated from: $\sigma_m(CH_2C{\equiv}CH)$ + average of $[\sigma_m(OCH_2X) - \sigma_m(CH_2X)]$, in which X = halogen.

For the 2,4,5-trisubstituted series shown in Table 5 in which the 2- and 4-positions are fixed as 2-F-4-Cl and the 5(*meta*)-position substituent is varied, Eq. (17) was formulated (Sato et al., unpubl.).

$$pI_{50}(\text{protox.}) = -1.771(\pm 1.339)\sum\sigma^{0} - 0.085(\pm 0.034)[\Delta L(m)]^{2} + 0.750(\pm 0.287)\Delta L(m) - 0.220(\pm 0.153)\Delta B_5(m) + 9.010(\pm 0.713)$$

$$n = 17, r = 0.885, s = 0.189, F = 10.88 \qquad (17)$$

Substituents at the 5-position included in Eq. (17) are mostly esters and ethers, the situation being somewhat similar to that for Eq. (16). The colinearity between ΔL and ΔB_5 is, however, higher (r = 0.84) than that in Eq. (16). The optimum ΔL(*m*) value is 4.4 which is lower than but similar to that in Eq. (16).

Table 5. Protox inhibition and physicochemical parameters of *N*-(2-F-4-Cl-5-X-phenyl)-tetrahydrophthalimides

						pI_{50}				
							Eq. (17)		Eq. (18)	
No.	X[a]	$\Sigma\sigma^{0}$ [b]	$\Delta L(m)$[b]	$\Delta B_5(m)$[b]	I(2-F)[c]	Obs.[d]	Cal.	Dev.	Cal.	Dev.
1	H	0.44	0.00	0.00	1.00	8.43	8.23	0.20	7.95	0.48
2	OMe	0.50	1.92	2.07	1.00	8.52	8.80	−0.28	8.65	−0.13
3	$OCHF_2$	0.75	1.92	2.61	1.00	7.96	8.24	−0.28	8.03	−0.07
4	$OCH_2C{\equiv}CH$	0.62	4.52	2.07	1.00	8.97	9.11	−0.14	9.28	−0.31
5	$OCH(Me)C{\equiv}CH$	0.62	4.52	3.35	1.00	9.00	8.83	0.17	8.90	0.10
6	O-*cyc*-Pent(3-Me)	0.69[e]	4.10	4.58	1.00	8.60	8.43	0.17	8.35	0.25
7	OCH_2COO-*n*-Pent*	0.56[f]	8.70	6.57	1.00	8.96	6.65	2.31	6.76	2.20
8	COOH*	0.81	1.85	1.66	1.00	6.42	8.31	−1.89	8.17	−1.75
9	COOMe	0.80	2.67	2.36	1.00	8.45	8.47	−0.02	8.39	0.06
10	COOEt	0.81	3.89	3.41	1.00	8.58	8.46	0.12	8.44	0.14
11	COO-*n*-Pr	0.80	4.71	3.83	1.00	8.41	8.40	0.01	8.44	−0.03
12	COO-*i*-Pr	0.80	3.91	2.43	1.00	8.86	8.69	0.17	8.76	0.10
13	COO-*n*-Bu	0.80	5.94	4.85	1.00	8.05	7.98	0.07	8.05	0.00
14	COO-*i*-Bu	0.80	4.71	4.71	1.00	8.35	8.20	0.15	8.17	0.18
15	COO-*s*-Bu	0.80	4.71	3.83	1.00	8.31	8.40	−0.09	8.44	−0.13
16	COO-*i*-Pent	0.80	5.95	4.85	1.00	7.89	7.97	−0.08	8.05	−0.16
17	COOAllyl	0.80	4.90	4.12	1.00	8.38	8.32	0.06	8.35	0.03
18	COOCH(Me)COOMe	0.80	5.75	4.71	1.00	7.80	8.06	−0.26	8.12	−0.32
19	$COOCH_2CH_2CN$	0.80	6.16	3.50	1.00	8.20	8.21	−0.01	8.41	−0.21

[a] Asterisked compounds are not included in the analysis.
[b] See footnote [b] in Table 1. Σ means that σ^0 values of 2-F and 4-Cl are added. STERIMOL parameters of some substituents were calculated by the program of Verloop (1983).
[c] Indicator variable for 2-F compounds.
[d] Sato et al., unpubl.
[e] Value for *m*-O-*cyc*-Pent.
[f] Value for *m*-OMe.

The 2-F-4-Cl-5-OCH_2CO_2Pent (no. 7 in Table 5) and 2-F-4-Cl-5-COOH (no. 8) compounds are not included in the correlation. The activity of the COOH compound which is much lower than and that of the CO_2Pent which is much higher than predicted could be rationalized by the ionization and the excessive length (nine heavy atoms), respectively, similar to outliers in Eq. (16).

Because Eqs. (16) and (17) look alike, they were combined into a single correlation equation to analyze the 5(*meta*)-substituent effects leading to Eq. (18) sharing features in common with Eqs. (16) and (17) (Sato et al., unpubl.). The results are shown in Tables 4 and 5.

$$\begin{aligned} pI_{50}(\text{protox.}) = &-1.825(\pm 0.966)\sum\sigma^0 - 0.093(\pm 0.027)[\Delta L(m)]^2 \\ &+ 0.926(\pm 0.230)\Delta L(m) - 0.301(\pm 0.095)\Delta B_5(m) \\ &+ 0.335(\pm 0.200)\,I(2\text{-F}) + 8.417(\pm 0.444) \\ &n = 35, r = 0.870, s = 0.230, F = 18.02 \end{aligned} \qquad (18)$$

In Eq. (18), I(2-F) is an indicator variable which takes the value of unity for compounds included in Eq. (17) but otherwise zero. This term corresponds to the "total" effect of the 2-F substituent except the "ordinary" electronic effect represented by $\sigma^0(para) \equiv \sigma^0(ortho)$. The electronic effect of 2-F is counted in the $\Sigma\sigma^0$ term in Eq. (18). From the quadratic $\Delta L(o)$ terms in Eq. (15) and $\Delta L(F) = 0.59$, the total steric effect of the 2-F substituent can be estimated as being $0.48 = 1.64\Delta L(F) - 1.39[\Delta L(F)]^2$. The value is similar to the coefficient of the I(2-F) term. This observation could be regarded as validating the correlations represented by Eqs. (15)–(18), although the fact that the $\Sigma\sigma^0$ term is insignificant in Eq. (15) whereas not in Eqs. (16)–(18) should be rationalized. In Eq. (15), compounds are mostly mono-substituted. As the number of substituents increases in the phenyl moiety, the averaged $\Sigma\sigma^0$ value of substituents included in each of the series shifts from 0.18 ($\pm$0.30) in Eq. (15) (n = 18) to 0.68 ($\pm$0.13) in Eq. (18) (n = 35), while the coefficient of the $\Sigma\sigma^0$ term becomes more negative. This suggests that the higher electron withdrawal from the substituted phenyl moiety is more unfavorable to the protox inhibitory activity. In Eq. (15), the variety and the number of substituents are not enough. In Eqs. (16) and (17), the variations in the structural type of the *meta* substituent are not great. To obtain a clearer picture, examinations should hopefully be continued.

4.2.3 Quantitative Correlation Between Protoporphyrinogen Oxidase Inhibition and Herbicidal Activity

For a set of mono- and di-substituted compounds in Table 6, almost the same as those included in Eq. (15), both the protox and growth inhibitory activities are available. The growth inhibitory activity against the sawa millet seedlings and green microalgae was expressed by Eqs. (19) and (20), respectively.

$$\begin{aligned} pI_{50}(\text{Ech.}) &= 0.838(\pm 0.153)pI_{50}(\text{protox}) - 0.823(\pm 0.554)\sum\sigma^0 \\ &\quad - 0.258(\pm 0.206)\Delta B_5(p) - 0.602(\pm 0.941) \\ &\quad n = 19, r = 0.950, s = 0.327, F = 46.44 \end{aligned} \tag{19}$$

$$\begin{aligned} pI_{50}(\text{Sce.}) &= 1.150(\pm 0.315)pI_{50}(\text{protox}) - 1.486(\pm 1.224)\sum\sigma^0 \\ &\quad - 0.483(\pm 0.410)\Delta B_5(p) - 0.718(\pm 1.868) \\ &\quad n = 16, r = 0.928, s = 0.469, F = 24.99 \end{aligned} \tag{20}$$

Similarity between Eqs. (19) and (20) is obvious. They are roughly equivalent with the difference in corresponding terms between Eqs. (1) and (15) and between Eqs. (2) and (15), respectively. The size of the pI_{50} (protox) term is close to unity showing that the growth inhibition is ultimately due to the protox inhibition in spite of the difference of plant species from which the

Table 6. Protox inhibition and herbicidal activity of *N*-(substituted phenyl)-tetrahydrophthalimides

					pI_{50}(Ech.)			pI_{50}(Sce.)		
No.	X	pI_{50}(Protox)[a]	$\Sigma\sigma^0$	$\Delta B_5(m)$	Obs.[b]	Cal.[c]	Dev.	Obs.[b]	Cal.[d]	Dev.
1	H	5.80	0.00	0.00	3.81	4.26	−0.45	5.25	5.95	−0.70
2	2-Me	5.89	−0.12	0.00	4.64	4.43	0.21	6.40	6.23	0.17
3	2-Cl	5.02	0.27	0.00	3.49	3.38	0.11	5.22	4.65	0.57
4	2-F	5.64	0.17	0.00	4.39	3.98	0.41	6.01	5.51	0.50
5	3-Me	5.89	−0.07	0.00	3.78	4.39	−0.61	5.40	6.16	−0.76
6	3-Cl	5.60	0.37	0.00	3.92	3.79	0.13	5.15	5.17	−0.02
7	4-$CONH_2$	4.97	0.36	2.07	2.93	2.73	0.20	–[e]	3.46	–
8	4-NH_2	4.80	−0.38	0.97	3.39	3.48	−0.09	–[e]	4.90	–
9	4-F	6.37	0.17	0.35	4.96	4.51	0.45	6.76	6.18	0.58
10	4-NO_2	6.40	0.82	1.44	3.76	3.72	0.04	6.00[f]	4.73	1.27
11	4-CF_3	6.55	0.53	1.61	3.43	4.04	−0.61	4.75	5.25	−0.50
12	4-SMe	6.68	−0.02	2.26	4.32	4.43	−0.11	5.92	5.90	0.02
13	4-OMe	7.00	−0.16	2.07	4.90	4.86	0.04	6.30	6.57	−0.27
14	4-Cl	7.60	0.27	0.80	5.25	5.34	−0.09	7.00	7.23	−0.23
15	4-Br	7.67	0.26	0.95	5.63	5.37	0.26	7.46	7.26	0.20
16	4-Me	6.08	−0.12	1.04	4.34	4.32	0.02	6.00	5.95	0.05
17	4-OCH_2Ph(*p*-Cl)	8.30	−0.23	2.50	6.11	5.90	0.21	8.40	7.96	0.44
18	2-F,4-Cl	8.43	0.44	0.80	5.93	5.89	0.04	8.03	7.93	0.10
19	2-F,4-Br	8.60	0.43	0.95	5.84	6.01	−0.17	7.92	8.07	−0.15

[a] Böger and Wakabayashi, unpubl.
[b] Sato et al., unpubl.
[c] By Eq. (19).
[d] By Eq. (20).
[e] Not measured.
[f] Not included in the analysis.

enzyme preparation originated. The $\Sigma\sigma^0$ and $\Delta B_5(p)$ terms in Eqs. (19) and (20) could correspond to the effects in which the difference is "most" significant between enzyme and whole plant/cellular activities, probably representing effects on the processes occurring during the transport. This suggests that the electronic effect of substituents and the steric effect of *p*-substituents are likely to be those exhibited in the transport processes. Note that the 4-OCH_2Ph(4-Cl) compound (no. 17 in Table 6) is not the outlier in Eqs. (19) and (20). The outlying behaviors of this compound from Eqs. (1) and (2) are supposed to be compensated by the outlying factor in Eq. (15). In Eq. (20), the 4-NO_2 compound (no. 10) is not included. The reason is not clear. The microalgae activity of this compound may be estimated as being too high.

4.3 QSAR Studies of *N*-Phenyl-Triazolin(thi)ones and -Tetrahydroindazoles

Unfortunately, the title compound series are almost the only precedents to which the QSAR procedure has been applied in the FMC Corporation more or less systematically among cyclic imide-like peroxidizers other than the tetrahydrophthalimides and analogs described earlier (Sect. 4.2). Although the efforts of the FMC scientists should be highly appreciated in demonstrating the versatility of the classical QSAR (Hansch and Leo 1995), their analyses would better be amended, hopefully, to give physicochemically more reasonable results. In this section, their analyses are compared with those we reanalyzed (Sato et al., unpubl.).

4.3.1 Herbicidal QSAR for Effects of Substituents in *N*-Phenyl Moiety of *N*-Phenyltriazolinones

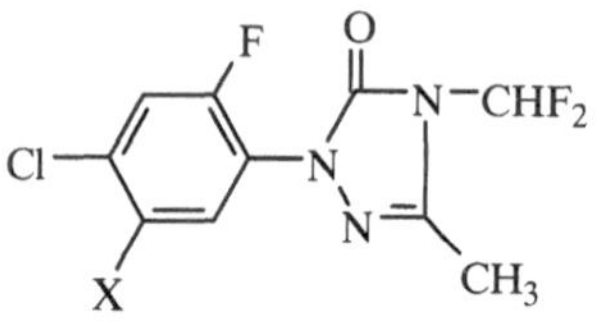

(**10**)

Theodoridis of the FMC Corporation recently published the QSAR analysis for the hydroponically measured herbicidal activity against cucumber seedlings of this important series of compounds (**10**), the 1-substituted phenyl-4*H*-1,2,4-triazolin-5-ones (Theodoridis 1997). Sulfentrazone (**10**: X = $NHSO_2Me$), a pre-emergence soybean herbicide, and carfentrazone-ethyl (**10**: X = $CH_2CHClCO_2Et$), a post-emergence cereal herbicide, belong to this series. Their selection procedure of the *meta* substituent X is similar to that used by Lyga and coworkers described in Section 4.2.1, so that the substituent physicochemical properties in terms of σ, π, and STERIMOL parameters cover a broad range and are independent of each other. For a set of 14 compounds (nos. 1–14) shown in Table 7, Eq. (21) was formulated.

$$\begin{aligned} pI_{50}(\text{Cuc.}) &= 7.02(\pm 2.351)B_1 - 2.57(\pm 0.775)B_1^{\,2} \\ &\quad - 0.18(\pm 0.045)[\pi(X/PhH)]^2 + 2.51 \\ &\quad n = 14, r = 0.884, s = 0.216 \end{aligned} \qquad (21)$$

The predictability of Eq. (21) was examined for five additional compounds (nos. 15–19) in which the substituent X is the *para* substituted phenoxy $[OC_6H_4(4\text{-}Y)]$ group. Except for compounds in which Y = Cl (no. 17) and $—OCH_2CO_2Et$ (no. 19), Eq. (21) predicted the hydroponic herbicidal activity well. The π value in Eq. (21) is, however, that from the monosubstituted benzene series. As described earlier (Sect. 4.2.1), this does not seem to be entirely correct.

In our reexamination, we included substituted phenoxy compounds except for compound (no. 19) the activity of which is outlyingly higher than that

Table 7. Herbicidal activity and physicochemical parameters of *N*-phenyl-triazolinones (**10**)

					pI_{50}		
No.	X[a]	ΔL[b]	ΔB_1[b]	ΔB_5[b]	Obs.[c]	Cal.[d]	Dev.
1	$OCH_2C{\equiv}CH$	4.52	0.35	2.07	7.6	7.56	0.04
2	$OCH_2CH{=}CH_2$	4.16	0.35	3.42	7.5	7.32	0.18
3	OMe	1.92	0.35	2.07	7.5	7.01	0.49
4	OH	0.68	0.35	0.93	7.2	6.89	0.31
5	CH_2OMe	2.72	0.52	2.40	7.1	7.01	0.09
6	$NHSO_2Et$	4.01	0.35	3.47	7.1	7.28	−0.18
7	OCOMe	2.68	0.35	2.67	7.1	7.10	0.00
8	Me	0.81	0.52	1.04	7.0	6.77	0.23
9	H	0	0	0	6.8	7.14	−0.34
10	$NHSO_2Me$	2.00[e]	0.35	3.12[e]	6.7	6.90	−0.20
11	OPh	2.45	0.35	4.89	6.6	6.78	−0.18
12	Cl	1.46	0.80	0.80	6.5	6.72	−0.22
13	Br	1.76	0.95	0.95	6.5	6.65	−0.15
14	Ph	4.22	2.11[e]	0.71[e]	6.3	6.29	0.01
15	$OPh(4\text{-}NHSO_2Et)$	3.68[e]	0.35	9.64[e]	6.6	6.46	0.14
16	OPh(4-OMe)	4.02	0.35	6.24	6.7	6.94	−0.24
17	OPh(4-Cl)	3.14	0.35	6.31	6.7	6.75	−0.05
18	$OPh(4\text{-}NO_2)$	3.04	0.35	6.31	6.8	6.73	0.07
19	$OPh(4\text{-}OCH_2CO_2Et)$*	7.95	0.35	8.98	9.0	7.44	1.56

[a] Compound 19 (asterisked) is not included in the analysis.
[b] Unless noted, either from Hansch et al. (1995) or calculated by STERIMOL program (Verloop 1983).
[c] Theodoridis (1997).
[d] By Eq. (22).
[e] Modified as indicated in the text.

predicted by Eq. (21). Among various combinations of substituent parameters, we observed Eq. (22) to show the highest quality, although some assumptions had to be made for the conformation of certain substituents as shown below. The correlation results are shown in Table 7.

$$pI_{50}(\text{Cuc.}) = 0.211(\pm 0.123)\Delta L - 0.780(\pm 0.334)\Delta B_1 - 0.123(\pm 0.066)\Delta B_5 + 7.135(\pm 0.299)$$
$$n = 18, r = 0.818, s = 0.240, F = 9.46 \quad (22)$$

The correlation coefficient, r, of Eq. (22) is not so high, but the equation is highly significant over 99.8%. The low r value is due to the fact that the variations in the pI_{50} value are not so broad, being from 6.5 to 7.6. No additional parameter term was significant over the 95% level. The B_1 parameter is the "minimum width" STERIMOL value for the perpendicular minimum distance from the L axis to the tangential surface plane of substituents. Equation (22) was able to predict the activity of compound no. 17.

In formulating Eq. (22), the STERIMOL ΔL and ΔB_5 values of $NHSO_2Me$ and OC_6H_4 (4-$NHSO_2Et$) groups in compounds (no. 10) and (no. 15), respectively, were calculated for a conformation somewhat different from that defined originally as being fully extended and staggered. In the $NHSO_2Me$ group, the dihedral angle between ·—N and S—C bonds (· denotes an atom located at the end of the rest of the molecule) was taken to be 0 degree instead of 180 degrees. In this way, the steric "repulsion" between ·—N and two S=O bonds would be minimized. In the $NHSO_2Et$ group, the original definition was revived so that the dihedral angle between ·—N and S—C bonds is 180 degrees to avoid a greater repulsion between the ·—N bond and the chain-end CH_2CH_3 group. In the OC_6H_4(4-$NHSO_2Et$) substituent, the dihedral angle between ·—O and N—S bonds on the different sides of the benzene ring was assumed to be 0 degree. Probably, an excessive length of the *para* Y substituent of the *m*-OPh group in the fully staggered and extended direction from the ·—O bond is unfavorable to the activity until the length of the Y substituents reaches that of six heavy atomic bonds in compound (no. 19).

The phenyl moiety of the *m*-Ph compound (no. 14) would be out of the core benzene ring-plane especially because of its *ortho*-substituted biphenyl structure (Suzuki 1967). Thus, the "thickness" of the Ph substituent in the direction perpendicular to the core benzene ring was approximated by the width, but the "width" in the direction coplanar to the core benzene ring was taken as the thickness parameters so that the original ΔB_1 and ΔB_5 values were exchanged to represent the "effective" conformational situation for the *m*-Ph group.

These conformational modifications seem to be validated by formulating the correlation equation by deleting conformationally "flexible" three $NHSO_2R$ compounds (nos. 6, 10, 15) and conformationally "rigid" *m*-Ph derivative (no. 14) in addition to the outlyingly potent compound (no. 19). With 14 compounds, the corresponding combination of substituent parameter terms is shown in Eq. (23) without modifying the original definition of STERIMOL parameters.

$$\begin{aligned} pI_{50}(\text{Cuc.}) &= 0.239(\pm 0.160)\Delta L - 0.881(\pm 0.739)\Delta B_1 \\ &\quad - 0.149(\pm 0.100)\Delta B_5 + 7.202(\pm 0.464) \\ &\quad n = 14, r = 0.792, s = 0.261, F = 5.62 \end{aligned} \tag{23}$$

Although the correlation coefficient is lower, Eq. (23) is practically identical with Eq. (22). The best combination for the same 14 compounds was that with ΔL, $(\Delta B_1)^2$ and $(\Delta B_5)^2$, r being 0.890 and s being 0.195 (equation not shown). Because the colinearities between ΔB and $(\Delta B)^2$ values are high ($r = 0.950$ for ΔB_1 and $r = 0.973$ for ΔB_5), the best correlation equation was also conceivable to simulate Eq. (23) sufficiently. The conformation of substituents other than Ph seems to be such that in which the minimum width, B_1, is in the direction perpendicular to and the maximum width, B_5, is in the direction coplanar to the core benzene ring. Equations (22) and (23) suggest that "slim and tall" substituents are favorable to the activity.

Equation (24) with use of the hydrophobic term of π(X/imide) was observed to show a better correlation than Eq. (23). The stereoelectronic features of the triazolinone structure were estimated to be similar to those of succinimido group in estimating the π value of X substituents. The procedure of the estimation will be described in the following Section 4.3.2.

$$\begin{aligned} pI_{50}(\text{Cuc.}) = & -1.101(\pm 1.011)\sum\sigma^0 - 0.215(\pm 0.140)\pi(X/\text{imide}) \\ & + 0.177(\pm 0.108)\Delta L + 7.426(\pm 0.613) \\ & n = 14, r = 0.859, s = 0.219, F = 9.37 \end{aligned} \qquad (24)$$

The counterpart of Eq. (24) for the set of 18 compounds included in Eq. (22) showed, however, a poorer correlation (r = 0.745, s = 0.264; equation not shown). In this counterpart equation, the π(X/imide) term was insignificant at the 95% level. The correlation equation including hydrophobic term(s) was unstable so that it was inappropriate to be selected as far as the set of compounds shown in Table 7 is concerned.

A possible rationalization for the very potent activity of compound (no. 19) is that there could be some additional site(s) or much better orientations for the binding of the entire substituent side chain with ten units in terms of the heavy atomic bond in total from the *N*-phenyl moiety similar to compound (no. 32) in Table 1, compound (no. 12) in Table 4, and compound (no. 7) in Table 5.

4.3.2
Herbicidal QSAR for Effects of Substituents in *N*-Phenyl Moiety of *N*-Phenyltriazolinethiones

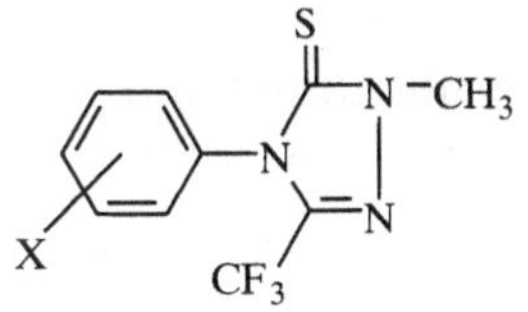

(11)

Simmons and coworkers analyzed hydroponic herbicidal activity of a number of the 4-substituted phenyl-1,2,4-triazoline-2*H*-3-thiones (**11**) (Simmons et al. 1992). They first examined structure-activity relationship of mono- and multichloro substituted compounds with the use of the modified Free-Wilson procedure (Fujita and Ban 1971) to identify "best" combinations of substituent positions with which substituent effects should be explored more vigorously. Thus, a variety of 2,4-di-, 2,4,5-tri-, and 3,4,5-tri-substituted compounds were synthesized and bioassayed. They also applied the sequential simplex optimization procedure (Plummer 1992) to mono-substituted derivatives to identify likely substituent types required for the higher activity at each of *ortho*-, *meta*, and *para*-positions. Their principle of substituent selection further includes that the physicochemical substituent parameters such as π, F and R, and MR vary as widely as possible.

As observed in Table 8, the potency variations seem to be broadest in *para* derivatives similar to those in *N*-phenyltetrahydrophthalimides. They formulated first the QSAR correlation equation for un- and *para*-mono-substituted derivatives (nos. 1–14) as shown in Eq. (25).

$$\begin{aligned} pI_{50}(\text{Cuc.}) &= 1.67(\pm 0.38)\pi(\text{X/PhH}) - 0.67(\pm 0.17)[\pi(\text{X/PhH})]^2 \\ &\quad - 1.18(\pm 0.22)\sigma + 4.34 \\ &\quad n = 14, r = 0.889, s = 0.28, F = 17.6 \end{aligned} \tag{25}$$

The correlation of Eq. (25) seems to be acceptable. Utilizing Eq. (25) as the basic model, Simmons and his group formulated Eq. (26) for un-, *ortho*-, *meta*-, *para*-mono-, and 2,4-di-substituted derivatives.

$$\begin{aligned} pI_{50}(\text{Cuc.}) &= 0.92(\pm 0.16)\pi(\text{X/PhH})(p) - 0.33(\pm 0.07)[\pi(\text{X/PhH})(p)]^2 \\ &\quad - 0.86(\pm 0.14)\sigma(p) + 1.02(\pm 0.15)B_1(o) - 1.05(\pm 0.15)I(o) \\ &\quad + 3.81 \\ &\quad n = 40, r = 0.921, s = 0.33, F = 45.0 \end{aligned} \tag{26}$$

In Eq. (26), $B_1(o)$ is the minimum width parameter for *ortho* substituents and $I(o)$ is an indicator variable which takes the value of unity for the presence of substituents at the *ortho* position as well as for the unsubstituted compound. The $\pi(\text{X/PhH})(p)$ term is used for the hydrophobicity specific to the *para* substituent only even in compounds substituted at positions other than *para*.

With further additions of 2,5-, 3,5-, and 2,6-dichlorosubstituted and 3,5-dichloro-4-substituted and 2,4,5-trisubstituted derivatives, Eq. (27) was formulated.

$$\begin{aligned} pI_{50}(\text{Cuc.}) &= 0.90(\pm 0.17)\pi(\text{X/PhH})(p) - 0.39(\pm 0.09)[\pi(\text{X/PhH})(p)]^2 \\ &\quad - 0.82(\pm 0.14)\sigma(p) + 0.98(\pm 0.14)B_1(o) - 1.12(\pm 0.19)I(o) \\ &\quad + 1.18(\pm 0.17)\pi(\text{X/PhH})(5) + 3.98 \\ &\quad n = 69, r = 0.918, s = 0.43, F = 60.4 \end{aligned} \tag{27}$$

In Eq. (27), to accommodate the trisubstituted derivatives occupied at the 5-position, another position-specific hydrophobicity term with $\pi(\text{X/PhH})(5)$ was needed.

The position-specific hydrophobic and steric effects could be understandable, but the requirement of $I(o) = 1$ for unsubstituted compounds should

Table 8. Herbicidal activity and physicochemical parameters of *N*-phenyl-triazolinethiones

No.	X	σ^0 [a]	π(X/imide) [b]	$\Delta L(o)$ [a]	$\Delta B_1(o)$ [a]	pI_{50} Obs. [c]	pI_{50} Cal. [d]	pI_{50} Dev.
1	H	0	0	0	0	4.1	4.34	−0.24
2	4-Cl	0.27	0.92	0	0	5.1	4.95	0.15
3	4-COPh	0.50	1.50	0	0	<4.0	4.54	>0.54
4	4-OPent	−0.31	1.90	0	0	5.3	5.59	−0.29
5	4-NMe$_2$	−0.48	0.04	0	0	5.5	5.23	0.27
6	4-*i*-Pr	−0.16	1.39	0	0	5.6	5.71	−0.11
7	4-OEt	−0.14	0.43	0	0	5.3	5.25	0.05
8	4-O-*i*-Pr	−0.17	0.86	0	0	5.5	5.67	−0.17
9	4-OBu	−0.14	1.55	0	0	5.3	5.60	−0.30
10	4-Me	−0.12	0.49	0	0	5.4	5.28	0.12
11	4-NEt$_2$	−0.53	0.96	0	0	6.7	6.32	0.38
12	4-NHPh	−0.33	1.39	0	0	6.2	6.00	0.20
13	4-NMe(*i*-Pr)	−0.53	0.68	0	0	6.1	6.16	−0.06
14	4-NPr$_2$	−0.53	1.82	0	0	6.1	6.05	0.05
15	4-NHEt	−0.49	0.04	0	0	5.5	5.25	0.25
16	3-Cl	0.37	0.99	0	0	5.0	4.81	0.19
17	3-NMe$_2$	−0.16	0.27	0	0	5.1	5.07	0.03
18	3-OPent	0.10	2.20	0	0	4.7	4.48	0.22
19	3-O-*i*-Pr	0.10	1.05	0	0	5.1	5.28	−0.18
20	3-Me	−0.07	0.53	0	0	4.6	5.24	−0.64
21	3-C≡CH	0.21	0.76	0	0	5.2	4.97	0.23
22	2-Cl	0.28	0.56	1.46	0.80	4.7	4.71	−0.01
23	2-F	0.16	0.04	0.59	0.35	4.2	4.18	0.02
24	2-Me	−0.12	0.21	0.81	0.52	4.6	5.00	−0.40
25	2-OMe	−0.16	−0.08	1.92	0.35	3.8	3.82	−0.02
26	2-C≡CH	0.23	0.49	2.60	0.60	4.0	3.95	0.05
27	2-CF$_3$	0.53	0.60	1.24	0.99	4.8	4.62	0.18

[a] See footnote [b] in Table 1.
[b] See footnote [c] in Table 2.
[c] Simmons et al. (1992).
[d] By Eq. (29).

reasonably be rationalized in Eqs. (26) and (27). Corresponding terms are almost equivalent between Eqs. (26) and (27), but changed much from Eq. (25). As described above, the substituent π values of this system should differ from corresponding π(X/PhH) values. Although the quality of correlation of Eqs. (26) and (27) seems to be sufficiently high, the use of $\sigma(p)$ value only for *para* substituents, so that any electronic effect of substituents at other positions is not considered, is unlikely to be justified physicochemically.

We reexamined their data with use of π(X/imide) values estimated from the log P of *N*-phenylsuccinimides (**9**) as we did earlier (see Sect. 4.3.1). The thionotriazolinyl structure of this series of compounds seems to differ considerably from that of succinimides. Nevertheless, they have similar stereoelectronic characters as follows. As described in Section 4.2.1, the succinimido group is of electron-withdrawing substituents, σ_m and σ_p values being 0.34 and 0.31, respectively. These values are close to those of the diacetylamino group with $\sigma_m = 0.35$ and $\sigma_p = 0.33$, as well as to those of the *N*-methylacetylamino group with $\sigma_m = 0.26$ and $\sigma_p = 0.31$ (Hansch et al. 1995). This could mean that σ values of nitrogen heterocyclic substituents are approximated by the "averaged" values of those for two structures formed by "splitting" the ring and methylating the nitrogen. The thionotriazolinyl structure could be split and methylated to give the *N*-methyl-trifluoroacetylamino ($\sigma_m = 0.41$, $\sigma_p = 0.39$) (Hansch et al. 1995) and *N,N'*-dimethylthioureido (estimated $\sigma_m = 0.32$, $\sigma_p = 0.40$) groups. The avaraged values are $\sigma_m = 0.37$ and $\sigma_p = 0.40$ which are not far from those of the succinimido group. In these estimations, the endocyclic C=N bond was simulated by the C=O bond and the *N*-methylation was expected to make the σ_m and σ_p values of non-methylated acylamino groups about 0.1 and 0.25 higher, respectively. This was deduced empirically from σ values of pairs of acylamino and N-methylacylamino groups (Hansch et al. 1995).

The thionotriazolinyl group with the CF_3 substituent in structure **11** is in fact broader around the nitrogen atom, with which the phenyl moiety is attached, than the succinimido group. In terms of the E_s value estimated by the procedure of Austel and coworkers (1979), however, the steric effect, perhaps considering the intramolecular situation more significantly, is very similar, the E_s value of succinimido and trifluoromethylthionotriazolinyl groups being −2.97 and −3.08, respectively. Thus, stereoelectronic features of the two nitrogen heterocycles toward the phenyl moiety are similar so that the use of π value from the *N*-phenylsuccinimides for the *N*-phenyltriazolinethiones is likely to be justified.

Equation (28) was formulated for un- and *para*-substituted compounds.

$$\begin{aligned} pI_{50}(\text{Cuc.}) = & -2.122(\pm 0.536)\sigma^0 + 2.153(\pm 0.673)\pi(\text{X/imide}) \\ & -1.008(\pm 0.355)[\pi(\text{X/imide})]^2 + 4.272(\pm 0.309) \\ & n = 14, r = 0.958, s = 0.200, F = 37.5 \end{aligned} \tag{28}$$

It is evident that the uses of π(X/imide) instead of π(X/PhH) and σ^0 instead of the ordinary σ have improved the correlation of the corresponding Equation (25).

Equation (29) is for un- and *ortho-*, *meta-*, and *para*-mono-substituted compounds.

$$\begin{aligned} pI_{50}(\text{Cuc.}) = & -1.694(\pm 0.457)\sigma^0 + 1.896(\pm 0.574)\pi(\text{X/imide}) \\ & -0.797(\pm 0.277)[\pi(\text{X/imide})]^2 + 0.959(\pm 0.754)\Delta B_1(o) \\ & -0.505(\pm 0.286)\Delta L(o) \pm 4.340(\pm 0.279) \\ & n = 26, r = 0.941, s = 0.266, F = 30.8 \end{aligned} \tag{29}$$

No outlier from members shown in Table 8 is observed in Eq. (29). The corresponding terms between Eqs. (28) and (29) are almost equivalent, showing the robustness and reliability of the correlation. There is an optimum hydrophobicity at about 1.05 in terms of π(X/imide). Similar to *N*-phenyltetrahydrophthalimdes, the sign of σ^0 term is negative, electron-donating substituents being favorable to the activity. To accommodate the *ortho* substituted compounds, two STERIMOL parameter terms are required. The positive ΔB_1 term means that the greater the minimum width, i.e., the greater the width as well as the more symmetric the shape of substituents, the higher the activity. This, along with the negative ΔL term, showing that "shorter" substituents are more favorable, coincides with the fact that Cl (no. 22) and CF_3 (no. 27) are most favorable among *ortho* substituents in Table 8. The negative σ^0 term and the optimum π(X/imide) value about 1.2 agree with the observation that the *p*-NEt_2 substituent [$\sigma^0 = -0.53$, $\pi(X/imide) = 0.96$] makes the activity highest among mono-substituted derivatives. Should the number and variety of substituents be increased, steric substituent effects at *meta* and *para* position similar to those shown in Eqs. (1), (15), and (16) may be shown in their QSAR. The pI_{50} value of the *p*-COPh compound (no. 3) which is unmeasurably low (<4.0) was predicted to be 4.5 by Eq. (29). The predictability does not seem to be too low.

To incorporate the 2,4-di, and 3,4,5- and 2,4,5-trisubstituted compounds in a correlation equation of a type similar to Eq. (29) was unsuccessful. The estimation of the total π(X/imide) values for multi-substituted groups was especially difficult. As indicated above, the π value of a certain substituent varies depending upon its environmental situations (Nakagawa et al. 1992). The intramolecular interactions among substituents and nitrogen-heterocyclic group work stereoelectronically to vary the extents of the relative hydrogen-bonding solvation between 1-octanol and water phases. For mono-substituted derivatives, the π(X/imide) value was estimated with the consideration of the interaction between the substituent and the heterocycle. For di- and tri-substituted derivatives, this type of interaction to affect the hydrophobic parameter should also be considered between substituents. Because of the lack of the necessary physicochemical parameters to estimate the interaction between hydrogen-bondable substituents included in the series, especially those located vicinally, the "accurate" substituent π values were difficult to estimate. Thus, further reanalysis was discontinued.

4.3.3
Herbicidal QSAR for Effects of Substituents in Pyrazole Moiety of *N*-Phenyltetrahydroindazoles

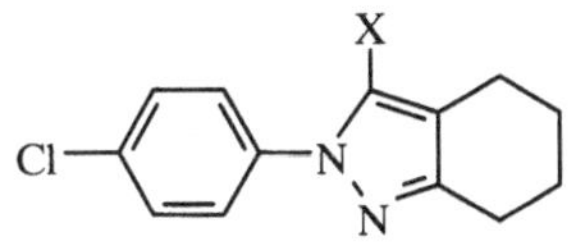

(12)

Lyga and coworkers analyzed the hydroponically measured herbicidal activity against cucumber seedlings of *N*-4-chlorophenyltetrahydroindazoles (**12**) (Lyga et al. 1994). They selected the substituents X so that the physicochemical parameters are as orthogonal as possible. Their analysis for 12 compounds (nos. 1–12) in Table 9 afforded Eq. (30).

$$pI_{50}(\text{Cuc.}) = -0.69(\pm 0.09)[\pi(X/PhH)]^2 + 1.56(\pm 0.37)B_1 + 3.65$$
$$n = 12, r = 0.933, s = 0.33, F = 30.5 \quad (30)$$

They examined predictability of Eq. (30) for additional four compounds (nos. 13–16) having a high B_1 value and the π value close to zero. Two compounds (no. 14) and (no. 16) showed the activity predicted by Eq. (30) but the other two did not.

We reexamined their analysis using a set of π values estimated for 3-substituted pyridazines (3PD) (Yamagami et al. 1990), in which the substituent is located at the position next to the two consecutive "aromatic" nitrogen atoms similar to the X on the pyrazole ring shown in structure **12**. For 15 compounds including additionally assayed compounds, Eq. (31) was formulated.

Table 9. Herbicidal activity and physicochemical parameters of *N*-phenyl-tetrahydroindazoles (**12**)

					pI_{50}		
No.	X[a]	$\sigma^0(m)$[b]	$\pi(X/3PD)$[c]	ΔL[b]	Obs.[d]	Cal.[e]	Dev.
1	Cl	0.37	1.04	1.46	6.7	6.32	0.38
2	SMe	0.12	0.91	2.24	6.0	5.68	0.32
3	SO_2Me	0.70	−0.52	2.05	5.5	5.25	0.25
4	NH_2	−0.08	−0.20	0.72	4.7	4.82	−0.12
5	CHF_2	0.31	0.56	1.24	6.2	6.21	−0.01
6	CF_3	0.48	1.35	1.24	5.9	6.41	−0.51
7	H	0.00	0.03	0.00	5.2	5.34	−0.14
8	Ph	0.05	1.74	4.22	3.8	3.94	−0.14
9	C≡CH	0.27	1.33	2.60	6.2	5.63	0.57
10	*i*-Pr	−0.07	1.22	2.05	5.1	5.33	−0.23
11	Br	0.38	1.17	1.76	6.2	6.21	−0.01
12	CN	0.62	0.29	2.17	5.9	6.29	−0.39
13	$CH(Me)CH_2OH$*	−0.05	0.29	2.05	4.5	5.19	−0.69
14	Me	−0.07	0.41	0.81	5.8	5.55	0.25
15	3-Pyridyl	0.17	0.59	4.25	4.6	4.63	−0.03
16	OMe	0.06	0.38	1.92	5.3	5.50	−0.20

[a] Compound 13 (asterisked) is not included in the analysis.
[b] See footnote [b] in Table 1.
[c] Estimated by correlation equation in Yamagami et al. (1990).
[d] Lyga et al. (1994).
[e] By Eq. (31).

$$\begin{aligned} pI_{50}(\text{Cuc.}) = {} & 1.704(\pm 0.858)\sigma^0(m) + 1.334(\pm 0.722)\pi(X/3PD) \\ & - 0.761(\pm 0.535)[\pi(X/3PD)]^2 - 0.082(\pm 0.042)\Delta L^2 \\ & + 5.299(\pm 0.392) \\ & n = 15, r = 0.920, s = 0.354, F = 13.8 \end{aligned} \tag{31}$$

In Eq. (31), compound (no. 13) which shows a much lower activity than predicted by Eq. (30) is not included. It behaves also as an outlier from Eq. (31). Compound (no. 15) which is not predicted well by Eq. (30) is incorporated in Eq. (31). The outlying behavior of compound (no. 13) may be due to the presence of the hydroxyl group in the substituent side chain. Equation (31) appears to confirm that electron-withdrawing short substituents with the hydrophobicity, about 0.9, in terms of $\pi(X/3PD)$ such as Cl [$\pi(X/3PD) = 1.04$] is among the best in this series.

4.4
Steric and Electrostatic Molecular Similarity Analyses of Cyclic Imide-type Peroxidizers

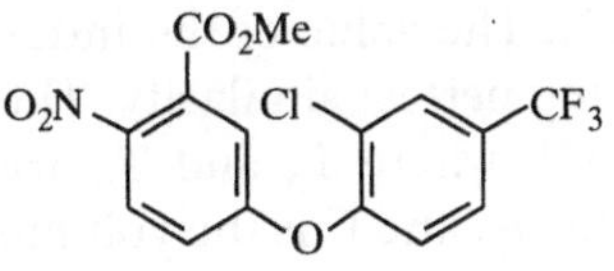

(13)

Peroxidizers inhibit competitively the protox, which catalyzes the oxidation of protoporphyrinogen IX (protogen) to protoporphyrin IX (proto IX). Hence, peroxidizing herbicides are to simulate the substrate at the active site. The molecular similarity between protogen and herbicidal diphenyl ethers was examined by Nandihalli et al. (1992; Nandihalli and Duke 1994). They compared the three-dimensional structure between protogen and acifluorfen (**13**), and suggested that acifluorfen is superimposable onto the B- and C-rings moiety of protogen. Scientists from Nippon Soda and Tamagawa University have performed similar molecular similarity analyses for cyclic imides and related peroxidizers. In this section, their studies, in which one of the present authors (A. N.) has been deeply involved, are presented.

4.4.1
Molecular Similarity Analysis of Cyclic Imide-type Herbicides

Hagiwara et al. (1993) of Nippon Soda found that a series of Δ^2-1,2,4-thiadiazoline derivatives having a 5-phenylimino group and a 3,4-fused ring structure (**14**) show a potent herbicidal activity. The phytotoxic symptoms and effects of benzene ring substituents on the activity were similar to those observed for the cyclic imide and urea herbicides such as chlorophthalim (**2**: X = 4-Cl) and azafenidin (**15**: X = 2,4-Cl_2-5-propargyloxy). Their structures are, however, not entirely similar among one another, differing in the heterocyclic ring system and the position of the "anilino" nitrogen, being exocyclic in compounds **14** but endocyclic in compounds **2** and **15**. The Nippon Soda group studied their structure-activity relationship in terms of the stereoelectronic features by using semi-empirical molecular orbital calculations and molecular similarity indices (Hagiwara and Nakayama 1994; Nakayama et al. 1995).

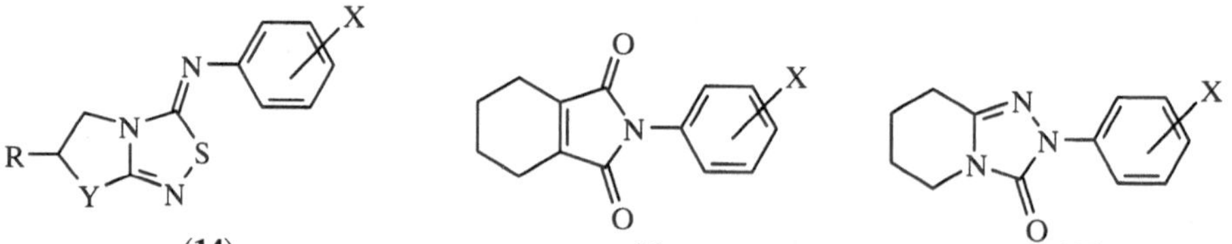

(**14**) (**2**) (**15**)

The electrostatic similarity index of molecules A and B is defined by Eq. (32), where ε_A and ε_B are electrostatic potential of molecules A and B, respectively, at each of the defined points outside the two molecules superimposed, and τ is the unit space (Hodgkin and Richards 1987). The value of the index varies in the range of -1 to 1, with $R_{AB} = 1$ indicating perfect similarity. The shape similarity index (S_{AB}) was defined by Eq. (33), where T_A and T_B are volume of the individual molecules A and B, respectively, and C is the volume commonly shared by the two molecules at the superimposition (Meyer and Richards 1991).

$$R_{AB} = \int \varepsilon_A \varepsilon_B d\tau \Big/ \left(\int \varepsilon_A{}^2 d\tau \int \varepsilon_B{}^2 d\tau \right)^{1/2} \quad (-1 \le R_{AB} \le 1) \tag{32}$$

$$S_{AB} = C/(T_A T_B)^{1/2} \quad (0 \le S_{AB} \le 1) \tag{33}$$

The compounds **I** (**14**: R = Me, Y = CH_2), **II** (**2**), and **III** (**15**) in Fig. 1 with X = 2-F-4-Cl-5-propargyloxy in common were selected for the analysis, since they exhibited an almost equivalent level of herbicidal activity. The molecular electrostatic potential values around the heterocyclic moiety in these compounds were calculated by the MNDO-PM3 procedure (Stewart 1989a,b), and shown in Fig. 2 as the pattern in contour maps. The conformational analysis for the rotation of the benzene ring in these molecules indicated that they share a stable conformation in common with respect to the torsion angle of the benzene and heterocyclic rings (Hagiwara and Nakayama 1994). Based on such information, the best superimposition for each pair of molecules was achieved by the simplex optimizing procedure (Nelder and Mead 1965) to maximize the electrostatic similarity index (R_{AB}) as shown in Fig. 1. In these superimposition

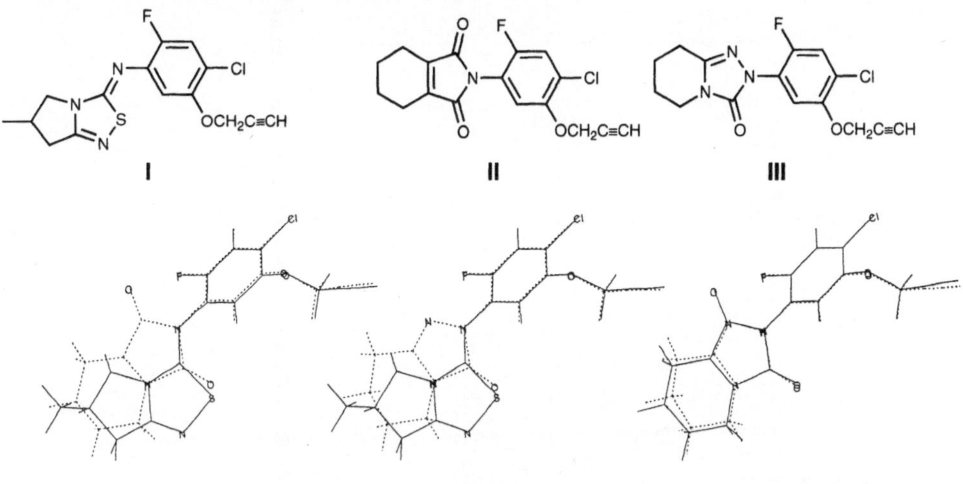

	I and II	I and III	II and III
R_{AB}	0.841	0.831	0.928
S_{AB}	0.861	0.850	0.889

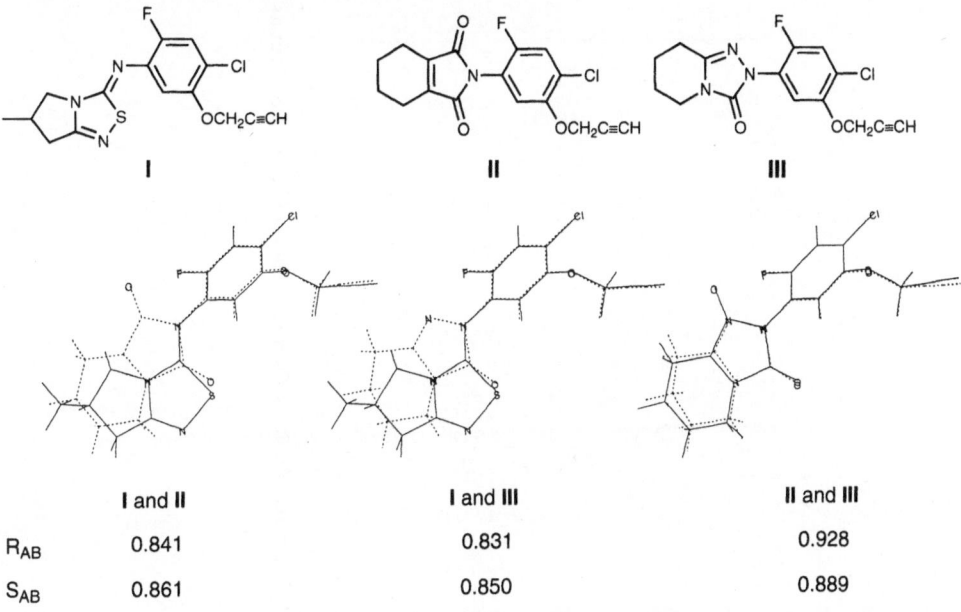

Fig. 1. Models for superimposed couples among three compounds, I, II and III, and the values of similarity indices (R_{AB} and S_{AB}) for each superimposition. In computer-drawn structures, C—H bonds are indicated by a *short line*. The same follows in Figs. 2–5. (Reproduced with permission from the Pesticide Science Society of Japan)

models, the R_{AB} values are in the range of 0.83–0.93, indicating significant similarity among these compounds in terms of electrostatic property. The values of S_{AB} are also significant enough to account for the similarity in the molecular shape. This means that the superimposition procedure to maximize the electrostatic similarity index is enough to obtain a significantly high shape similarity within the set of three compounds. In each of the superimposition models in Fig. 1, the substituted phenyl and the condensed alicyclic moieties are closely overlapped between two molecules. A position-specific hydrophobicity or steric effect of the condensed alicyclic moiety in the molecules could be important for the activity. This position-specific effect is also understood because the presence of the methyl group on the alicyclic ring of compound **I** not only enhances the herbicidal activity, but also improves significantly the superimposability with the alicyclic moieties of compounds **I** and **II**. The heterocyclic five-membered rings in the molecules are not always superimposed well to each other as shown in pairs of **I** and **II**, and **I** and **III**. The sulfur atom in the thiadiazoline compound **I** corresponds with the carbonyl oxygen in compounds **II** and **III**. The other carbonyl group in compound **II** and the nitrogen atom at the 1-position of compound **III** are not superimposed onto any atom in compound **I**, and the nitrogen atom at 2-position in compound **I** is not on any atom in compounds **II** and **III**. This suggests that two different ways of fusion of the alicyclic structure with the heterocyclic moiety are

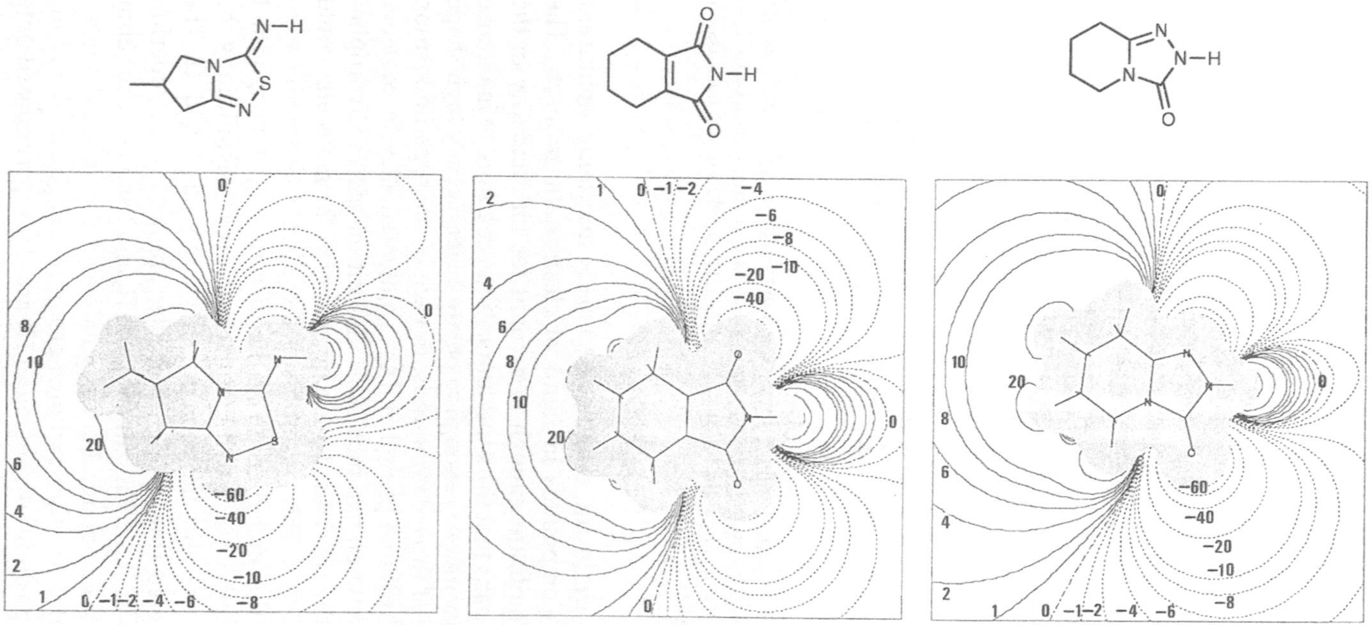

Fig. 2. Molecular electrostatic potential contours of the fused heterocyclic ring moiety in protox inhibitors. Contour values are in 10^{-3} atomic units (au). (Reproduced with permission from the Pesticide Science Society of Japan)

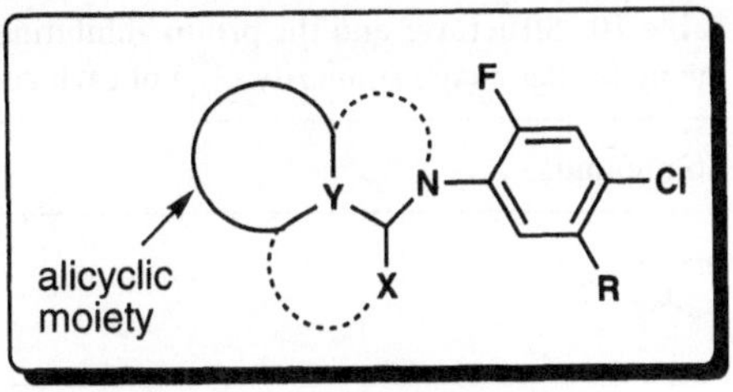

Fig. 3. Hypothetical structural requirements for imide-type peroxidizers. (Reproduced with permission from the Pesticide Science Society of Japan)

X : hetero atom (O or S)

Y : sp^2 carbon or planar nitrogen

------ : bond to fuse the ring system

possible to give similar molecular structures in terms of steric and electrostatic properties. In other words, the role of the heterocyclic five-membered rings in the molecules is to fix the positions of the benzene ring and the alicyclic moiety.

According to these results, hypothetical structural requirements are represented in Fig. 3, where X is a hetero atom such as oxygen and sulfur, Y is an sp^2 carbon or a planar nitrogen. At the position of Y, the sp^2 carbon of compound **II** and the planar nitrogen in compounds **I** and **III** are superimposed. The planar configuration around Y is suggested to be important to fuse the alicyclic moiety with the heterocyclic ring coplanarly. The broken lines represent two ways of ring fusion to make the entire fused-ring system rigid, in which the "anilino" nitrogen is exocyclic or endocyclic depending upon the way of the ring fusion. The bioisosteric or bioanalogous requirements among dissimilar backbone structures for the cyclic imides and related peroxidizers could be understood through this model on the basis of molecular similarity.

4.4.2
Molecular Similarity of Cyclic Imide-type Peroxidizers with Protoporphyrinogen IX

The molecular similarity of the substrate of the target enzyme with inhibitors such as cyclic imides is also of interest to clarify the process of molecular recognition at the active site of peroxidizers. Uraguchi, Wakabayashi, and others of Tamagawa University, jointly with one of the present authors (A.N.) (Uraguchi et al. 1997), investigated the molecular similarity between the protox substrate (protogen) and each member of a set of imide-type inhibitors listed in Table 10. The energy-minimum structures of protogen were surveyed by the conformational analyses using the MNDO-PM3 calculations (Stewart 1989a,b). The most stable conformation is shown in Fig. 4. In this conformation, the four pyrrole rings are not coplanar but twisted around the bonds

Table 10. Structures and the protox inhibiting activity (pI_{50}) of cyclic imides and relatives, and the molecular shape similarity (S_{AB}) of each compound with the "half-protogen"

Compounds	X_n	pI_{50} [a]	S_{AB} [a]	$\log[S_{AB}/1-S_{AB}]$
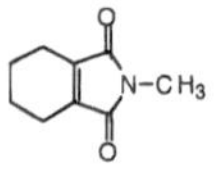		4.41	0.62	0.21
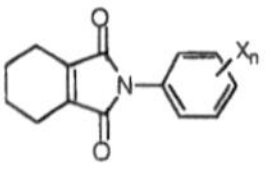	H	5.80	0.75	0.48
	4-Cl	7.60	0.80	0.60
	3-$OCH_2C{\equiv}CH$,4-Cl	9.40	0.85	0.75
	3-CO_2-*i*-Pr,4-Cl	9.00	0.84	0.72
	3-CO_2-*n*-Bu,4-Cl	8.04	0.83	0.69
	3-$CO_2CH_2CO_2Me$,4-Cl	8.80	0.84	0.72
	3-CO_2-*n*-Pent,4-Cl	8.12	0.81	0.63
	3-CO_2-*n*-Oct,4-Cl	7.60	0.76	0.50
	2-F,4-Cl,5-$OCH_2C{\equiv}CH$	8.74	0.85	0.75
	2-F,4-Cl,5-CO_2-*i*-Pr	8.86	0.84	0.72
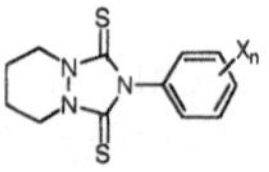	4-Br	8.10	0.78	0.55
	4-Cl	8.17	0.78	0.55
	2-F,4-Cl	8.14	0.78	0.55
	2-F,4-Cl,5-O-*i*-Pr	8.92	0.82	0.66
	2-F,4-Cl,5-$OCH_2C{\equiv}CH$	9.05	0.83	0.69
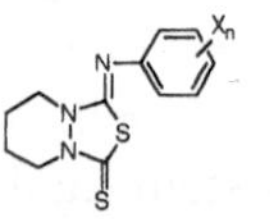	4-Br	6.13	0.70	0.37
	4-Cl	5.96	0.70	0.37
	2-F,4-Cl	5.96	0.69	0.35
	2-F,4-Cl,5-O-*i*-Pr	6.21	0.73	0.43
	2-F,4-Cl,5-$OCH_2C{\equiv}CH$	6.96	0.74	0.45
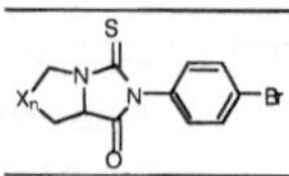	$-CH_2-$	7.77	0.76	0.50
	$-CH_2CH_2-$	8.29	0.79	0.58
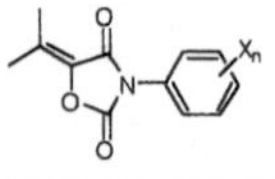	4-Cl	7.66	0.74	0.45
	2-F,4-Cl,5-O-*cyc*-Pent	8.28	0.78	0.55
	2-F,4-Cl,5-OCH(Me)$C{\equiv}CH$	8.57	0.81	0.63

[a] Uraguchi et al. (1997).

connecting with the CH_2 moiety in four bridges. The nitrogens in the A and C rings and those in the B and D rings are directed oppositely.

The 3-D structural comparison of the most stable conformer of protogen with the imide-type inhibitors was first carried out on the display geometrically using the imide compound (**2**: X = 3-propargyloxy-4-Cl) as the reference, which is most active among the series of 26 compounds listed in Table 10. Among a number of superimposition patterns, the one in which the benzene ring and the cyclic imide moiety of the peroxidizer molecule are overlaid with the C- and D-rings of protogen, respectively, as shown in Fig. 5 seemed to be best. In this model, the 2-carboxyethyl group on the C-ring is matched to the

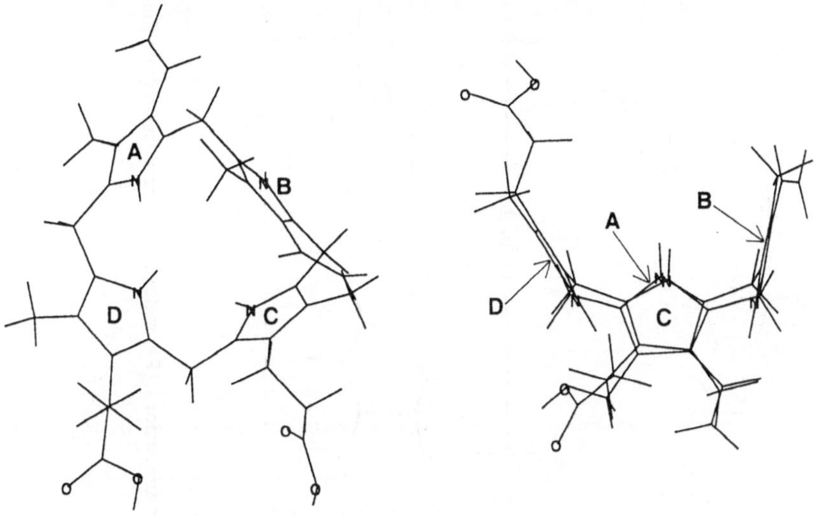

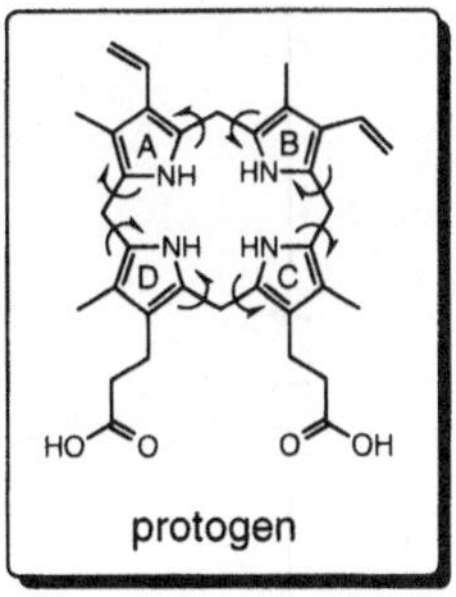

Fig. 4. The most stable structure of protogen obtained by conformational analysis. *Left* Top view; *right* side view. (Reproduced with permission from the Pesticide Science Society of Japan)

propargyloxy group on the benzene ring of the inhibitors. The steric dimensions of the reference compound in terms of the maximum length of the whole molecule (11.54 Å) and the length of the propargyloxy group (5.21 Å) correspond to the dimensions of protogen such as the diameter (10.85 Å) and the length of the 2-carboxyethyl group (5.61 Å), respectively.

A hypothetical model molecule for the "half-protogen" containing only the C- and D-rings moiety was built in which the coordinates of each atom in the most stable conformation of protogen are retained. The molecular shape similarity (S_{AB}) of this "half-protogen" with each of the inhibitors in Table 10 was calculated for the superimposed model obtained by minimizing the root mean square (RMS) of distances between corresponding atomic positions as defined in Fig. 6. The correlation between the shape similarity index and the protox inhibiting activity (pI_{50}) of the 26 compounds was shown by Eqs. (34) and (35), and plotted in Fig. 7. In Eq. (35), the index S_{AB} was transformed in the logistic scale, $\log[S_{AB}/(1 - S_{AB})]$, not to restrict the range of variations from zero to unity. The logistic scale may be more reasonable than S_{AB} itself to correlate with the activity in terms of the pI_{50} scale varying beyond the range from zero to unity. The electrostatic similarity between the half-protogen and

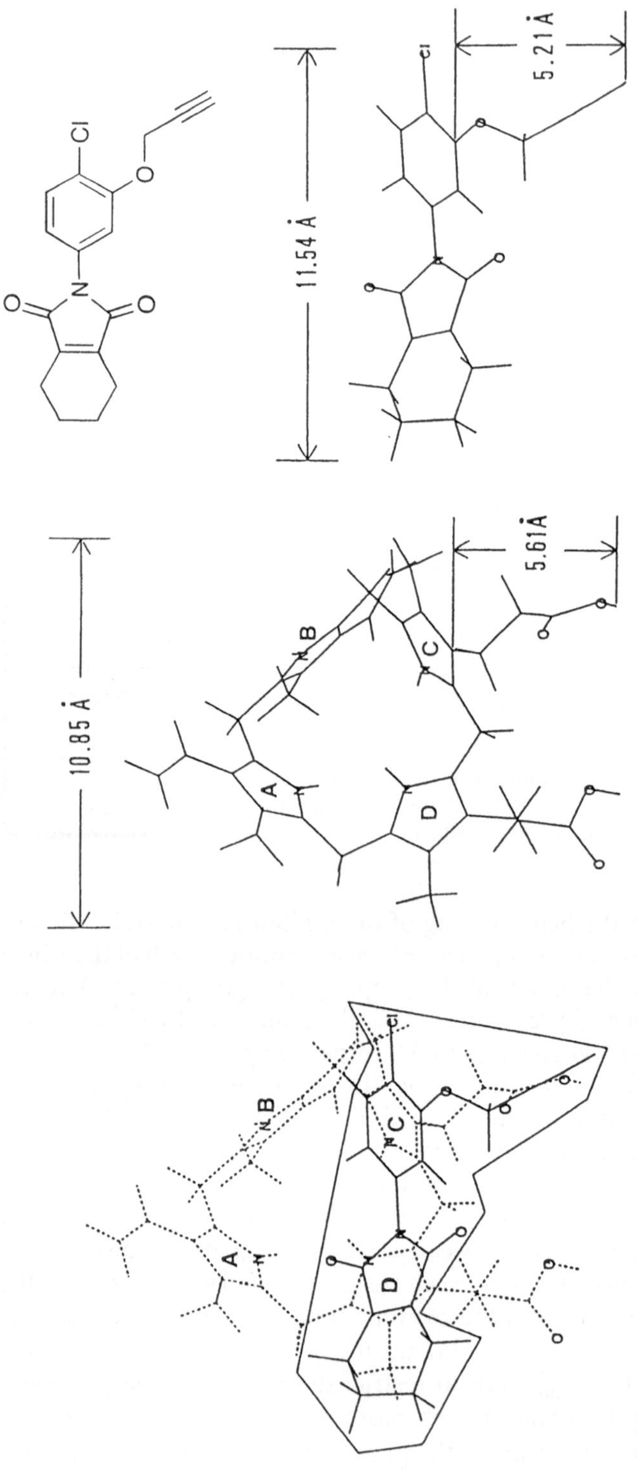

Fig. 5. Superimposition (*left*) of protogen (*middle*) and a cyclic imide (*right*) with molecular steric dimensions. (Reproduced with permission from the Pesticide Science Society of Japan)

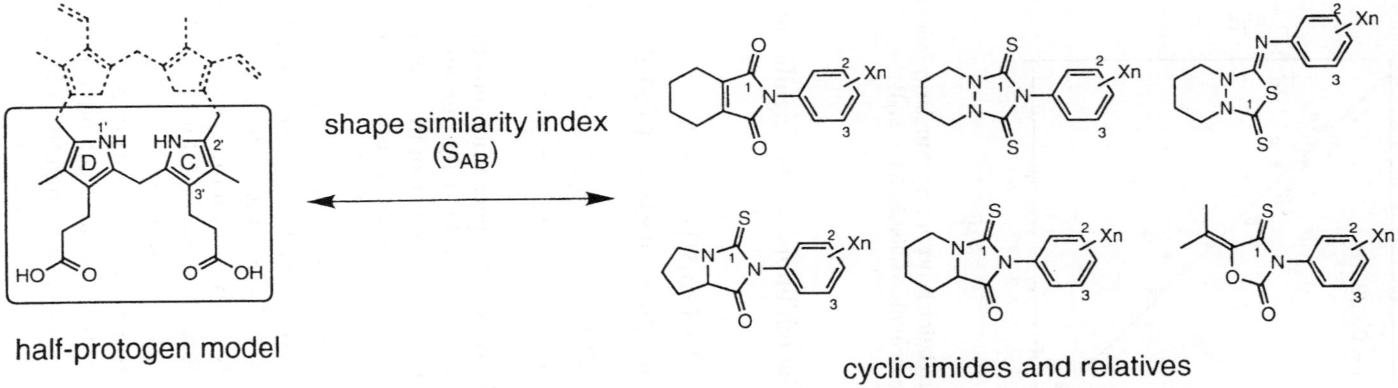

Fig. 6. Structure of the half-protogen model and various cyclic imide-type compounds and relatives subjected to structure-studies. Atoms labeled *1*, *2*, and *3* in the inhibitors were superimposed onto positions of *1′*, *2′*, and *3′* in the half-protogen, respectively, by minimizing the RMS (root mean square) value of distances between corresponding paired atoms

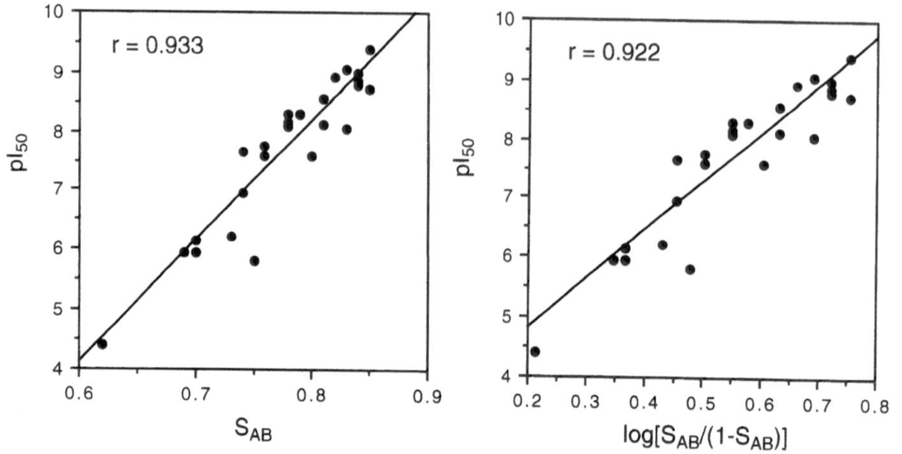

Fig. 7. Correlation between shape-similarity index (S_{AB}) and protox inhibiting activity (pI_{50}). *Left* pI_{50} vs. S_{AB}; *right* pI_{50} vs. $\log[S_{AB}/(1 - S_{AB})]$

each of the inhibitors was not significant in correlating the enzyme inhibition activity.

$$pI_{50} = 20.34(\pm 3.31)S_{AB} - 8.10$$
$$n = 26, r = 0.933, s = 0.463, F_{1,24} = 160.53 \quad (34)$$

$$pI_{50} = 8.22(\pm 1.45)\log[S_{AB}/(1 - S_{AB})] + 3.15$$
$$n = 26, r = 0.922, s = 0.498, F_{1,24} = 135.90 \quad (35)$$

The above result suggests not only that herbicidal cyclic imides and relatives may mimic the C- and D-rings moiety of protogen at the active site of protox, but also that the similarity in the molecular shape may be predominant in the mechanism for peroxidizers to recognize the active site at the (sub)molecular level. Nandihalli et al. (1992; Nandihalli and Duke 1994; Reddy et al. 1995) proposed that acifluorfen (**13**) is superimposable onto the B- and C-rings moiety of protogen. Thus, the diphenyl ethers may mimic half the protogen containing B- and C-rings, whereas the cyclic imides may mimic the C- and D-rings. Recently, a mutant species of *Scenedesmus acutus* highly resistant against chlorophthalim (220-fold) was shown to have only a much lower resistance against oxyfluorfen (9-fold) (Watanabe et al., unpubl.). This implies that the cross-resistance relationship between herbicidal cyclic imides and diphenyl ethers is rather weak. Such phytotoxic features could depend on the difference in the molecular recognition of protox inhibitors between diphenyl ethers and cyclic imides. The procedure for the conformational analysis of protogen and inhibitors by Nandihalli et al. was not the same as that used by Uraguchi et al. (1997). Further studies are needed to understand the common as well as the different features of cyclic imides and diphenyl ethers with respect to the molecular similarity with the substrate.

4.5 Overview of QSAR Studies and Their Implications in Molecular Design

Besides imide-type compounds and diphenyl ethers, some series of compounds simply with an amide bond have been shown to be peroxidizers (Chaps. 2 and 3). There are QSAR examples for compounds **16** (Matsunari et al. 1998) and **17** (Osabe et al. 1992a–c). The light-dependent herbicidal activity against *Echinochloa oryzicola* has been clearly analyzed with a certain combination of physicochemical parameters of variable substituents, X_n and R_n, in each of series **16** and **17**. The QSAR results shared a few features in common, but it was difficult to find comparable patterns between them as well as with those of imide-type peroxidizers, as discussed in Sections 4.2 and 4.3. The integrated discussions about substituent effects with imide-type peroxidizers are not easily accessible. Thus, their QSAR are not included here.

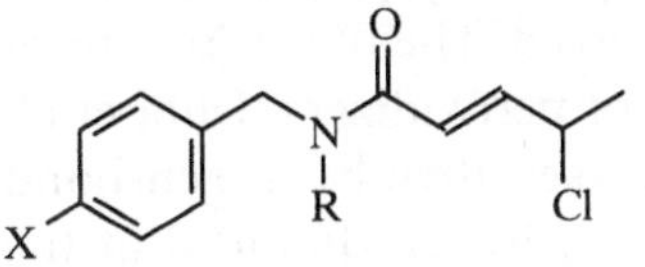

(16)

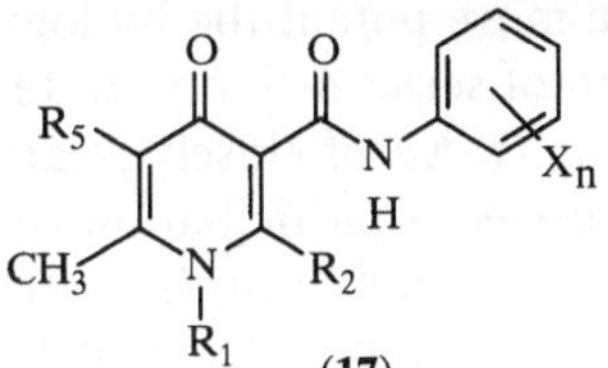

(17)

As mentioned earlier, QSAR studies (Sects. 4.2 and 4.3) are mainly for those of substituent effects in the *N*-phenyl moiety, while molecular similarity analyses (Sect. 4.4) deal with compound series in which the nitrogen-heterocyclic moiety is considerably changed. In this section, the above analyses are overviewed. Some possible guiding principles for designing novel candidate structures with substituent variations are also proposed.

4.5.1 *N*-Phenyltetrahydrophthalimides and Analogs

For the title series compounds studied by the Mitsubishi, Tamagawa, and Konstanz group described in Sections 4.2.1–4.2.3, the QSAR results from various subseries are consistent with each other. The QSAR pattern is almost independent from differences in the test biological systems and conditions between the sawa millet seedlings (7 days) and green microalgae (24h) in the light-dependent growth inhibition potency [Eqs. (1)–(12)]. Moreover, the enzyme protox prepared from corn etioplasts is inhibited with a QSAR pattern similar to the millet seedling and algal cellular growth [Eqs. (1), (2), (15), and

(18)]. The most important physicochemical substituent property is the steric features representable by the STERIMOL parameters of substituents at each of the positions. It seems peculiar that the hydrophobicity term is of no significance in correlations on the seedling [Eq. (1)] and cellular [Eq. (2)] as well as enzyme [Eq. (16)] levels. This would not mean that the hydrophobicity is not required for the peroxidizing activity, because the highly ionized and thus the more hydrophilic compounds exhibit the lower growth inhibition. The (sub)molecular hydrophobicity above a certain level should be required, but variations in the hydrophobicity do not critically work to modify the bioactivities in tetrahydrophthalimides used here under respective test conditions.

For the pI_{50}(protox) values of 11 *para* substituted compounds from the data included in Table 3, Nicolaus et al. (1993) suggested a linear dependence on the hydrophobicity in terms of the π(X/PhH) value of the substituents ($n = 11$, $r = 0.83$, $s = 0.64$). The use of the π(X/PhH) value is inappropriate and the quality of the correlation is much lower than that of Eq. (15). Thus, Eq. (15) probably supersedes that formulated by Nicolaus et al. (1993).

In Eqs. (1)–(4) for the growth inhibition of compounds series **2** and **3**, the negative $\Sigma\sigma^0$ term is significant. The greater the electron-donating property of substituents, the more potent the biological activity. In Eqs. (5)–(12) for the growth inhibition of series **4–7**, no $\Sigma\sigma^0$ term is involved. The σ^0 (or $\Sigma\sigma^0$) term is unstable among QSAR of closely related series of peroxidizers. Ohta et al. (1980) have suggested a participation of geometry-sensitive hydrogen-bond formation of the carbonyl group(s) with possible hydrogen donor(s) at the active site for series compounds of which QSAR requires the negative σ^0 term. The greater the electron-donating property of aromatic substituents, the more enhanced the hydrogen-bonding as well as the activity.

Recently, Sato et al. (1991) found that series **2** compounds are hydrolyzed to give equilibrium mixtures with corresponding amide-acids under test conditions for the sawa millet activity in the presence of millet seedlings. They also observed that, in most cases, the growth inhibitory activity of amide-acids is lower than that of the corresponding dicarboximides. For compound series **3**, the hydrolysis leading to the equilibrium mixtures similar to that for series **2** has been observed under non-biological conditions, but, for series **4–7**, not observable significantly (Wakabayashi 1997, pers. comm.). The greater the electron-withdrawing character of aromatic substituents, the greater the extent of hydrolysis giving the lower activity. To protect from the hydrolysis, the electron-donating effect of substituents is favorable to the activity. The significance in the negative $\Sigma\sigma^0$ term in Eqs. (1) and (3) is understandable on this basis. Under conditions for the short term microalgal test, the hydrolysis does not occur significantly. Thus, the electronic effect of aromatic substituents on the microalgal activity should be less important than that on the sawa millet activity. This does not, however, seem to be reflected explicitly. Comparisons between Eqs. (1)–(4), (19), and (20) indicate that, even in the microalgal activity, the $\Sigma\sigma^0$ term is comparable with that in the sawa millet activity. The effect of the hydrolysis-cyclization equilibrium on the activity against different plant species should be investigated more comprehensively.

In addition to the protecting effect from the hydrolytic equilibrium mixture formation, the electron-donating substituents could inherently be favorable for the protox inhibition as shown in Eqs. (16)–(18). As described in Section 4.2.2, the average of the σ^0 value of substituents included in Eq. (15) is close to zero. The protox inhibition is worked out within a short time not enough for the hydrolysis-cyclization equilibration. Equations (19) and (20) indicate that the electron-donating property of substituents is far more important in growth inhibitory activities than in the enzyme inhibition, at least for compound series **2**. In the growth inhibition, the effect on the hydrolysis protection would be added to the effect on the inherent enzyme inhibition.

One of the serious drawbacks in the above QSAR studies of the tetrahydrophthalimides is the colinearity among STERIMOL parameters. However, the fact that the growth inhibitory activity against two plant species and the target enzyme inhibition are consistently illustrated, overcomes the drawback to a great extent.

The fact that the *ortho* position in the phenyl moiety of the imide-type peroxidizers has been most frequently substituted by F [$\Delta L = 0.59$, cf. $\Delta L(Cl) = 1.46$] is understandable from the steric requirement [$\Delta L(opt) = 0.75$ in Eqs. (1) and (2)] of substituents at this position. For the *para* position, Cl seems to be the first choice in peroxidizers of various *N*-heterocyclic series. The optimum ΔL at the *para* position in Eqs. (1) and (2) is 2.78, which is close to that of OEt (2.74), but its high ΔB_5 value (2.36) makes the Cl substituent ($\Delta L = 1.46$, $\Delta B_5 = 0.80$) one of the best *para* substituents.

Often used substituents at the 5(*meta*)-position are propargyloxy and 1-butyn-3-yloxy groups. Their ΔL and ΔB_5 values are 4.52 and 2.07 for $OCH_2C{\equiv}CH$ and 4.52 and 3.35 for $OCH(CH_3)C{\equiv}CH$, respectively. The optimum ΔL value for the protox inhibition is estimated as being about 5.0 from Eq. (18). The $\sigma^0(m)$ value of these groups is estimated to be 0.18 (see footnote [f] in Table 4). The lower σ^0 and ΔB_5 values working together with the near optimum quadratic ΔL terms, the protox inhibition of the 4-Cl-3-propargyloxy (no. 20 in Table 4) and the 2-F-4-Cl-5-propargyloxy and 2-F-4-Cl-5-(1-butyn-3-yl)oxy (nos. 4 and 5 in Table 5) compounds are highest among compounds included in Eq. (18).

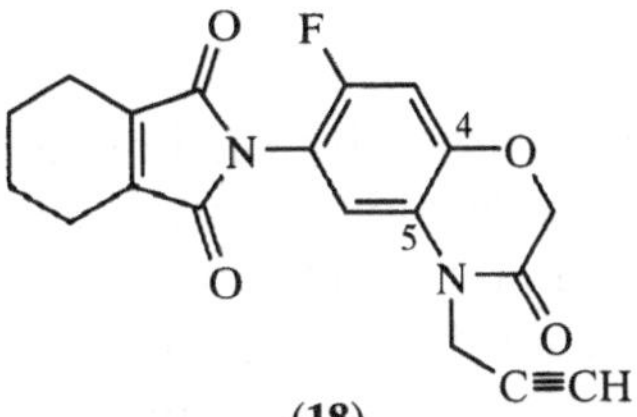

(**18**)

It is interesting to note that, in the recently commercialized flumioxazin (**18**), the steric features of the propargylamino "group" at the 5-position of the benzene ring are very similar to those of the propargyloxy group. Besides the ΔL value close to the optimum, the linear propargyl group has an advantage to make the ΔB_5 value lower. The steric features of the "alkoxy" substituent at the 4-position are also favorable to the activity. In "ordinary" 4,5-disubstituted structures, the maximum width of the 4-substituents would be directed in the less hindered space closer to the 3-position. In the 4,5-fused 1,4-oxazin system, however, the situation is different. Being cyclized with the end of the "substituent" at the 5-position, the ΔB_5 value of the 4-"alkoxy group" should be replaced by the ΔB_1 value. With this "reversed" ΔB_5 value (0.35: ΔB_1 for alkoxy groups), the effect of the "cyclized" alkoxy group, simulated by the "*endo*-OMe", at the *para* position is estimated as being about 10 times higher than that of Cl. In flumioxazin, the electron-withdrawing carbonyl group in the oxazin ring could be deleted without losing the activity (see Fig. 8 of Chap. 2).

The above type of cross examinations of the QSAR results are believed not only to rationalize the past molecular design principles but also to promote the motivation of synthetic chemists toward "rational" trials.

For the hydroponic growth inhibition of cucumber seedlings of the FMC series of compounds (**8**), the electron-withdrawing and hydrophobic properties of *para* substituents are significant. The QSAR pattern is quite different from those for compounds from the Mitsubishi-Tamagawa-Konstanz group. The participation of the hydrophobic term may be due to the use of the hydroponic cucumber assay system. The positive σ^0 term in Eq. (14) appears to conflict with those discussed above. One of the possible rationalizations would be the fact that, for this type of 2,4,5-trisubstituted compounds, the amide-acids formed by the hydrolysis during the 10-day assay period are more active than the original dicarboxyimides. For the short-term protox inhibition of 2,4,5-trisubstituted compounds, the negative $\Sigma\sigma^0$ term intrinsically important to the protox inhibition is significant in Eq. (18). The experimental results of Sato et al. (1991) suggest that the amide-acid is slightly but significantly more active than the imide for the 2-F-4-Cl-5-propargyloxy compounds in the growth inhibition (7 days) of sawa millet seedlings. Systematic QSAR examination for the amide-acids should also be a future project.

4.5.2
Other Compound Series

Including the last mentioned compound series **8** in the preceding section, the QSAR studies from the FMC Corporation described in Section 4.3 are based upon the hydroponically measured growth inhibition of cucumber seedlings for 10 days. The difference in the test conditions from the sawa millet and green microalgal assays may be one of the origins of the significant contribution of the quadratic hydrophobic terms to the QSAR correlation equations for compound series **11** and **12**. The significance of the quadratic hydrophobic terms was also the case in the amide compound series **16** and **17**, even though the biological test system uses the sawa millet seedlings. Thus, the participation of the (sub)molecular hydrophobicity may also be dependent on the structural features of peroxidizers.

As described in respective sections, the original QSAR correlation equations from the FMC Corporation were able to be rewritten with the use of hopefully "improved" physicochemical substituent parameters. For the *N*-phenyltriazolinethiones (**11**), the original Eq. (25) as well as the revised Eq. (29) are with the significant negative σ^0 (σ) term. The effect would be to protect the compounds from the possible hydrolysis similar to the effect observed in Eqs. (1) and (2) for the tetrahydrophthalimides (**2**), the exact mechanism of hydrolytic "detoxification" being uncertain. For the *N*-phenyltetrahydroindazoles (**12**), the revised Eq. (31) shows the contribution of the positive σ^0 term. This observation could mean that the electron-withdrawing substituents may protect from the possible ring splitting which could be initiated rate-determiningly by the protonation of the 1-N atom and subsequent attack of nucleophile(s) against the partly positive-charged 3-C atom leading to phenylhydrazine derivatives.

The QSAR from the FMC Corporation which is represented only with the steric constant terms is Eq. (22) for *N*-phenyltriazolinones (**10** in Sect. 4.3.1). In formulating Eq. (22), we made assumptions about the conformation of the 5(*m*)-Ph, -$NHSO_2Me$ and -OC_6H_5(4-$NHSO_2Et$) groups (nos. 10, 14, and 15 in Table 7, respectively). The procedure was validated by Eq. (23) without including three compounds occupied by the above substituents. Substituents included in Table 7 and Eq. (22) seem rather rich in variety. As a matter of the fact, however, the ΔL value of substituents, except for that in compound (no. 19), does not exceed the possible optimum value. The optimum ΔL for 5(*m*) substituents was assumed from Eq. (18) for the protox inhibition of the 3,4-di- and 2,4,5-trisubstituted phenyltetrahydrophthalimides as being about 5.0. Thus, only the positive ΔL term seems to appear in Eq. (22). The *p*-substituted phenoxy groups (nos. 15–18), as the *meta* substituents attached to the core phenyl moiety, are by no means too lengthy in terms of ΔL. They work very well along with such simple substituents as halogens (nos. 12 and 13) in representing Eq. (22).

It is worth examining the substituent features of outlyingly very active compounds. The ΔL value of the substituent (no. 19) is too great, being much above the optimum estimated in the analogous compound series **2**. Nevertheless, the activity of the compound (no. 19) is very high. In fact, it is about 75 times higher than that expected from Eq. (22). The number of heavy atomic bonds along the longest "chain" of this substituent is counted as 11. There are observations similar to this. The substituent, 3-CO_2Oct (no. 12 in Table 4), also makes the activity of the 3-CO_2Oct-4-Cl-phenyltetrahydrophthalimide about 900 times higher than that predicted by Eq. (18). The "length" of the substituent is 10 in terms of the heavy atomic bond number. The substituent, 5(*m*)-OCH_2CO_2Pent (no. 7 in Table 5) in the recently launched flumiclorac-pentyl

(**19**) is a further example. Its activity is about 160 times higher than expected by Eq. (18). The chain "length" in terms of the bond number is 9. Still further example is provided by the 4-$OCH_2C_6H_4$(4-Cl) substituent in compound (no. 32 in Table 1). The activity is very high, about 10^{10} times higher than that predicted by Eqs. (1) and (2). The chain "length" of this substituent is 7 heavy-atomic bonds.

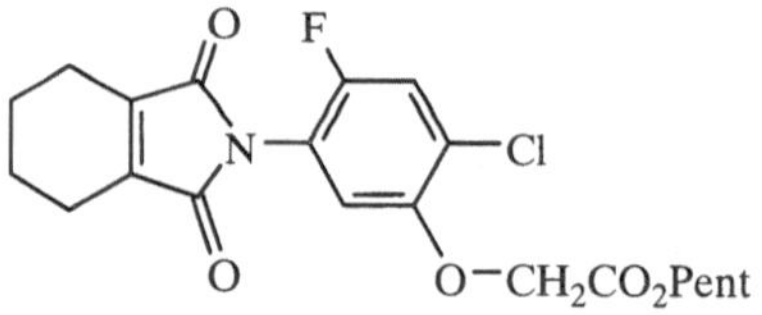

(**19**)

Beyond the optimum ΔL of substituents, the activity of compounds should be decreased, other things being equal. In the imide-type peroxidizers, however, there should be a certain threshold for the revival of the activity. Depending upon the optimum ΔL value at each of substituent positions, the threshold length seems to vary correspondingly. The ΔL(opt) for *meta* substituents is estimated as being about 5.0 Å (5 bonds) from Eq. (18), while that for *para* substituents is about 2.8 Å (3 bonds) from Eq. (1). The total bond "length" of substituents in which the bond number exceeds the threshold in highly potent compounds is 9–11 for *meta* and 7 for *para* substituents.

The above arguments are believed to demonstrate an instructive example in which outlying behaviors of certain compounds from ordinary members included in the QSAR of related series compounds have been analyzed to reveal a "common" origin. Because the outlying behavior is to enhance the biological activity much higher than that predicted by the original QSAR in the present series of peroxidizers, the common origin could be used as a kind of "rule" especially for the molecular design. The seemingly "too" lengthy substituents could be introduced at *meta* and *para* positions of the phenyl moiety to give highly potent peroxidizers. The conditions required here would be that, besides the substituent length being higher than the threshold, the physicochemical features should match the structure of *N*-heterocycles. Beyond the threshold, the hydrophobicity of substituents might work to increase the activity. It should be mentioned that this kind of finding could only be obtained by the comparative QSAR examinations, although the rule or guiding principle for the molecular design is semi-quantitative in this case.

In Figs. 6–23 of Chapter 2, Hirai compiled various imide-type peroxidizers. The structure of the nitrogen heterocyclic moiety ranges from five-membered dicarboxiimides and pyrazoles via a variety of di-, tri-, and tetra-azolin(thi)ones with and without the oxa modification to six-membered uracil and thimidine analogs. In the phenyl moiety, the most popular substitution pattern at the 2- and 4-positions is 2-F-4-Cl, as mentioned above. The most widely modified is the substituent at the 5(*meta*)-position. Quite often the end of the 5-position substituent is cyclized with the 4-position to give condensed bicyclic structures such as flumioxazin (**18**). This observation is believed to mean that there are a number of chances to disclose favorable novel structures by elaborating the 4- and 5-substituents longer than the threshold as well as the 4,5-fused ring structures. The regions into which these substituents and fused rings are to be extended/located are relatively not restricted sterically as the QSAR analyses have suggested. In this respect, the effects of lengthy substituent at the 4- and 5-positions and those of 4,5-fused rings should be analyzed quantitatively with rationally designed substituent/substructure variations perhaps with branching, insertion of heteroatom(s), and/or cyclization.

4.5.3 Structural Modification and Molecular Design of Nitrogen Heterocyclic Moiety

Molecular similarity analyses described in Section 4.4 indicated that the steric and/or stereoelectrostatic features are critical for the effective heterocyclic structure in peroxidizers in general. This would be true for novel heterocycles to be designed. Unfortunately, the variations in structural features included in the analyses are limited so that the analyses are not sufficient for detailed as well as concrete information about the molecular design of heterocyclic structures. As mentioned earlier (Sect. 4.5.2), there are a great variety of structures in the heterocyclic moiety of imide-type peroxidizers. In overviewing the compilation of Hirai (Chap. 2) for the heterocyclic structures, we notice that the structural variation patterns are very similar to those observed in the alcoholic moiety in the synthetic pyrethroids of tetramethrin-type. In Fig. 8, the heterocyclic skeletal structures shared by peroxidizers and tetramethrin analogs are shown. No common mode of action exists between the two series. Pyrethroids are insecticides working as the Na channel opener of the nervous system (Kobayashi et al. 1989). Neverthless, there is a close similarity between the two series in substructural variation patterns of the heterocyclic moiety.

When various structural moieties or substructural units are interchangeable within a certain frame structure of bioactive compounds without losing the activity, these substructural units have been called bioisosteric. Because the term *bioisosterism* inherently involves the concept of isosterism or isometricity, Floersheim et al. (1992) proposed replacing the term with *bioanalogy* which could apply to cases in which the structural dimensions are not close but more or less drastically varied. Therefore, such a (sub)structural relationship as shown in Fig. 8 indicates that the substructural bioanalogy is not limited within a single category of bioactive compounds but spread over the difference in pharmacology. Similar bioanalogous structural variations have been observed in various bioactive compound series of different pharmacology (Fujita 1996). Examples shared by peroxidizers have been further found

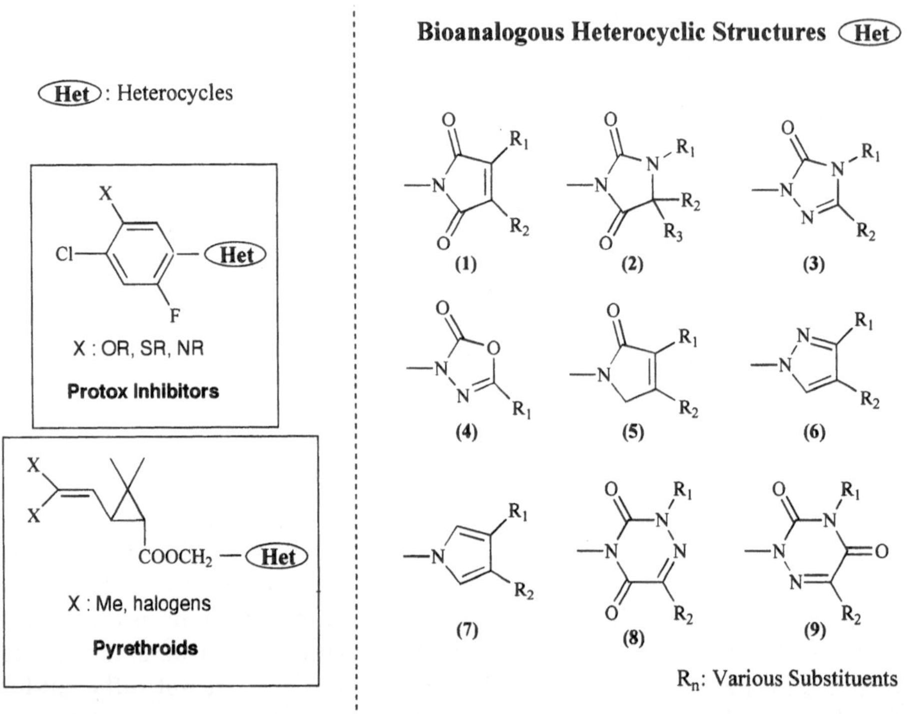

Fig. 8. Bioanalogous heterocyclic structures shared by protox inhibitors and tetramethrin-type pyrethroids. Example structure and the literature source are: for protox inhibitors, Hirai (Chap. 2, this Volume); for pyrethroids, **1** tetramethrin: R_1–R_2=-$(CH_2)_4$-, X=Me (Kato et al. 1964), **2** imiprothrin: R_1=-$CH_2C{\equiv}CH$, R_2=R_3=H, X=Me (Itaya and Hirano 1979), **3**: R_1=-$CH_2C{\equiv}CH$, R_2=Me, X=Me and Cl (Kishida and Yano 1982), **4**: R_1=Pr, X=Cl (Ueda et al. 1965), **5**: R_1–R_2= -(CH=CH)-$_2$, X=Me (Nishimura et al. 1988), **6**: R_1=-CH_2OMe, R_2=H, X=Cl (Ohsumi et al. 1981), **7**: R_1–R_2=-$(CH_2)_4$-, X=Cl (Ohsumi et al. 1981), **8**: R_1=-$CH_2C{\equiv}CH$, R_2=H, X=Cl (Mizutani et al. 1981), **9**: R_1=-$CH_2C{\equiv}CH$, R_2=H, X=Cl. (Mizutani et al. 1981)

in nonacidic pyrine-type antiinflammatory/analgesic agents (Sugiura et al. 1977) shown in Fig. 9 as well as in fipronil series insecticides (Wittle et al. 1995).

Thus, by keeping close attention on substructural modification examples in pyrethroid alcohols, fipronil-type insecticides, and pyrine analgesics and other bioactive compounds having heterocyclic moiety which have been accumulated so far as well as just newly disclosed or patented, it could be possible to gain invaluable insights into how to molecular-design novel heterocyclic moiety for peroxidizers. The bioanalogous relationship found in other categorical series compounds could be utilizable in designing novel heterocyclic skeletal structure of peroxidizers. Starting from the novel skeletal structure, structural optimization should be further explored by examining "best" patterns of substitution and physicochemical properties of substituents. At this stage, the

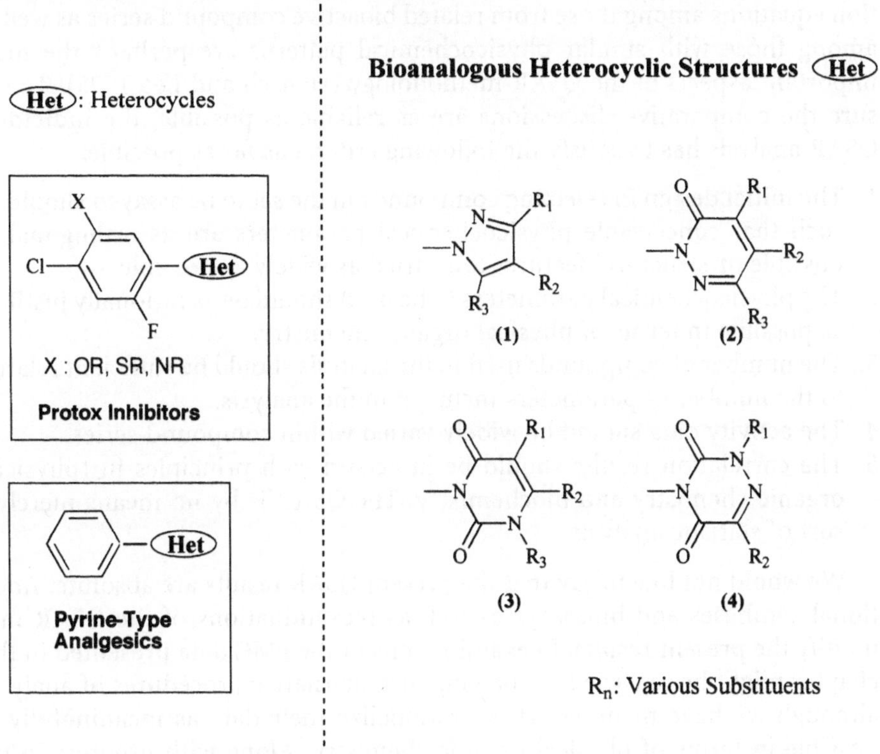

Fig. 9. Bioanalogous heterocyclic structures shared by protox inhibitors and pyrine-type antiinflammatory/analgesic agents. Example structure and literature source are: for protox inhibitors, Hirai (Chap. 2 this Vol.); for analgesics, **1:** R_1=$CONMe_2$, R_2=H, R_3=OEt (Sugiura et al. 1977), **2:** R_1=$NH(CH_2)_3NMe_2$, R_2=H, R_3=OEt (Suzuki et al. 1972), **3:** R_1=Me, R_2=NMe_2, R_3=Allyl (Senda et al. 1972), **4:** R_1=Me, R_2=$(CH_2)_2NMe_2$. (Saikawa et al. 1971)

QSAR procedure is that to rationally reach the "best" compounds hopefully with minimum detour.

4.6 Concluding Remarks

In this chapter, we showed a number of QSAR and similarity analysis examples of imide-type peroxidizers. The reason why we exhibited examples rather collectively was to demonstrate that structure-activity data for any congeneric series of compounds including peroxidizers could be analyzed by the QSAR procedures rationally, if reasonable physicochemical substituent (or substructural) parameters can be selected and utilized. Another was to show how to examine comparatively similar and dissimilar features in QSAR results for related compound series to extract "comprehensive" information about possible mechanisms for the potency variations and "rules" for the structural modification. The comparative examination and lateral validation of correlation equations among those from related bioactive compound series as well as among those with similar physicochemical patterns are perhaps the most important aspects of the QSAR methodology (Hansch and Leo 1995). To ensure the comparative discussions are as reliable as possible, the individual QSAR analysis has to satisfy the following criteria as far as possible:

1. The initial design in selecting compounds in the set to be assayed should be such that conceivable physicochemical parameters are as orthogonal as possible or structural features are varied as widely as possible.
2. The physicochemical parameters to be used should be as rationally justified as possible in terms of physical organic chemistry.
3. The number of compounds used in the analysis should be sufficient relative to the number of parameters included in the analysis.
4. The activity data should be widely varied within compound series.
5. The correlation results should be in accord with principles in (physical) organic chemistry and biochemistry. The QSAR is by no means merely a sort of statistic analysis.

We would not like to say that the present QSAR results are absolute. Additional syntheses and bioassays as well as reexaminations of the QSAR may modify the present results. Reexaminations of the FMC data presented in this chapter might be regarded as showing just alternative procedures of analysis, although we have made efforts to rationalize their data as meaningfully as possible in terms of physical organic chemistry. Along with progress in the knowledge of the receptor (protoporphyrinogen oxidase) as well as with improvement/elaboration in how to represent the biological results, we would like to expect further development of the QSAR studies in this important field of peroxidizing herbicide research.

Acknowledgments. The authors would like to express heartfelt thanks to Professors Ko Wakabayashi and Yukiharu Sato and members of their laboratory at Tamagawa University, Dr. Yoshiaki Nakagawa of Kyoto University and Dr. Kenji Hagiwara of Nippon Soda Co. Without their help, this chapter would never have been completed.

References

Austel V, Kutter E, Kalbfleisch W (1979) A new easily accessible steric parameter for structure-activity relationships. Arzneim Forsch 29:585–587

Boesch R, Metivier J (1963) FP 957151 RP

Charton M (1981) Electrical effect substituent constants for correlation analysis. Prog Phys Org Chem 13:119–251

Draber W, Fujita T (1992) Rational approches to structure, activity, and ecotoxicology of agrochemicals. CRC Press, Boca Raton

Floersheim P, Pombo-Villar E, Shapiro G (1992) Isosterism and bioisosterism case studies with muscarinic agonists. Chimia 46:323–334

Fujita T (1983) Substituent effects in the partition coefficient of disubstituted benzenes: bidirectional Hammett-type relationships. Prog Phys Org Chem 14:75–113

Fujita T (1996) Similarities in bioanalogous structural transformation patterns among various bioactive compound series. Biosci Biotech Biochem 60:557–566

Fujita T, Ban T (1971) Structure-activity study of phenethylamines as substrates of biosynthetic enzymes of sympathetic transmitters. J Med Chem 14:148–152
Fujita T, Nishioka T (1976) The analysis of the *ortho* effect. Prog Phys Org Chem 12:49–89
Fujita T, Iwasa J, Hansch C (1964) A new substituent constant, pi, derived from partition coefficients. J Am Chem Soc 86:5175–5180
Hagiwara K, Nakayama A (1994) Molecular similarity of peroxidizing herbicides: bioisosterism in Δ^2-1,2,4-thiadiazolines and related heterocyclic compounds. J Pesticide Sci 19:111–117
Hagiwara K, Saitoh K, Iihama T, Hosaka H (1993) Synthesis and herbicidal activity of fused Δ^2-1,2,4-thiadiazolines. J Pesticide Sci 18:309–318
Hansch C, Fujita T (1995) Classical and three-dimensional QSAR in agrochemistry. ACS symposium series 606. Am Chem Soc, Washington, DC
Hansch C, Leo A (1995) Exploring QSAR fundamentals and applications in chemistry and biology. Am Chem Soc, Washington, DC
Hansch C, Leo A, Hoekman D (1995) Exploring QSAR hydrophobic, electronic, and steric constants. Am Chem Soc, Washington, DC
Hansch C, Leo A, Taft RW (1991) A survey of Hammett substituent constants and resonance and field parameters. Chem Rev 91:165–195
Hansch C, Unger SH, Forsythe AB (1973) Strategy in drug design. Cluster analysis as an aid in the selection of substituents. J Med Chem 16:1217–1222
Hodgkin EE, Richards WG (1987) Molecular similarity based on electrostatic potential and electric field. Int J Quantum Chem: Quantum Biol Symp 14:105–110
Itaya N, Hirano M (1979) Jpn. Kokai Tokkyo Koho JA 54-9269
Kato T, Ueda K, Fujimoto K (1964) New insecticidally active chrysanthemates. Agr Biol Chem 28:914–915
Kishida H, Yano T (1982) Jpn. Kokai Tokkyo Koho JA 57-158765
Kobayashi T, Masutani H, Nishimura K, Fujita T (1989) Quantitative structure-activity studies of pyrethroids 16. Pestic Biochem Physiol 35:58–69
Kohno H, Hirai K, Hori M, Sato Y, Böger P, Wakabayashi K (1993) New peroxidizing herbicides: activity compared with X-ray structure. Z Naturforsch 48c:334–338
Lyga JW, Patera RM, Theodoridis G, Halling BP, Hotzman FW, Plummer MJ (1991) Synthesis and quantitative structure-activity relationships of herbicidal *N*-(2-fluoro-5-methoxyphenyl)-3,4,5,6-tetrahydrophthalimides. J Agric Food Chem 39:1667–1673
Lyga JW, Patera RM, Plummer MJ, Halling BP, Yuhas DA (1994) Synthesis, mechanism of action, and QSAR of herbicidal 3-substituted-2-aryl-4,5,6,7-tetrahydroindazoles. Pestic Sci 42:29–36
Matsui K, Kasugai H, Matsuya K, Aizawa H (1973) Jpn. Tokkyo Koho JP 48-11940 Mitsubishi Kasei Kogyo
Matsunari K, Yoshida F, Nakamura Y, Fujita T (1998) Quantitative structure-activity relationship of herbicidal *N*-alkyl-*N*-(4-substituted-benzyl)-4-chloro-2-pentenamides against *Echinochloa oryzicola*. J Pestic Sci 23 (in press)
Meyer AM, Richards WG (1991) Similarity of molecular shape. J Comput Aided Mol Design 5:427–439
Mizutani M, Tsushima K, Sanemitsu Y, Hirano M (1981) Europ Pat 0033163 (Jap Prior 29. 01. 80)
Nakagawa Y, Izumi K, Oikawa N, Sotomatsu T, Shigemura M, Fujita T (1992) Analysis and prediction of hydrophobicity parameters of substituted acetanilides, benzamides and related aromatic compounds. Environ Toxicol Chem 11:901–916
Nakayama A, Hagiwara K, Hashimoto S, Hosaka H (1995) Quantitative structure-activity and molecular modeling studies of novel fungicides and herbicides having 1,2,4-thiadiazoline structures. In: Hansch C, Fujita T (eds) Classical and three-dimensional QSAR in agrochemistry. ACS symposium series 606. Am Chem Soc, Washington, DC, pp 213–228
Nandihalli UB, Duke SO (1994) Structure-activity relationships of protoporphyrinogen oxidase inhibiting herbicides. In: Duke SO, Rebeiz CA (eds) Porphyric pesticides: chemistry, toxicology, and pharmaceutical applications. ACS symposium series. 559. Am Chem Soc, Washington, DC, pp 133–146

Nandihalli UB, Duke MV, Duke SO (1992) Quantitative structure-activity relationships of protoporphyrinogen oxidase-inhibiting diphenyl ether herbicides. Pestic Biochem Physiol 43:193–211

Nelder JA, Mead R (1965) A simplex method for function minimization. Comput J 7:308–313

Nicolaus B, Sandmann G, Böger P (1993) Molecular aspects of herbicide action on protoporphyrinogen oxidase. Z Naturforsch 48c:326–333

Nishimura K, Kitahaba T, Ikemoto Y, Fujita T (1988) Quantitative structure-activity studies of pyrethroids 14. Pestic Biochem Physiol 31:155–165

Ohsumi T, Hirano M, Itaya N, Fujita Y (1981) A new pyrethroid of high insecticidal activity. Pestic Sci 12:53–58

Ohta H, Jikihara T, Wakabayashi K, Fujita T (1980) Quantitative structure-activity study of herbicidal *N*-aryl-3,4,5,6-tetrahydrophthalimides and related cyclic imides. Pestic Biochem Physiol 14:153–160

Osabe H, Morishima Y, Goto Y, Masamoto K, Nakagawa Y, Fujita T (1992a) Quantitative structure-activity relationships of light-dependent herbicidal 4-pyridone-3-carboxanilides I. Effect of benzene ring substituents at the anilide moiety. Pestic Sci 34:17–25

Osabe H, Morishima Y, Goto Y, Masamoto K, Nakagawa Y, Fujita T (1992b) Quantitative structure-activity relationships of light-dependent herbicidal 4-pyridone-3-carboxanilides II. Substituent effects of anilide and pyridone moieties. Pestic Sci 34:27–36

Osabe H, Morishima Y, Goto Y, Fujita T (1992c) Quantitative structure-activity relationships of light-dependent herbicidal 4-pyridone-3-carboxanilides III. 3-D (comparative molecular field) analysis including light-dependent diphenyl ether herbicides. Pestic Sci 35:187–200

Plummer EL (1992) Sequential simplex optimization strategies in agrochemical/drug design. In: Draber W, Fujita T (eds) Rational approches to structure, activity, and ecotoxicology of agrochemicals. CRC Press, Boca Raton, pp 3–39

Reddy KN, Nandihalli UB, Lee HJ, Duke MV, Duke SO (1995) Predicting activity of protoporphyrinogen oxidase inhibitors by computer-aided molecular modeling. In: Reynolds CH, Holloway MK, Cox HK (eds) Computer-aided molecular design applications in agrochemicals, materials, and pharmaceuticals. ACS symposium series 589. Am Chem Soc, Washington, DC, pp 211–224

Saikawa I, Osada T, Hori T, Maeda T (1971) Jpn. Tokkyo Koho JP 46-28025

Sato Y, Kojima T, Goto T, Oomikawa R, Watanabe H, Wakabayashi K (1991) Hydrolysis and phytotoxic activity of cyclic imides. Agric Biol Chem 55:2627–2681

Senda S, Hirota K, Banno K (1992) Pyrimidine derivatives and related compounds 15. J Med Chem 15:471–476

Simmons KA, Dixson JA, Halling BP, Plummer EL, Plummer MJ, Tymonko JM, Schmidt RJ, Wyle MJ, Webster CA, Baver WA, Witkowski DA, Peters GR, Gravella WD (1992) Synthesis and activity optimization of herbicidal substituted 4-aryl-1,2,4-triazole-5(1*H*)-thiones. J Agric Food Chem 40:297–305

Stewart JJP (1989a) Optimization of parameters for semiempirical method I. Method. J Comput Chem 10:209–220

Stewart JJP (1989b) Optimization of parameters for semiempirical method II. Applications. J Comput Chem 10:221–264

Sugiura S, Ohno S, Ohtani O, Izumi K, Kitamikado T, Asai H, Kato K, Hori M, Fujimura H (1977) Syntheses and antiinflammatory and hypnotic activity of 4-alkoxy-3-(*N*-substituted carbamoyl)-1-phenylpyrazoles 4. J Med Chem 20:80–84

Suzuki H (1967) Electronic absorption spectra and geometry of organic molecules. Academic Press, New York, pp 279–285

Suzuki Y, Ito M, Hayashi M, Yamagami I (1972) Pharmacological studies on TM3, a new analgesic agent. Folia Pharmacol Japan 68:442–459

Takayama C, Fujinami A (1979) Quantitative structure-activity relationships of antifungal *N*-phenylsuccinimides and *N*-phenyl-1,2-dimethylcyclopropanedicarboximides. Pestic Biochem Physiol 12:163–171

Theodoridis G (1997) Structure-activity relationships of herbicidal aryltriazolinones. Pestic Sci 50:283–290

Ueda K, Mizutani T, Yoshioka H, Fujimoto K, Okuno Y (1965) Jpn. Kokai Tokkyo Koho JA 65-46032

Uraguchi R, Sato Y, Nakayama A, Sukekawa M, Iwataki I, Böger P, Wakabayashi K (1997) Molecular shape similarity of cyclic imides and protoporphyrinogen IX. J Pesticide Sci 22:314–320

Verloop A (1983) The sterimol approach: further development of the method and new applications. In: Miyamoto J, Kearney PC (eds) Pesticide chemistry, human welfare and environment vol 1. Pergamon Press, Oxford, pp 339–344

Verloop A, Hoogenstraaten W, Tipker J (1976) Development and application of new steric substituent parameters in drug design. In: Ariëns EJ (ed) Drug design vol 7. Academic Press, New York, pp 165–207

Wakabayashi K, Matsuya K, Jikihara T, Suzuki S, Matsunaka S (1978) Mechanism of action of cyclic imide-type herbicides. Abstr Vol 4th Int Congr Pestic Chem. Zürich IV:655

Wakabayashi K, Matsuya K, Ohta H, Jikihara T (1979) Structure-activity relationship of cyclic imide herbicides. In: Geissbühler H (ed) Advances in pesticide science part 2. Pergamon Press, Oxford, pp 256–260

Wittle AJ, Fitzjohn S, Mullier G, Pearson DPJ, Perrior TR, Taylor R, Salomon R (1995) The use of computer-generated electrostatic surface map for design of new "GABA-ergic" insecticides. Pestic Sci 44:29–31

Yamagami C, Takao N, Fujita T (1990) Hydrophobicity parameter of diazines (1). Quant Struct Act Relat 9:313–320

CHAPTER 5

Structure-Activity Relationships of Diphenyl Ethers and Other Oxygen-Bridged Protoporphyrinogen Oxidase Inhibitors

Franck E. Dayan[1], Krishna N. Reddy[2], and Stephen O. Duke[1]

Contents

5.1 Introduction

The first commercial inhibitors of protoporphyrinogen oxidase (Protox) were diphenyl ether (DPE) molecules (Matsunaka 1976; see Chapter 1). Nitrofen was introduced in the early 1960s and was followed by several other, more potent Protox inhibitors (Dayan and Duke 1996, 1997). The molecular site of action of this class of herbicides was unknown until 1989 (Matringe et al. 1989). Thus, all quantitative structure-activity relationship (QSAR) studies conducted before this time were done solely on the correlation of structure with biological activity in intact tissues (e.g., Lambert et al. 1983). In some of these

[1] USDA-ARS, NPURU, National Center for the Development of Natural Products, University of Mississippi, P. O. Box 8048, Mississippi 38677, USA
[2] USDA-ARS, Southern Weed Science Laboratory, P. O. Box 350, Stoneville, Mississippi 38776, USA

Peter Böger, Ko Wakabayashi
Peroxidizing Herbicides

studies, the "activity" portion of the QSAR study may not have been completely or even at all associated with Protox inhibition, in that diphenyl ether compounds have several other potential secondary sites of action (Kunert et al. 1987). In 1992, we published the first QSAR study of DPE herbicides in which inhibition of Protox was correlated with molecular parameters (Nandihalli et al. 1992). This earlier work and subsequent published work on the QSAR of Protox inhibitors have largely been performed at a two dimensional level (Nandihalli and Duke 1994; Reddy et al. 1995). This chapter reviews this previous work and extends QSAR studies with DPEs to three-dimensional descriptors that more precisely describe the requisites for activity of Protox inhibitors at the molecular level. The scope of this chapter will be limited to the Protox-inhibiting herbicides that are bicyclic molecules connected by an oxygen bridge, although several second generation Protox inhibitors have been developed on non-bridged bicyclic structures (e.g., phenyl triazolinones, isoxazole carboxamides and phenyl imides). The QSAR analysis of several classes of first and second generation Protox inhibitors has recently been reviewed by Reddy et al. (1997).

5.2 Diphenyl Ether Protoporphyrinogen Oxidase Inhibitors

Nitrofen is a very weak inhibitor of Protox (Nandihalli et al. 1992), and it is likely to have an additional mode of action (Kunert et al. 1987). Nonetheless, many *p*-nitro diphenyl ethers (e.g., acifluorfen, oxyfluorfen) are potent competitive inhibitors of Protox (Nandihalli et al. 1992), and nearly a dozen of them are or have been commercialized (Anderson et al. 1994). Commercial DPE herbicides, along with some other experimental examples, sorted in order of increasing potency are shown in Table 1. These herbicides cause rapid photobleaching and light-dependent desiccation of foliage. The gross effects observed on the foliage of DPE-treated plants include leaf cupping, crinkling, and bronzing (Johnson et al. 1978). The first detectable damage is cellular leakage, followed sequentially by loss of photosynthetic capacity, evolution of ethylene, ethane and malondialdehyde, and finally bleaching of chloroplast pigments – all characteristics of photodynamic membrane lipid peroxidation (Kenyon et al. 1985).

5.2.1 Substituent Effects on Phenyl Ring A

In the absence of any substitution on the phenyl ring A (**21**, **23**), the inhibitory potency is very low (Table 2). Adding a CH_3 group to the 3-carbon (R_2) (**20**) increases the herbicidal activity by 3.5-fold. Adding electron-withdrawing substituents, e.g., two Cl atoms (**11**) or a Cl and a CF_3 (**12**), to the 2-carbon position (R_1) and 4-carbon position (R_3) of the ring enhances the inhibition of Protox by

Table 1. Chemical structures and biological activities of DPE herbicides.

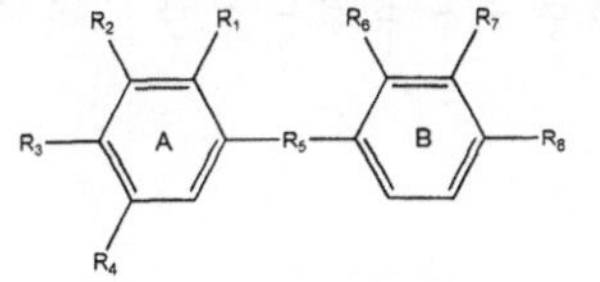

Compounds 1–24

Compounds	Herbicide	R_1	R_2	R_3	R_4	R_5	R_6	R_7	R_8
1	PPG-1055	Cl	H	CF_3	H	O	H	$C(CH_3){=}NOCH_2COOH$	NO_2
2	PPG-1013	Cl	H	CF_3	H	O	H	$C(CH_3){=}NOCH_2COOCH_3$	NO_2
3	MT-124	Cl	H	CF_3	H	O	H	—O—Furan	NO_2
4	Oxyfluorfen	Cl	H	CF_3	H	O	H	OC_2H_5	NO_2
5	Benzofluorfen	Cl	H	CF_3	H	O	H	$COOCH_2\ COOH$	NO_2
6	RH-4638 (*R*)	Cl	H	CF_3	H	O	H	$OCH(CH_3)COOC_2H_5$	NO_2
7	Lactofen	Cl	H	CF_3	H	O	H	$COOCH(CH_3)COOC_2H_5$	NO_2
8	Acifluorfen-Me	Cl	H	CF_3	H	O	H	$COOCH_3$	NO_2
9	MC-15608	Cl	H	CF_3	H	O	H	$COOCH_3$	Cl
10	RH-1460	Cl	H	CF_3	H	S	H	$COOCH_3$	NO_2
11	Nitrofen	Cl	H	Cl	H	O	H	H	NO_2
12	Nitrofluorfen	Cl	H	CF_3	H	O	H	H	NO_2
13	Bifenox	Cl	H	Cl	H	O	H	$COOCH_3$	NO_2
14	Fluorodifen	NO_2	H	CF_3	H	O	H	H	NO_2
15	Fomesafen	Cl	H	CF_3	H	O	H	$CONHSO_2CH_3$	NO_2
16	RH-4639 (*S*)	Cl	H	CF_3	H	O	H	$OCH(CH_3)COOC_2H_5$	NO_2
17	RH-5348	Cl	H	H	CF_3	O	H	$COOCH_3$	NO_2
18	Acifluorfen	Cl	H	CF_3	H	O	H	COOH	NO_2
19	RH-8827	Cl	H	CF_3	H	SO	H	$COOCH_3$	NO_2
20	RH-8327	H	CH_3	H	H	O	H	H	NO_2
21	Aclonifen	H	H	H	H	O	Cl	NH_2	NO_2
22	RH-8826	Cl	H	CF_3	H	SO_2	H	$COOCH_3$	NO_2
23	RH-0211	H	H	H	H	O	H	H	NO_2
24	RH-5349	Cl	H	H	CF_3	O	H	COOH	NO_2

about 12000-fold. When the CF_3 of acifluorfen-methyl (**8**) is moved from the *para* to the *meta* carbon as in **17**, the enzyme inhibiting capacity is reduced by 100-fold. A similar trend is observed between the free acid form of acifluorfen (**18**) and **24** (Table 2).

5.2.2 Substituent Effects on Phenyl Ring B

Among the 24 compounds presented in Table 1, there are 12 structures (**1–8, 12, 15, 16, 18**) which differ solely in their substitution at *meta* carbon, R_7. These compounds have exhibited moderate to high Protox inhibition and herbicidal activities. Protox inhibition generally increases with higher lipophilicity, and the herbicidal activity increases with higher surface area and molecular electrostatic potential (MEP) of the substituents at the *meta* carbon. Substitution

Table 2. Observed and predicted Protox pI_{50} activities of the DPE herbicides from Table 1

Compounds	Herbicide	Observed	Multilinear regression		Partial least square regression			
			Nitrophenyl aligned		Nitrophenyl aligned		Trifluoromethylphenyl aligned	
			Predicted	Residual	Predicted	Residual	Predicted	Residual
1	PPG-1055	8.097	6.967	−1.130	7.220	−0.877	7.338	−0.759
2	PPG-1013	7.921	7.721	−0.200	8.039	0.118	7.699	−0.222
3	MT-124	7.824	6.848	−0.976	7.275	−0.549	7.022	−0.802
4	Oxyfluorfen	7.602	7.921	0.319	7.264	−0.338	6.954	−0.648
5	Benzofluorfen	7.523	7.658	0.135	6.888	−0.635	7.808	0.285
6	RH-4638	7.456	7.444	−0.012	6.174	−1.282	6.317	−1.139
7	Lactofen	7.398	8.907	1.509	8.455	1.057	7.733	0.335
8	Acifluorfen-methyl	7.398	6.991	−0.407	7.048	−0.350	6.855	−0.543
9	MC-15608	7.398	7.886	0.488	9.451	2.053	6.383	1.015
10	RH-1460	7.347	7.244	−0.103	5.912	−1.435	6.316	−1.031
11	Nitrofen	7.301	6.738	−0.563	6.387	−0.914	6.867	−0.434
12	Nitrofluorfen	7.222	6.995	−0.267	7.664	0.442	7.824	0.621
13	Bifenox	7.222	6.863	−0.359	5.803	−1.419	5.922	−1.300
14	Fluorodifen	6.745	6.191	−0.554	5.076	−1.669	5.781	−0.964
15	Fomesafen	6.553	6.468	−0.085	5.689	−0.864	5.914	−0.639
16	RH-4639	5.456	-	-	7.489	2.033	7.375	1.919
17	RH-5348	5.409	5.310	−0.099	4.994	−0.415	6.31	0.901
18	Acifluorfen	5.398	5.399	0.001	6.424	1.026	6.407	1.009
19	RH-8827	4.553	5.770	1.217	2.818	−1.735	4.663	0.110
20	RH-8378	4.222	4.602	0.380	4.764	0.542	4.650	0.428
21	Aclonifen	3.979	4.347	0.368	4.236	0.257	4.032	0.053
22	RH-8826	3.939	4.469	0.530	5.768	1.829	5.298	1.359
23	RH-0211	3.658	3.696	0.038	4.438	0.780	4.179	0.521
24	RH-5349	3.377	3.472	0.095	4.522	1.145	3.285	−0.092
			$r^2 = 0.76$		$r^2 = 0.96$		$r^2 = 0.98$	
			q^2 = none		$q^2 = 0.567$		$q^2 = 0.70$	

with CCH_3NOCH_2COOH (**1**) increases Protox-inhibitory activity by 7.5-fold and substitution with $COOCH_3$ (**8**) increased Protox activity by 1.5-fold compared to no substitution (**12**). Similarly, Hayashi (1990) found that herbicidal activity is greatly affected by a broad range of substitutions at the *meta* position in a series of 22 compounds represented by 2-chloro-4-(trifluoromethyl)-3′-(substituted)-4′-nitrodiphenyl ethers. Furthermore, the herbicidal activity increased in the order of amide > thioester > ester substitution.

5.2.3
Effects of Ether, Sulfide, Sulfoxide, or Sulfone Bridges

The enzyme-inhibiting activity is similar whether the two phenyl rings were connected by an ether bridge (**18**) or a sulfide bridge (**10**), but the latter molecule is herbicidally inactive. This is probably due to rapid metabolism of the sulfide bridge relative to the diphenyl ether counterparts. The enzyme-inhibiting activity is greatly reduced at the molecular site of action and herbicidal activity is lost when diphenyls are connected via sulfone (**22**) or sulfoxide (**19**) bridges.

5.2.4
Two-Dimensional or Classical QSAR Analysis

Early QSAR analysis of these herbicides utilized molecular modeling techniques to examine the relationships between herbicidal activities and structural properties of 24 DPE analogs that included a pair of enantiomers (Nandihalli et al. 1992). This work led to development of several equations including Eq. 1 for 23 out of the 24 DPE herbicide analogs (Table 1):

$$\begin{aligned}\log \text{Protox } I_{50} &= 11.95 - 2.84(\log P) - 12.04(\text{partial charge on C5}) \\ &\quad - 1.82(S_{\text{LUMO}}) - 0.61(\mu) \\ \text{n} &= 23,\ F_{(4,18)} = 28.7,\ \text{s} = 0.65,\ \text{r}^2 = 0.86,\ \mathbf{16}\ \text{was an outlier,}\end{aligned} \qquad (1)$$

where Protox I_{50} is the μM concentration of the herbicide required to produce 50% Protox inhibition, $\log P$ is lipophilicity descriptor, C5 represents the carbon of the phenyl ring to which R_4 is attached (refer to figure in Table 1) and S_{LUMO} is superdelocalizability of lowest unoccupied molecular orbital, μ is dipole moment. Statistics: n is the number of compounds, F statistic is shown with degrees of freedom in subscript to test the null hypothesis, s is standard error, and r^2 is the coefficient of determination.

The S enantiomer (**16**) was an outlier and was excluded from the regression analyses. The S enantiomer (**16**) was 100-fold less active than the R form (**6**), although both had relatively similar "whole molecule" descriptor values and could not be distinguished using 2-D QSAR method. The use of more powerful molecular modeling techniques such as 3-D QSAR to elucidate steric and electrostatic differences between enantiomers is discussed in Section 5.2.5.

The Protox activity of DPE analogs was dependent on lipophilic and electronic properties [Eq. (1)]. The most potent compounds (I_{50} values $< 0.06\,\mu M$) had $\log P$ values greater than 4.26, whereas the $\log P$ values of the ten analogs whose I_{50} ranged from 0.18 to $420\,\mu M$ were mostly less than 4.0. $\log P$ (lipophilicity descriptor) is an indicator of the ability of a compound to move through the biomembranes in order to reach the active site. The fact that $\log P$ was one of the parameters in Eq. (1) suggests that the active site of Protox is embedded in a membrane. This is consistent with the fact that the Protox is a membrane-bound protein (Matringe et al. 1992; Jacobs and Jacobs 1993, Lee et al. 1993). There were also clear trends between I_{50} values and net charges at the 5-carbon (C5) of the ring A, S_{LUMO}, and dipole moment (μ). The compounds with very low activity (I_{50} values $>28.0\,\mu M$) had negative partial charges on the C5 atom, while the remaining structures in the set had mostly positive values. This difference in partial charge distribution was attributed to the fact that this atom had either no substitution or an electronegative CF_3 group. The calculated I_{50} values using Eq. (1) deviated somewhat from the observed I_{50} values (Table 2). This may be because (1) the descriptors accounted for only 86% of the variation in the Protox I_{50} data, (2) the observed I_{50} range was 52 500-fold larger, and (3) there was some colinearity among the descriptors ($r = -0.17$ to $+0.40$) (Nandihalli et al. 1992).

5.2.5 Three-Dimensional QSAR Analysis

Recent advances in computer modelling have led to the development of several methods enabling 3-D QSAR. In particular, workers at Tripos Inc. (St. Louis, Missouri) have developed a method called Comparative Molecular Field Analysis (CoMFA) (Cramer et al. 1988). The benefit of 3-D QSAR relative to classical 2-D QSAR is that the influence of relative position in space and orientations of substituents on a structure can be taken into account. As a result, such methods may better accommodate sets of compounds that include enantiomeric structures with differing biological activities that cannot be differentiated by 2-D techniques (Nandihalli et al. 1994).

5.2.5.1 *Methodology*

To begin 3-D QSAR, structures are built, generally using fragments present in a library, minimized and optimized. Structures are then aligned using a template (usually a substructure common to all the analogs being investigated), by pairing homologous atoms. CoMFA relies on the building of a lattice around the molecules. Though the resolution can be customized, we used the default parameters consisting of a grid with a width of 2-Å. Steric and electrostatic potentials are then determined by positioning a hypothetic probe along the lattice. The default probe consists of an sp3 carbon carrying a positive charge, but other probes such as H^+, NH_3^+ and O^- are also available. GRID, a software developed by Goodford (1985), can be used similarly.

Molecular mechanics and force fields (steric and electrostatic) have successfully been used in QSAR studies because interactions between herbicides and their receptor binding sites usually involve non-covalent binding. CoMFA was developed on the belief that biological activity was correlated with differences in intensity and relative spatial distribution of molecular fields. From empirical experiences gathered during the last two decades, steric and electrostatic potentials appear to be the most meaningful fields in QSAR studies. Hydrophobic interactions may also play an important role in QSAR. Hydrophobic "fields" can be included in CoMFA by using the HINT software developed by Kellogg et al. (1991).

The large number of highly correlated descriptors generated by CoMFA precludes the use of simple multivariate linear regression (MLR). There must be thousands of nominally significant linear regressions within a single CoMFA table purely by accident. The vast array of data found in the CoMFA table requires the use of partial least square (PLS) analysis. This statistical method is more computer demanding, but it yields a more robust answer. Unlike MLR, which maximizes commonality of a single parameter to biological activity (yields a prediction that best "reproduces" biological activity), PLS attempts to maximize commonality of an ensemble of descriptors to biological activity (yields an equation that best "predicts" biological activity). This is achieved by cross-validation (leave-one-out) by which a molecule is "removed" from the data set and the system attempts to predict the biological activity of the omitted molecule based on the other analogs. The analysis is repeated until each molecule in the data set has been omitted once.

This method allows the detection of parameters that most reliably predict activity from the entire CoMFA table. PLS analysis yields several r^2 values. The r^2 value evaluating the overall predictiveness of the analysis is called cross-validated r^2, and is referred to as q^2. Conventional r^2 is also provided as an indication of how much of the variation in the data set was accounted by the equation. Since it is easier to match the biological activity than actually predict it, it is not surprising that q^2 is always smaller than r^2.

We applied these more recent methods of analysis to the set of 24 DPEs described above (Table 1). All the analogs were built on the template of the X-ray crystal structure of acifluorfen obtained from the Cambridge Structure Database (Kennard et al. 1987). The orientation around the ether bond was set to a common starting point defined by an angle of 40.3 degrees between the nitrophenyl plane and the opposite O-phenyl bond and an angle of 57.2 degrees between the trifluoromethylphenyl plane and the opposite O-nitrophenyl bond (Fig. 1). Structural side chains of the analogs were constructed in trans configu-

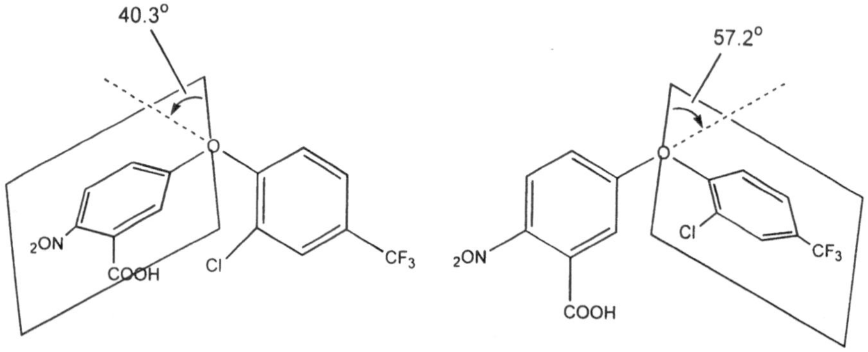

Fig. 1. Initial conformation of all the DPE analogs around the oxygen bridge before being submitted for minimization

ration. Finally, the molecules where minimized to 0.005 kcal/mol using MAXIMIN2, and Huckel-Gasteiger charges were calculated.

The structure of acifluorfen was diplayed using the "BEST VIEW" feature of Sybyl, and the analogs were then aligned on this template. Because alignment of the molecules is crucial to ensure maximum signal to noise ratio during the partial least square (PLS) analysis, we applied two alignment rules for CoMFA analyses. The first consisted of fitting three carbons from the trifluoromethylphenyl rings (Fig. 2A) and the second consisted of fitting three carbons from the nitrophenyl ring (Fig. 2B). PLS analyses were done on both sets of alignment with and without log*P* values of each of the compounds. Experimentally calculated log*P* (Nandihalli et al. 1992) was included using CoMFA standard scaling factors. We selected inhibition of Protox as our biological activity thereby minimizing the effect of uptake, translocation and metabolism. This allowed us to focus more precisely on the interaction between the ligands (herbicides) and the yet to be characterized receptor on Protox.

5.2.5.2
Results

For brevity, we limited our 3-D QSAR investigation to correlating structural characteristics with inhibition of Protox. Including log*P* greatly improved the predictiveness of the PLS analysis. The q^2 for the nitrophenyl and trifluoromethylphenyl aligned data sets was 0.367 and 0.454 without including log*P*, respectively, and 0.567 and 0.700 including log*P*. The improved q^2 obtained by including log*P* values indicates that whole molecule properties such as lipophilicity, an indicator of the ability of compounds to cross or partition into membrane, can be crucial.

With the assumptions that herbicide analogs interact with the same binding site and that differences in biological activity are attributable to differences in

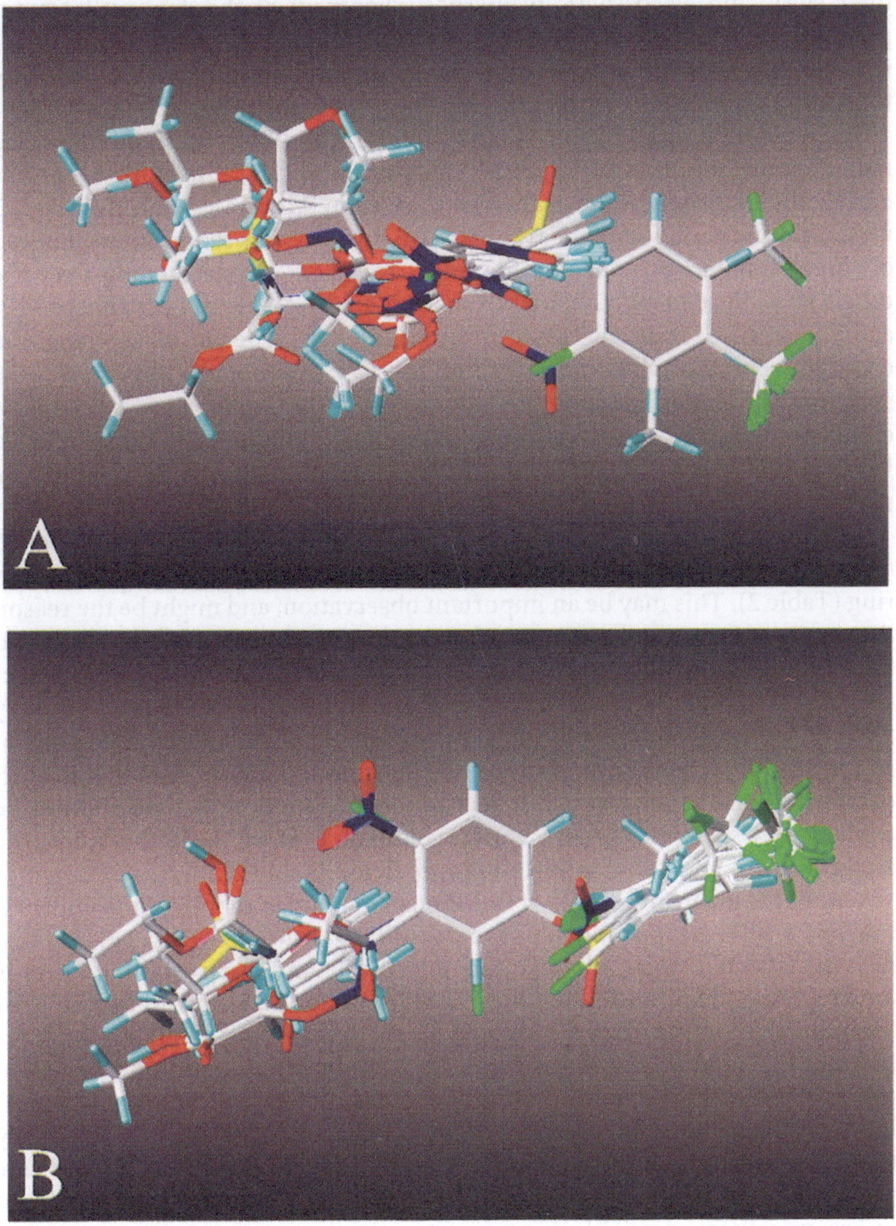

Fig. 2A,B. Spatial distribution of the DPE analogs and their respective side chains when aligned on the trifluoromethylphenyl ring (**A**) and nitrophenyl ring (**B**)

their steric and electrostatic properties, alignment of the data set becomes paramount. In the case of our set of diphenyl herbicides, we elected to create two alignment groups because little information is known about their binding to Protox, except that they all apparently compete for the same site.

It should be mentioned that results from MLR and PLS cannot be compared meaningfully in terms of relative r^2. For example, the r^2 of both equations derived by PLS is higher than the one derived by MLR (Table 2). However, if we look at the variation in the residuals, MLR provided less variations. Instead of direct comparison, one can identify advantages of one method over the other. First, the equation derived from MLR analysis does not actually predict activity. Instead, it establishes a relationship only in terms of how well the equation reproduces the data. Second, the two enantiomers appeared to be structurally identical at a 2-D level; thus, one of the enantiomers was considered an outlier, and was simply omitted from the statistical analysis (Table 2). PLS analysis of the CoMFA table actually indicates how well the data can be predicted.

The predictiveness of the biological activity was greater for the set aligned along the trifluoromethylphenyl ring than the set aligned with the nitrophenyl ring (Table 2). This may be an important observation, and might be the reason that many of the second generation oxygen-bridged Protox inhibitors retained the trifluoromethylphenyl group instead of the nitrophenyl ring (refer to Section 5.3). This suggests that the plane of the trifluoromethylphenyl ring may play an important role in binding to Protox. Another attractive feature of 3-D QSAR is its ability to compare enantiomeric pairs of inhibitors. Such compounds are similar at a 2-D level and differences in activity cannot be explained by such traditional means (Nandihalli et al. 1992). In this study, the enantiomers **6** and **16** were effectively predicted (Table 2).

One of the great advantages of CoMFA is its ability to represent the information in a 3-D format that highlights regions of the molecules where positive and negative effects associated with increased steric and electrostatic potentials are important. Figs. 3A,B illustrates CoMFA steric and electrostatic maps for Protox activity of the data set aligned on the trifluoromethylphenyl ring, respectively. The structure of acifluorfen is placed within the CoMFA results to assist in the visualization of the CoMFA maps. The steric contours (Fig. 3A) indicate areas where increased steric bulk is positively (green) or negatively (yellow) correlated with potency. The electrostatic contours (Fig. 3B) indicate areas where higher biological activity is correlated with greater positive (blue) or greater negative (red) electrostatic potentials.

Based on the analysis of the 24 DPE analogs available to us, the result of CoMFA indicates that increasing bulk around the carboxylic acid also increases activity (green region in molecular space beyond the carboxylic group of the template displayed in Fig. 3A). Increasing the electron density in the area of the trifluoromethyl side chain or the chloro substituents has a positive effect on activity (red areas in Fig. 3B). On the other hand, increasing the positive electrostatic potential in the area of the carboxylic acid increased activity (blue

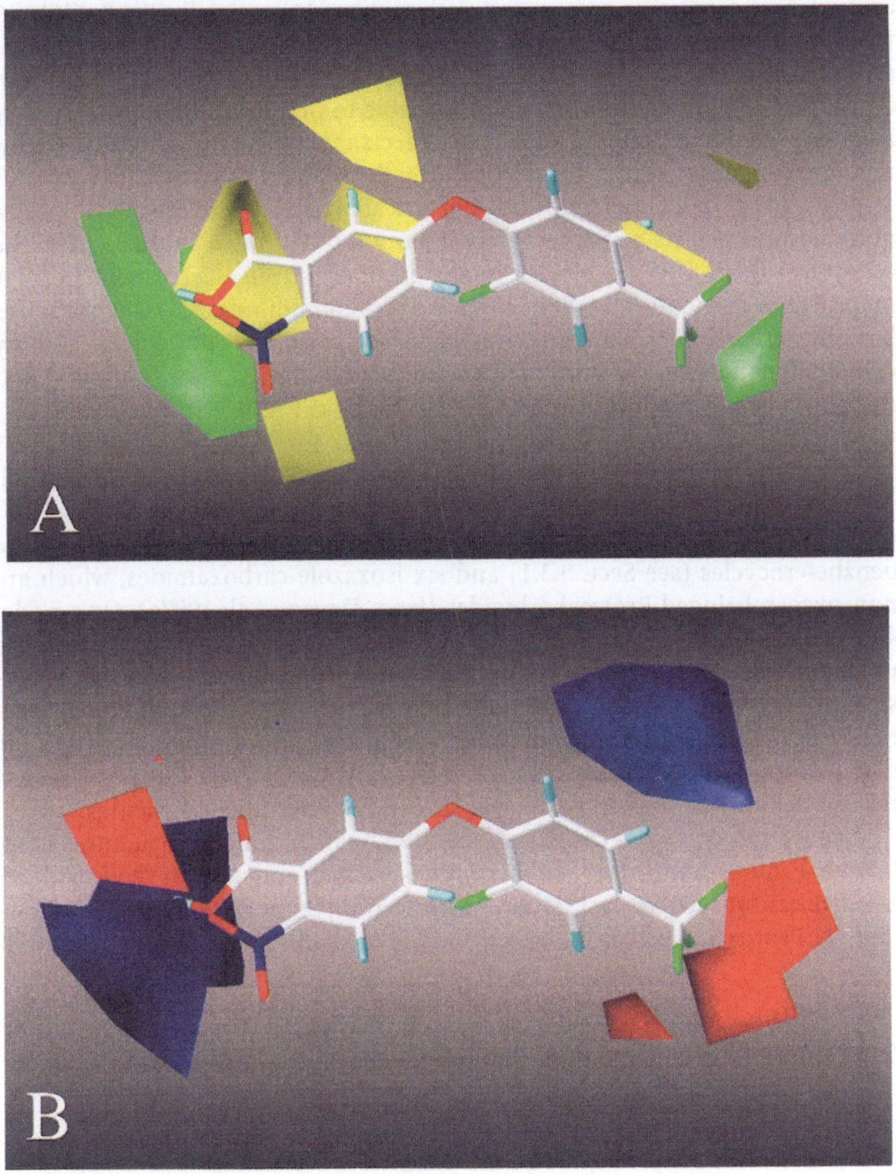

Fig. 3A,B. CoMFA maps generated by PLS analysis of the DPE analogs aligned on the trifluoromethylphenyl ring. Structure of acifluorfen is shown within the CoMFA maps as point of reference. **A** Increased biological activity is correlated with increased (*green*) or decreased (*yellow*) bulk on the molecule. **B** Increased biological activity is correlated with increased positive (*blue*) or negative (*red*) electrostatic charges on the molecule.

area around carboxylic acid in Fig. 3B). Unlike classical 2-D QSAR, CoMFA was able to differentiate the R and S enantiomers (**6** and **16**) in their spatial arrangement. Using this 3-D QSAR technique, **16** (S enantiomer) was not considered an outlier and the model was able to predict its biological activity relatively well. The steric CoMFA map reflects this by the presence of a yellow region in the molecular space occupied by the bulky S side chain.

Admittedly, one of the limitations of this analysis is the relatively small size of the sample set. As a result, the contribution of the nitro substituent to the activity of these DPE herbicides is not detected in the electrostatic CoMFA map. This is due to the fact that only one of the compounds (**9**) out of the series has other than a nitro substituent. None of the variation in biological activity was attributed to the presence of the nitro group. This in no way indicates that the nitro substituent is not important for biological activity.

Nonetheless, we attempted to validate our model by submitting other Protox inhibitors for which Protox I_{50} values were known. Using our model, cyperine, a natural DPE with low levels of Protox inhibition (pI_{50} = 5.0) was predicted to have a pI_{50} of 5.42. Table 3 contains the data of six DPE-type benzheterocycles (see Sect. 5.3.1) and six isoxazole carboxamides, which are non-oxygen bridged Protox herbicides (from Dayan et al. 1997a). Our model successfully approximated the activity of DPE-type Protox inhibitors, but was unable to predict the activity of the isoxazole carboxamides. While this analysis is an improvement on other models derived from 2-D QSAR (which could not be used to predict activity of other related compounds not included in the data set from which the models were derived), it is clear that the usefulness of this model may be limited to oxygen-bridged phenyl ether herbicides.

A more robust model could be generated by including the structures of all known Protox inhibitors – both oxygen-bridged and non-oxygen-bridged. Unfortunately, such analysis is not possible at this time because of the following: (1) Protox inhibition (I_{50} values) is not available for most of these com-

Table 3. Comparison of observed and predicted activities of six oxygen-bridged and six non-oxygen-bridged Protox inhibitors

	Oxygen-bridged phenyl ether				Isoxazole carboxamide (non-DPE)		
Id[a]	Observed	Predicted	Residual	Id[b]	Observed	Predicted	Residual
2	7.160	6.990	−0.17	1	8.22	5.71	−2.51
6	6.600	6.410	−0.19	2	8.890	5.540	−3.35
7	7.260	7.580	0.32	3	4.600	5.720	1.12
11	7.460	7.060	−0.4	4	6.260	5.660	−0.6
15	6.830	6.500	−0.33	5	8.550	3.060	−5.49
16	6.190	7.620	1.43	6	7.750	3.470	−4.28

[a] Id number is from numbering sytem used in Lee et al. (1995).
[b] Id number is from numbering system used in Dayan et al. (1997a).

pounds; (2) currently available biological activity data are not expressed in a standard form (e.g., the range in herbicide concentrations and the rating scales used); and (3) compounds have not been tested on the same weed species and therefore activities cannot be directly compared between analysis.

5.3 Other Oxygen-Bridged Protoporphyrinogen Oxidase Inhibitors

This chapter would not be complete without including at least a brief description of other oxygen-bridged compounds that appear to act by inhibiting Protox. It should be pointed out that the activity of many of these compounds has not been tested directly on Protox. Furthermore, direct comparison between the different classes is difficult because of the lack of standardized testing methods. Some activities are reported as ED_{50} or ED_{80}, and compounds were tested on various crop and weed species.

In heterocyclic phenyl ethers, one of the phenyl rings of ether is replaced by an aromatic heterocycle. Pyrazolyl, pyridyl, and furyl rings have been investigated as the heterocyclic component (Anderson et al. 1994; Clark 1994, 1996). While typical DPE such as acifluorfen have a *p*-nitro substituent on one of ring and a trifluoromethyl on the other ring, the new generation of Protox inhibitors tends to favor keeping the ring with the *p*-trifluoromethyl substitutions. Apparently, the trifluoromethylphenyl ring may have a more significant effect on activity than the nitrophenyl ring since better predictiveness was obtained by aligning on the trifluoromethylphenyl ring than on the other.

Several novel heterocyclic phenyl ethers have been reported recently by several groups. As a whole, these phenyl ethers have been synthesized by linking the X and Y substituents of a ring to form a heterocyclic ring. Heterocyclic phenyl ethers are highly active both pre- and postemergence on a wide variety of weeds. A few recently reported examples include 6-aryloxy-1*H*-benzotriazoles (Condon et al. 1995), aryloxyindolin-2-(3*H*)-ones (Karp et al. 1995), 5-aryloxybenzisoxazoles (Wepplo et al. 1995), 6-aryloxyquinoxalin-2,3-diones (Anderson et al. 1994), benzheterocycles (Lee et al. 1995), and benzoxazines (Sumida et al. 1995). Structure-activity relationships of benzheterocycles and benzoxazines are briefly discussed below (Sects. 5.3.1 and 5.3.2).

5.3.1 Benzheterocycles

Lee et al. (1995) described the herbicidal and Protox inhibitory activities of an array of 18 compounds belonging to the phenoxy-substituted benzheterocycle class (Fig. 4). Except for one totally inactive compound, Protox I_{50} values were within a relatively narrow range (25–650 nM). Several of the compounds were

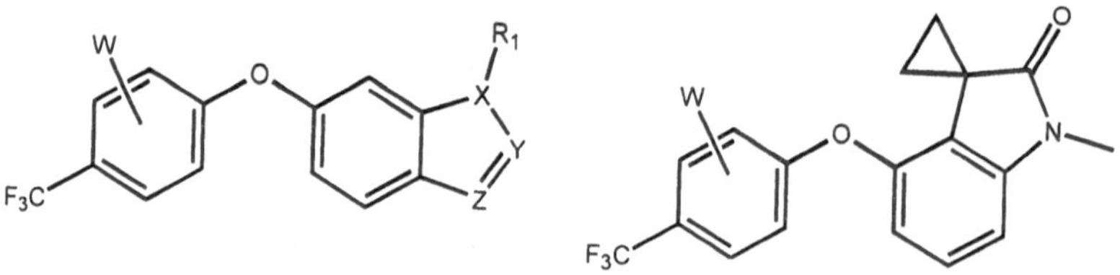

Fig. 4. Structural characteristics of benzheterocyclic analogs

quite phytotoxic in darkness, suggesting that some members of this chemical family may have another site of action. However, some generalizations on structure-activity relationships were suggested. For good Protox inhibition, chlorine and fluorine at the 2- and 6-positions, respectively, of the phenyl ring are required. The position of the phenoxy group on the indolin-2-(3*H*)-one is critical. Analogs that are 5- and 6-aryloxy are active, whereas 4-aryloxy analogs are inactive. Heterocyclic structures other than indole provided good activity.

5.3.2
Benzoxazines

Sumida et al. (1995) reported the peroxidizing herbicidal activity of 20 derivatives of 7-[2-chloro-4-(trifluoromethyl)phenoxy]-2-methyl-2*H*,4*H*-1,4-benzoxazin-3-one, a new chemical family of diphenyl ether herbicides (Fig. 5). Compounds with no substitution at the R_1 position on the nitrogen atom of the benzoxazine moiety were not phytotoxic. Methyl, ethyl, and propargyl substitutions on the nitrogen atom of the benzoxazine moiety exhibited herbicidal activity, but isopropyl substitution resulted in complete loss of herbicidal activity. Addition of fluorine to the 6-position (labeled X on Fig. 5) of the phenoxy ring greatly increased the herbicidal activity. Substituting the benzoxazine functionality with either benzothiazine or quinazoline groups decreased activity.

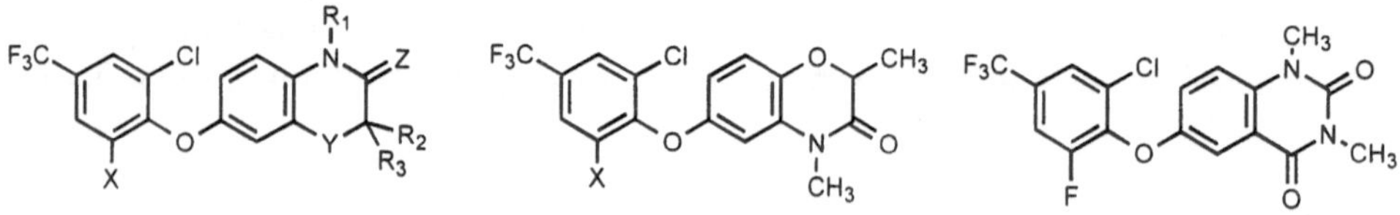

Fig. 5. Structural characteristics of benzoxazines

5.3.3
Benzisoxazoles

Welppo et al. (1995) reported that benzisoxazole phenyl ethers (Fig. 6) were herbicidally active. Unlike typical DPE, these compounds have a high level of

Fig. 6. Structural characteristics of benzisoxazoles

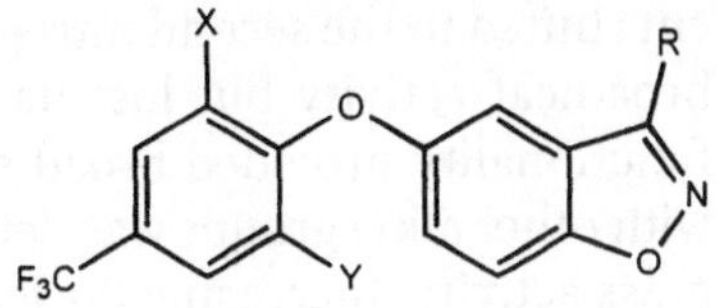

both preemergence and postemergence activity. Benzisoxazole phenyl ethers are very phytotoxic compounds that can control certain weeds at rates as low as 1 g/ha. Unfortunately, this chemical group has shown very little crop selectivity, although they appear to be more active on broadleaf than grass weeds.

As is customary for DPE's, the most active substitution corresponded to a 2-Cl, 6-F pattern on the phenoxy ring, and replacing one or both of these halogens with nitro groups completely deactivated the structure. Alkyl ether substitution (R) that did not contain carboxylate functionalities on the benzisoxazole ring (i.e., —O—CH(Cl)CH_3 or —O—CH_2CH=CH_2) were not active. Alkyl substitution directly on the benzisoxazole ring did not affect activity as much as long as the groups contained alkyl or alkene ester functionalities.

5.3.4 Aryloxybenzotriazoles

Condon et al. (1995) have reported the SAR for 31 aryloxybenzotriazoles (Fig. 7). The basic structure (as shown in Fig. 7) with an oxygen bridge connected to C6 controlled pigweed, ragweed, velvetleaf, morningglory, barnyardgrass, green foxtail, crabgrass, and proso millet at rates as low as 2 g/ha. Shifting the oxygen bridge to C5 resulted in loss of activity on the tested weeds. The backbone structure was based on the trifluoromethyl substituted phenoxy ring (Y). Other substituents tested on Y, such as NO_2 and CH_3CO, were not as active. The most active substitutions on the aryloxy ring were based on the 2-F, 6-Cl pattern that is common to most of the second generation Protox inhibitors. Replacing fluoro with chloro in position 2 decreased grass activity. The presence of the trifluoromethyl group in *para* position seems to be a recurring pattern too.

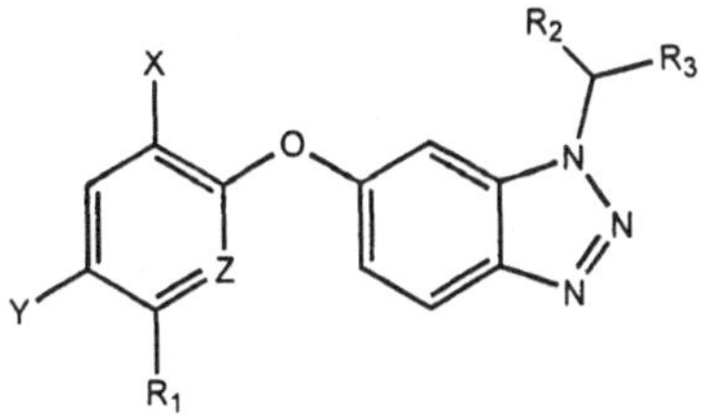

Fig. 7. Structural characteristics of aryloxybenzotriazoles

Regiospecificity analysis determined that substitutions on N1 were more active than on the other nitrogen, although a regioisomer with the N substituent shifted to the second nitrogen of the benzotriazole ring retained most of its broadleaf activity but lost its grass weed activity. The presence of an ester functionality provided broad spectrum weed control. Replacing ester groups with other alkyl groups was detrimental to the overall activity, and particularly grass activity. Increasing the length of the ester chain also had a negative effect on the activity of the benzotriazoles. Replacing the ester functionality with groups of similar or lower oxidation states (e.g., aldehyde, hydroxyl, amide) maintained good broadleaf activity but resulted in an overall decrease in grass activity. The presence of three nitrogens (as shown in Fig. 7) provided most activity, and analogs with either two or one nitrogen were incrementally less active. The diazole derivative lost activity against grasses, while retaining activity against broadleaf weeds, while the azole derivative controlled pigweed but not the other broadleaf weeds tested.

5.3.5
Aryloxyindolinones

As with most diphenyl ether-based herbicides, aryloxyindolinone derivatives (Fig. 8) have higher postemergence activity than preemergence activity and are more phytotoxic to dicotyledonous than to monocotyledonous plants (Karp et al. 1995). However, one of these compounds was selective toward rice while providing excellent control of barnyardgrass and flatsedge. Substitutions on both the phenyl and indolinone rings as well as several regioisomeric forms of these compounds were investigated.

Fluoro substitution (X) on the phenyl ring yielded compounds that were more active than the non-fluorinated analogs. The pattern of substitution of the most active analogs was similar to that of several other DPE-derived compounds presented in this chapter. Fluoro and chloro halogen substitutions in *ortho* positions accompanied by trifluoromethyl group in *para* position of the phenyl ring were the most active. The presence of relatively small hydrophobic groups like methyl and ethyl on nitrogen increased the activity of the compounds relative to the N—H derivatives. The presence of small side chains on the C3 carbon also increased the activity. Comparison of spirocyclic side chain size on that position confirmed that larger hydrophobic groups have a negative effect on activity. The smallest spirocyclic chain (two carbons) was the most active. The remaining compounds consisted of N substitutions based on the smallest spirocyclic derivative. Smaller non-polar groups were generally more

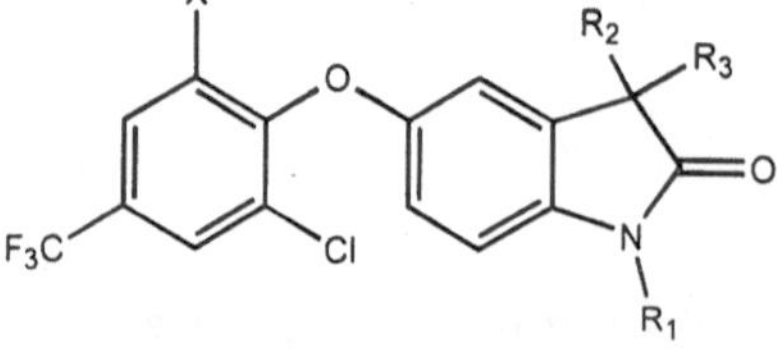

Fig. 8. Structural characteristics of aryloxyindolinones

active than the analogs with larger hydrophobic groups. Activity was affected by regioisomeric substitutions. 5-Aryloxy indolinones were more active than the 6-regioisomer. The 4-regioisomer did not have any activity.

5.3.6 Pyrazoles

Several series of pyrazolyl phenyl ethers including *p*-nitro (Rogers and Moedritzer 1990; Moedritzer et al. 1992) and *p*-trifluoromethyl derivatives (Clark 1996) have been described. Pyrazole phenyl ethers were first reported to inhibit Protox by Sherman et al. (1991). Two sets of isomers tested in this study differed in activity. These isomers were designed on the *N*-substituted (R_1) phenoxy backbone. Carboxylated derivatives on R_2 were much less active than the analogs lacking a carboxylate group.

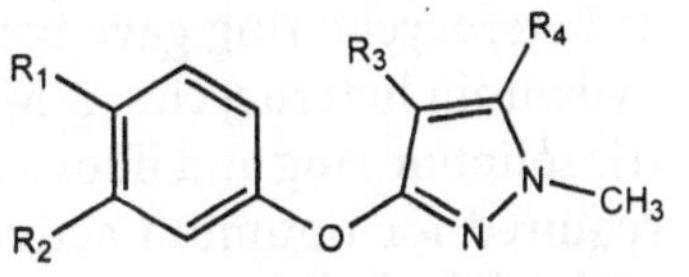

Fig. 9. Structural characteristics of pyrazoles

Clark (1996) recently reported a QSAR study of 50 pyrazole phenyl ethers (Fig. 9). These compounds were derived from the apparently preferred trifluoromethyl-substituted (R_1) phenyl backbone. The most active pyrazoles were the 5-sulfonyl (R_4), 3′amido (R_2) analogs. Replacing the anilide hydrogen by a methyl group did not affect activity. In contrast to the earlier report of Sherman et al. (1991), the carboxyl derivatives at position R_3 were more active than the methoxy analogs. A major difference was that the carboxylate groups were ethylated in this study whereas they were free acids in the previous study. Replacing the trifluoromethyl group with the more polar trifluoromethoxy moeity yielded derivatives that were not active.

5.3.7 Phenoxyphenoxy Triazolinones

Phenoxyphenoxy triazolinones (Fig. 10) are included in this chapter because they can be seen as the fusion of two successful chemical groups of Protox inhibitors. These molecules are derived from the phenyl triazolinone class of Protox inhibitors which include the commercially available sulfentrazone and carfentrazone herbicides (Dayan et al. 1997b; Dayan et al. 1997c). This backbone does not include an ether bridge between the two rings. However, phenoxyphenoxy triazolinones have been modified by adding a phenoxy side chain and have some of the characteristics of "classical" DPE.

These highly active (at 7–15 g/ha) polycyclic compounds apparently mimic the three ring-propionate portion of Protogen (Theodoridis et al. 1995). These compounds have a characteristic heterocyclic ring in addition to two substi-

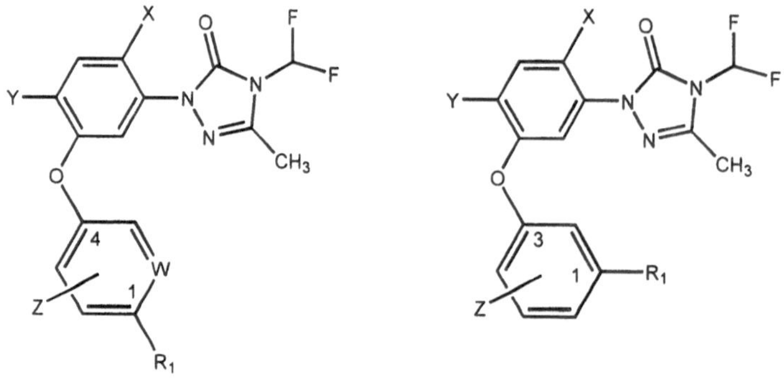

Fig. 10. Structural characteristics of phenoxyphenoxy triazolinones

tuted aryl rings with an oxygen bridge as in diphenyl ethers. Triazolin-5-one as heterocyclic ring gave better activity than either tetrahydrophthalimide or hydantoin heterocycles. 3-Methyl and 4-N-difluoromethyl substitution on the triazolinone ring and fluoro and chloro substitution on both phenyl rings were required for optimum activity. Activity was greatly reduced if either one or both of the halide groups were removed. Presence of an ester side chain (R_1) on the phenyl ring farther from the heterocycle improved herbicidal activity. Adding a hydrogen, amino, or nitro group in this ring had a negative effect. Several acid derivatives of the ester side chain were investigated. The best control of both broadleaf and grass weeds was obtained with a side chain of oxypropionate ester derivative. Triazolinone phenoxyphenoxy oxypropionate derivatives have good crop selectivity for rice.

5.4 Conclusions and Projections

We have surveyed the existing literature on the SAR and QSAR of diphenyl ether and closely related Protox inhibitors. QSAR modeling of Protox inhibitors has become more refined and predictive as molecular modeling and statistical software have advanced, along with more powerful and faster computers. Steric, electronic, chemical, and physicochemical parameters can be quickly calculated and correlated accurately with biological activities. In this chapter, we have applied these more robust methods of QSAR analysis to a set of 24 DPE Protox inhibitors. Activity of suspected Protox inhibitors can be predicted with some confidence with these methods, even differentiating between enantiomers that differ significantly as inhibitors due to steric dissimilarity.

Such capabilities may have been used in the design of new Protox inhibitors. Because of private sector priorities, the published work on QSAR studies of Protox inhibitors most likely represent only a small fraction of the research in this area. Since good inhibitor activity at the molecular target site does not generally translate directly into good herbicidal activity at the whole plant level, much of the industrial QSAR work is based on relationships to activity at the whole plant level. Nonetheless, in the case of Protox inhibitors, there have generally been acceptable correlations (e.g., Nandihalli et al. 1992; Reddy et al. 1995). However, finding more potent Protox inhibitors has not been a major objective of the herbicide industry, as many compounds with strong activity are already known. Discovery of potent compounds with good crop selectivity to major crops other than soybeans and rice has been the greatest challenge. Applying the QSAR approach to this problem, which is not likely to be resolved in terms of selectivity of Protox itself (Matsumoto et al. 1994), may be useful.

The available QSAR information about Protox inhibitors should provide some clues to the chemical nature of their precise intramolecular binding site(s). The plant plastid Protox gene has reportedly been cloned and sequenced (Ward and Volrath 1995), but the enzyme's secondary and tertiary structures, and details of the catalytic site remain unknown. Clearly, the next step in QSAR analysis of Protox inhibitors will involve more detailed modeling of the interactions between Protox and its inhibitors at the binding site(s).

References

Anderson RJ, Norris AE, Hess FD (1994) Synthetic organic chemicals that act through the porphyrin pathway. In: Duke SO, Rebeiz CA Porphyric pesticides: Chemistry, toxicology, and pharmaceutical applications. Am Chem Soc Symp Ser 559:18–33

Clark RD (1994) Synthesis of protoporphyrinogen oxidase inhibitors. In: Duke SO, Rebeiz CA Porphyric pesticides: Chemistry, toxicology, and pharmaceutical applications. Am Chem Soc Symp Ser 559:34–47

Clark RD (1996) Synthesis and QSAR of herbicidal 3-pyrazolyl α,α,α-trifluorotolyl ethers. J Agric Food Chem 44:3643–3652

Condon ME, Alvarado SI, Arthen FJ, Birk JH, Brady TE, Crews AD Jr., Marc PA, Karp GM, Lavanish JM, Nielsen DR, Lies TA (1995) 6-Aryloxy-1*H*-benzotriazoles: synthesis and herbicidal activity. In: Baker DR, Fenyes JG, Basarab, GS Synthesis and chemistry of agrochemicals IV. Am Chem Soc Symp Ser 584:122–135

Cramer RD III, Patterson DE, Bunce JD (1988) Comparative molecular field analysis (CoMFA). Effect of shape of binding of steroids to carrier proteins, J Am Chem Soc 110:5959–5967

Dayan FE, Duke SO (1996) Porphyrin-generating herbicides, Pestic Outlook 7:22–27

Dayan FE and Duke SO (1997) Phytotoxicity of protoporphyrinogen oxidase inhibitors: Phenomenology, mode of action and mechanisms of resistance. In: Roe RM, Burton JD, Kuhr RJ Herbicides activity: Toxicology, biochemistry and molecular biology I S O Press, Amsterdam, pp 11–35

Dayan FE, Duke SO, Reddy KN, Hamper BC, Leschinsky KL (1997a) Effects of isoxazole herbicides on protoporphyrinogen oxidase and porphyrin physiology. J Agric Food Chem 45:967–975

Dayan FE, Weete JD, Duke SO, Hancock HG (1997b) Soybean (*Glycine max*) cultivar differences in response to sulfentrazone. Weed Sci 45:634–641

Dayan FE, Duke SO, Hancock HG, Weete JD (1997c) Selectivity and mode of action of carfentrazone-ethyl, a novel phenyl triazolinone herbicide. Pestic Sci 51:65–73

Goodford PJ (1985) A computational procedure for determining energetically favorable binding sites on biologically important macromolecules. J Med Chem 28:849–857

Hayashi Y (1990) Synthesis and selective herbicidal activity of methyl (E,Z)-[[[1[5-[2-chloro-4-(trifluoromethyl)phenoxy]-2-nitrophenyl]-2-methoxyethylidene]amino]oxy]acetate and analogous compounds. J Agric Food Chem 38:839–844

Jacobs JM and Jacobs NJ (1993) Porphyrin accumulation and export by isolated barley (*Hordeum vulgare* L.) plastids: Effects of diphenyl ether herbicides. Plant Physiol 101:1181–1188

Johnson WO, Kollman GE, Swithenbank C, Yih RY (1978) RH-6201 (Blazer): A new broad spectrum herbicide for postemergence use in soybean. J Agric Food Chem 26:285–286

Karp GM, Condon ME, Arthen FJ, Birk JH, Marc PA, Hunt DA, Lavanish JM, Schwindeman JA (1995) Aryloxyindolin-2(3*H*)-ones: synthesis and herbicidal activity. In: Baker DR, Fenyes JG, Basarab GS Synthesis and chemistry of agrochemicals IV. Am Chem Soc Symp Ser 584:136–148

Kellogg GE, Semus SF, Abraham DJ (1991) HINT: A new method of empirical hydrophobic field calculation for CoMFA. J Comput-Aided Mol Design 5:2723–2729

Kennard CHL, Smith G, Hari T (1987) The crystal of acifluorfen {5-[2-chloro-4-(trifluoromethyl)-phenoxy]-2-nitrobenzoic acid}. Austr J Chem 40:1131–1135

Kenyon WH, Duke SO, Vaughn KC (1985) Sequences of effects of acifluorfen on physiological and ultrastructural parameters in cucumber cotyledon discs. Pestic Biochem Physiol 24:240–250

Kunert KJ, Sandmann G, Böger P (1987) Modes of action of diphenyl ethers. Rev Weed Sci 3:35–55

Lambert RG, Sandmann G, Böger P (1983) Correlation between structure and phytotoxic activities of nitrodiphenyl ethers. Pestic Biochem Physiol 19:309–320

Lee HJ, Duke MV, Birk JH, Yamamoto M, Duke SO (1995) Biochemical and physiological effects of benzheterocycles and related compounds. J Agric Food Chem 43:2722–2727

Lee HJ, Duke MV, Duke SO (1993) Cellular localization of protoporphyrinogen-oxidizing activities of etiolated barley (*Hordeum vulgare* L.) leaves: Relationship to mechanism of action of protoporphyrinogen oxidase-inhibiting herbicides. Plant Physiol 102:881–889

Matringe M, Camadro J-M, Block M, Joyard J, Scalla R, Labbe P, Douce R (1992) Localization within chloroplasts of protoporphyrinogen oxidase, the target enzyme for diphenylether-like herbicides. J Biol Chem 267:4646–4651

Matringe M, Camadro J-M, Labbe P, Scalla R (1989) Protoporphyrinogen oxidases as molecular target for diphenyl ether herbicides. Biochem J 260:231–235

Matsumoto H, Lee JJ, Ishizuka K (1994) Variation in crop response to protoporphyrinogen oxidase inhibitors. Am Chem Soc Symp Ser 559:120–132

Matsunaka S (1976) Diphenyl ethers. In: Kearney PC, Kaufman DD Herbicides: Chemistry, degradation, and mode of action, vol 2, Dekker, New York, pp 709–739

Moedritzer K, Allgood SG, Charumilind P, Clark RD, Gaede BJ, Kurtzweil ML, Mischke DA, Parlow J, Rogers MD, Singh RK, Sikes GL, Weber RK (1992) Novel pyrazole phenyl ether herbicides. In: Baker DR, Fenyes JG, Steffens JJ Synthesis and chemistry of agrochemicals III. Am Chem Soc Symp Ser 504:147–160

Nandihalli UB and Duke SO (1994) Structure-activity relationships of protoporphyrinogen oxidase inhibiting herbicides. In: Duke SO, Rebeiz CA Porphyric pesticides: Chemistry, toxicology, and pharmaceutical applications. Am Chem Soc Symp Ser 559:133–146

Nandihalli UB, Duke MV, Duke SO (1992) Quantitative structure-activity relationships of protoporphyrinogen oxidase-inhibiting diphenyl ether herbicides. Pestic Biochem Physiol 43:193–211

Nandihalli UB, Duke MV, Ashmore JW, Musco VA, Clark RD, Duke SO (1994) Enantioselectivity of protoporphyrinogen oxidase-inhibiting herbicides. Pestic Sci 40:265–277

Reddy KN, Nandihalli UB, Lee HJ, Duke MV, Duke SO (1995) Predicting activity of protoporphyrinogen oxidase inhibitors by computer-aided molecular modeling. In: Reynolds CH, Holloway MK, Cox HK Computer-aided molecular design: Applications in agrochemicals, materials, and pharmaceuticals. Am Chem Soc Symp Ser 589:221–224

Reddy KN, Dayan FE, Duke SO (1997) QSAR analysis of protoporphyrinogen oxidase inhibitors. In: Devillers J Comparative QSARs, Taylor and Francis, Washington DC, pp 197–233

Rogers MD, Moedritzer K (1990) Substituted 4-(4-nitrophenoxy)pyrazoles and their use as herbicides Monsanto Co., US Patent 4,964,895

Sherman TD, Duke MV, Clark RD, Sanders EF, Matsumoto H, Duke SO (1991) Pyrazole phenyl ether herbicides inhibit protoporphyrinogen oxidase. Pestic Biochem Physiol 40:236–245

Sumida M, Niwata S, Fukami H, Tanaka T, Wakabayashi K, Böger P (1995) Synthesis of novel diphenyl ether herbicides. J Agric Food Chem 43:1929–1934

Theodoridis G, Poss KM, Hotzman FW (1995) Herbicidal 1-(2,4-dihalo-5-phenoxyphenyl)-4-difluoromethyl-4,5-dihydro-3-methyl-1,2,4-triazolin-5(1*H*)-one derivatives: synthesis and structure-activity relationships. In: Baker DR, Fenyes JG, Basarab GS Synthesis and chemistry of agrochemicals IV. Am Chem Soc Symp Ser 584:78–89

Ward ER, Volrath S (1995) Manipulation of protoporphyrinogen oxidase enzyme activity in eukaryotic organisms. International patent application WO 95/34659

Wepplo P, Birk JH, Lavanish JM, Manfredi M, Nielsen DR (1995) 5-Aryloxybenzisoxazole esters: synthesis and herbicidal activity. In: Baker DR, Fenyes JG, Basarab GS Synthesis and chemistry of agrochemicals IV. Am Chem Soc Symp Ser 584:149–160

CHAPTER 6

General Physiological Characteristics and Mode of Action of Peroxidizing Herbicides

Ko Wakabayashi[1] and Peter Böger[2]

Contents

6.1 Introduction

Scientists all over the world have been engaged in elucidating the mode of action of herbicides for years, and the herbicide target domains can be summarized as follows: deregulation of auxin-induced cell growth, interference with the microtubular cell-division system, the electron-transport system, inhibition of cellulose biosynthesis, amino acid synthesis, and destruction of structure and function of the plant-specific apparatus, the chloroplast. Although the modes of action are not yet fully understood even for many commercially available herbicides, most of the compounds introduced during the last 40 years act upon the chloroplast. More than 60% of modern herbicides interfere with its structure and function. Five major herbicidal plastidic targets have been established, as indicated in Table 1.

Many herbicides with biocidal side effects are being phased out, due to toxicological problems and environmental impacts. Such side activities have often been caused by high use rates of conventional herbicides. Even with

[1] Chair of Physiology and Biochemistry, Faculty of Agriculture, 6-1-1 Tamagawa University, Machida-shi, Tokyo 194, Japan
[2] Lehrstuhl für Physiologie und Biochemie der Pflanzen, Universität Konstanz, D-78457 Konstanz, Germany

Table 1. Herbicides acting upon the chloroplast

	Examples
Photosynthetic electron-transport system	
Electron-transport inhibitors	Atrazine, diuron, ioxynil, metribuzin, etc.
Energy-transfer inhibitors	Dinoseb, pentachlorophenol, some diphenyl ethers, etc.
Superoxide-forming herbicides	Diquat, paraquat, etc.
Photosynthetic membrane systems	
Peroxidizers inhibiting protoporphyrinogen-IX oxidase	Chlorophthalim, oxyfluorfen, etc.
Biosynthesis of photosynthetic pigments	
Bleaching herbicides interfering with chlorophyll biosynthesis	Pyrazolate, pyrazoxyfen, clomazone, etc.
Bleaching herbicides interfering with carotenogenesis	Clomazone, norflurazon, methoxyphenone, etc.
Ammonia metabolism and amino acid biosynthesis	
Inhibitors of glutamine synthetase involved in ammonia assimilation	Glufosinate, bialaphos
Inhibitors of enolpyruvylshikimate phosphate synthase	Glyphosate
Inhibitors of acetolactate synthase (branched amino acid synthesis)	Chlorsulfuron, imazapyr, pyrithiobac, metosulam, etc.
Fatty acid biosynthesis	
Acetyl-CoA carboxylase inhibitors	Fluazifop, sethoxydim, clethodim, etc.

compounds targeting the chloroplast, many conventional herbicides have been applied with use rates more than 1 kg a.i./ha. During the past 15 years, however, scientists have concentrated their efforts on molecular design of low use rate herbicides, considering the mechanisms of action obtained from biological and biochemical studies. Two inhibitor types targeting the chloroplast can essentially produce low use rate herbicides. One group are the herbicidal inhibitors interfering with acetolactate synthase, and the other group is represented by the "peroxidizers", inhibiting protoporphyrinogen-IX oxidase (Fig. 1). Mode of action studies on herbicides interfering with acetolactate synthase have been reviewed in detail (Stetter 1994). In this chapter, general physiological characteristics and mode of action of peroxidizing herbicides will be discussed based upon up-to-date information.

6.2 Physiological Response of Plants: Light-Dependent Herbicidal Action

p-Nitrodiphenyl ethers (DPEs) and cyclic imides (CyIs), both called peroxidizing herbicides of the first generation (1965–1980), are shown in Fig. 2. They induce a rapid wilting and browning of shoots after application, bleach

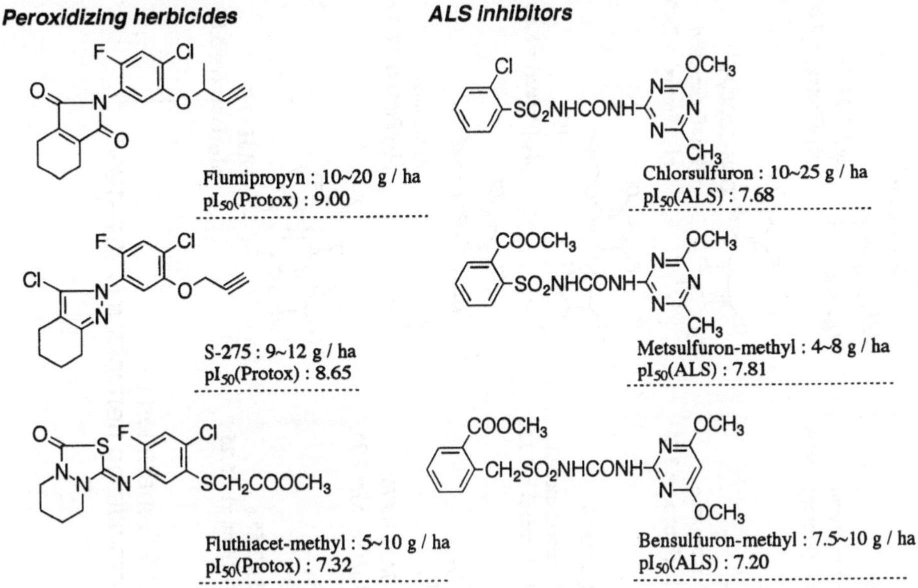

Fig. 1. Low use rate herbicides inhibiting acetolactate synthase (*ALS*) and protoporphyrinogen-IX oxidase

out the pigments and eventually result in growth retardation and death of plants (Wakabayashi et al. 1979). Light is required for their herbicidal action. The absorption of light energy essential for this herbicidal action has been misconceived to be mediated by carotenoids in earlier experiments. While green plants and yellow mutants with almost equal amounts of carotenoids were found to be susceptible to the herbicides, white mutants lacking carotenoids and chlorophylls were resistant to the light-dependent herbicidal action (Matsunaka 1969; Fadayomi and Warren 1976; Suzuki et al. 1976; Prendeville and Warren 1977; Wakabayashi et al. 1979). It has been subsequently considered that peroxidizing herbicides exhibit their light-dependent herbicidal action by interfering somewhere before the greening step in chlorophyll biosynthesis also in the yellow mutants, independently of carotenoid formation. Peroxidizing herbicides resulted in a stronger chlorophyll decrease compared to that of carotenoids in green plants by primarily inhibiting chlorophyll biosynthesis, which in turn affects carotenoid formation (Wakabayashi et al. 1986; Teraoka et al. 1987).

It should be noted that pI_{50}(*Echinochloa*) values, the negative logarithms of the molar concentration of peroxidizing herbicides which produces a 50% inhibition of *root growth* of sawa millet (*Echinochloa utilis*), surprisingly indicate a good correlation with the light-dependent herbicidal activity exhibited in pot tests using crabgrass, common purslane, hairy gallinsoga, barnyard grass and others (Suzuki et al. 1976; Wakabayashi et al. 1979).

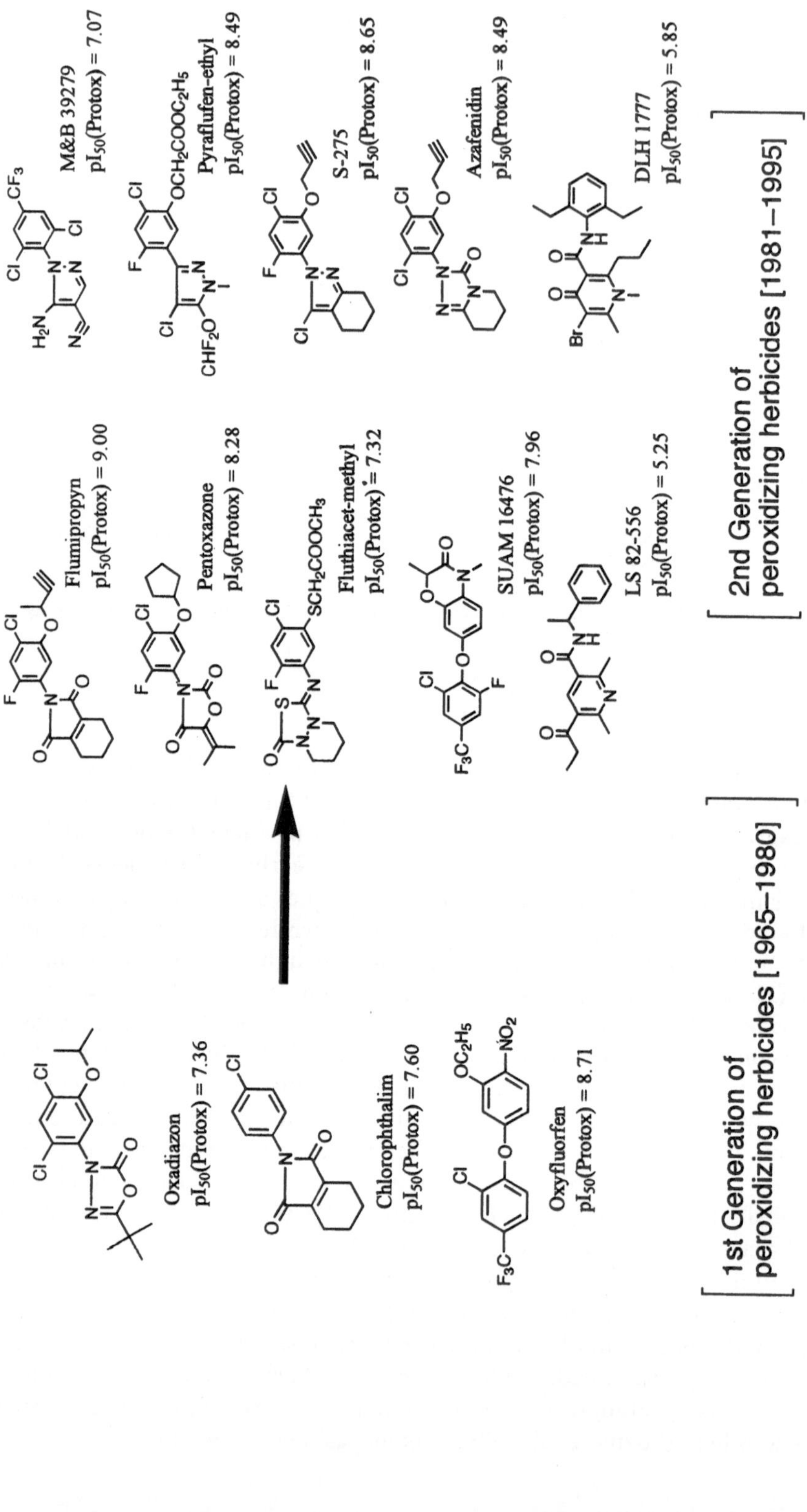

Fig. 2. Development of peroxidizing herbicides: two generations since 1965

6.3 Markers to Detect Peroxidizing Activity

Whether a given compound is a "peroxidizer" can be proven reliably with liquid suspension cultures of green microalgae. *Scenedesmus acutus*, as used in our laboratory, can be grown in light as well as in the dark, offering many possibilities to determine adequate peroxidative parameters with the intact cell. Figure 3 shows several typical markers. The first indication for a peroxidizer in action is the immediate halt of chlorophyll biosynthesis as demonstrated by part I, which was reported as the first inhibitory effect for peroxidizing compounds (Kunert and Böger 1981a; Wakabayashi et al. 1986). This is observed in heterotrophic dark cultures as well as in illuminated autotrophic ones (Nicolaus et al. 1989). However, in addition to inhibited chlorophyll biosynthesis, in the light a strong *degradation* of chlorophylls and carotenoids is observed. Photosynthesis is affected after a couple of hours (part II), and subsequently ethane is produced in the light (part III). Polyunsaturated fatty acids, major constituents of the acyllipids of thylakoids and cell membranes, are rapidly degraded leading to evolution of saturated short-chain hydrocarbons. With *Scenedesmus*, ethane is evolved and propane shows up besides ethane in treated soybean suspension cell cultures (Böger and Nicolaus 1993). Ethane determination with *Scenedesmus* can be performed with small sample volumes and is a convenient and quick assay for peroxidizing compounds applicable to greater series of chemicals using automatic sampling devices.

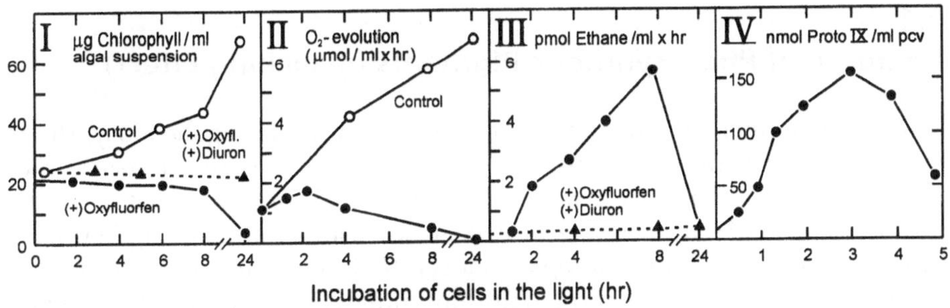

Fig. 3. Markers to measure peroxidative activity. A culture of *Scenedesmus acutus* was treated with oxyfluorfen, a peroxidizing diphenyl ether herbicide [2-chloro-1-(3-ethoxy-4-nitrophenoxy)-4-(trifluoromethyl)benzene]. Cells were grown autotrophically (in the light) except for data of dashed line curve in part I. These data belong to a glucose-supported *dark* culture to show *halt* of chlorophyll formation but without pigment degradation when a peroxidizer is present. Part III demonstrates that ethane formation is alleviated by diuron [3-(3,4-dichlorophenyl)-1,1-dimethylurea]. Also decrease of chlorophyll content (*I*) and proto-IX formation (*IV*) is stopped in the presence of a photosynthesis inhibitor besides a peroxidizing herbicide (not shown). Part II demonstrates loss of photosynthetic activity after a time lag.

A further parameter easily detectable with either intact algae cells or with higher plant seedlings is the accumulation of protoporphyrin IX (Matringe and Scalla 1988). Under moderate illumination a high level shows up in the cell within 2–3h after peroxidizer application and then gradually disappears apparently due to photooxidation (part IV of Fig. 3). In the cell the accumulated protoporphyrin IX (proto IX) is sensitized by light leading to activated oxygen species which in turn attack the unsaturated fatty acids of the membranes, degrading them, causing ion leakage (Orr and Hess 1981) and water loss of the cell. Since proto IX is readily photooxidized its accumulation proceeds best in low light, which may explain the improved kill efficiency of peroxidizing herbicides applied before the weeds in the field are exposed to strong sunlight, a fact well-known to extension experts. A typical sixth marker effect shown by autotrophic cells treated with peroxidizers is the alleviation of light-induced short-chain hydrocarbon formation by photosystem-II inhibitors like diuron (part III of Fig. 3).

The following text will deal closer with bleaching, ethane formation, the "diuron effect" and proto-IX accumulation. For measurement and assays of these parameters see Böger and Sandmann (1993). It should be noted that the physiological parameters mentioned are also observed with soybean cells or *Lemna minor* (duckweed) (Fig. 4).

On the basis of due consideration of the physiological markers obtained by treatment with peroxidizing herbicides, a herbicide-mediated radical peroxidation process has been proposed as mode of action of peroxidizing herbicides, being supported by the quantitative activity – activity correlations (see Fig. 5 and Sect. 6.10; Wakabayashi and Böger 1993; Böger and Wakabayashi 1995).

6.4 Decrease of Photosynthetic Pigments (Bleaching Effect)

"Bleaching herbicides" can exert their influence by inhibiting the biosynthesis of chlorophylls and/or carotenoids, or by causing peroxidative destruction of pigments (Sandmann et al. 1984a). It has been thus confirmed that when the green microalga *Scenedesmus acutus* is grown with a peroxidizing herbicide present, it interferes with *chlorophyll* biosynthesis and cytochrome formation in light or darkness while carotenoid biosynthesis is unaffected (Sandmann et al. 1984b; Wakabayashi et al. 1986; Teraoka et al. 1987).

This finding has given the earliest indication that the primary action of the peroxidizing herbicides is directed against a step in the common pathway of chlorophyll and cytochrome biosynthesis close to the branching point of protoporphyrin-IX formation, before the exact finding of protoporphyrinogen-IX oxidase inhibition by the herbicides became known (Matringe et al. 1989). Phytotoxic activity of the herbicides is due to interference with chlorophyll biosynthesis as a primary mechanism of action, which in turn inhibits β-carotene formation (see data of compounds (2) and (3) of Table

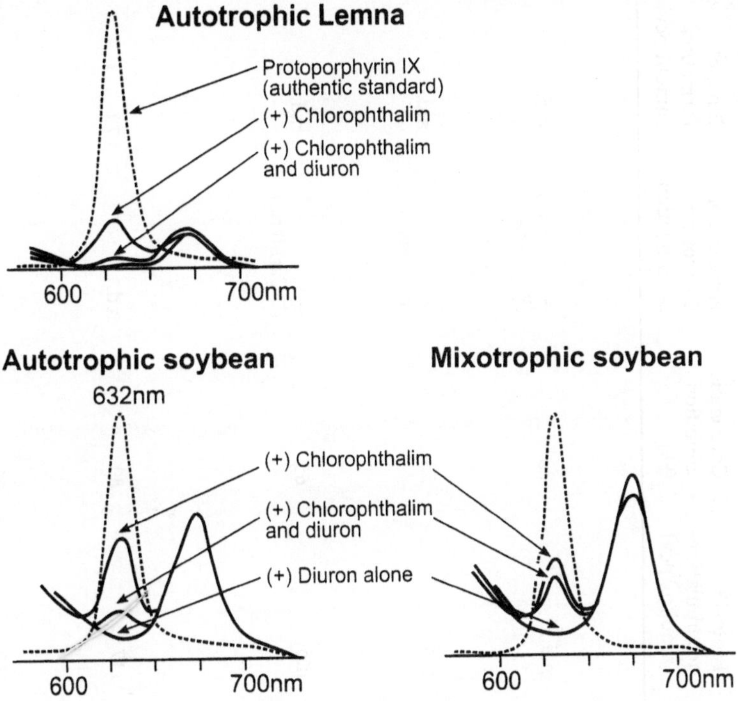

Fig. 4. Formation of protoporphyrin IX in autotrophic *Lemna minor*, autotrophic and mixotrophic soybean cells. Presence of protoporphyrin IX is shown by its fluorescence band which peaks at 632 nm. The peak of about 685 nm belongs to chlorophyll fluorescence. *Lemna* and cultured soybean cells were incubated in the light with peroxidizer chlorophthalim and diuron for about 20 h. Note that the antagonistic effect of diuron against peroxidation is absent in mixotrophic cells (cf. Sandmann et al. 1990b)

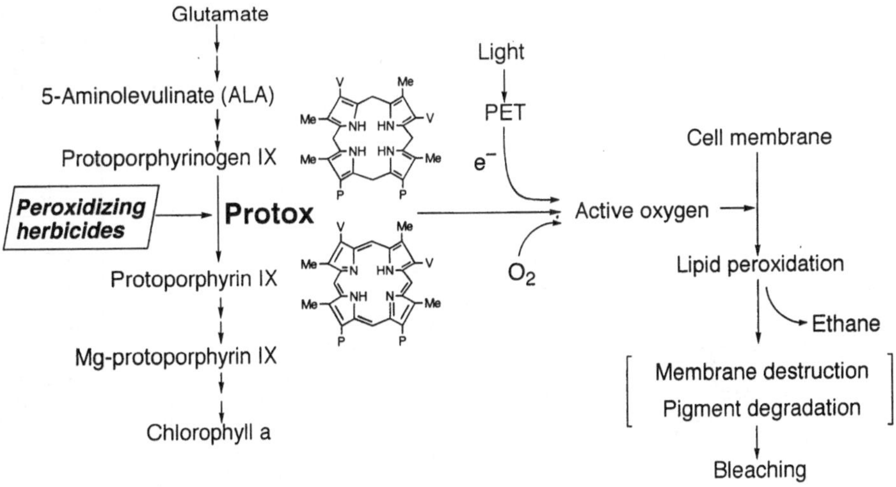

Fig. 5. Mode of action of peroxidizing herbicides: herbicide-mediated lipid peroxidation and destruction of photosynthetic membranes and pigments

Table 2. Influence of bleaching herbicides on pigments

Herbicides	Concentration (μM)	Chlorophyll inhibition (%)	Proto-IX[a] accumulation (nmol/ml pcv)	Carotenoid inhibition (%)	Phytoene or ζ-Carotene accumulation	Ethane[b] formation (nmol/l pcv)
Group A						
(1)[c] 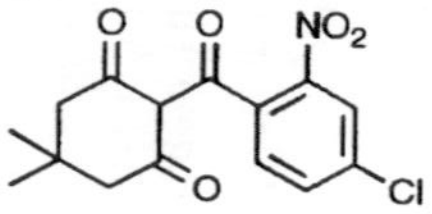	0.1	92	n.d.	82	+ (Phytoene)	n.d.
(2) 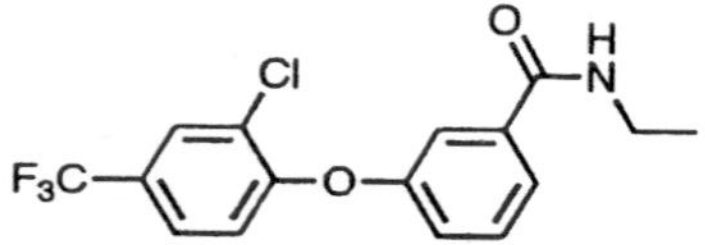	1.0	21	n.d.	53	+ (Phytoene)	n.d.
(3) 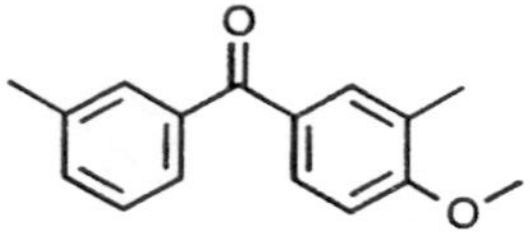	1.0	26	n.d.	65	+ (ζ-Carotene)	n.d.
Group B						
(4) 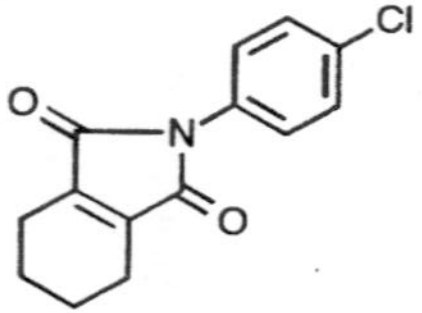	1.0	78	117	80	n.d.	7.6
(5) 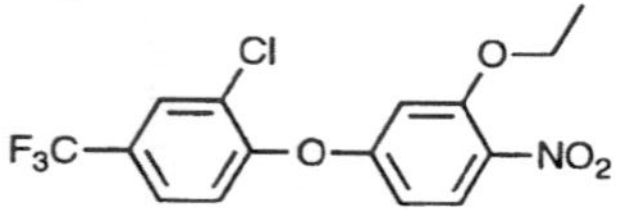	1.0	84	120	82	n.d.	9.3

n.d., not detected; +, detected.
[a]Proto-IX accumulation after 1-h incubation with compounds in the dark.
[b]Ethane formation after 17-h incubation in the light.
[c]Data of compound (1) (CH-309, benzoylcyclohexanedione) were obtained with cress seedlings (Sandmann et al. 1990*a*). Data with compounds (2)–(5) were obtained with heterotrophic *Scenedesmus*. *Scenedesmus* responded poorly in this type of herbicides. Difference in inhibition of chlorophyll and carotenoid formation depends on methodology of assay systems used.

2). Also, in experiments with *Nicotiana xanthi* cells, peroxidizing herbicides exhibit a strong inhibition of chlorophyll formation at a very low concentration (10^{-8} to 10^{-9}M), while carotenoid biosynthesis is not affected by these concentrations (Teraoka et al. 1987).

6.5 Ethane Formation: Peroxidative Destruction of Cell Membranes

Peroxidizing herbicides are now considered to essentially exhibit their phytotoxic activity by light-induced peroxidative destruction of photosynthetic membranes (Kunert and Böger 1981a; Orr and Hess 1981). In the light, in the presence of these herbicides, higher plants exhibit short-chain alkane formation (Kunert and Böger 1981a,b; Kunert 1984; Matringe et al. 1986; Gillham and Dodge 1987; Wakabayashi et al. 1988; Matsumoto et al. 1994) and membrane leakage (Orr and Hess 1981; Sato et al. 1988, 1994). Although this herbicide-induced in vivo peroxidation can also be measured by either malondialdehyde (Orr and Hess 1982) or degradation of ^{35}S-sulfolipid (Sandmann and Böger 1982b, 1983a), the convenient method measuring ethane released by decomposition of unsaturated fatty-acid hydroperoxides in the green unicellular microalga, *Scenedesmus acutus*, has provided a tool to assay and compare peroxidizing activity of the herbicides (Kunert and Böger 1981a,b; Wakabayashi et al. 1988). Thus, many bleaching herbicides and their relatives have been quantitatively checked for their peroxidizing activity by ethane formation and many of them have been confirmed to be peroxidizing herbicides. Furthermore, structure and peroxidizing activity have been discussed using this peroxidizing parameter (Lambert et al. 1983, 1984; Wakabayashi et al. 1988; Nicolaus et al. 1989; Watanabe et al. 1992; Hoshi et al. 1993; Kohno et al. 1993, 1995; Wakabayashi and Böger 1993; Ogino et al. 1994; Sato et al. 1994; Böger and Wakabayashi 1995; Ihara et al. 1995; Iida et al. 1995; Sumida et al. 1995, 1996; Wakabayashi and Böger 1995; Senoo et al. 1996; Kohno et al. 1995; Shouda et al. 1996).

Although ethane together with a small amount of ethylene is peroxidatively generated from the predominant thylakoid lipid (linolenates) of *Scenedesmus*, other kinds of short-chain volatile hydrocarbons are produced by the plants, which have different polyunsaturated fatty acid moieties of the acyllipids in their membranes. The chain length of the alkane generated depends on the ω-number which denotes the position of the double bond most distant from the carboxyl group of the fatty acid. Linolenic acid, an ω-3 polyunsaturated fatty acid, forms C_2-ethane, an ω-4 acid produces C_3-propane, an ω-5 species C_4-butane etc., according to the "ω-1 rule" (Sandmann and Böger 1982a, 1983b).

The peroxidative ethane formation is light-dependent (Kunert and Böger 1981a). No immediate inhibition by peroxidizing herbicides is found on photosynthesis and respiration except for concentrations far exceeding the pI_{50}-

values for growth inhibition. *Scenedesmus* cells cultured in the dark with glucose and incubated with the herbicides never exhibit ethane formation, indicating lack of peroxidation in the dark (Wakabayashi et al. 1988). It should be noted that although peroxidizing herbicides reduce the chlorophyll content of dark-grown cells, the pigment decrease is stronger in autotrophic light than in heterotrophic dark cultures. A three- to four fold difference in bleaching activity is found. This is a much lower figure than expected from general experience with peroxidizing herbicides in greenhouse pot tests. Obviously the chlorophyll decrease in the dark cultures of *Scenedesmus* reflects inhibition of chlorophyll biosynthesis alone and the decrease in the light cultures indicates a combination effect of inhibition of chlorophyll biosynthesis together with photooxidative destruction of chlorophylls (Sect. 6.10). In fact, pI_{50}(Chlorophyll, D) values, indicating the chlorophyll decrease in the dark cultures of *Scenedesmus* cells, correlate quite well with protoporphyrinogen-IX oxidase inhibition by peroxidizing compounds (Table 3). The pI_{50} (Chlorophyll, L) values, expressing the chlorophyll decrease in the autotrophic light cultures, indicate a significant correlation with two phytotoxic parameters, the protoporphyrinogen-IX oxidase inhibition [pI_{50}(Protox)] and the peroxidizing ethane formation [pKa(Ethane)] as well [see Eqs. (4), (5) and (7)].

Undoubtedly, an oxygen radical and/or activated oxygen is initiated to start herbicide-induced peroxidation in *Scenedesmus* cells (Kunert and Böger 1981a; Kunert and Dodge 1989). This line of thought has been supported by the strong evidence that peroxidation is counteracted by high endogenous levels and adequate regeneration of cellular radical quenchers, like α-tocopherol and/or ascorbate (see Chapt. 12). ESR signals of the peroxidative breakdown products can be found by illuminating isolated thylakoids from spinach with peroxidizing herbicides present, if spin-trap techniques are used (Lambert et al. 1984). This radical formation is also alleviated by the photosystem-II inhibitor (diuron) as well as by the antioxidant ethoxyquin. A report on formation of the diphenyl ether radicals in illuminated isolated chloroplasts cannot be realized. We realize an involvement of photosynthetic electron transport, activated oxygen and oxygen radicals in peroxidizing processes. As will be discussed in Section 6.7, the starting oxygen radicals for peroxidation are induced by sensitized protoporphyrin-IX or by an analogue fluorescing at 590 nm (Iwata et al. 1994). In the presence of a reductant like glutathione the superoxide radical may be formed, possibly leading to the •OH radical. This can be the starter radical in the peroxidation reaction of α-linolenate (Sandmann and Böger 1986).

6.6 Diuron Effect: Alleviation of Phytotoxic Symptoms

Diuron or other photosynthetic electron-transport inhibitors such as atrazine, metribuzin, etc., applied at concentrations high enough to completely block

Table 3. Quantitative correlations of phytotoxic parameters obtained with various peroxidizing compounds

Equations				Equation numbers
pI_{50}(*Echinochloa*)	= 0.775 pI_{50}(*Scenedesmus*, L) (±0.042)	−0.191	(n = 167, r = 0.944, s = 0.354)	(1)
pI_{50}(*Scenedesmus*, L)	= 0.966 pI_{50}(Chlorophyll, L) (±0.019)	+0.093	(n = 188, r = 0.991, s = 0.168)	(2)
pI_{50}(Chlorophyll, D)	= 1.036 pI_{50}(Protox) (±0.170)	−0.294	(n = 30, r = 0.921, s = 0.581)	(3)
pI_{50}(Chlorophyll, L)	= 0.553 pI_{50} (Protox) (±0.073)	+2.408	(n = 108, r = 0.827, s = 0.480)	(4)
pI_{50}(Chlorophyll, L)	= 0.882 pKa(Ethane) (±0.101)	+1.071	(n = 108, r = 0.859, s = 0.434)	(5)
pKa(Ethane)	= 0.495 pI_{50}(Protox) (±0.066)	+2.535	(n = 105, r = 0.823, s = 0.446)	(6)
pI_{50}(Chlorophyll, L)	= 0.266 pI_{50}(Protox) + 0.661 pKa(Ethane) (±0.063) (±0.095)	+0.471	(n = 108, r = 0.918, s = 0.338)	(7)
pI_{50}(Carotenoid, L)	= 1.231 pKa(Ethane) (±0.196)	−0.341	(n = 23, r = 0.943, s = 0.483)	(8)

Activity – activity correlations in this Table are revised (Wakabayashi and Böger 1993; Böger and Wakabayashi 1995). Data refer to *Scenedesmus acutus*, *Echinochloa utilis* (sawa millet) and maize protoporphyrinogen-IX oxidase (isolated from etiolated seedlings). pI_{50}(*Echinochloa*) indicates root growth inhibition by the compounds assayed. The alga (*Scenedesmus*) was grown in heterotrophic liquid dark cultures (D) or cultivated autotrophically in the light (L). Thus, pI_{50}(Chlorophyll, D) refers to the cellular chlorophyll decrease in the dark, reflecting chlorophyll biosynthesis inhibition. Values of pI_{50} (*Scenedesmus*, L), pI_{50}(Chlorophyll, L) and pI_{50}(Carotenoid, L) represent growth inhibition, chlorophyll decrease (inhibition of chlorophyll biosynthesis plus degradation of chlorophyll) and carotenoid decrease in the light-cultured cells, respectively. pI_{50}(Protox) indicates inhibition of protoporphyrinogen-IX oxidase by the compounds assayed. pKa(Ethane)-value ("activity value") is the molar concentration giving half of the maximum of light-induced ethane formation (Lambert et al. 1983). In some papers this figure has been called I_{50}(Ethane) or pI_{50}(Ethane), respectively.

photosynthetic electron flow, alleviate the light-induced phytotoxic action of peroxidizing herbicides in autotrophic *Scenedesmus* cells, namely, ethane formation, chlorophyll decrease, and protoporphyrin-IX accumulation. This alleviation of peroxidation by photosystem-II inhibitors (the "diuron effect") is essentially a safening effect in autotrophic cells. This inhibition of peroxidative activity has been observed in autotrophic *Scenedesmus* and tissue of higher plants (Matringe et al. 1986; Gillham and Dodge 1987; Nicolaus et al. 1989; Sandmann et al. 1990b); but no influence of diuron or metribuzin on peroxidation was reported for cucumber cotyledons (Orr et al. 1983).

The diuron effect disappears in glucose-supplemented heterotrophic cultures (Sandmann et al. 1990b) or in cucumber cotyledon pieces (Nurit et al. 1988). The supplemented glucose or storage material in the cells substitute for photosynthesis, and no diuron effect shows up. NADPH, ATP and organic carbon precursors have to be provided by photosynthesis or sugar breakdown since they are required for synthesis of 5-amino-levulinate (ALA), the starting metabolite and limiting compound for tetrapyrrole biosynthesis.

Peroxidizing activity is synergized by ALA (Rebeiz et al. 1987; Watanabe et al. 1992), while gabaculine (3-amino-2,3-dihydrobenzoic acid) and 4,6-dioxoheptanoic acid prevent formation of protoporphyrin-IX by inhibiting glutamate-1-semialdehyde aminotransferase or by inhibiting ALA dehydratase, respectively (Lydon and Duke 1988; Sandmann et al. 1990b; Watanabe et al. 1992). Peroxidative bleaching can also be substantially reduced by application of peroxidizing herbicides and gabaculine together.

6.7 Protoporphyrin-IX Accumulation

Peroxidizing compounds inhibit the protoporphyrinogen oxidase in the biosynthetic pathway (protox, EC 1.3.3.4) which oxidizes protoporphyrinogen (protogen; Matringe and Scalla 1988; Matringe et al. 1989). The protogen precursors uro- and coproporphyrinogen are not oxidized (Duke et al. 1994). There are several sites in the cell housing enzymes with protox activity as indicated in Fig. 6. The protox enzymes of the plastid are instrumental for processing of protogen into proto IX, which is the precursor for chlorophyllide *a* and heme (see Jacobs and Jacobs 1994 for overview). Plant mitochondria depend on protogen produced in the plastid. They also contain a membrane-bound protox and the subsequent enzymes to produce the heme moiety for cytochromes. The endoplasmic reticulum (microsomes) also include a high protox activity (Retzlaff and Böger 1996). Additionally, a protogen-oxidizing enzyme located in the plant plasma membrane was claimed (Jacobs et al. 1991; Dayan and Duke 1996), but this location could not be confirmed (for maize) in our laboratory. A soluble protogen-oxidizing peroxidase has been reported (Yamato et al. 1994) as well as a soluble enzyme from chloroplasts of photomixotrophic tobacco cells (Yoshida et al. 1993).

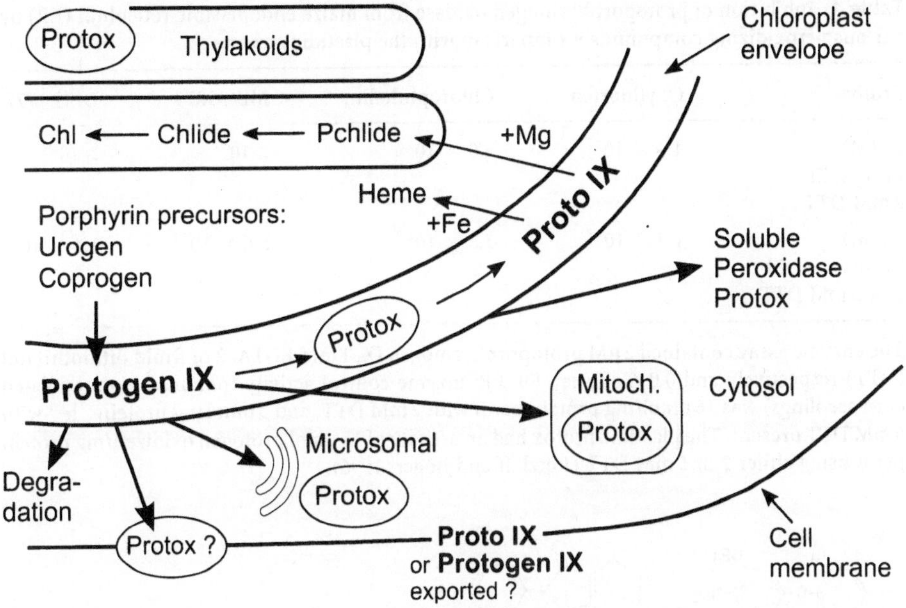

Fig. 6. Oxidation of protoporphyrinogen IX (protogen IX) inside the plant cell. Enzymic oxidation may occur at six sites as indicated by protox (protoporphyrinogen oxidase). Membrane-bound enzymes of plastid and mitochondrium are sensitive to peroxidizers while the indicated cytosolic soluble peroxidase (protox) and the microsomal one are not. Whether a protox is generally present in the plasma membrane is still unclear. Biosynthesis of protogen precursors occurs in the plastid only. Urogen and coprogen = uroporphyrinogen and coproporphyrinogen, respectively; *Chl* chlorophyll; Chlide and Pchlide = immediate chlorophyll *a* precursors

The role of the last four enzymes is not yet clear. Their protox activities may be due to an unspecific side effect or – in case of the microsomal enzyme – may be involved in an additional heme-producing pathway. It looks that plastidic and mitochondrial protox are the major biosynthetic enzymes, and these are most sensitive to various peroxidizing chemicals. The other protox enzymes are less susceptible as could be shown in detail for an endoplasmic reticulum samples prepared from maize microsomes which exhibited a strong protogen oxidation activity. Sensitivity of this microsomal enzyme against several peroxidizers was found 100 to more than 10 000 times less than the sensitivity of the plastidic protox (Table 4). The paradox of the accumulation of protoporphyrin IX as product of an inhibited enzyme reaction is explained by both non-enzymic oxidation of protogen (which proceeds at alkaline pH) and by enzymic oxidation performed by the protox activities found in the cytosol (Lee et al. 1993). Apparently the non-enzymic oxidation is much slower than the enzymic one. This is evident from our finding that formation of proto IX in isolated maize etioplasts is strongly reduced by inhibiting the (plastidic)

Table 4. Inhibition of protoporphyrinogen oxidase from maize endoplasmic reticulum (ER) by various peroxidizing compounds. Comparison with the plastidic enzyme

Protox	Oxyfluorfen	Chlorophthalim	MB-39279	DLH 1777
I_{50} (M) Protox, ER 2 mM DTT	4.4×10^{-7}	4.9×10^{-5}	$>10^{-4}$	$>10^{-4}$
I_{50} (M) Protox, plastid 2 or 5 mM DTT	1.3×10^{-9}	2.3×10^{-8}	8.6×10^{-8}	1.4×10^{-6}

The enzyme assay contained 2 μM protoporphyrinogen IX, 1 mM EDTA, 2 or 5 mM dithiothreitol (DTT) respectively and 0.03% Tween 80. ER enzyme control activity (obtained from etiolated corn seedlings) was 16 nmol/mg protein per h with 2 mM DTT, and 3 nmol mg protein^{-1}h^{-1} with 5 mM DTT present. The plastidic protox had an activity of 15 nmol protogen oxidized/mg protein per h using either 2 or 5 mM DTT (Retzlaff and Böger, 1996).

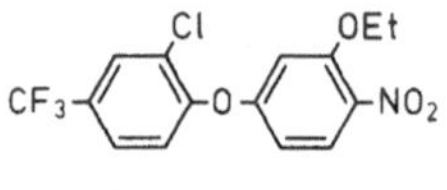

Oxyfluorfen

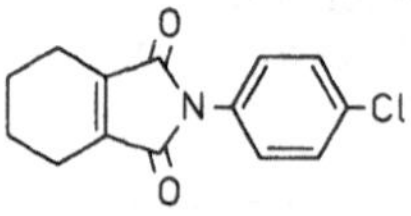

Chlorophthalim

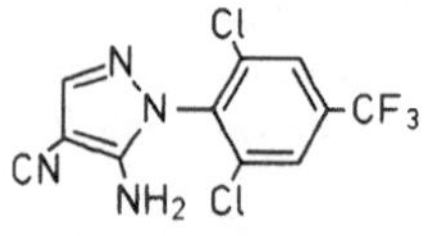

MB-39279

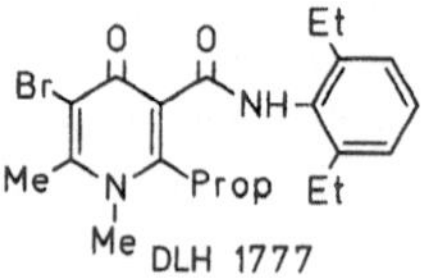

DLH 1777

protox with oxyfluorfen during a 2-h incubation time (Retzlaff and Böger 1996). This should not be the case if autoxidation of protogen proceeds rapidly. Although most of the protogen oxidation occurs in the cytosol (Lehnen et al. 1990), proto-IX formation also proceeds *inside* the chloroplasts as we have shown with liverwort cell cultures treated with acifluorfen methyl (Sumida et al. 1996b).

Once the biosynthetic protox is inhibited, protogen is overproduced in the plastid since feedback control by the endproduct of chlorophyll biosynthesis is missing. Apparently protochlorophyllide *a* is the regulating endproduct (Kotzabasis et al. 1989; Becerril et al. 1992). Protogen is exported to the cytosol and proto IX can accumulate in treated plants up to 20 nmol/g fresh weight while a few nanomoles per gram are already considered as phytotoxic and lethal (Dayan and Duke 1997).

Most authors agree that the accumulating colored tetrapyrrole is proto IX. This assumption is based on fluorescence characteristics and – in part – on absorbance spectroscopy. The existence of proto IX, however, was proven by

NMR and mass spectroscopy in *one* case only with herbicide-treated *Bumilleriopsis*. This eukaryotic microalga excretes the tetrapyrrole into the culture medium giving it the deep red color of proto IX and it can subsequently tolerate strong peroxidizers (Sandmann and Böger 1988). Acifluorfen-methyl-treated liverwort cells accumulate a fluorescent tetrapyrrole (the "590 fluorescent pigment") which exhibits a fluorescence maximum at 590 nm instead of 632 nm for proto IX. This may be an oxidized or modified proto IX, but structural elucidation failed so far due to the poor yield of a pure preparation (Iwata et al. 1994).

As noted above, the membrane-bound biosynthetic protoporphyrinogen oxidase exhibits a high sensitivity towards peroxidizing compounds with I_{50} values down to 10^{-10} M. However, as shown by Table 5, the sensitivity differs markedly between plastidic and mitochondrial protox depending on the chemical class of inhibitors. For instance, the arylpyrazole (similar to pyraflufen-ethyl, first line of Table 5) shows strong inhibition of maize

Table 5. Comparison of I_{50} values (molar) for protoporphyrinogen oxidase

Peroxidizing compound	Maize etioplasts	Yeast mitoch.	Rat liver mitoch.	Ratio I_{50} Yeast/Maize	Ratio I_{50} Rat/Maize
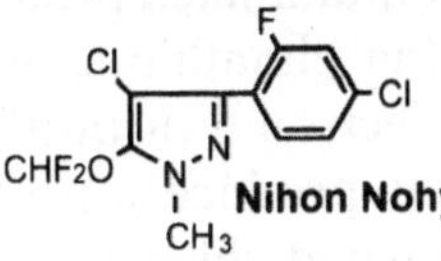 Nihon Nohyaku type	3.2×10^{-9}	2.1×10^{-9}	6.4×10^{-7}	0.65	200
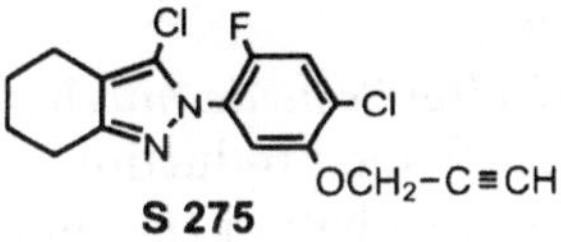 S 275	7.4×10^{-10}	4.7×10^{-8}	1.1×10^{-9}	63	1.5
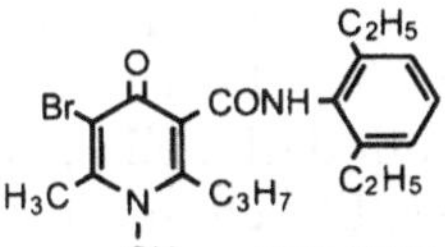 DLH 1777	1.4×10^{-6}	3.7×10^{-6}	2.0×10^{-8}	2.6	0.014
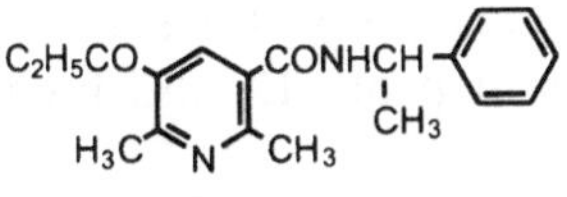 LS 82556	5.6×10^{-6}	2.7×10^{-7}	2.5×10^{-6}	4.8	0.5
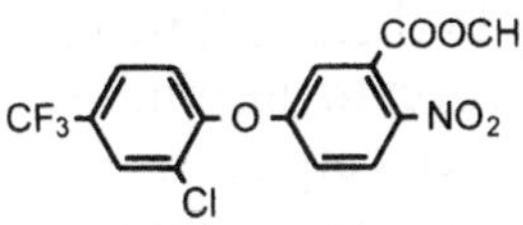 Acifluorfen-methyl	5.6×10^{-9}	1.3×10^{-8}	3.7×10^{-8}	2.3	6.6

etioplast protox but the I_{50} value is 200-fold less for the mitochondrial rat liver enzyme. DLH 1777 exhibits the opposite. Moderate inhibition of the plastid protox but good inhibition of the liver enzyme (see last column in Table 5). This finding may have some bearing to toxicology. As a working hypothesis it may be assumed that the arylpyrazole is less toxic to mammals than the DLH 1777 compound.

6.8 Reasons for the Phytotoxic Effect

Why do so many peroxidizers exhibit an excellent phytotoxicity with a low use rate in the field? The reasons are summarized as follows:

1. The binding constant to the target enzyme (and the inhibition constant) is very low. That is, for half-inhibition of the protox in the cell up to some 10g a.i./ha are often sufficient (see Böger and Wakabayashi 1995).
2. Proto IX acts catalytically to produce active oxygen species in the light. Although proto IX is photooxidized eventually it may be safely assumed that one proto-IX molecule will produce several radicalic oxygens. The "radical productivity"of proto IX, however, still has to be ascertained. It is unclear which type of activated oxygen is formed to initiate the attack on the fatty acids of the biological membrane. Generally, light-activated proto IX leads to singlet oxygen. We could show, however, that illuminated proto IX can reduce, e.g., nitro compounds in the light using glutathione as electron donor (unpubl. results). This implies that an "electron transport" through the sensitized tetrapyrrole is possible which may produce superoxide anions (O_2^-) under certain conditions. Also illuminated chlorophyll can generate superoxide anions (You and Fong 1986).
3. A substantial amount of ethane is evolved in the *dark* after *Scenedesmus* has been pre-illuminated with oxyfluorfen present. This ethane production in the dark depends on the length of the previous light incubation (unpubl. results). Apparently this is due to stable radical(s), most probably hydroperoxides originating from unsaturated fatty acids (cf. Böger and Sandmann 1990 and the breakdown scheme therein). Both dark and light ethane formation could be alleviated by radical quenchers like ethoxyquin (1,2-dihydro-6-ethoxy-2,2,4-trimethylquinoline).
4. For reasons of point 2, only a part of the (sensitive) plastid protox activity has to be inhibited still leading to small amounts of proto IX and high phytotoxicity.
5. The binding constant of an active peroxidizer is much smaller than the k_m of the substrate protoporphyrinogen (see Sect. 6.9). So there is little substrate competition at the common binding site with respect to the inhibitor. As shown by Becerril and Duke (1989) with cucumber cotyledon discs, leakage of cellular membranes correlates with the content of accumulated proto IX in a certain range. We could show with our model alga

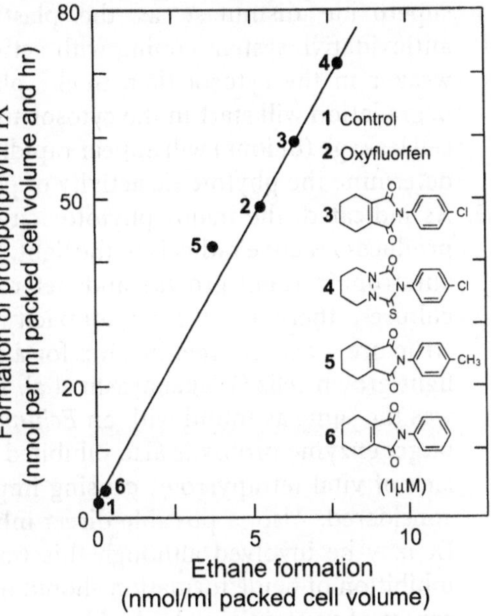

Fig. 7. Correlation between short-term formation of protoporphyrin IX and ethane evolution. Autotrophic *Scenedesmus* cultivated for 1 h in the presence of 1 μM cyclic imides of different peroxidative activity. Ethane formation was determined after 17-h incubation of *Scenedesmus* under strong white light (350 μE/m^2 per s; cf. Watanabe et al. 1992)

Scenedesmus that the *maximum* accumulation of proto IX reached after a 5 to 10-h incubation time did not correlate with the phytotoxic effects. Only the smaller proto-IX amount (about one tenth or one fifth of the maximum level) attained after a short-term herbicide treatment was appropriate for a reliable data set for structure – activity considerations as indicated by Fig. 7 (Watanabe et al. 1992).

6. It should also be mentioned that during long-term accumulation, proto-IX degradation will occur both by photooxidation (see again Watanabe et al. 1992) or – at least in certain species – by enzymic degradation of protogen IX (Jacobs et al. 1996). Using acifluorfen, very different proto-IX accumulation was found in leaf discs of several plant species (e.g. mustard, cucumber, morning glory) after a certain period of treatment. The protox of the three species, however, had a similar sensitivity to acifluorfen (Sherman et al. 1991). The question whether different uptake or degradation of the herbicide took place is not yet clear. These questions are important for selectivity; regrettably information on this topic is meager. It is also not clear whether (a part) of protogen or proto IX may penetrate the cell wall or may be transported (cf. Fig. 6).
7. Peroxidizers exhibit a rapid "burn-down" effect. Quickly the leaf bleaches, becoming necrotic and brown in the light within 1–2 days. This is in sharp contrast to, e.g., glyphosate or photosynthesis inhibitors. After application it takes a much longer time before their action is clearly visible.
8. As pointed out, most of the proto IX is produced in the cytosol. The cytosol does not contain an active membrane-bound ascorbate peroxidase and

superoxide dismutase as the plastid (Polle 1996). Accordingly, the antioxidative system coping with activated oxygen and peroxy radicals is weaker in the cytosol than in the plastid (cf. Chapter 12). Peroxidative degradation will start in the cytosol affecting the plasma membrane. Thus, cell leakage (of ions) will appear rapidly which can be used to quantitatively determine the phytotoxic activity of peroxidizing compounds.

9. As indicated, the major phytotoxicity is due to a *product* formed which produces reactive radicals in the light. However, as was shown in detail with autotrophic (light-grown) and heterotrophic (dark-grown) *Scenedesmus* cultures, there is also phytotoxicity in the *dark* exhibiting the same structure – activity relationship for a series of cyclic imides as found with light-grown cells (Wakabayashi et al. 1988). Furthermore, this relationship was the same as found with an *Echinochloa* root tip test. Conceivably, the target enzyme protox is also inhibited in the dark. Under these conditions lack of vital tetrapyrroles causing impaired heme biosynthesis should be considered. Also, a possible direct inhibitory effect of accumulated proto IX may be involved although this was never assayed. Anyway, the dark inhibition of heme formation should not be neglected when the phytotoxic causes of peroxidizing herbicides are assessed.

A low use rate is the driving force in modern herbicide development. This requires highly active compounds. In our case, peroxidizing herbicides lead to rapid cell degradation and burning which prevents the induction of enzymes to strengthen the antioxidative radical-quencher system of the plant or the formation of breakdown enzymes for tetrapyrroles. Accordingly, poor selectivity is generally encountered with peroxidizing herbicides. The rapid burning also prevents substantial transport of the herbicide in the plant. Regrowth is often observed starting from the apical meristem of (grass) weeds after postemergent treatment.

6.9
Some Information on Protoporphyrinogen Oxidase

Peroxidizers act as reversible, competitive inhibitors with respect to the substrate proto-porphyrinogen as has been demonstrated for compounds of different chemical classes (Camadro et al. 1991, 1994; Nandihalli et al. 1992; Nicolaus et al. 1995). The inhibition constants (I_{50}) of active compounds are in the range of 10^{-8} to 10^{-9} M (the binding constants k_b are about ten times less) while the k_m for protogen was found as 0.21 μM for the purified maize protox, as shown by Fig. 8 (below). This discrepancy explains lack of effective substrate competition with the inhibitor as mentioned in Section 6.7. Figure 8 also demonstrates the strict competition of the substrate with oxyfluorfen, an inhibition type which is preserved after a more than 800-fold enrichment of the enzyme. Since competition and binding reversibility (Nicolaus et al. 1995) are

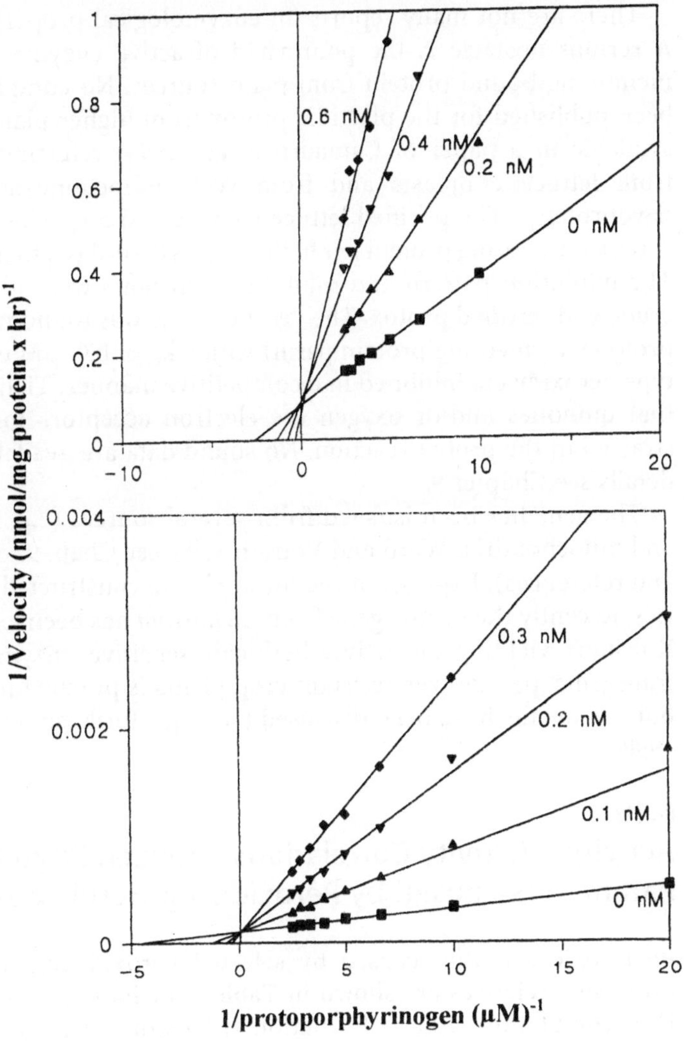

Fig. 8. Competitive inhibition of plastidic protoporphyrinogen oxidase by peroxidizing compounds, here exemplified by oxyfluorfen (0.1–0.6 nM concentration range). A crude preparation from maize (above) and a purified maize protox below were used. Specific activities were 12 and 9600 nmol proto IX formed/mg protein per h, respectively. (C. Adomat, this laboratory, unpubl.)

characteristic features, a displacement was observed of a labeled inhibitor (e.g. ^{3}H-acifluorfen or ^{14}C-thidiazimin) by the unlabeled form or by other peroxidizers like oxyfluorfen. Moreover, protoporphyrinogen leads to competitive displacement of the inhibitor (Nicolaus 1993). The in-vitro inhibition assays are run with dithiothreitol present (DTT, 2–5 mM) to reduce the autoxidation of the substrate and to stabilize the reaction. Also, DTT competes with substrate and inhibitor in our experiments, indicating an influence of redox agents on inhibition.

There are not many reports on enzymological properties of plant protox. A serious obstacle is the poor yield of active enzyme when purifying the membrane-bound protein from plant sources. No complete purification has been published for the plastidic protox from higher plants. Some details are available in a paper of Camadro et al. (1994) referring to purified protox from lettuce etioplasts and from yeast mitochondria. Both are 55-kDa flavoproteins. The purified lettuce enzyme had a specific activity of 9.3 μmol protox formed/mg protein per h, the k_m was 0.3 μM (cf. that for maize in Fig. 8). The inhibition pattern and inhibition constants were identical both for the crude and purified protox. The yeast enzyme was found more active (40 μmol proto IX formed/mg protein per h) with a k_m of 0.07 μM only. Diphenyl-ether type peroxidizers inhibited in a competitive manner. Tietjen (1991) suggested that quinones and/or oxygen are electron acceptors for the hydrogen abstracted in the protox reaction. No sound data are available as yet. For more details see Chapter 9.

The gene has been isolated from several sources (e.g. from maize plastids and mitochondria: Ward and Volrath 1995; see Chap. 9 for additional species and references). Expression vectors are being constructed in several laboratories. Recently the protox gene from *Cichorium* has been expressed in *E. coli* in Konstanz yielding an active, herbicide-sensitive enzyme. Development of transgenic, peroxidizer-resistant crop plants is pursued in several companies but few details have been disclosed (see e.g. Horikoshi et al. 1998; Choi et al. 1998).

6.10 Activity – Activity Correlations Between Phytotoxic Parameters Caused by Peroxidizing Herbicides

Phytotoxic activities caused by selected peroxidizing herbicides and their promising relatives are shown in Table 6, including protox inhibition, proto IX-accumulation, ethane formation (peroxidizing activity), chlorophyll decrease (bleaching effect) and growth inhibition. The activity – activity correlations, comparing a group of compounds in different biological or biochemical test systems, sometimes provide useful forerunning information concerning mechanism of action studies and QSAR research of pesticides. Eight phytotoxic parameters obtained from peroxidizing herbicides can be correlated with high significance to the light-dependent herbicidal activity in greenhouse pot tests (herbicidal activity), namely the root growth inhibition of *Echinochloa* [pI_{50} (*Echinochloa*)], growth inhibition of *Scenedesmus acutus* [pI_{50} (*Scenedesmus*)], chlorophyll decrease in light or in darkness [pI_{50}(Chlorophyll, L) or pI_{50}(Chlorophyll, D)], the protoporphyrinogen-IX oxidase inhibition [pI_{50}(Protox)], photo-peroxidative ethane formation [pKa(Ethane)] and eventually the carotenoid decrease in the light [pI_{50}(Carotenoid)], (Eqs. (1)–(8) of Table 3) (Wakabayashi et al. 1988; Watanabe et al. 1992; Wakabayashi and Böger 1993; Böger and Wakabayashi 1995). Since all the parameters except the

Table 6. Phytotoxic activities of peroxidizing herbicides and candidates

Formulas		pI_{50} (*Scenedesmus*)	pI_{50} (Chlorophyll, L)	pI_{50} (Ethane)	Proto IX[a] (nmol/ml pcv)	pI_{50} (Protox)
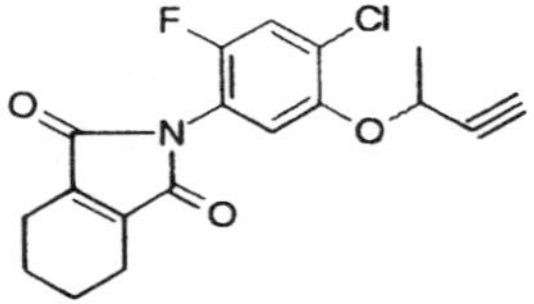	Flumipropyn	7.36	7.53	6.98	74.8	9.00
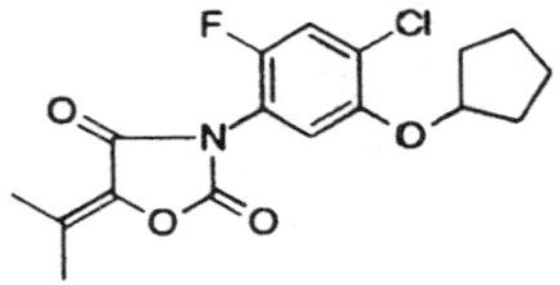	Pentoxazone	6.83	6.88	6.20	54.9	8.28
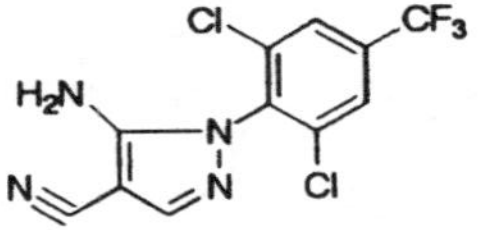	M&B 39279	7.15	7.42	6.85	62.2	7.07
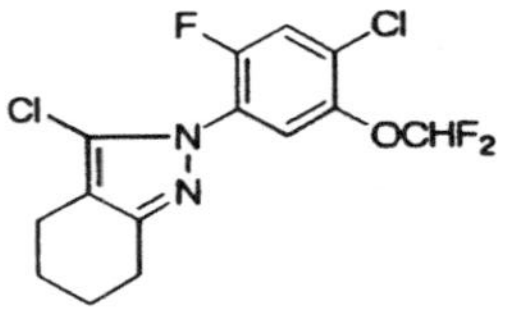	Shionogi 1	7.48	7.47	6.62	61.1	8.55
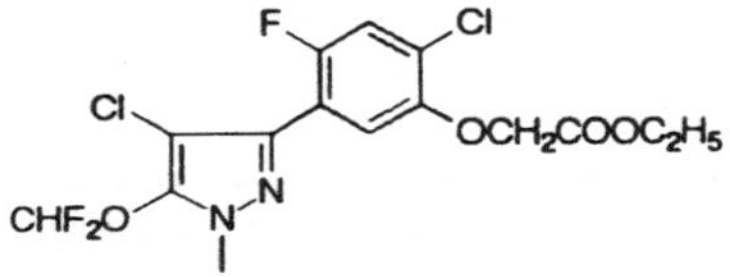	Pyraflufen-ethyl	8.14	8.33	7.23	–	8.49

Table 6. (*Continued*)

Formulas		pI_{50} (*Scenedesmus*)	pI_{50} (Chlorophyll, L)	pI_{50} (Ethane)	Proto IX[a] (nmol/ml pcv)	pI_{50} (Protox)
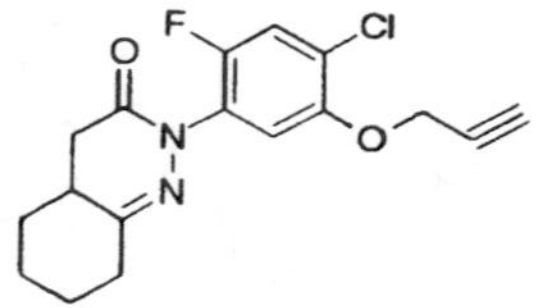	EXP 9301	6.41	6.78	5.81	44.6	8.96
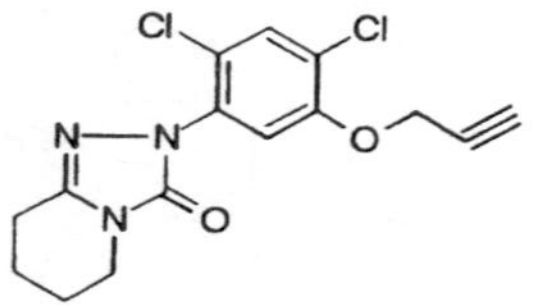	Azafenidin	7.65	7.80	7.11	–	8.49
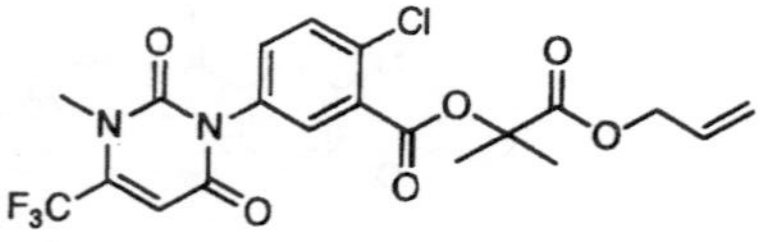	CGA-276854	7.06	7.30	6.48	–	7.97
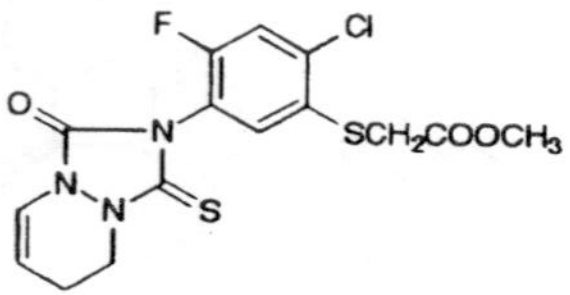	BW-66	6.37	5.65	6.10	48.0	7.21
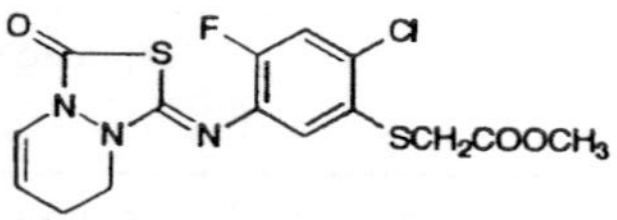	NCl-876648	6.86	6.90	6.82	70.2	8.80

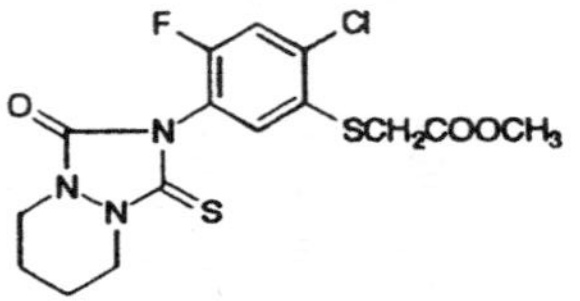	BW-110	6.66	6.57	6.40	66.5	8.08
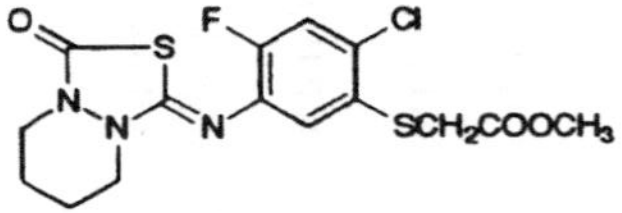	Fluthiacet-methyl	6.50	5.57	6.02	52.8	7.32
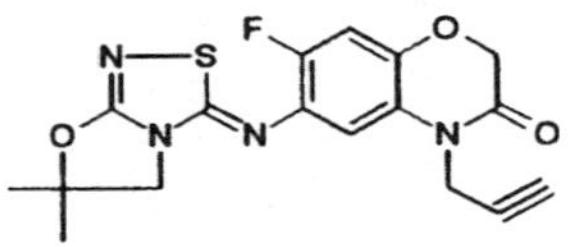	Thidiazimin	5.78	5.78	5.06	25.8	8.52
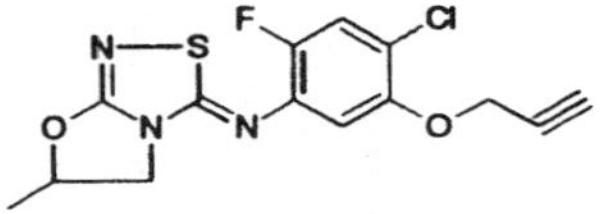	NS-1552	7.60	7.25	6.36	55.8	8.56
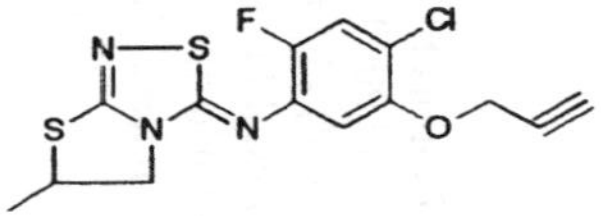	NS-1556	7.52	7.32	6.43	57.5	8.77
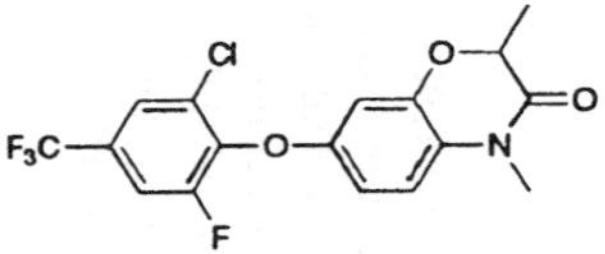	SUAM 16476	6.80	6.87	6.98	65.6	7.96
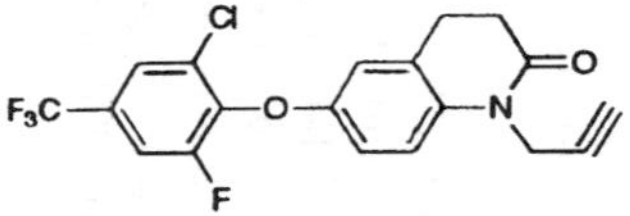	SUAM 19233	6.84	6.89	7.07	67.3	7.92

[a] Short-term proto IX formation in *Scenedesmus* (see Table 3). See Table 3 also for determination of the data.
Recently Cinidon-ethyl (BAS 615H) was introduced, a cyclic-imide type peroxidizing herbicide (ethyl(Z)-2-chloro-3-[2-chloro-5-dioxo-4,5,6,7-tetrahydroisoindol-2-yl)phenyl]acrylate).

pI_{50}(Protox) are data "in cell-level" experiments, the Eqs. (3), (4), (6) and (7) of Table 3 may be improved by adding physico-chemical lipophilicity parameters such as log P and others, taking into account absorption, translocation or distribution of the herbicides. Often lipophilicity plays an important role, when different compartments of a phytotoxic system or data obtained from cellular and cell-free assays are compared.

The light-dependent herbicidal activity exhibited by peroxidizing herbicides correlates quite well with root growth inhibition of *Echinochloa* and growth inhibition of *Scenedesmus acutus* cells [Eq. (2) in Table 3]. Equation (2) quantitates the correlation with an extremely high probability between growth inhibition and chlorophyll decrease by peroxidizing herbicides and their relatives, when applied to autotrophically grown *Scenedesmus* cells in the light. As mentioned previously, peroxidizing herbicides also decrease the chlorophyll content in heterotrophically dark-grown *Scenedesmus*, while peroxidative ethane formation is never observed. This finding is verified by Eq. (3), so that the chlorophyll decrease in the dark reflects protoporphyrinogen-IX oxidase inhibition. Chlorophyll decrease in the light does not correlate too well with the protoporphyrinogen-IX oxidase inhibition and the photo-peroxidative ethane formation, respectively. Chlorophyll decrease, however, exhibits a quite significant correlation with the two phytotoxic variables combined [cf. Eq. (7) with Eqs. (4) and (5)]. Equation (7) can be improved by adding the physicochemical log P parameter (unpubl. result of our group). The decrease of the carotenoid content in autotrophically grown *Scenedesmus* cells in the light can be understood as a follow-up effect after inhibition of chlorophyll biosynthesis and photo-peroxidative destruction of thylakoid membranes [see Eq. (8)].

QSAR research of peroxidizing herbicides and their relatives is discussed in Chapters 4 and 5. Especially in QSAR studies of cyclic imides (CyI), fore-running data obtained from the activity – activity correlations have contributed to identifying significant structure – activity relationships. These could be related to all phytotoxic parameters measured with *Echinochloa* and *Scenedesmus* like chlorophyll bleaching, growth, protox inhibition, and – most importantly – to herbicidal activity. Surprisingly the same physicochemical variables (STERIMOL parameters, L and B_5) could be applied for all phytotoxic data.

References

Becerril JM, Duke SO (1989) Protoporphyrin IX content correlates with activity of photobleaching herbicides. Plant Physiol 90:1175–1180

Becerril JM, Duke MV, Duke SO (1992) Light control of porphyrin accumulation in acifluorfen-treated *Lemna paucicostata*. Physiol Plant 86:6–16

Böger P, Nicolaus B (1993) Ethane formation by peroxidizing herbicides. In: Böger P, Sandmann G (eds) Target assays for modern herbicides and related phytotoxic compounds. Lewis Publ, Boca Raton, pp 51–60

Böger P, Sandmann G (1990) Modern herbicides affecting typical plant processes. In: Bowers WS, Ebing W, Martin D, Wegler R (eds) Chemistry of plant protection, vol 6. Springer, Berlin-Heidelberg-New York, pp 174–216

Böger P, Sandmann G (1993) Target assays for modern herbicides and related phytotoxic compounds. Lewis Publ, Boca Raton

Böger P, Wakabayashi K (1995) Peroxidizing herbicides (I): mechanism of action. Z Naturforsch 50c:159–166

Camadro JM, Matringe M, Scalla R, Labbe P (1991) Kinetic studies on protoporphyrinogen oxidase inhibition by diphenyl ether herbicides. Biochem J 277:17–21

Camadro JM, Matringe M, Brouillet N, Thomé F, Labbe P (1994) Characterization of plant and yeast protoporphyrinogen oxidase. In: Duke SO, Rebeiz CA (eds) Porphyric pesticides. ACS Symp Ser 559. Am Chem Soc, Washington, DC, pp 81–90

Choi KW, Han O, Lee HJ, Yun YC, Moon YH, Kim M, Kuk YI, Han SU, Guh JO (1998) Generation of resistance to the diphenyl ether herbicide, oxyfluorfen, via expression of the *Bacillus subtilis* protoporphyrinogen oxidase gene in transgenic tobacco plants. Biosci Biotechnol Biochem 62:558–560

Dayan FE, Duke SO (1996) Porphyrin-generating herbicides. Pestic Outlook 7:22–27

Dayan FE, Duke SO (1997) Phytotoxicity of protoporphyrinogen oxidase inhibitors: phenomenology, mode of action and mechanisms of resistance. In: Roe RM, Burton JD, Kuhr RJ (eds) Herbicide activity, toxicology, biochemistry and molecular biology. IOS Press, Amsterdam, pp 11–35

Duke SO, Nandihalli UB, Lee HJ, Duke MV (1994) Protoporphyrinogen oxidase as the optimal herbicide site in the porphyrin pathway. In: Duke SO, Rebeiz CA (eds) Porphyric pesticides. ACS Symp Ser 559. Am Chem Soc, Washington, DC, pp 191–204

Fadayomi O, Warren GF (1976) The light requirement for herbicidal activity of diphenyl ethers. Weed Sci 24:598–600

Gillham DJ, Dodge AD (1987) Studies into the action of the diphenyl ether herbicides acifluorfen and oxyfluorfen I: activation by light and oxygen in leaf tissue. Pestic Sci 19:19–24

Horikoshi M, Mametsuka K, Hirooka T (1998) Molecular breeding of photobleaching herbicide-resistant plant (II). Abstr. B 112, Annu. Meeting Pestic. Soc. Japan, Matsue (in Japanese)

Hoshi T, Koizumi K, Sato Y, Wakabayashi K (1993) Hydrolysis and phytotoxic activity of N-Aryl-3,4,5,6-tetrahydrophthalimides. Biosci Biotech Biochem 57:1913–1915

Ihara T, Iida T, Takasuka S, Kohno H, Sato Y, Nicolaus B, Böger P, Wakabayashi K (1995) Peroxidizing phytotoxic activity of 1,3,4-thiadiazolidines and 1,2,4-triazolidines. J Pestic Sci 20:41–47

Iida T, Senoo S, Sato Y, Nicolaus B, Wakabayashi K, Böger P (1995) Isomerization and peroxidizing phytotoxicity of thiadiazolidine-thione compounds. Z Naturforsch 50c:186–192

Iwata S, Sumida M, Nakayama N, Tanaka T, Wakabayashi K, Böger P (1994) Porphyrin synthesis in liverwort cells induced by peroxidizing herbicides. In: Duke SO, Rebeiz CA (eds) Porphyric pesticides. ACS Symp Ser 559, Am Chem Soc, Washington, DC, pp 147–160

Jacobs JM, Jacobs NJ (1994) Factors affecting protoporphyrin accumulation in plants treated with diphenyl ether herbicides. In: Duke SO, Rebeiz CA (eds) Porphyric Pesticides. ACS Symp Ser 559, Am Chem Soc, Washington, DC, pp 105–119

Jacobs JM, Jacobs NJ, Sherman TD, Duke SO (1991) Effect of diphenyl ether herbicides on oxidation of protoporphyrinogen to protoporphyrin in organellar and plasma membrane enriched fractions of barley. Plant Physiol 97:197–203

Jacobs JM, Wehner JM, Jacobs NJ (1994) Porphyrin stability in plant supernatant fractions: implications for the action of porphyrinogenic herbicides. Pestic Biochem Physiol 50:23–30

Jacobs JM, Jacobs NJ, Duke SO (1996) Protoporphyrinogen destruction by plant extracts and correlation with tolerance to protoporphyrinogen oxidase-inhibiting herbicides. Pestic Biochem Physiol 55:77–83

Kohno H, Hirai K, Hori M, Sato Y, Böger P, Wakabayashi K (1993) New peroxidizing herbicides: activity compared with X-ray structure. Z Naturforsch 48c:334–338

Kohno H, Ogino C, Iida T, Takasuka S, Sato Y, Nicolaus B, Böger P, Wakabayashi K (1995) Peroxidizing phytotoxic activity of pyrazoles. J Pesticide Sci 20:137–143

Kotzabasis K, Breu V, Dörnemann D (1989) The inhibitory effect of 4,5-dioxovalerate on 5-aminolevulinate dehydratase and its implication in the regulation of light-dependent chlorophyll formation in pigment mutant C-2A′ of *Scenedesmus obliquus*. Biochim Biophys Acta 977:309–314

Kunert KJ (1984) The diphenyl-ether herbicide oxyfluorfen: a potent inducer of lipid peroxidation in higher plants. Z Naturforsch 39c:476–481

Kunert KJ, Böger P (1981a) The bleaching effect of the diphenyl ether oxyfluorfen. Weed Sci 29:169–173

Kunert KJ, Böger P (1981b) The diphenyl ether herbicide oxyfluorfen: action of antioxidants. J Agric Food Chem 32:725–728

Kunert KJ, Dodge AD (1989) Herbicide-induced radical damage and antioxidative systems. In: Böger P, Sandmann G (eds) Target sites of herbicide action. CRC Press, Boca Raton, pp 45–63

Lambert R, Sandmann G, Böger P (1983) Correlation between structure and phytotoxic activities of nitrophenyl ethers. Pestic Biochem Physiol 19:309–320

Lambert R, Kroneck PMH, Böger P (1984) Radical formation and peroxidative activity of phytotoxic diphenyl ethers. Z Naturforsch 39c:486–491

Lee HJ, Duke MV, Duke SO (1993) Cellular localization of protoporphyrinogen-oxidizing activities of etiolated barley (*Hordeum vulgare* L.) leaves. Plant Physiol 102:881–889

Lehnen LP Jr, Sherman TD, Becerril JM, Duke SO (1990) Tissue and cellular localization of acifluorfen-induced porphyrins in cucumber cotyledons. Pestic Biochem Physiol 37:239–248

Lydon J, Duke SO (1988) Porphyrin synthesis is required for photobleaching activity of the *p*-nitrosubstituted diphenyl ether herbicides. Pestic Biochem Physiol 31:74–83

Matringe M, Dufour JL, Lehrminier J, Scalla R (1986) Characterization of the experimental herbicide LS 82-556. Pestic Biochem Physiol 26:150–159

Matringe M, Scalla R (1988) Studies on the mode of action of acifluorfen-methyl in non-chlorophyllous soybean cells. Plant Physiol 86:619–622

Matringe M, Camadro JM, Labbe P, Scalla R (1989) Protoporphyrinogen oxidase as a molecular target for diphenyl ether herbicides. Biochem J 260:231–235

Matsumoto H, Lee JJ, Ishizuka K (1994) Variation in crop response to protoporphyrinogen oxidase inhibition. In: Duke SO, Rebeiz CA (eds) Porphyric pesticides. ACS Symp Ser 559, Am Chem Soc, Washington, DC, pp 120–132

Matsunaka S (1969) Acceptor of light energy in photoactivation of diphenyl ether herbicides. J Agric Food Chem 17:171–175

Nandihalli UB, Duke MV, Duke SO (1992) Quantitative structure – activity relationships of protoporphyrinogen oxidase-inhibiting diphenyl ether herbicides. Pestic Biochem Physiol 43:193–211

Nicolaus B (1993) Zur peroxidativen Wirkungsweise von Herbiziden. PhD Thesis, Konstanz, Germany

Nicolaus B, Sandmann G, Watanabe H, Wakabayashi K, Böger P (1989) Herbicide-induced peroxidation: influence of light and diuron on protoporphyrin IX formation. Pestic Biochem Physiol 35:192–201

Nicolaus B, Sandmann G, Böger P (1993) Molecular aspects of herbicide action on protoporphyrinogen oxidase. Z Naturforsch 48c:326–333

Nicolaus B, Johansen JN, Böger P (1995) Binding affinities of peroxidizing herbicides to protoporphyrinogen oxidase from corn. Pestic Biochem Physiol 51:20–29

Nurit F, Ravanel P, Tissut M (1988) The photodependent effect of LS 82556 and acifluorfen in cucumber (*Cucumis sativus* L.) cotyledons. Pestic Biochem Physiol 31:67–73

Ogino C, Hoshi T, Iida T, Koura S, Ogawa H, Kohno H, Sato Y, Takai M, Wakabayashi K (1994) Peroxidizing phytotoxic activity of thiadiazolidine and triazolidine compounds. J Pesticide Sci 18:369–373

Orr GL, Hess FD (1981) Characterization of herbicidal injury by acifluorfen-methyl in excised cucumber (*Cucumis sativus* L.) cotyledons. Pestic Biochem Physiol 16:171–178

Orr GL, Hess FD (1982) Mechanism of action of the diphenyl ether herbicide acifluorfen-methyl in excised cucumber (*Cucumis sativus* L.) cotyledons. Plant Physiol 69:502–507

Orr GL, Elliott CM, Hogan ME (1983) Activity in vivo and redox states in vitro of nitro- and chloro-diphenyl ether herbicide analogs. Plant Physiol 73:939–944

Polle A (1996) Mehler reaction: friend or foe in photosynthesis. Botan Acta 109:84–89

Prendeville GN, Warren GF (1977) Effect of four herbicides and two oils on leaf-cell membrane permeability, Weed Res 17:251–258

Rebeiz CA, Montazer-Zouhoor A, Mayasich JM, Tripathy BC, Wu SM, Rebeiz CC (1987) Photodynamic herbicides and chlorophyll biosynthesis modulators. In: Heitz JR, Downum KR (eds) Light activated pesticides. ACS Symp. Ser 339, Am Chem Soc, Washington, DC, pp 295–328

Retzlaff K, Böger P (1996) An endoplasmic reticulum plant enzyme has protoporphyrinogen IX oxidase activity. Pestic Biochem Physiol 54:105–114

Sandmann G, Böger P (1982a) Volatile hydrocarbons from photosynthetic membranes containing different fatty acids. Lipids 17:35–41

Sandmann G, Böger P (1982b) Formation and degradation of photosynthetic membranes determined by ^{35}S-labeled sulfolipid, Plant Sci Lett 24:347–352

Sandmann G, Böger P (1983a) Comparison of the bleaching activity of norflurazon and oxyfluorfen. Weed Sci 31:338–341

Sandmann G, Böger P (1983b) Peroxidative formation of C_3-hydrocarbons from ω-4 polyunsaturated fatty acid (16:3, ω4) in the alga *Bumilleriopsis*. Lipids 18:37–41

Sandmann G, Böger P (1986) Site of herbicide inhibition at the photosynthetic apparatus. In: Staehelin LA, Arntzen CJ (eds) Encyclopedia of plant physiology. New series, vol 19. Photosynthesis III. Springer, Berlin, Heidelberg, New York, pp 595–602

Sandmann G, Böger P (1988) Accumulation of protoporphyrin IX in the presence of peroxidizing herbicides. Z Naturforsch 43c:699–704

Sandmann G, Clarke IE, Bramley PM, Böger P (1984a) Inhibition of phytoene desaturase – the mode of action of certain bleaching herbicides. Z Naturforsch 39c:443–449

Sandmann G, Reck H, Böger P (1984b) Herbicidal mode of action on chlorophyll formation. J Agric Food Chem 32:868–872

Sandmann G, Böger P, Kumita I (1990a) A typical inhibition of phytoene desaturation by 2-(4-chloro-2-nitrobenzoyl)-5,5-dimethylcyclohexane-1,3-dione. Pestic Sci 30:353–355

Sandmann G, Nicolaus B, Böger P (1990b) Typical peroxidative parameters verified with mungbean seedlings, soybean cells and duckweed. Z Naturforsch 45c:512–517

Sato R, Nagano E, Oshio H, Kamoshita K (1987) Diphenylether-like physiological and biochemical actions of S-23142, a novel *N*-phenyl imide herbicide. Pestic Biochem Physiol 28:194–200

Sato R, Nagano E, Oshio H, Kamoshita K (1988) Activities of the *N*-phenyl imide S-23142 in carotenoid-deficient seedlings of rice and cucumber. Pestic Biochem Physiol 31:213–220

Sato Y, Hoshi T, Iida T, Ogino C, Nicolaus B, Wakabayashi K, Böger P (1994) Isomerization and peroxidizing phytotoxicity of thiadiazolidine herbicides. Z Naturforsch 49c:49–56

Senoo S, Iida T, Shouda K, Sato Y, Nicolaus B, Böger P, Wakabayashi K (1996) Enzyme-modified phytotoxic structure of thiadiazolidine compounds. Z Naturforsch 51c:518–526

Sherman TD, Becerril JM, Matsumoto H, Duke MV, Jacobs JM, Jacobs NJ, Duke SO (1991) Physiological basis for differential sensitivities of plant species to protoporphyrinogen oxidase-inhibiting herbicides. Plant Physiol 97:280–287

Shouda K, Iida T, Uchida A, Kohno H, Sato Y, Nicolaus B, Böger P, Wakabayashi K (1996) Peroxidizing phytotoxicities of 1,2-alkylene-1,2,4-triazolidines and 3,4-alkylene-1,3,4-thiadiazolidines. J Pestic Sci 21:187–193

Stetter J (1994) Herbicides inhibiting branched-chain amino acid biosynthesis – recent developments. In: Stetter J (ed) Chemistry of plant protection, vol 10. Springer, Berlin Heidelberg New York

Sumida M, Niwata S, Fukami H, Tanaka T, Wakabayashi K, Böger P (1995) Synthesis of novel diphenyl ether herbicides. J Agric Food Chem 43:1929–1934

Sumida M, Kohno H, Shouda K, Fukami H, Tanaka T, Wakabayashi K (1996a) Protoporphyrinogen-IX oxidase inhibition of new peroxidizing diphenyl ethers. J Pestic Sci 21:317–321

Sumida M, Niwata S, Tanaka T, Furono T, Nakanishi M, Wakabayashi K, Böger P (1996b) Accumulation of protoporphyrinogen IX induced by acifluorfen methyl. Z Naturforsch 51c:174–178

Suzuki S, Watanabe H, Jikihara T, Ohta H, Wakabayashi K (1976) Structure - activity relationship of cyclic imide herbicides. In: Kataoka T (ed) Proceedings of the 5th Asian Pacific Weed Science Society Conference, Asian Pacific Weed Science Society, Tokyo, pp 176–182

Teraoka T, Sandmann G, Böger P, Wakabayashi K (1987) Effect of cyclic imide herbicides on pigment formation in plants. J Pestic Sci 12:499–504

Tietjen KG (1991) Quinone activation of protoporphyrinogen oxidase of barley plastids. Pestic Sci 33:467–471

Wakabayashi K, Böger P (1993) Peroxidizing herbicides: mechanism of action and molecular design. In: Mitsui T, Matsumura F, Yamaguchi I (eds) Pesticides/environment: molecular biological approaches. Pesticide Science Society of Japan, Tokyo, pp 239–253

Wakabayashi K, Böger P (1995) Peroxidizing herbicides (II): structure - activity relationship and molecular design. Z Naturforsch 50c:591–601

Wakabayashi K, Matsuya K, Ohta H, Jikihara T (1979) Structure - activity relationship of cyclic imide herbicides. In: Geissbühler H (ed) Advances in pesticide science, part 2. Pergamon Press, Oxford, pp 256–260

Wakabayashi K, Matsuya K, Teraoka T, Sandmann G, Böger P (1986) Effect of cyclic imide herbicides on chlorophyll formation in higher plants. J Pestic Sci 11:635–640

Wakabayashi K, Sandmann G, Ohta H, Böger P (1988) Peroxidizing herbicides: comparison of dark and light effect. J Pestic Sci 13:461–471

Ward ER, Volrath S (1995) Manipulation of protoporphyrinogen oxidase enzyme activity in eukaryotic organisms. PCT Int Appl WO 95/34659, Dec 21, 1995, PCT/IB95/00452, p 112

Watanabe H, Ohori Y, Sandmann G, Wakabayashi K, Böger P (1992) Quantitative correlation between short-term accumulation of protoporphyrin IX and peroxidative activity of cyclic imides. Pestic Biochem Physiol 42:99–109

Yamato S, Katagiri M, Ohkawa H (1994) Purification and characterization of a protoporphyrinogen-oxidizing enzyme with peroxidase activity and light-dependent herbicide resistance in tobacco cultured cells. Pestic Biochem Physiol 50:72–82

Yoshida S, Ichinose K, Wang JM, Asami T, Che FS (1993) Characterization of protoporphyrinogen oxidase in chloroplasts. In: Mitsui T, Matsumura F, Yamaguchi I (eds) Pesticides/environment: molecular biological approaches. Pesticide Science Society of Japan, Tokyo, pp 227–237

You JL, Fong FK (1986) Superoxide photogeneration by chlorophyll a in water/acetone. Electron spin resonance studies of radical intermediates in chlorophyll a photoreaction in vitro. Biochem Biophys Res Commun 139:1124–1129

CHAPTER 7

Metabolic Modification of Isoimide Type Peroxidizing Compounds Catalyzed by An Isoenzyme of Glutathione *S*-Transferase

YUKIHARU SATO[1]

Contents

7.1 Introduction

Peroxidizing herbicides cause membrane destruction in plants by inhibiting the membrane-bound protoporphyrinogen oxidase (protox, EC 1.3.3.4) competitively (Nicolaus et al. 1995). This inhibition induces the accumulation of protoporphyrin IX (Lydon and Duke 1988; Matringe and Scalla 1988; Sandmann and Böger 1988; Wittkowski and Halling 1988) which is sensitized by light with subsequent radical formation leading to degradation of cellular constituents with evolution of ethane (Böger and Sandmann 1990; Böger and Wakabayashi 1995).

Differences in protoporphyrin IX (proto IX) accumulation were correlated with the herbicidal injury obtained (Sherman et al. 1991; Nandihalli et al. 1992; Watanabe et al. 1992). Tolerance against peroxidizers requires an antioxidative radical-scavenging system or a metabolic breakdown of the herbicide (Böger and Sandmann 1993; Cole 1994). It is becoming obvious that soybean and peanuts are tolerant to diphenyl ethers due to detoxification caused by making a conjugation with glutathione *S*-transferase (GST) (Eastin 1971; Frear et al. 1983).

To establish a suitable weed control by peroxidizing herbicides, the selectivity of each peroxidizer should be found. Herbicide selectivity is determined by several distinct factors, such as rates of absorption into the plant and subsequent translocation, subcellular localization, variation in target site sensitivity

[1] Tamagawa University, Graduate School of Agricultural Science, Physiology and Biochemistry Division, 6-1-1 Tamagawa Gakuen, Machida-shi, Tokyo 194

Peter Böger, Ko Wakabayashi
Peroxidizing Herbicides

and the metabolic detoxification to less toxic substances. Because protox is present in all photosynthetic plants, it is necessary to study the differential metabolism of the peroxidizing herbicides which may influence their interaction with the target site in plants.

Hoshi et al. (1993) and Sato et al. (1997) found that some *N*-aryl-3,4,5,6-tetrahydroisophthalimides were converted into corresponding *N*-aryl-3,4,5,6-tetrahydrophthalimides in the presence of *Echinochloa utilis* seedlings, and equine glutathione *S*-transferase (GST) with a reduced glutathione (GSH). One type of 5-arylimino-3,4-tetramethylene-1,3,4-thiadiazolidines showed their herbicidal activity after isomerization into their corresponding isomers, 4-aryl-1,2-tetramethylene-1,2,4-triazolidines in the culture of *E. utilis* and *Scenedesmus acutus*, and equine GST with GSH (Sato et al. 1994a,b, 1995; Senoo et al. 1996). However, another type of 5-arylimino-3,4-tetramethylene-1,3,4-thiadiazolidines were not isomerized into their corresponding isomers in the same conditions (Iida et al. 1995; Sato et al. 1995). Nicolaus et al. (1996a,b) revealed that this isomerization was caused by a GST isoenzyme isolated from corn (*Zea mays* var. Anjou) with some SH compounds, especially with GSH. Shimizu et al. (1995) reported that isourazole type herbicide, fluthiacet-methyl [5-(4-chloro-2-fluoro-5-methoxycarbonylmethylthiophenylimino)-3,4-tetramethylene-1,3,4-thiadiazolidin-2-one] was converted into its urazole type compound [4-(chloro-2-fluoro-5-methoxycarbonylmethylthiophenyl)-1,2-tetramethylene-1,2,4-triazolidin-3-one-5-thione] in the presence of the partially purified GST from velvetleaf (*Abutilon theophrasti* Medic). Uchida et al. (1997) discussed the structural necessity of thiadiazolidine compounds for isomerization.

In this chapter, the isomerization of cyclicisoimide type peroxidizing compounds to the corresponding cyclic imide types is described.

7.2 Intrinsic Active Structure of Cyclic Imide Type Peroxidizing Compounds

N-aryl-3,4,5,6-tetrahydrophthalimides (**1**, **3**, **5**, **7** and **9** in Table 1) and their hydrolysates *N*-aryl-3,4,5,6-tetrahydrophthalamic acids (**2**, **4**, **6**, **8** and **10** in Table 1) are peroxidizing compounds. They have close phytotoxic activities, and have the same mode of action by analyzing the quantitative structure-activity relationship (QSAR) (Wakabayashi et al. 1979). Sato et al. (1991) investigated their phytotoxic activities against *E. utilis* and *S. acutus*, and the relationship of interconversion between the imides and amide acids during bioassay, to explain their similar phytotoxic activities.

In almost all cases, the imides showed stronger activity than the corresponding amide acids. During a 7-day culture of *Echinochloa*, three types of interconversion between the imides and amide acids were found: type I (**1** and **2**, and **3** and **4**) – the imides were hydrolyzed to amide acids, and the amide

Table 1. Phytotoxicities of imides and amide acids

Compounds	pI_{50}					Proto IX[e] (nmol/ml pcv per hour)
	Ech.[a]	*Sce.*[b]	Chl.[b]	Ethane[c]	Protox[d]	
N-(4-Chlorophenyl)-3,4,5,6-tetrahydrophthalimide (**1**)	5.82	7.06	7.07	6.04	7.60	68.2
N-(4-Chlorophenyl)-3,4,5,6-tetrahydrophthalamic acid (**2**)	5.55	5.78	5.40	–	–	17.6
N-(4-Chloro-2-methylphenyl)-3,4,5,6-tetrahydrophthalimide (**3**)	5.61	6.18	5.08	–	–	57.8
N-(4-Chloro-2-methylphenyl)-3,4,5,6-tetrahydrophthalamic acid (**4**)	5.36	5.00	4.26	–	–	0.8
N-(4-Chloro-2-fluorophenyl)-3,4,5,6-tetrahydrophthalimide (**5**)	6.72	7.43	7.61	6.62	8.43	63.7
N-(4-Chloro-2-fluorophenyl)-3,4,5,6-tetrahydrophthalamic acid (**6**)	5.91	6.77	6.42	–	–	27.1
N-(4-Chloro-2-fluoro-5-propargyloxyphenyl)-3,4,5,6-tetrahydrophthalimide (**7**)	7.98	7.80	7.80	6.85	8.74	74.4
N-(4-Chloro-2-fluoro-5-propargyloxyphenyl)-3,4,5,6-tetrahydrophthalamic acid (**8**)	8.07	7.20	7.00	–	–	29.2
N-{4-(4-Chlorobenzyloxy)phenyl}-3,4,5,6-tetrahydrophthalimide (**9**)	6.28	8.40	8.52	7.95	8.30	83.5
N-{4-(4-Chlorobenzyloxy)phenyl}-3,4,5,6-tetrahydrophthalimide (**10**)	5.40	7.70	7.60	–	–	54.8

[a] pI_{50}(*Ech.*) = $-\log$[molar I_{50} of root growth of *Echinochloa utilis*]. This value indicates phytotoxic activity, since a good correlation has been found between the pI_{50}(*Ech.*) value and phytotoxic activity against various kinds of weeds.
[b] pI_{50}(*Sce.*) and pI_{50}(Chl.) refer to the 50% growth inhibition of *Scenedesmus acutus* and 50% decrease of chlorophyll content in algal cell, respectively.
[c] pI_{50}(Ethane) = $-\log I_{50}$(Ethane). This value indicates 50% formation of ethane in *S. acutus* cell (Lambert et al. 1983).
[d] pI_{50}(Protox) = $-\log_{50}$(Protox). This refers to 50% inhibition of isolated maize protoporphyrinogen oxidase.
[e] Protot IX indicates the short-term accumulation of protoporphyrin IX after 1 h incubation of *S. acutus* in presence of a compound (10^{-6} M). This value correlates well with pI_{50}(Chl.), indicating peroxidative destruction of photosynthetic pigments.

Table 2. Phytotoxicity of *N*-(4-bromophenyl)-3,4,5,6-tetrahydroisophthalimide and *N*-(4-bromophenyl)-3,4,5,6-tetrahydrophthalimide

Compounds	pI_{50}			
	Ech.[a]	Chl.[b]	Ethane[c]	Protox[d]
N-(4-Bromophenyl)-3,4,5,6-tetrahydroisophthalimide (11)	6.09	5.65	6.40	7.37
N-(4-Bromophenyl)-3,4,5,6-tetrahydrophthalimide (12)	6.04	5.91	6.47	7.55

[a-d] See footnotes to Table 1.

acids were cyclized to the imide and hydrolyzed to the corresponding aniline; type II (**5** and **6**, and **7** and **8**) – the imides were hydrolyzed to amide acids and further to the corresponding aniline, and the amide acids were cyclized to imides and hydrolyzed to the corresponding aniline; type III (**9** and **10**) – the imide was not hydrolyzed to an amide acid in 7 days of culture, and a small amount of the amide acid was cyclized to the imide and hydrolyzed. The introduction of fluorine at the 2-position, and of a propargyloxy group at the 5-position on the 4-halophenyl ring increased the hydrolysis of both the imides and amide acids. This result is the first evidence indicative of a bioactivation through isomerization of cyclic imide compounds.

Hoshi et al. (1993) and Sato et al. (1997) synthesized *N*-(4-bromophenyl)-3,4,5,6-tetrahydroisophthalimide (**11**), *N*-(4-chlorophenyl)-3,4,5,6-tetrahydro-iso-phthalimide (**13**), *N*-(4-chloro-2-methylphenyl)-3,4,5,6-tetrahydroiso-phthalimide (**14**), *N*-(4-chloro-2-fluorophenyl)-3,4,5,6-tetrahydroisophthalimide (**15**), *N*-(4-chloro-2-fluoro-5-propargyloxyphenyl)-3,4,5,6-tetrahydro-isophthalimide (**16**) and *N*-[4-(4-chlorobenzyloxy)phenyl]-3,4,5,6-tetrahydro-phthalimide (**17**), and investigated the phytotoxic action and hydrolytic conversion of isoimides using *E. utilis* and *S. acutus* bioassay.

All the isoimide analogs assayed had strong herbicidal activity; the pI_{50}(*Ech.*) of **11**, **13–17** was 6.09, 5.90, 5.63, 6.25, 8.55 and 6.37, respectively. As shown in Table 2, the isoimide **11** caused almost the same type of phytotoxic activities as the corresponding imide **12**. The amide acids were less active than isoimides. Other isoimides (**13–17**) caused analogous phytotoxic symptoms to that of the isoimide **11**. This result demonstrates that all the isoimides tested are peroxidizing herbicides.

The isoimides are the isomers of the imide, and their phytotoxic activities against the root growth of *E. utilis* are close to those of imides and amide acids. To inspect this herbicidal action of isoimides, conversion of isoimides to imides and/or amide acids in the culture medium for *Echinochloa* was investigated. As shown in Table 3, after 7 days of incubation, about 70–80% of isoimides **13–16** hydrolyzed to the amide acids **2**, **4**, **6** and **8**, about 7–20% of isoimides was further hydrolyzed to the corresponding anilines and 3,4,5,6-

Table 3. Conversion of isoimides in *Echinochloa utilis* culture after 7 days of incubation (in molar percent)

Compounds	Isoimide (%)	Imide (%)	Amide acid (%)	Corresponding aniline (%)	3,4,5,6-Tetrahydrophthalic acid (%)	Half-life time (day)[a]
13	0.0	20.0	68.9	11.1	11.1	0.8
14	0.0	11.7	82.0	6.3	6.3	0.3
15	0.0	6.5	83.7	9.8	9.8	0.0
16	0.0	8.4	79.2	12.4	12.4	0.6
17	16.0	13.6	68.4	2.0	2.0	2.4

[a] Half-life time of isoimides.

tetrahydrophthalic acid, about 6–12% of isoimides was isomerized to imides **1**, **3**, **5** and **7** presumably via the amide acids, and of the starting isoimides no more remained. However, 16% of isoimide **17** still remained without any change, although 13.6% of **17** was isomerized to **9**, 68.4% of **17** was hydrolyzed to **10**, and 2% of **17** was further hydrolyzed to the corresponding aniline and 3,4,5,6-tetrahydrophthalic acid. The slower hydrolysis rate of the isoimide **17** may be due to the bulky *p*-substituent on the *N*-aryl ring.

Conversion of *N*-(4-bromophenyl)-3,4,5,6-tetrahydroisophthalimide (**11**) into corresponding imide and amide acid was investigated under three conditions as shown in Fig. 1 by Sato et al. (1997). As shown in Fig. 1a, with the reaction in buffer without GST, the decrease of isoimide **11** and the increase of both imide **12** and 4-bromophenyl-3,4,5,6-tetrahydrophthalamic acid (amide acid) began 3 min after start of the reaction. A rapid accumulation of the amide acid started 10 min after start. At the end of incubation (2 h), the percent of the amounts of the isoimide, the imide, the amide acid and 4-bromoaniline was 5.1, 52.3, 39 and 3.6%, respectively. The changes in amount of the four compounds during the conversion experiment with equine GST were almost the same as found in buffer solution without GST (Fig. 1b). Figure 1c shows the conversion of isoimide **11** with GST and GSH present. Isoimide **11** began to convert into imide **12** at the beginning of the incubation, and 49.4% of the isoimide was isomerized to the imide already 1 min after start. The maximum isomerization (92.7%) of the isoimide was observed 10 min after start, and at this point the amide acid appeared in the test solution and decrease of the imide began. At the end of incubation, the percent of the amounts of the isoimide, the imide, the amide acid and 4-bromoaniline was 0, 56.8, 42.0 and 1.2%, respectively. These results indicate that GST with GSH isomerized the isoimide to the imide rapidly and directly. Without enzyme and cofactor, however, the isoimide was converted into the imide by hydrolysis and recyclization in buffer at pH 6.8, because the hydrolysis of the isoimide and the accumulation of both imide and amide acid started at almost the same time.

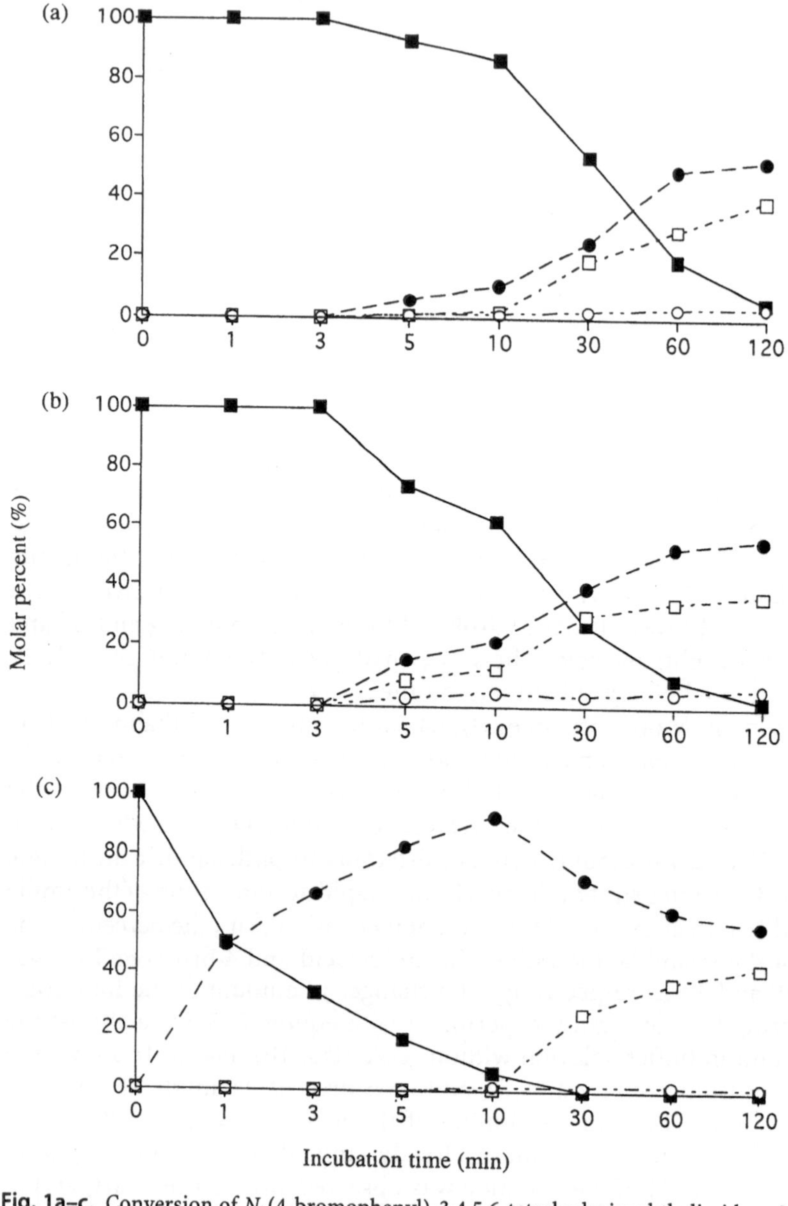

Fig. 1a–c. Conversion of *N*-(4-bromophenyl)-3,4,5,6-tetrahydroisophthalimide. **a** In potassium phosphate buffer (pH 6.8). **b** With GST present in the buffer. **c** With GST and GSH present in the buffer. *N*-(4-Bromophenyl)3,4,5,6-tetrahydroisophthalimide [1] (■); *N*-(4-Bromophenyl)-3,4,5,6-tetrahydrophthalimide [2] (●); *N*-(4-Bromophenyl)3,4,5,6-tetrahydrophthalamic acid (□); 4-bromoaniline (○). (Reproduced with permission from the Pesticide Science Society of Japan)

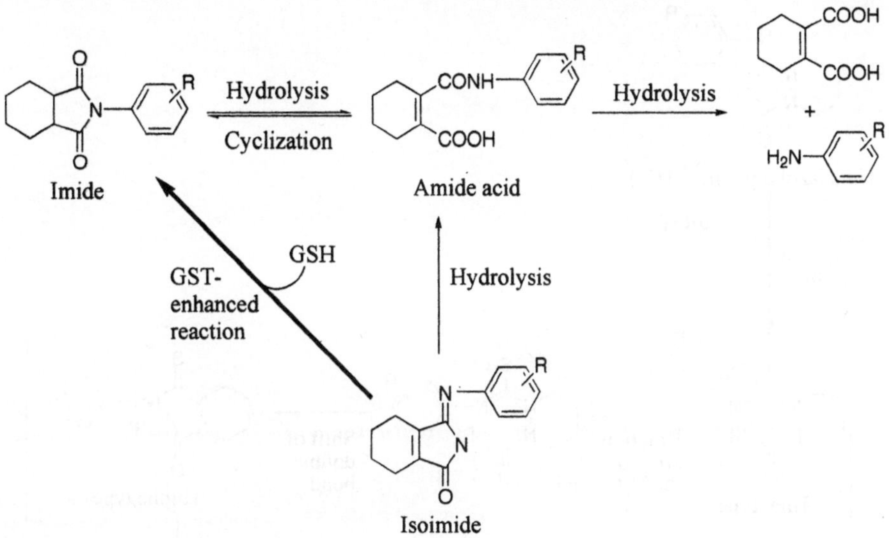

Fig. 2. Conversion of an isoimide into imide, amide acid, corresponding aniline, and 3,4,5,6-tetrahydrophthalic acid

These results indicate that (1) the intrinsic active structures among *N*-aryl-3,4,5,6-tetrahydrophthalimides, *N*-aryl-3,4,5,6-tetrahydrophthalamic acids and *N*-aryl-3,4,5,6-tetrahydroisophthalimides are *N*-aryl-3,4,5,6-tetrahydrophthalimides, (2) *N*-aryl-3,4,5,6-tetrahydroisophthalimides are isomerized into corresponding imides in the presence of equine GST and GSH and (3) GST and GSH acted only on the isoimide, not on the amide acid (Fig. 2).

The enzymatic isomerization of isoimides into imides occurs by the same mechanism as that of thiadiazolidines into triazolidines (see Fig. 3).

7.3
Isomerization of Thiadiazolidines to Triazolidines

Sato et al. (1994a,b, 1995) and Iida et al. (1995) prepared 5-arylimino-3,4-tetramethylene-1,3,4-thiadiazolidin-2-ones (thiadiazolidin-ones **18**, **22**, **26**, **30**, **34**, **38**, **42** and **46**), 4-aryl-1,2-tetramethylene-1,2,4-triazolidin-3-one-5-thiones (triazolidin-one-thiones **19**, **23**, **27**, **31**, **35**, **39**, **43** and **47**), 5-arylimino-3,4-tetramethylene-1,3,4-thiadiazolidine-2-thiones (thiadiazolidine-thiones **20**, **24**, **28**, **32**, **36**, **40**, **44** and **48**) and 4-aryl-1,2-tetramethylene-1,2,4-triazolidin-3,5-dithiones (triazolidine-dithiones **21**, **25**, **29**, **33**, **37**, **41**, **45** and **49**). Biological activities of these compounds were determined by light-induced ethane evolution, protoporphyrin IX formation, chlorophyll decrease in *S. acutus* under the light and *Echinochloa* root growth inhibition (Table 4). As discussed by Watanabe et al. (1992) for a series of cyclic imides, all these parameters

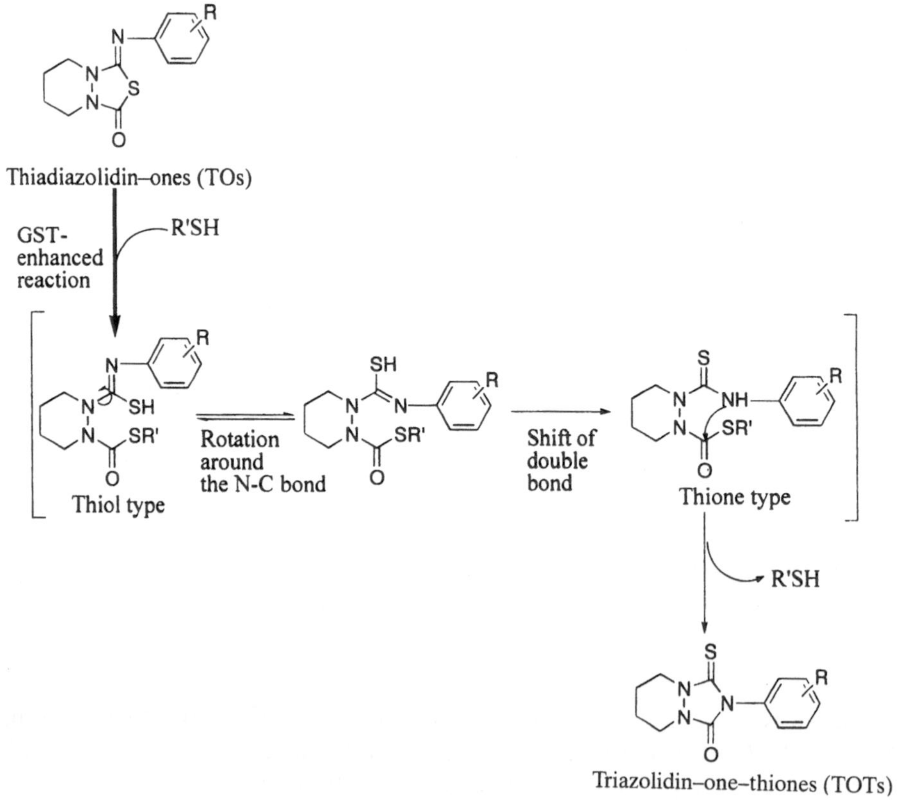

Fig. 3. Proposed mechanism of isomerization of thiadiazolidin-ones by GST and GSH

correlate quantitatively, indicative of the phytotoxic activity of the peroxidizing compound examined. The short-term accumulation of protoporphyrin IX in autotrophic *Scenedesmus* cells suggests that the chlorophyll biosynthesis pathway is interrupted by both thiadiazolidin-ones and triazolidin-one-thiones. This was confirmed by determining their inhibitory activity on isolated maize protoporphyrinogen oxidase (Table 4). Inhibition of chlorophyll biosynthesis at this step leads to subsequent accumulation of protoporphyrin IX, the product of protoporphyrinogen oxidase.

Proto IX is either formed by non-enzymatic protoporphyrinogen oxidation with molecular oxygen, or catalyzed by inhibitor-insensitive protoporphyrinogen-oxidizing enzymes present in the plasma membrane or in microsomes (Jacobs et al. 1990; Lee et al. 1993; Nandihalli and Duke 1993; Retzlaff and Böger 1996). The proto IX accumulation observed by treatment of algae cells with thiadiazolidines and triazolidines is caused by protox inhibition. The same holds for light-induced ethane evolution with autotrophic *Scenedesmus* cells as a marker for thylakoid degradation.

No marked differences in the pI_{50} values between thiadiazolidin-ones and triazolidin-one-thiones have been found when using the *Echinochloa* root test

Table 4. Peroxidizing activities of thiadiazolidines and triazolidines

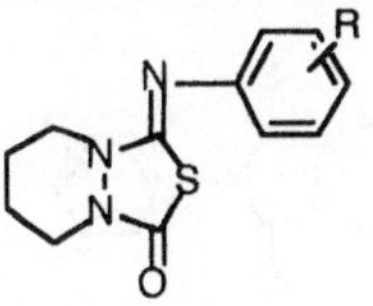	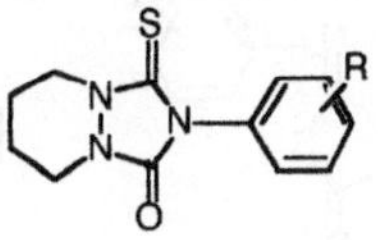	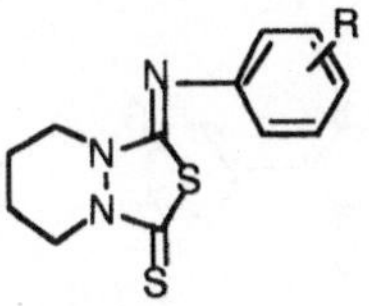	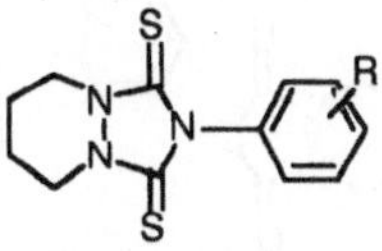
Thiadiazolidin-ones (ThOs)	Triazolidin-one-thiones (TrOTs)	Thiadiazolidine-thiones (ThTs)	Triazolidine-dithiones (TrDTs)

Compounds	pI$_{50}$				Proto IX[e] (nmol/ml pcv per hour)
	Ech.[a]	Chl.[b]	Ethane[c]	Protox[d]	
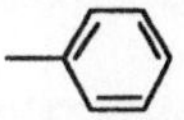					
18 (ThOs)	4.42	5.12	4.76	–	12.1
19 (TrOTs)	4.49	4.95	4.70	–	14.6
20 (ThTs)	4.82	4.86	5.36	<4.50	10.8
21 (TrDTs)	5.45	5.20	5.69	5.49	13.2
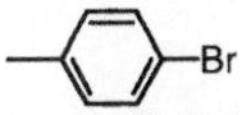					
22 (ThOs)	6.60	6.57	6.43	5.30	27.8
23 (TrOTs)	6.79	6.46	6.57	7.90	73.7
24 (ThTs)	6.61	6.83	6.26	6.13	25.3
25 (TrDTs)	7.18	7.43	7.33	8.14	67.9
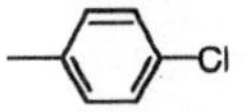					
26 (ThOs)	6.47	6.68	6.53	–	65.1
27 (TrOTs)	6.85	6.65	6.56	8.10	66.8
28 (ThTs)	6.39	6.63	6.26	5.96	20.5
29 (TrDTs)	7.05	7.36	7.30	8.17	50.3
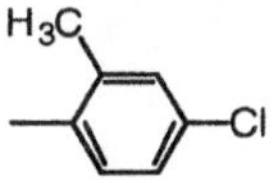					
30 (ThOs)	5.53	6.36	6.30	5.50	34.5
31 (TrOTs)	5.89	6.78	6.53	6.70	82.2
32 (ThTs)	5.54	6.56	6.29	5.12	20.3
33 (TrDTs)	5.65	7.07	7.07	6.82	63.3
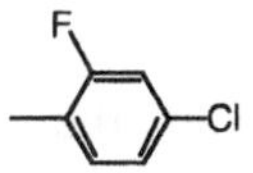					
34 (ThOs)	6.83	6.12	6.41	7.22	61.1
35 (TrOTs)	6.93	7.21	6.71	8.23	66.5
36 (ThTs)	6.80	7.17	7.04	5.96	31.5
37 (TrDTs)	7.04	7.74	7.56	8.14	70.8

Table 4. (*Continued*)

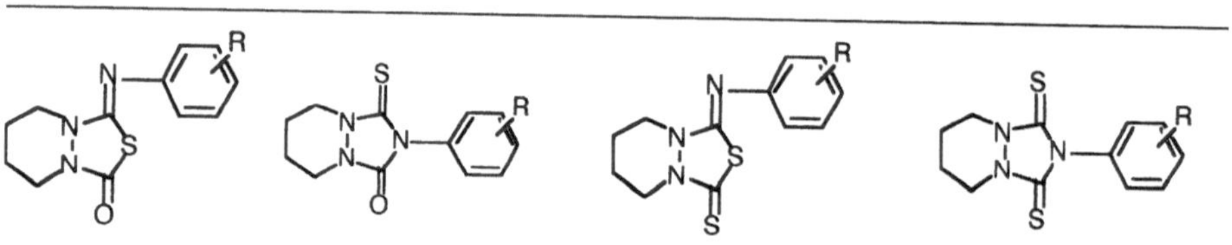

Thiadiazolidin-ones (ThOs)	Triazolidin-one-thiones (TrOTs)	Thiadiazolidine-thiones (ThTs)	Triazolidine-dithiones (TrDTs)

Compounds	pI_{50} Ech.[a]	Chl.[b]	Ethane[c]	Protox[d]	Proto IX[e] (nmol/ml pcv per hour)
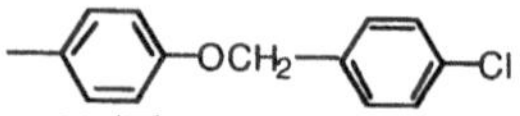					
38 (ThOs)	5.37	7.33	7.23	5.16	79.3
39 (TrOTs)	6.98	7.34	7.46	7.85	85.0
40 (ThTs)	5.26	7.70	6.92	6.09	30.7
41 (TrDTs)	6.06	7.81	7.64	7.92	80.3
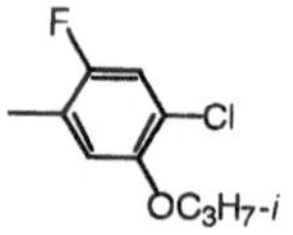					
42 (ThOs)	7.56	7.47	7.26	6.13	80.6
43 (TrOTs)	7.56	7.42	7.98	–	92.8
44 (ThTs)	7.55	7.95	6.78	6.21	30.7
45 (TrDTs)	7.91	8.28	7.17	8.92	60.2
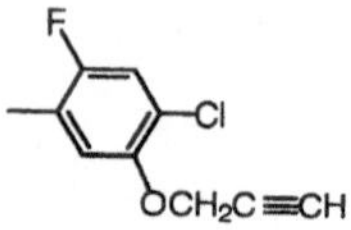					
46 (ThOs)	8.09	8.30	7.64	–	75.1
47 (TrOTs)	8.28	8.43	7.50	–	67.3
48 (ThTs)	8.23	8.26	7.10	6.96	63.5
49 (TrDTs)	8.77	8.77	8.14	8.05	87.2

[a–e] See footnotes to Table 1.

in Table 4, although triazolidin-one-thiones are a little stronger than thiadiazolidin-ones in all cases. The same holds for the ethane assay with *Scenedesmus*. These are long-term experiments. When assaying for short-term proto IX formation, however, a substantial difference was found, the thiadiazolidines being the less active forms. This difference of inhibition is even stronger in the cell-free enzymological protox inhibition assay.

These findings may be due either to a different uptake velocity of both compound types into the cells, or to a conversion of a less active form into an active one taking place during the bioassays with *Echinochloa* and

Table 5. Conversion of thiadiazolidines and triazolidines by *Echinochloa* seedling during a 7-day incubation period (in molar percent)

Compound present at start	Thiadiazolidine-form	Triazolidine-form
Thiadiazolidine-one (ThOs)		
18	1.9	98.1
22	4.1	95.9
26	5.3	94.7
30	26.2	73.8
34	4.6	95.4
38	100	0
42	34.6	65.4
46	15.6	84.4
Triazolidine-one-thione (TrOTs)		
19,23,27,31,35,39,43,47	No conversion	
Thiadiazolidine-thione (ThTs)		
20	97.6	n.d.
24	97.4	n.d.
28	92.9	n.d.
32	83.9	n.d.
36	83.7	n.d.
40	94.8	n.d.
44	95.6	n.d.
48	90.5	n.d.
Triazolidine-dithiones (ThDTs)		
21,25,29,33,37,41,45,49	No conversion	

n.d., Not detected.

Scenedesmus. Indeed, a unidirectional conversion occurs between thiadiazolidin-ones and triazolidin-one-thiones as shown in Table 5 using 7-day-old *Echinochloa* seedlings. No other peaks except the sets of thiadiazolidines and triazolidines used were detected in quantitative HPLC analysis. Starting the conversion experiment with thiadiazolidin-ones, 98.1% of **18**, 95.9% of **22**, 94.7% of **26**, 73.8% of **30**, 95.4% of **34**, 65.4% of **42** and 84.4% of **46** was converted to the corresponding triazolidin-one-thiones (**19**, **23**, **27**, **31**, **35**, **43** and **47**) respectively. Thiadiazolidin-one **38** was not converted to triazolidin-one-thione **39**. Triazolidin-one-thiones **19**, **23**, **27**, **31**, **35**, **43** and **47**, however, did not change to corresponding thiadiazolidines at all under these conditions. These findings indicate that thiadiazolidin-ones exhibit their herbicidal activities after they isomerized to triazolidin-one-thione structures.

To define whether this conversion proceeds biologically or chemically, only buffer solution was used, omitting *Echinochloa* seedlings. No structural changes of thiadiazolidin-ones and triazolidin-one-thiones were observed independent of the pH applied. These results indicate that the conversion of thiadiazolidin-ones to triazolidin-one-thiones does not occur chemically. In a spinach homogenate, 37% of **22** was converted to triazolidin-one-thione **23**, but – as expected – **23** did not convert to the thiadiazolidin-one **22**, and thiadiazolidin-one **38** and triazolidin-one-thione **39** were not converted. In a boiled spinach homogenate, thiadiazolidin-ones **22** and **38** did not convert to the corresponding triazolidin-one-thiones. These results indicate that isomerization of the thiadiazolidin-ones to triazolidin-one-thiones is catalyzed by an enzyme. As indicated by Fig. 3 it appears that thiadiazolidin-ones are hydrolyzed enzymatically to an unstable intermediate (compounds in parentheses) which can rapidly change non-enzymatically to the corresponding triazolidine derivative. Lack of isomerization of **38** can be explained by a specificity of the isomerizing enzyme for binding of certain of thiadiazolidin-ones only. The structure of **38** may not fit into a binding niche therefore excluding itself as a substrate for isomerization.

Table 6 shows the conditions of interconversion of thiadiazolidines into triazolidines. It is known that GST exhibits its conjugating activity in the

Table 6. Isomerization of thiadiazolidin-ones (ThOs) and thiadiazolidine-thiones (ThTs) to triazolidin-one-thiones (TrOTs) and triazolidine-dithiones (TrDTs), respectively (in molar percent in test solution)

Additions	Compounds					
	22 (ThOs)	⇒	**23** (TrOTs)	**24** (ThTs)	⇒	**25** (TrDTs)
Echinochloa utilis[a]	4.1		95.9	97.4		n.d.
Scenedesmus acutus[b]	66.9		33.1	93.1		n.d.
Spinach homogenate[c]	38.9		61.1	–[e]		–
Heated spinach homogenate[d]	100		0	–		–
GST	98.4		1.6	96.0		n.d.
GST + GSH	2.4		97.6	95.4		n.d.
GST + DTT	3.8		96.2	98.2		n.d.
GST + L-cysteine	20.5		79.5	92.8		n.d.
GST + thioglycollate	71.1		23.9	92.4		n.d.
GST + ethyl mercaptan	55.5		40.5	98.5		n.d.
GST + *S*-methyl-L-cysteine	97.2		2.8	96.0		n.d.
GST + methionine	96.1		3.9	94.9		n.d.
GSH	98.8		n.d.	–		–
DTT	99.4		n.d.	–		–
L-Cysteine	98.3		n.d.	–		–
Thioglycollate	99.1		n.d.	–		–
Ethyl mercaptan	98.7		n.d.	–		–

Triazolidines were not isomerized into thiadiazolidines in the same conditions.
n.d., not detected.
[a] In the culture medium in which *E. utilis* was grown (dark, 27 °C, 7 days).
[b] In the culture medium of *S. acutus* (dark, 22 °C, 20 h).
[c] Assayed at 27 °C for 24 h in the dark.
[d] Spinach homogenate was heat-denatured for 30 min at 60 °C (dark, 27 °C, 24 h).
[e] No data.

presence of GSH. Both thiadiazolidin-ones and thiadiazolidine-diones did not convert into the corresponding triazolidines without any cofactors. Thiadiazolidin-one **22** converted into the corresponding triazolidin-one-thione **23** in the presence of GSH and DTT, *S*-methyl-L-cysteine or L-methionine. These results show that equine GST needs SH compounds when it acts as a thiadiazolidine-converting enzyme.

Jablonkai et al. (1997) studied the non-enzymatic conversion of thiadiazolidin-one **22** into triazolidin-one-thione **23** both in aprotic solvent with various —SH, —OH and —NH nucleophiles, and in aqueous solution using a buffer other than the non-nucleophilic phosphate. Their results suggested that the basicity and the nucleophilicity of cofactors are of prime importance in the isomerization. They proposed a possible mechanism of the isomerization of **22** into **23** starting a nucleophilic attack by the anionic form of the thiol at the carbonyl group in **22**. This result supports the proposed mechanism of isomerization of thiadiazolidin-one into triazolidin-one-thione shown in Fig. 3.

7.4 Identification of Isomerase

Nicolaus et al. (1996a,b) and Sato et al. (1995) identified this thiadiazolidine-isomerase as GST II isoform from maize etioplast grown with or without safener. Corn seedlings (*Zea mays* var. Anjou) were grown in vermiculite for 6 days in darkness at 30 °C with watering every second day. Naphthalic anhydride (0.1 mM) was applied as safener during germination. The purification scheme of isomerase using thiadiazolidin-one **22** for bioassay is shown in Table 7. After homogenization and ammonium sulfate fractionation, 45–75% ammonium sulfate precipitate was chromatographed on a hydroxyapatite SC column. The isomerase fraction, not retained by the column, was subjected to FPLC system with a Mono Q Sepharose anion exchange column. Two isomerase activity, fraction 1 and 2, was eluted with a combination of stepwise and discontinuous sodium chloride gradients (Fig. 4). Fraction 1 was eluted at 0.1 M NaCl, and fraction 2 was eluted at 0.25–0.3 M NaCl. Fraction 1 contained about 50–65% of the activity of fraction 2. GST I is present constitutively in etiolated corn with some increase after treatment of seedlings with safeners. It elutes at lower salt concentrations than other isoforms (II, III, IV) from anionic exchange columns and exists as a homodimer of 2 × 29 kDa (Mozer et al. 1983; Moore et al. 1986; Timmerman 1989; Fuerst et al. 1993; Jepson et al. 1994). After desalting, two fractions were further purified by GSH-Sepharose 4B affinity chromatography, respectively.

The purity, molecular weight and qualitative amounts of different peptides of the isomerase fractions obtained by affinity chromatography were analyzed by SDS-PAGE (Fig. 5) according to Laemmli (1970). Fraction 2 contained the main thiadiazolidine-isomerase activity. There were differences in protein content and apparent molecular weight of the respective peptides between

Table 7. Typical purification thiadiazolidine-isomerase activity from naphthalic anhydride (NA)-treated and NA-non-treated corn seedlings (after Nicolaus et al. 1996a)

Fractions	Total activity (units) NA-treated	NA-non-treated	Specific activity (units/mg protein) NA-treated	NA-non-treated	Purification (x-fold) NA-treated	NA-non-treated
45–75% Ammonium sulfate precipitate	7780	19 740	6	11	1	1
Hydroxyapatite	9000	15 840	71	18	12	6
Mono Q						
(1) Fraction 1	460	840	28	23	5	2
(2) Fraction 2	1290	5 600	37	56	6	5
(1) GSH-affinity	158	322	350	428	58	38
(2) GSH-affinity	276	500	13 200	25 250	2200	23 600
(2) SB-affinity	182	303	27 600	60 600	4600	5 660

A unit is defined as 1 nmol triazolidin-one-thione **23** formed per hour. Numbers in parenthesis denote the active fraction obtained after separation on Mono Q chromatography.

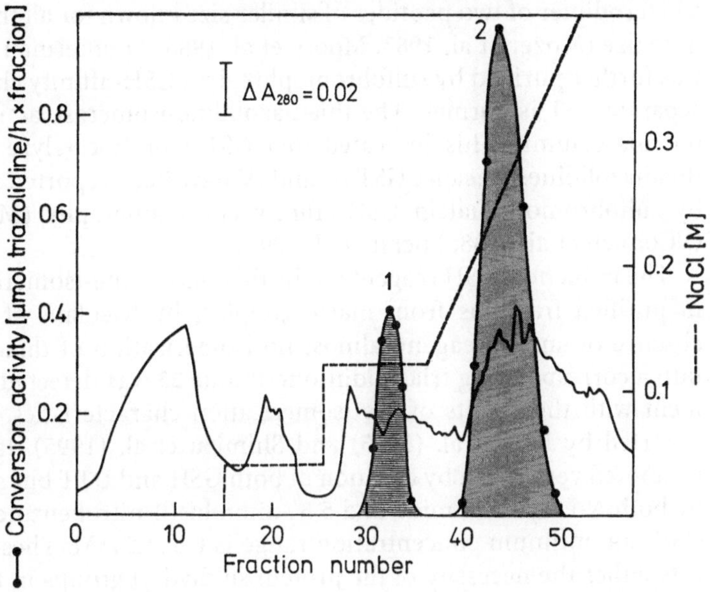

Fig. 4. Elution pattern of isomerase activity from Mono Q-FPLC. 1 Fraction 1; 2 Fraction 2. (Nicolaus et al. (1996a); reproduced with permission from Verlag der Zeitschrift für Naturforschung)

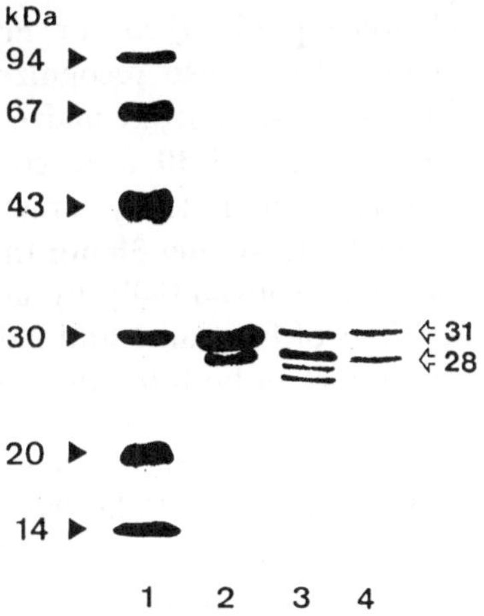

Fig. 5. SDS-PAGE of isomerase fractions obtained from chromatography. *Lane 1* Maker protein; *Lane 2* fraction 1 after GSH-affinity chromatography [(1) GSH-affinity in Table 7]; *Lane 3* fraction 2 after GSH-affinity chromatography [(2) GSH-affinity in Table 7]; *Lane 4* isomerase fraction from sulfobromophthalein affinity chromatography [(2) SB-affinity in Table 7]. (Reproduced with permission from Verlag der Zeitschrift für Naturforschung)

fraction 1 and 2 by SDS-PAGE. Four dominant peptides of which two were present in a similar quantity were detected in fraction 2, with apparent molecular weights from 25 to 31 kDa. Similar apparent molecular weights were obtained for the peptides of fraction 1, but one peptide of about 30 kDa was dominant. A purification of nearly 95% is estimated for the latter fraction. The native apparent molecular weight of isomerase activity from fraction 2 was estimated as 60 ± 3 kDa (data not shown) by gel filtration. This apparent molecular weight shows that the isomerase exists either as a homodimer or as a heterodimer of two peptides of similar size known for all the GSTs identified in maize (Mozer et al. 1983; Moore et al. 1986; Timmerman 1989). Fraction 2 was further purified by sulfobromophthalein GSH-affinity chromatography to separate GST isoformes. The thiadiazolidine-isomerase activity was retained on this column. This indicated that GST I or II catalyze isomerization of thiadiazolidines, because GST III and IV have been reported not to be retained by sulfobromophthalein GSH-affinity chromatography (Mozer et al. 1983; O'Connell et al. 1988; Fuerst et al. 1993).

The influence of SH reagent on the thiadiazolidine-isomerase was examined in purified fractions from maize etioplast by Nicolaus et al. (1996a,b). In absence of an SH reagent, almost no isomerization of thiadiazolidin-one **22** into a corresponding triazolidin-one-thione **23** was detected. This is in agreement with the results of the isomerization characteristics in crude extracts reported by Iida et al. (1995) and Shimizu et al. (1995). Isomerization was promoted very much by addition of both GSH and DTT but strongly inhibited by both *N*-ethylmaleimide and 5,5′-dithiobis(2-nitrobenzoic acid). In case of GSH, an optimum concentration range is 0.5–1.0 mM. These chemicals indicate either the necessity of the protein sulfhydryl groups in the isomerization process or the necessity for thiols as a cofactor.

Isomerization of thiadiazolidines in vivo was dependent on the 5-arylimino moiety (Sato et al. 1994a,b), and on the thiadiazolidine ring structure (Iida et al. 1995). Nicolaus et al. (1996a,b) re-examined whether a structure-activity relationship observed for *Echinochloa* seedlings, spinach homogenate or bovine GST is also recognized with the fractions purified by GSH-affinity chromatography from maize etioplasts (Table 8). The three thiadiazolidin-ones **22**, **26** and **30** were converted rapidly by fraction 2 while fraction 1 converted them slowly. Fraction 2 was 5- to 30-fold more active than the bovine GST. Neither **38** nor the thiadiazolidine-thiones **24**, **28**, **32** and **40** were converted substantially by all three enzyme samples assayed. The specific activities of fraction 1 and 2 differed, but the pattern of the conversion rates was similar in both fractions and are in complete agreement with the in vivo data reported by Iida et al. and Sato et al. previously.

GSTs catalyze the nucleophilic attack of GSH to electrophilic substrates (Timmerman 1989; Cole 1994). Currently, four isoforms of GST are recognized in maize, namely GST I, II, III and IV. Their properties are summarized in Table 9 based on relative activities to conjugate the chloroacetamide herbicides, such as metolachlor, and 1-chloro-2,4-dinitrobenzene (CDNB). They exist as polypeptide dimers, each polypeptide having an apparent molecular weight of 20–30 kDa (Jepson et al. 1994), which agrees with the protein pattern obtained for both fractions 1 and 2 of thiadiazolidine-isomerase activity. The isomerization of thiadiazolidines could be either a new function of a known GST isoform or the activity of a new isoform not yet detected. To examine this assumption, [^{14}C]atrazine, [^{14}C]metazachlor and CDNB were chosen to assay GST activity in isomerase fractions (Table 10). Both fractions contained GST activity, but their activity for CDNB, metazachlor and thiadiazolidin-one **22**

Table 8. Comparison of thiadiazolidine-isomerase specificity of two Mono Q fractions from corn seedlings with bovine GST (after Nicolaus et al. 1996a)

Compounds	Specific activity (unit/mg protein)		
	Fraction 1	Fraction 2	Bovine GST
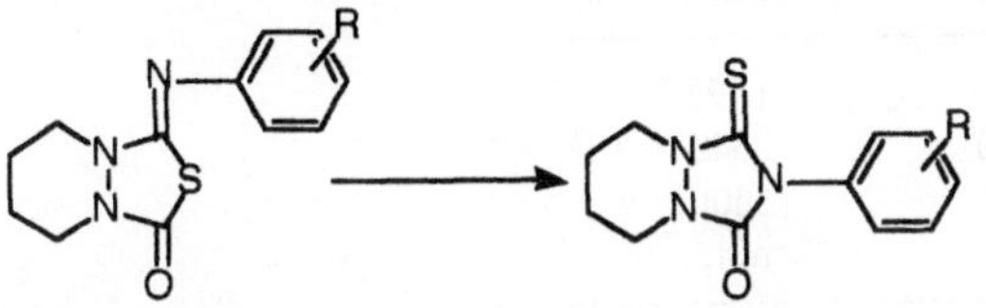			
22 (R = 4-Br)	306	62930	2121
26 (R = 4-Cl)	306	77980	1889
30 (R = 2-CH_3, 4-Cl)	194	26970	2721
38 (R = 4-OCH_2-C_6H_4Cl-*p*)	3	3564	308
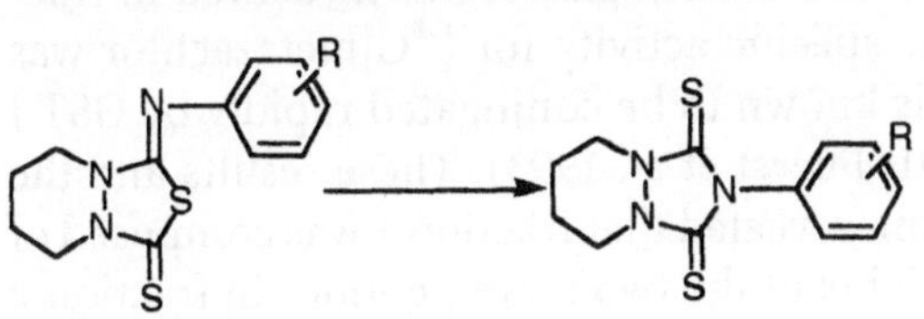			
24 (R = 4-Br)	3	4774	43
28 (R = 4-Cl)	3	2332	155
32 (R = 2-CH_3, 4-Cl)	2	n.d.	23
40 (R = 4-OCH_2-C_6H_4Cl-*p*)	1	n.d.	178

n.d., Not detected.
A unit is defined as 1 nmol corresponding triazolidines formed per hour.
Fraction 1 and 2 refer to the active fraction from Mono Q chromatography obtained from NA-treated corn seedlings.

Table 9. Classification of glutathione *S*-transferase in maize

GST Isoforms	Properties
I[a]	Constitutive, with affinity for CDNB and chloroacetamide. A homodimer of 29-kDa subunits.
II[a]	Safener-induced, with greater affinity for chloroacetamide. A heterodimer of 27-kDa and 29-kDa subunits.
III[b]	Constitutive, with high affinity for chloroacetamide, possibly some inducibility. A homodimer of 26-kDa subunits.
IV[c]	Constitutive and inducible, with high affinity for chloroacetamide and none towards CDNB. A homodimer of 27-kDa subunits.

[a] Mozer et al. (1983).
[b] O'Connell et al. (1988).
[c] Irzyk and Fuerst (1993).

Table 10. Specificities of thiadiazolidine-isomerase fraction from NA-treated corn for GST substrates (after Nicolaus et al. 1996a)

Substrates	Specific activity (nmol/mg protein per hour)	
	Fraction 1	Fraction 2
Thiadiazolidin-one **22**	327	13227
CDNB	4896×10^3	2642×10^3
[^{14}C]Metazachlor	126	5400
[^{14}C]Atrazine	n.d.	n.d.

n.d., not detected.

differed markedly. A higher rate of CDNB conjugation was measured in fraction 1 than in fraction 2, while the specific activity for [^{14}C]metazachlor was much higher in fraction 2. CDNB is known to be conjugated rapidly by GST I (Mozer et al. 1983; Dean et al. 1991; Fuerst et al. 1993). These results and the elution pattern from Mono Q column revealed that fraction 1 was composed of GST I. The apparent molecular weights of the two main peptides in fraction 2 are different from those reported for GST III and IV which have molecular weights of 26 and 27kDa, respectively (Moore et al. 1986; O'Connell et al. 1988; Miller et al. 1994). Isomerase activity was bound to sulfobromophthalein GSH-affinity column, but GST III and IV are not retained on this column.

Safener treatment increases the activity of GST isoforms constitutively present in non-treated plants and induces additional isoforms, such as GST II. GST II was only detectable by CDNB conjugation after safener treatment (Mozer et al. 1983). Fraction 2 showed less CDNB conjugation in comparison to fraction 1, but specific CDNB-conversion activity of fraction 2 remained constant with further purification. GST II was shown to utilize the chloroacetamide and was identified as a heterodimer of 27- and 29-kDa polypeptides (Moore et al. 1986; O'Connell et al. 1988; Fuerst et al. 1993; Jepson et al. 1994).

Fifteen N-terminal amino residues of the 31-kDa peptide were sequenced as Met-Ala-Pro-Met-Lys-Leu-Tryp-Gly-Ala-Val-Met-Ser-Trp-Asn-Val. This sequence was completely homologous to GST I as reported (Shah et al. 1986), exactly as has been shown for the 29-kDa peptide of GST II (Moore et al. 1986; Timmerman 1989; Jepson et al. 1994). The 28-kDa peptide of fraction 2 was resistant to Edman degradation because of N-terminal blockage as the second peptide of GST II purified from plants.

As shown from the above results, the thiadiazolidine-isomerase is a GST II isomer. Bioactivation (isomerization) of thiadiazolidines caused by GST II isomers is completely different from the typical detoxification reaction by GST. In isomerization, GSH is not conjugated to the herbicide as intermediate but

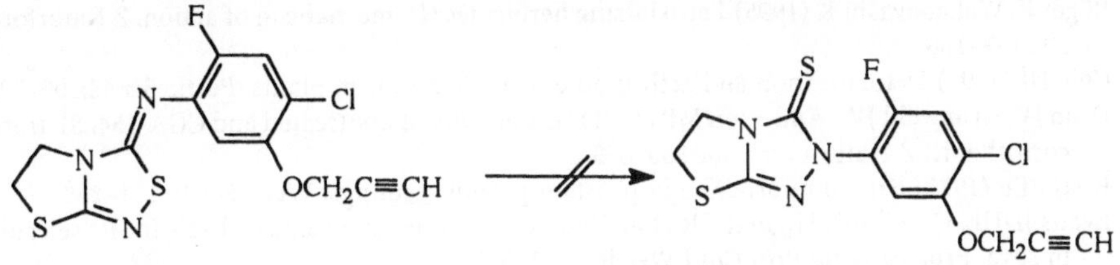

Fig. 6. Non-isomerization of Δ^2-1,2,4-thiadiazolidine derivative

acts as a cofactor (Sato et al. 1994a,b; Iida et al. 1995). One isoenzyme of GST is mainly responsible for thiadiazolidine isomerization, having characteristics as those described for GST II. Although safener treatment increased isomerase activity, a constitutive conversion activity in untreated plants can only be detected by using high-affinity substrates like thiadiazolidines.

It is very interesting that weeds and crop have different sensitivities to activate peroxidizing compounds, depending on the presence of the respective GST isoform. Thiadiazolidine type peroxidizers may be developed as pro-herbicides which are converted more rapidly in weeds than in the crop, possibly in combination with a safener.

7.5 Isomerization of Other Peroxidizers

There are many other types of peroxidizing herbicides in use these days. No isomerization was observed for oxyfluorfen, acifluorfen-methyl and 3-[4-chloro-2-fluoro-5-(2-propynyloxy)phenylimino]-5,6-dihydro-6,6-dimethyl-3H-thiazolo[2,3-c][1,2,4]-thiadiazole (**50**) (Sato Y, Hoshi T, Ohki A, Wakabayashi K 1994, pers. comm.). Oxyfluorfen and acifluorfen-methyl are the diphenyl ether type compound and different from thiadiazolidines structures. Fomesafen [5-(2-chloro-4-trifluoromethylphenoxy)-*N*-methane-sulfonyl-2-nitrobenzamide] having a diphenyl ether type structure was metabolized to less active homoglutathione conjugate in soy bean leaves (Evans et al. 1987). Compound **50** (Hagiwara et al. 1993) is a series of Δ^2-1,2,4-thiadiazolidine derivatives having a 5-phenylimino group and a fused ring structure. No isomerization of **50** into **51** under biological condition with GST is observed (see Fig. 6; Hagiwara and Nakayama 1994).

References

Böger P, Sandmann G (1990) Modern herbicides affecting typical plant processes. In: Bowers WS, Ebing W, Martin D, Wegler R (eds) Chemistry of plant protection, vol. 6 Springer, Heidelberg, New York, Berlin, pp 173–176

Böger P, Sandmann G (1993) Pigment biosynthesis and herbicide interaction. Photosynthetica 28:481–493

Böger P, Wakabayashi K (1995) Peroxidizing herbicides (I), mechanism of action. Z Naturforsch 50c:159–166

Cole DJ (1994) Detoxification and activation of agrochemicals in plants. Pestic Sci 42:209–222

Dean JV, Gronwald JW, Anderson MP (1991) GST activity in nontreated and CGA-154281-treated corn shoots. Z Naturforsch 46c:850–855

Eastin EF (1971) Fate of fluorodifen in resistant peanut seedlings. Weed Sci 19:261–265

Evans JDHL, Cavell BD, Hignett RR (1987) Fomesafen – metabolism as a basis for its selectivity in soya. Proc Br Crop Prot Conf Weeds, pp 345–352

Frear DS, Swanson HR, Mansager ER (1983) Acifluorfen metabolism in soybean: diphenyl ether bond cleavage and the formation of homoglutathione, cysteine, and glucose conjugates. Pestic Biochem Physiol 20:299–310

Fuerst EP, Irzyk GP, Miller KD (1993) Partial characterization of GST isozymes induced by the herbicide safener benoxacor in corn. Plant Physiol 102:795–802

Hagiwara K, Nakayama A (1994) Molecular similarity of peroxidizing herbicides: bioisosterism in Δ^2-1,2,4-thiadiazolines and related heterocyclic compounds. J Pestic Sci 19:111–117

Hagiwara K, Saitoh K, Iihama T, Hosaka H (1993) Synthesis and herbicidal activity of fused Δ^2-1,2,4-thiadiazolines. J Pestic Sci 18:309–318

Hoshi T, Koizumi K, Sato Y, Wakabayashi K (1993) Hydrolysis and phytotoxic activity of *N*-aryl-3,4,5,6-tetrahydroisophthalimides. Biosci Biotech Biochem 57:1913–1915

Iida T, Senoo S, Sato Y, Nicolaus B, Wakabayashi K, Böger P (1995) Isomerization and peroxidizing phytotoxicity of thiadiazolidine-thione compounds. Z Naturforsch 50c:186–192

Irzyk GP, Fuerst EP (1993) Purification and characterization of a glutathione *S*-transferase from benoxacor-treated maize (*Zea mays*). Plant Physiol 102:803–810

Jablonkai I, Kömives T, Böger P, Sato Y, Wakabayashi K (1997) Chemical catalysis of the isomerization of a peroxidizing herbicidal thiadiazolidine. Proc Br Crop Prot Conf Weeds, pp 771–776

Jacobs JM, Jacobs NJ, Borotz SE, Guerinot ML (1990) Effects of the photobleaching herbicides, acifluorfen-methyl, on protoporphyrinogen oxidation in barley organelles, soybean root mitochondria, soybean root nodules, and bacteria. Arch Biochem Biophys 280:369–375

Jepson I, Lay VJ, Holt DC, Bright SWJ, Greenland AJ (1994) Cloning and characterization of maize herbicide safener-induced cDNAs encoding subunits of glutathione *S*-transferase isoforms I, II and IV. Plant Mol Biol 26:1855–1866

Laemmli UK (1970) Cleavage of structural proteins during assembly of the head of bacteriophage T4. Nature 227:680–685

Lambert R, Sandmann G, Böger P (1983) Correlation between structure and phytotoxic activities of nitrodiphenyl ethers. Pestic Biochem Physiol 19:309–320

Lee HJ, Duke MV, Duke SO (1993) Cellular localization of protoporphyrinogen-oxidizing activities of etiolated barley (*Hordeum vulgare* L.) leaves. Relationship to mechanism of action of protoporphyrinogen oxidase-inhibiting herbicides. Plant Physiol 102:881–889

Lydon J, Duke SO (1988) Porphyrin synthesis is required for photobleaching activity of the *p*-nitrosubstituted diphenyl ether herbicides. Pestic Biochem Physiol 31:74–83

Matringe M, Scalla R (1988) Effects of acifluorfen-methyl on cucumber cotyledons: protoporphyrin accumulation. Pestic Biochem Physiol 32:164–172

Miller KD, Irzyk GP, Fuerst EP (1994) Benoxacor treatment increases GST activity in suspension cultures of *Zea mays*. Pestic Biochem Physiol 48:123–134

Moore RE, Davies MS, O'Connell KM, Harding EI, Wiegand RC, Tiemeier DC (1986) Cloning and expression of a cDNA encoding a corn glutathione transferase in *E. coli*. Nucleic Acids Res 14:7227–7235

Mozer TJ, Tiemeier DC, Jaworski EG (1983) Purification and characterization of corn GST. Biochemistry 22:1068–1072

Nandihalli UB, Duke SO (1993) The porphyrin pathway as a herbicide target side. In: Duke SO, Menn JJ, Plimmer JR (eds) Pest control with enhanced environmental safety ACS Symp Ser 524, Am Chem Soc, Washington, DC, pp 62–78

Nandihalli UB, Sherman TD, Duke MV, Fisher JD, Mausco VA, Becerril JM, Duke SO (1992) Correlation of protoporphyrinogen oxidase inhibition by *o*-phenyl-pyrrolidino and piperidino-carbamates with their herbicidal effects. Pestic Sci 35:227–235

Nicolaus B, Johansen JN, Böger P (1995) Binding affinities of peroxidizing herbicides to protoporphyrinogen oxidase inhibiting diphenyl ether herbicides. Pestic Biochem Physiol 51:20–29

Nicolaus B, Sato Y, Wakabayashi K, Böger P (1996a) Isomerization of peroxidizing thiadiazolidine herbicides is catalyzed by glutathione *S*-transferase. Z Naturforsch 51c:342–354

Nicolaus B, Sato Y, Wakabayashi K, Böger P (1996b) Activation of proherbicides by glutathione *S*-transferase. Nato advanced research workshop. Regulation of enzymatic systems detoxifying xenobiotics in plants, p 32

O'Connell KM, Breaux EJ, Fraley RT (1988) Different rates of metabolism of two chloracetanilide herbicides in Pioneer 3320 corn. Plant Physiol 86:359–363

Retzlaff K, Böger P (1996) An endoplasmic reticulum plant enzyme has protoporphyrinogen IX oxidase activity. Pestic Biochem Physiol 54:105–114

Sandmann G, Böger P (1988) Accumulation of protoporphyrin IX in the presence of peroxidizing herbicides. Z Naturforsch 43c:699–704

Sato Y, Kojima T, Goto T, Oomikawa R, Watanabe H, Wakabayashi K (1991) Hydrolysis and phytotoxic activity of cyclic imides. Agric Biol Chem 55:2677–2681

Sato Y, Hoshi T, Nicolaus B, Wakabayashi K, Böger P (1994a) Isomerization and peroxidizing phytotoxicity of thiadiazolidine herbicides. Z Naturforsch 49c:49–65

Sato Y, Hoshi T, Nicolaus B, Wakabayashi K, Böger P (1994b) Intrinsic phytotoxic structures of cyclic imide class of peroxidizing herbicides. Abstract paper, 8th international congress of pesticide chemistry, Washington, DC, p 749

Sato Y, Iida T, Senoo S, Nicolaus B, Wakabayashi K, Böger P (1995) Enzymatic conversion of thiadiazolidine-type peroxidizing herbicides into more active triazolidines. Proc 15th Asian Pacific Weed Sci So Conf 1:193–198

Sato Y, Böger P, Wakabayashi K (1997) The enzymatic activation of peroxidizing cyclicisoimide: a new function of glutathione *S*-transferase and glutathione. J Pestic Sci 22:33–36

Senoo S, Iida T, Shouda K, Sato Y, Nicolaus B, Böger P, Wakabayashi K (1996) Enzyme-modified phytotoxic structure of thiadiazolidine compounds. Z Naturforsch 51c:518–526

Shah DM, Hironaka CM, Wiegand RC, Harding EI, Krivi GG, Tiemeier DC (1986) Structural analysis of a corn gene coding for GST involved in herbicide detoxification. Plant Mol Biol 6:203–211

Sherman TD, Becerril JM, Matsumoto H, Duke MV, Jacobs JM, Jacobs NJ, Duke SO (1991) Physiological basis for differential sensitivities of plant species to protoporphyrinogen oxidase inhibiting herbicides. Plant Physiol 97:280–287

Shimizu T, Hashimoto N, Nakayama I, Nakao T, Mizutani H, Unai T, Yamaguchi M, Abe H (1995) A novel isourazole herbicide, fluthiacet-methyl, is a potent inhibitor of protoporphyrinogen oxidase after isomerization by gluthathione *S*-transferase. Plant Cell Physiol 36:625–632

Timmerman KP (1989) Molecular characterization of corn GST isozymes involved in herbicide detoxification. Physiol Plant 77:465–471

Uchida A, Iida T, Sato Y, Böger P, Wakabayashi K (1997) Isomerization of 3,4-dialkyl-1,3,4-thiadiazolidines and 3,4-alkylene-1,3,4-thiadiazolidines by glutathione *S*-transferase. Z Naturforsch 52c:345–350

Wakabayashi K, Matsuya K, Ohta H, Jikihara T (1979) Structure-activity relationship of cyclic imide herbicides. In: Geissbühler H (ed) Advances in pesticide science, part 2. Pergamon Press, Oxford, pp 256–260

Watanabe H, Ohori Y, Sandmann G, Wakabayashi K, Böger P (1992) Quantitative correlation between short term accumulation of proto IX and peroxidative activity of cyclic imides. Pestic Biochem Physiol 42:99–102

Wittkowski DA, Halling BP (1988) Accumulation of photodynamic tetrapyrroles induced by acifluorfen-methyl. Plant Physiol 87:632–637

CHAPTER 8

The Metabolic Pathway of Tetrapyrrole Biosynthesis

BERNHARD GRIMM[1]

Contents

[1]Institut für Pflanzengenetik und Kulturpflanzenforschung Gatersleben (IPK), Corrensstr. 3, D-06466 Gatersleben, Germany

Peter Böger, Ko Wakabayashi
Peroxidizing Herbicides

8.1 Introduction

The aim of this chapter is to provide an overview of the metabolic pathway of tetrapyrrole biosynthesis. Tetrapyrroles play a pivotal role in many biochemical processes. They participate as prosthetic groups or chromophores in light harvesting (chlorophyll(Chl)-binding proteins), respiration or phosphorylation (cytochromes), removal of reactive oxygen species or detoxification (cytochrome P450, catalase, peroxidase), nitrogen fixation (leghemoglobin), oxygen transport (hemoglobin), storage (myoglobin) and light perception (phytochrome). All tetrapyrroles are synthesised in a branched pathway, in which various end products are formed in different amounts. The most abundant cyclic tetrapyrroles are Chl and heme which are characterised by a chelated magnesium and iron, respectively. Protoheme is an intermediate in the formation of the open chain and iron free phycobilins which are the chromophores of the phycobiliproteins in cyanobacteria and red algae, and of phytochromobilin for phytochrome in plants and algae. Uroporphyrinogen (Urogen) III, which is a metabolic intermediate in the main pathway leading to heme and chlorophyll synthesis, is also the precursor of a branch pathway leading to the synthesis of cobalt containing vitamin B_{12} (cobalamins), siroheme or the nickel chelating corrins (coenzyme F 430).

Tetrapyrrole biosynthesis in eukaryotic and prokaryotic photosynthetic organisms has been reviewed by several authors in the last few years (Beale 1993; Smith and Griffiths 1993; Chadwick and Ackrill 1994; Jordan 1994; von Wettstein et al. 1995; Reinbothe and Reinbothe 1996; Porra 1997). These authors provide either a general survey of the pathway or selectively emphasise particular biochemical, regulatory or genetic aspects. The present chapter will focus on the most recent developments in research on tetrapyrrole biosynthesis in higher plants unless an introduction to the subject requires results obtained with other organisms. The chapter will recall the current views on each enzymatic step and will focus on novel aspects which have arisen out of the cloning of genes involved in tetrapyrrole synthesis and studies on their regulation. The last section will summarise some prospects for genetic engineering of the metabolic pathway. The author will describe the biochemical analysis of transgenic plants with deregulated gene expression for specific enzymes of tetrapyrrole biosynthesis. These plants simulate potential herbicidal effects. By the aid of transgenic plants with a single deregulated enzymatic step suitable targets are predictable for potential herbicides.

8.2 Metabolic Pathway of Tetrapyrrole Biosynthesis

A simplified flow diagram of the metabolic pathway is given in Fig. 1. The pathway can be subdivided into three sections: 5-aminolevulinate (ALA) syn-

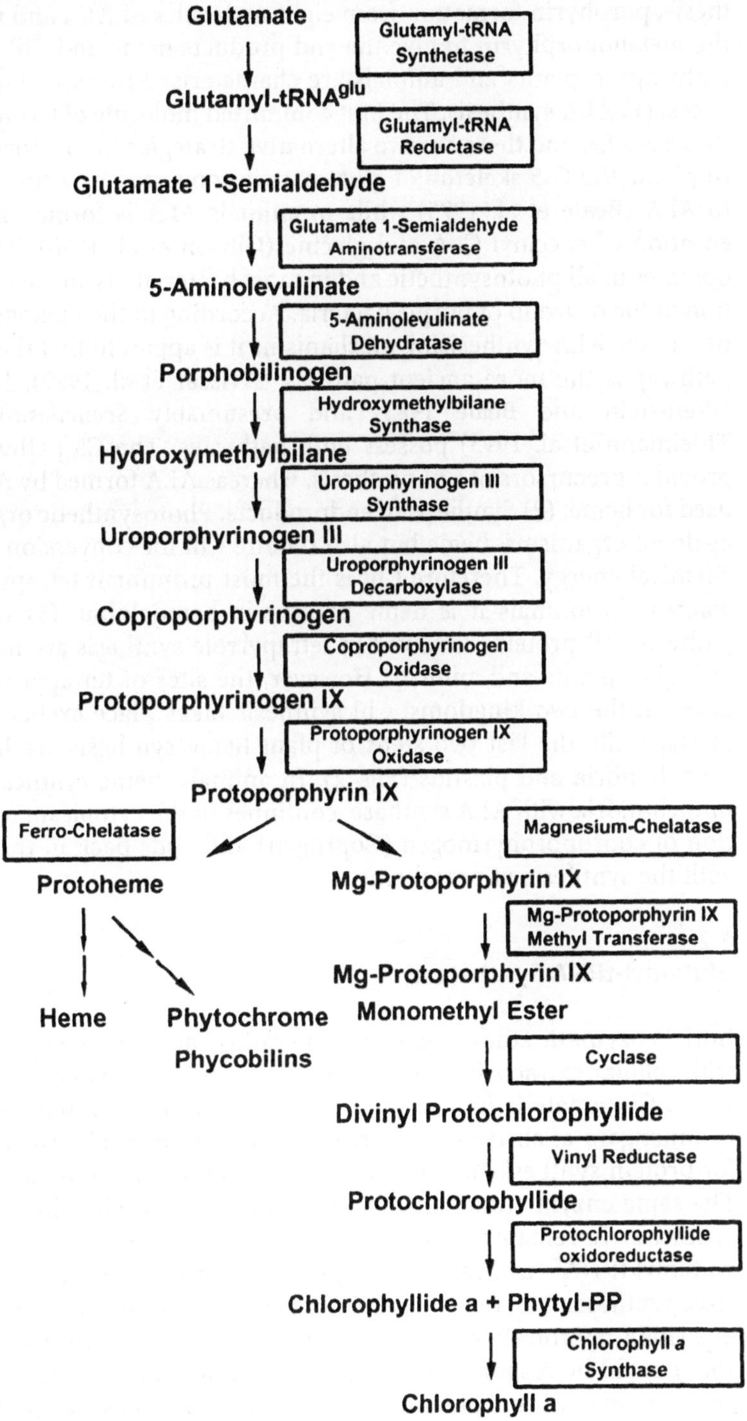

Fig. 1. Tetrapyrrole biosynthetic pathway in plants

thesis, porphyrin formation from eight molecules of ALA and modification of the metalloporphyrin to give the end products heme and Chl. The metabolic pathways in plants and animals are characterised by three important differences: (1) ALA synthesis. The first committed molecule of tetrapyrrole biosynthesis is ALA and there are two alternative strategies for the formation of ALA. In plants the C-5 skeleton of glutamate is converted by a three-step pathway to ALA (Beale et al. 1975) while in animals ALA is formed from the condensation of succinyl CoA and glycine (Gibson et al. 1958). The C5 pathway operates in all photosynthetic and in most bacterial organisms with the exception of the α-group of purple bacteria. According to the phylogenetic distribution of the ALA synthesising mechanisms it is apparent that the ubiquitous C5 pathway is the more ancient pathway (Avissar et al. 1989). *Euglena gracilis* (Weinstein and Beale 1983) and presumably *Scenedesmus* (Drechsler-Thielmann et al. 1993) possess both pathways. The C5 pathway in *Euglena* provides precursors of Chl synthesis, whereas ALA formed by ALA synthase is used for heme. (2) Synthesis of endproducts. Photosynthetic organisms utilise, as do all organisms, heme but also require Chl for conversion of radiation to chemical energy. Therefore Chl is the most prominent tetrapyrrole in plants whereas in animals it is heme present in hemoglobin. (3) Location of the pathway. All proteins involved in tetrapyrrole synthesis are nuclear-encoded in higher plants and animals. However, the sites of tetrapyrrole metabolism differ in the two kingdoms. Chl synthesis takes place exclusively in chloroplasts, while the last two steps of plant heme synthesis are located in both mitochondria and plastids (Fig. 2). In animals, heme synthesis starts in the mitochondria with ALA synthase, continues in the cytoplasm up to the formation of coproporphyrinogen (Coprogen) and ends back in the mitochondria with the synthesis of protoheme.

8.2.1 Glutamyl-tRNA Synthetase

Initial enzymatic studies on plants operating the C5 pathway were carried out with cellular extracts from greening barley seedlings (Gough and Kannangara 1977). Glutamate is first activated by means of an unusual cofactor, $tRNA^{glu}$ (Kannangara et al. 1984). This reaction, which resembles the aminoacylation for protein synthesis, is catalysed by glutamyl-tRNA synthetase (EC 6.1.1.17). The same enzyme functions simultaneously in protein and tetrapyrrole biosynthesis. It remains a challenging question in tetrapyrrole research how glutamyl $tRNA^{glu}$ is diverted from protein synthesis and channelled to tetrapyrrole synthesis. The $tRNA^{glu}$ sequence is encoded in the plastid genome as part of a tRNA operon (Berry-Lowe 1987). Specific bases of the tRNA (Schön et al. 1986) could be responsible for the recognition and binding of the synthetase and the next enzyme glutamyl-tRNA reductase to ensure an immediate GSA formation by the adjoining enzyme (Willows et al. 1995). It cannot be excluded

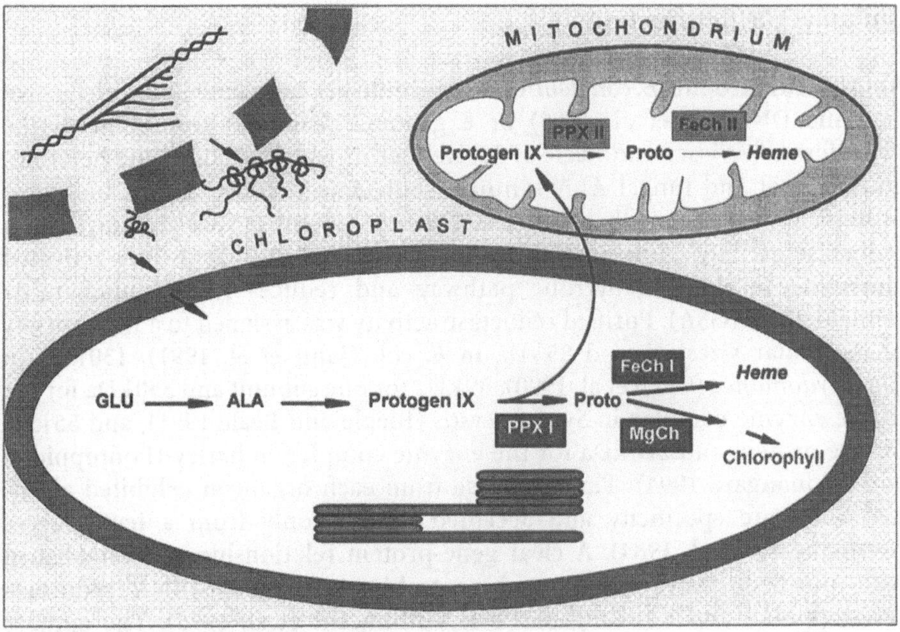

Fig. 2. Cellular organisation of tetrapyrrole biosynthesis. All enzymes involved are nuclear-encoded. Early steps of the pathway from glutamate (*Glu*) via 5-aminolevulinate (*ALA*) to protoporphyrinogen IX (*Protogen IX*) are exclusively located in plastids. Chlorophyll synthesis proceeds with Mg chelatase (*MgCh*) and following enzymes in the plastids. The two final steps of the heme synthesising branch protoporphyrinogen oxidase (*PPX*) and ferro chelatase (*FeCh*) occur in plastids and mitochondria

that the highly modified anticodon could also fulfil this function. Although the $tRNA^{glu}$ of an *Euglena gracilis* mutant is charged with glutamate by the glutamyl-tRNA synthetase, a substitution of C56 $\rightarrow$ U in the T loop of the $tRNA^{glu}$ prevents the coupling to glutamyl-tRNA reductase and leads to a very low ALA synthesising rate (Stange-Thomann et al. 1994). The plastidal glutamyl-tRNA synthetase was purified first from barley and has a molecular mass of 54–58 kDa (Bryuant and Kannangara 1987; Andersen 1992). This enzyme charges three tRNA isoacceptors with glutamate. Two of them are $tRNA^{gln}$ and a third acceptor is $tRNA^{glu}$ which binds to both glutamate codons and participates in ALA and protein synthesis (Kannangara et al. 1984). A full-length barley cDNA clone was obtained by an amplification of a DNA fragment using degenerate primers of internal peptides of purified glutamyl-tRNA synthetase (Andersen 1992). This cDNA sequence and a full-length cDNA clone from tobacco (Giesen and Grimm, unpubl.) encode 65-kDa (barley) and 67-kDa (tobacco) precursor proteins, respectively, with aminoterminal extensions for plastid translocation.

8.2.2 Glutamyl-tRNA Reductase

An ALA auxotrophic *E. coli hemA* mutant could be complemented with its own genomic DNA (Li et al. 1989) or a genomic fragment from *Chlorobium vibrioforme* (Avissar and Beale 1990) neither of which encoded the heterologous animal and fungal ALA synthase, but instead coded for an unknown protein. It was shown that the *E. coli hemA* mutant lacked glutamyl-tRNA reductase activity (Avissar and Beale 1989a). This protein directs activated glutamate to the tetrapyrrolic pathway and reduces it to glutamate 1-semialdehyde (GSA). Purified reductase activity was assigned to a wide variety of molecular sizes: 45 and 85kDa in *E. coli* (Jahn et al. 1991), 130kDa in *Chlamydomonas* (Chen et al. 1990), 39kDa for one subunit and 350kDa for the native enzyme complex in *Synechocystis* (Rieble and Beale 1991), and 55kDa for the protein and 270kDa for the enzyme complex in barley (Pontoppidan and Kannangara 1994). The reductase from each organism exhibited a limited substrate specificity and accepted $tRNA^{glu}$ only from a few sources (Kannangara et al. 1984). A clear gene-protein relationship was established with purified native barley and recombinant glutamyl-tRNA reductase (Pontoppidan and Kannangara 1994; Vothknecht et al. 1996). The authors showed unequivocally that the reductase is encoded by the homologous *hemA* gene. Physiological concentrations of heme inhibited enzyme activity and a chromophore of the cytochrome c-type is apparently associated with the protein. Both findings point to a significant heme-dependent regulatory mechanism for glutamyl-tRNA reductase (Rieble and Beale 1991; Pontoppidan and Kannangara 1994). cDNA sequences encoding the reductase were identified from *Arabidopsis* (Ilag et al. 1994), barley (Bougri and Grimm 1996) or cucumber (Tanaka et al. 1996).

8.2.3 Glutamate 1-Semialdehyde Aminotransferase

The transfer of an amino group of the C2 of GSA to C1 of the product ALA is catalysed by GSA aminotransferase (EC 5.4.3.8). The protein was purified from the stroma fraction of greening barley plastids, *Synechococcus* PCC 6301(Grimm et al. 1989), *Chlorella vulgaris* (Avissar and Beale 1989b) and subsequently from various sources. The active enzyme forms a homodimer and contains the vitamin B_6-derivatives pyridoxamine phosphate or pyridoxal phosphate. By means of oligonucleotides derived from partial peptide sequences of the barley enzyme a cDNA fragment was amplified and finally a full-length cDNA sequence obtained (Grimm 1990). Further plant cDNA sequences were identified in *Chlamydomonas* (Matters and Beale 1994), *Arabidopsis* (Ilag et al. 1994), soybean (Sangwan and O´Brian 1993) and tobacco (Höfgen et al. 1994). The absorption characteristics of the B_6-dependent GSA aminotransferase enabled it to be biochemically characterised, principally with the recombinant *Synechococcus* (Grimm et al. 1991) and native pea enzymes (Nair et al. 1991). The spectral (Smith et al. 1991a) and steady state kinetics (Smith et al. 1991b) were analysed and identified the enzymic mechanism as being consistent with the Ping-Pong-Bi-Bi mechanism common to aminotransferases. Lysine272/265 of the cyanobacterial or *E. coli* enzyme, respectively, was identified as the pyridoxal phosphate-binding site of GSA aminotransferase and thus channels the overall transamination towards ALA (Grimm et al. 1992). The X-ray structure was recently obtained from the crystallized *Synechococcus* GSA aminotransferase (Hennig et al. 1997) and substantiates the biochemical data.

8.2.4
5-Aminolevulinate Dehydratase

The asymmetric condensation of two molecules of ALA to the first monopyrrole, porphobilinogen, is catalysed by ALA dehydratase (EC 4.2.1.24). Animal, bacterial and plant enzyme sequences derived from nucleotide sequences show a significant similarity, but differ in their molecular size and metal requirement for catalysis (Jaffe 1995). Animal ALA dehydratase requires Zn^{2+} while the plant enzyme additionally uses Mg^{2+} ions. In the catalytic centre a lysine residue forms a Schiff base with the 4-oxo group of the first ALA moiety which supplies the propionate side chain of porphobilinogen. A second ALA provides the acetate side chain of porphobilinogen (Jordan and Gibbs 1985). The first purification of the enzyme was achieved from spinach (Liedgens et al. 1980). Plant cDNA sequences encoding ALA dehydratase were isolated from pea (Boese et al. 1991), spinach (Schaumburg et al. 1992), *Glycine max* (Kaczor et al. 1994), tomato (Polking et al. 1995) and *Chlamydomonas* (Matters and Beale 1995). The Zn^{2+} binding domain of the animal protein sequence contains cysteine and histidine residues which are not present in the plant sequence. A change of the effector requirement from Mg^{2+}- to Zn^{2+}-dependence was achieved by three amino acid substitutions in the *Bradyrhizobium* enzyme (Chauhan and O´Brian 1995).

8.2.5
Hydroxymethylbilane (HMB) Synthase (Porphobilinogen Deaminase)

Hydroxymethylbilane (HMB) synthase (EC 4.1.3.8) catalyses the stepwise addition of four porphobilinogen molecules with the loss of a free amino group from each to form the linear tetrapyrrole HMB (preuroporphyrinogen). In addition to enzyme preparations from animal and bacterial sources, purification of the plant monomeric enzyme has been described from several sources: spinach (Higuchi and Bogarad 1975), *Euglena gracilis* (Sharif et al. 1989), *Scenedesmus obliquus* (Juknat et al. 1994) and pea (Spano and Timko 1991). The molecular mass ranges from 34 to 44 kDa. Chemical analysis has revealed that porphobilinogen molecules are repeatedly condensed onto an unusual cofactor, dipyrrolemethan (Hart et al. 1987), until a porphobilinogen hexamer is assembled. The tetrapyrrole is then hydrolysed and the cofactor remains bound to the enzyme (Jordan 1994). The enzymatic studies were supported by the crystallisation of the *E. coli* enzyme and the determination of the X-ray structure to 1.76 Å (Louie et al. 1992). The structural data are also summarised by Jordan (1994). Plant cDNA sequences encoding the HMB synthase have been isolated from pea and *Arabidopsis* (Witty et al. 1993, 1996).

8.2.6 Uroporphyrinogen III Synthase (Co-Synthase)

The instability of the linear HMB suggests an association between HMB synthase and the next enzyme in the pathway, Urogen III synthase (EC 4.2.1.75), to allow a direct transfer of the metabolite. Deficiency of the latter enzyme would lead to non-enzymatic formation of Urogen I. Urogen synthase catalyses the inversion of ring D followed by the ring closure of the linear tetrapyrrole to form Urogen III. The first enzyme purifications were described for extracts from spinach (Higuchi and Bogorad 1975) and *Euglena gracilis* (Hart and Battersby 1985). The mature enzyme is a monomer of approximately 30 kDa. By means of the recombinant *E. coli* Urogen synthase, an analysis of the properties of this enzyme was undertaken (Jordan et al. 1988; Crocket et al. 1991).

8.2.7 Uroporphyrinogen III Decarboxylase

Urogen decarboxylase (EC 4.1.1.37) catalyses the sequential removal of carboxyl groups from the four acetate side chains of Urogen. The source for the first partially purified plant enzymes were tobacco and *Euglena*, respectively (Chen and Miller 1974; Juknat et al. 1989). The sequential decarboxylation of the acetate residues follows a clockwise order of ring D, A, B and C of the macrocycle (Luo and Lim 1993). The first plant cDNA clones were identified from barley and tobacco by Mock et al. (1994). The deduced amino acid sequences of the enzymes show significant similarity to the cyanobacterial enzyme (73%) (Kiel et al. 1992). Previously proposed reaction mechanisms (Akhtar 1994) can be proved in future using the purified recombinant tobacco enzyme (Mock et al. 1994).

8.2.8 Coproporphyrinogen III Oxidase

Coprogen III oxidase (EC 1.3.3.3) catalyses the oxidative decarboxylation of the two propionate side chains of ring A and B to vinyl groups. The animal enzyme is localised in mitochondrial membranes (Elder and Evans 1978) and the yeast enzyme in the cytosol (Camadro et al. 1986). The activity of the plant enzyme has been studied in tobacco (Hsu and Miller 1970) and pea and spadices of cuckoo pint (Smith et al. 1993). The purified enzyme consists of two identical subunits of approximate 35–37 kDa. Plant cDNA sequences encoding Coprogen oxidase were obtained from soybean (Madsen et al. 1993), *Chlamydomonas* (Hill and Merchant 1994), tobacco and barley (Kruse et al. 1995). The expression of active recombinant Coprogen oxidase from tobacco resulted in an extremely photosensitive *E. coli* strain grown at 37°C in the light (Kruse et al. 1995a). Anaerobic organisms are endowed with an O_2-independent enzyme (Hem N) which uses alternative electron acceptors. This enzyme does not share any significant similarity to its aerobic counterpart. The anaerobic enzymes, e.g. from *Rhodobacter sphaeroides*, require ATP, NADP or NAD, Mg^{2-} and methionine (Tait 1972).

8.2.9 Protoporphyrinogen IX Oxidase

The last common step of the metabolic pathway to heme and Chl is the removal of six electrons from protoporphyrinogen IX (Protogen IX) catalysed by Protogen IX oxidase (EC 1.3.3.4). Because of its susceptibility to tetrapyrrole-based photodynamic herbicides, it has become an attractive subject for enzymological studies. A peculiarity of the previous studies was the fact that the site of enzyme inhibition and the site of substrate oxidation are in different cellular compartments (Becerril and Duke 1989; Matringe et al. 1989; Lehnen et al. 1990). For a detailed description of this enzyme I refer to other contributions in this volume. Enzyme activities were detected in plastidal and mitochondrial extracts (Jacobs and Jacobs 1987). The plastidal activity was determined in fractions of the envelope and the thylakoid membrane reflecting a possible mechanism for the intracellular distribution of Protogen to different compartments (Matringe et al. 1992). Enzyme purification was hampered because of the difficulty in solubilising active Protogen oxidase and recovering activity in a cell-free preparation. Most recently, plant cDNA clones has been isolated for different Protogen oxidase isoforms by functional complementation of the protoporphyrin IX (Proto IX) accumulating *E. coli hemG* mutant (Ward and Volrath 1995; Narita et al. 1996; Lermontova et al. 1997). The deduced peptide sequences are significantly similar to HemY, a gene product from *Bacillus subtilis* to which Protogen oxidase activity was first assigned (Hansson and Hederstedt 1992). The two isoforms from tobacco share only 27.2% identical amino acid residues. The isoenzyme with a putative transit sequence was exclusively imported into plastids. The other isoenzyme without an extended amino terminus was only targeted to mitochondria without a reduction in size (Lermontova et al. 1997).

There are a number of challenging questions that remain to be answered. These concern the regulation of Protogen allocation between plastids and mitochondria and the transfer mechanisms of Protogen from plastids to mitochondria, the source of heme for cytosolic heme-containing enzymes. A carrier function of plastidic Protogen oxidase has previously been suggested (Matringe et al. 1992). A similar membrane transport mechanism was suggested from studies on the animal Protogen oxidase, which is located in the inner mitochondrial membrane and faces the cytosol (Ferreira et al. 1988). For anaerobically grown micro-organisms another reaction mechanism has been proposed that ensures the transfer of electrons to other acceptors. *Desulfovibrio gigas* requires three proteins of 12, 18.5 and 57 kDa for Protogen oxidase activity (Klemm and Barton 1987).

8.2.10
Magnesium Protoporphyrin IX Chelatase

The Mg chelatase is the first committed enzyme of the Chl synthesising branch and catalyses the insertion of Mg^{2+} into Proto IX. The enzyme activity was determined first from intact plastids (Castelfranco et al. 1979; Fuesler et al. 1982) followed by attempts to assay Mg^{2+} chelation either by combining soluble and membrane fractions of lysed plastids (Walker and Weinstein 1991a), with a subplastidal membrane fraction without soluble components (Lee et al. 1992) or, more recently, with combined soluble plastid protein fractions (Kannangara et al. 1997). It is generally accepted that ATP is required for protein activation as well as Mg^{2+} insertion (Walker and Weinstein 1991a,b 1994). Disruption of three open reading frames *bchI, bchH* and *bchD* of the *Rhodobacter* 45-kB photosynthetic gene cluster (Burke et al. 1993, Bollivar et al. 1994a) led to accumulation of Proto IX. The recombinant peptide derived from these three genes from *Rhodobacter* and *Synechococcus* could be used to reconstitute Mg^{2+} chelatase activity (Gibson et al. 1995; Jensen et al. 1996b). The plant enzyme also consists of the three different subunits designated CHL I, CHL H and CHLD. The stoichiometry of the subunits in planta is unknown.

Plant genes encoding two subunits of Mg chelatase were identified by T-DNA tagged gene inactivation in *Arabidopsis* (Koncz et al. 1990) or by transposon tagging in *Antirrhinum* (Hudson et al. 1993). By means of this insertion mutagenesis the *ch42* gene, a *Chl I* homologue, was identified. The second approach helped to identify the *oli* gene which corresponds to the bacterial *bchH*. The genes encoding the three Mg chelatase subunits H, I and D are apparently effected in the barley *xantha f, h* and *g* mutants, respectively (Jensen et al. 1996a). Other plant cDNA sequences encoding CHL I were identified in soybean (Nakayama et al. 1995), tobacco (Kruse et al. 1997b) and homologous genes have been isolated from the plastid genome of *Euglena gracilis* (Orsat et al. 1992), and other algae, such as *Porphyra purpurea* (Reith and Munholland 1993) and *Cryptomonas* sp., (Douglas and Reith 1993). Coding sequences for CHL H have been identified in *Arabidopsis* (Gibson et al. 1996) and tobacco (Kruse et al. 1997b) and a cDNA clone encoding the third subunit, CHL D, was recently obtained (Papenbrock et al. 1997). Pairwise combination of the three subunits was tested for protein-protein interaction using the yeast two-hybrid system. Interaction between CHL D and CHL I and homodimerisation of CHL D could be demonstrated. The entire active tobacco Mg chelatase complex was reconstituted with all three recombinant tobacco subunits. The final functional evidence for Proto IX chelation was provided with all three tobacco subunits expressed in *Saccharomyces cerevisiae* (Papenbrock et al. 1997). It was suggested that CHL H binds Proto IX and CHL D functions in metal binding of the Mg^{2+} cation. Both CHL I and CHL D possess a nucleotide binding site and both peptides form an activating complex in the presence of ATP (Gibson et al. 1995; Willows et al. 1996).

8.2.11
Ferro Chelatase

The ferro chelatase (EC 4.99.1.1) channels the substrate Proto IX to the heme branch and, subsequently, to the phycobilins and phytochromobilins. Although the insertion of a ferrous ion into Proto IX seems to be a comparable reaction to Mg chelation, ferro chelatase does not show any structural similarity to the competing enzyme. Enzyme activities were measured in plastids and mitochondria (Porra and Lascelles 1968; Little and Jones 1976). Apart from its natural substrates the enzyme also accepts Zn^{2+} and Co^{2+} and different Proto IX derivatives in in vitro assays. The first plant cDNA sequences encoding ferro chelatase were obtained from *Arabidopsis* by functional complementation of the *hem15 Saccharomyces cerevisiae* mutant which has a lesion in ferro chelatase (Smith et al. 1994), and by functional complementation of the *visA* gene using cDNAs from barley and cucumber (Miyamoto et al. 1994). cDNA sequences for two isoforms of ferrochelatase were described. Isoform I precursor has the ability to be imported into plastids and mitochondria (Chow et al. 1998). Ferro chelatase II is targeted to plastids and is associated with the thylakoid and envelope membranes. Interestingly, Protogen oxidase is also localised in envelopes. A close proximity of both enzymes would facilitate a direct transfer of photosensitive Proto IX and a continuous export of heme into the cytosol.

8.2.12
S-Adenosyl -L-Methionine Mg Protoporphyrin IX Methyltransferase

The enzyme S-adenosyl-L-methionine Mg protoporphyrin IX methyltransferase (EC 2.1.1.11) catalyses methylation of the C6-propionic group of Mg protoporphyrin IX (Mg Proto IX). A methyl group from S-adenosyl-L-methionine is substituted for a hydrogen to form Mg protoporphyrin IX monomethylester (Mg Proto IX MME). The enzyme activity was measured first in *Zea mays* (Radmer and Bogorad 1967), *Euglena* (Ebbon and Tait 1969) and wheat (Ellsworth et al. 1974). It has been previously suggested that Mg chelation is obligatorily coupled to the methyltransferase reaction (Gorchein 1972). Enzyme assays with extracts of *E. coli* strains expressing the *Rhodobacter capsulatus* or *R. sphaeroides bchM* gene demonstrated that the 27.5-kDa BchM protein had methyltransferase activity (Bollivar et al. 1994a; Gibson and Hunter 1994).

8.2.13 Mg Protoporphyrin IX Monomethylester Cyclase

The enzyme Mg protoporphyrin IX monomethylester cyclase forms a fifth isocyclic ring in the Mg porphyrin macrocycle. Mg Proto MME is converted to divinyl protochlorophyllide (Pchlide) using NADPH and oxygen. The cyclic ring is made from the methylated propionic acid group at the 6th position of the substrate (Wong and Castelfranco 1985). The complex reaction consists of three sequential steps: hydroxylation of the methylpropionate at the 13-β-carbon atom, oxidation of the hydroxyl group to a carbonyl group and ligation of the α-carbon of the β-ketomethylpropionate to the γ-meso bridge C-15 carbon between the C and D pyrrole rings of the tetrapyrrole. Enzyme activity has been measured in chloroplast extracts of greening cucumber cotyledons (e.g. Chereskin et al. 1982; Fuesler et al. 1984; Wong and Castelfranco 1984), in intact chloroplasts (Nasrulhaq-Boyce et al. 1987) and in *Chlamydomonas* (Bollivar and Beale 1995). Experiments in which plastids were fractionated and cyclase activity reconstituted have indicated that the enzyme consists at least of a soluble and a membrane-bound component (Wong and Castelfranco 1984; Walker et al. 1991). The formation of the isocyclic ring requires oxygen. Therefore, the cyclase reaction must be subject to a different mechanism in anaerobic photosynthetic bacteria.

8.2.14 Vinyl Reductase

The 8-vinyl group on ring B of the divinyl Pchlide, which is the product of the cyclase, is converted by the next enzyme in the pathway, the vinyl reductase. Due to a wider substrate specificity of the enzyme, the vinyl group reduction can take place at various steps of Chl synthesis between Mg Proto MME and Chl. (Rebeiz et al. 1983). Monovinyl Pchlide was mainly found in dark-grown plants, but mono and divinyl Chl were also detected in plant tissue (Rebeiz and Lascelles 1982).

8.2.15 Protochlorophyllide Oxidoreductase

The membrane associated enzyme protochlorophyllide oxidoreductase (POR) (EC 1.6.99.1) binds NAPDH and Pchlide to form a photoreactive ternary enzyme-substrate complex. It catalyses the transreduction of a double bond in ring D to yield chlorophyllide (Chlide), a reaction that has an absolute requirement for light. Several catalytic intermediates can be depicted from the spectroscopically detectable enzyme complex in prolamellar bodies (Griffiths 1978), in in vitro reconstituted solubilised enzyme (Oliver and Griffiths 1982) and with recombinant POR A and B (Holtorf et al. 1995). The protein was purified from several sources, such as wheat, barley and pea. Plant cDNA clones were obtained from barley (Schulz et al. 1989), oat (Darrah et al. 1990), pea (Spano et al. 1992a), pine (Spano et al. 1992b), wheat (Teakle and Grifftiths 1993) and *Arabidopsis* (Armstrong et al. 1995). Sequencing in barley and *Arabidopsis* revealed cDNA sequences which encode two different isoenzymes, POR A and POR B (Armstrong et al. 1995; Holtorf et al. 1995). The gene products differ in their expression pattern and posttranslational plastid import mechanisms. The isoenzyme POR A is a short lived enzyme in greening tissue and predominantly present in the prolamellar body of angiosperm etioplasts. Processing and import of the POR A precursor depends on the presence of its substrate Pchlide (Reinbothe et al. 1995). POR B is present in etiolated seedlings and persistent in green tissue. Physiological reasons for the distinct functions of POR A and POR B during chloroplast development have been proposed by Reinbothe and Reinbothe (1996).

Most green algae, gymnosperms and photosynthetic bacteria are able to synthesise Chl in both the light and the dark (Bogorad 1950). The genes encoding the polypeptides essential for dark Pchlide oxidoreductase activity were found using mutants that arrested Chl synthesis in the dark and accumulated Pchlide (Roitgrund and Mets 1990; Yang and Bauer 1990; Choquet et al. 1992; Fujita et al. 1992; Suzuki and Bauer 1992; Li et al. 1993). The genes *FrxC/Chl L/bchL, BchN/chl N* and *bchB/Bch B* are required for operating the light-independent pathway of Pchlide reduction. The CHL L, N and B subunits reveal some similarity to the gene products from *Azetobacter vinelandii*, NifD, H and K, three subunits of the nitrogenase that catalyse the reduction of N_2 to NH_3.

8.2.16 Chlorophyll *a* Synthase

A propionate substituent at C-17 on ring D of the macrocycle is esterified with a long chain fatty alcohol, mainly phytol or geranylgeraniol by Chl *a* synthase. Etioplast preparations preferentially incorporate geranylgeranyl-pyrophosphate into Chlide rather than phytyl pyrophosphate (Rüdiger et al. 1980). A reductase subsequently catalyses the saturation of double bonds of geranylgeranyl to phytylated Chl (Schoch et al.1977). Phytyl pyrophosphate is preferentially esterified in green tissue (Soll et al. 1983). By gene disruption of the *Rhodobacter capsulatus* photosynthetic gene cluster the *bchG* gene was identified as encoding bacteriochlorophyll synthase (Bollivar et al. 1994c) and recombinant BchG has now been shown to have this enzymatic activity (Oster et al. 1997). Homologous plant sequences have been detected in both *Arabidopsis* (Gaubier et al. 1995) and *Chloroflexus auranticus* (Lopez et al. 1996).

8.2.17
Interconversion of Chlorophyll *a* and Chlorophyll *b*

The oxidation of the 7-methyl group of Chl to the 7-formyl group of Chl *b* was described in *Chlorella vulgaris* and maize (Schneegurt and Beale 1992; Porra et al. 1993). The substrate specificity of this enzyme is broad and the oxygenase accepts Pchlide, Chlide or Chl *a*. Chl *b* is converted to Chl *a* via a 7-hydroxymethyl Chl intermediate and this reaction is accompanied by the loss of light harvesting chlorophyll binding complex (LHC) apoproteins and formation of new reaction centre proteins (Ito et al. 1993). The release of Chl *b* from the antenna pigment proteins, conversion to Chl *a* and assembly with core complex proteins were recently demonstrated (Ohtsuka et al. 1997). An interesting question that still remains to be answered is how the control of the interconversion between both forms of Chl affects the organisation of the photosystems during adaptation to different light intensities.

8.3
Regulation of Tetrapyrrole Biosynthesis in Plants

8.3.1
General Considerations

Control of tetrapyrrole biosynthesis occurs in response to an endogenous developmental program and to environmental factors. The most important exogenous stimulus for the development of the photosynthetically active plant is light. The photoperiodic alteration of irradiation and the perception of specific wavelengths by different photoreceptors affect the rate of Chl formation. The photosynthetic apparatus consisting of the Chls and the pigment binding proteins is modulated in terms of size, activity of energy transduction and protective capacity against photooxidative damage in response to daily changes in light intensities. The phytochrome and different blue light receptors control cellular developmental and structural processes. The signal transduction pathways triggered by the photoreceptors await final elucidation. In general, these pathways result in gene activation and posttranslational modification for many cellular processes including tetrapyrrole synthesis.

Apart from the daily adjustments to environmental conditions a tight control of tetrapyrrole biosynthesis is also essential for other reasons. Additionally, in response to the developmental program, the need for different tetrapyrrolic endproducts requires a complex regulation to ensure their supply in appropriate proportions. This implies a coordinated allocation of metabolic intermediates at the branchpoints towards Chl or heme synthesis. Secondly, the control mechanism of tetrapyrrole biosynthesis has to guarantee a regular flow of substrate through the pathway to prevent hazardous accumulation of tetrapyrrolic intermediates and to avoid uncontrolled reactions which have been described as a result of chemically induced or hereditary lesions in the pathway (porphyrias) which lead to cellular destruction. The control of such a complex pathway requires many levels of regulation from transcriptional to posttranslational events with the aim of synchronising enzyme activities to guarantee an appropriate flow of metabolites in response to changing physiological and environmental conditions (Fig. 2). Posttranscriptional control includes the transport of the enzymes into plastids, the routing to their final destination inside the organelle, the assembly of complex protein-protein interactions between neighbouring enzymes or several subunits of an enzyme and activation and derepression of enzymatic steps. However, the substrate flow will actually be controlled at only a few limiting sites in the pathway while other non-limiting enzymes ensure a constant flow of metabolites. The sites of metabolic control can be assumed to be at the beginning of the pathway, at the branchpoints and at the end of the pathway. Finally, control of tetrapyrrole biosynthesis is required to synchronise the metabolic pathway with the synthesis of the apoproteins. The coordination of both processes is a prerequisite for the functional assembly of the photosynthetic apparatus (Herrin et al. 1992; Plumley and Schmidt 1995). Accumulation of different Chl-binding proteins appears to be a sequential process starting with the assembly of the core complex in the thylakoid membrane and a preferential accumulation of LHC II apoproteins relative to LHC I apoproteins (Anandan et al. 1993; Dreyfuss and Thornber 1994). This implies that Chl limitation leads to a preferential accumulation of core complexes of photosystem I and II.

Our own studies with Chl-deficient transgenic plants that are affected in the expression of GSA aminotransferase indicated that a lack of Chl does not necessarily leave plants in an early developmental state and, what is more, does not prevent synchronous assembly of entire photosynthetic units. Under low light conditions without a photoinhibitory danger, these plants tend to maintain their light harvesting antenna size even when Chl synthesis is suppressed (Härtel et al. 1997).

8.3.2 Gene Expression at Key Regulatory Steps

Three major steps have previously been reported to be limiting for tetrapyrrole metabolism: (1) ALA synthesis is accepted as the rate limiting step in the synthesis of tetrapyrroles. It is stimulated upon irradiation and restricted in the dark most likely by a feedback mechanism (Kannangara and Gough 1978b; Huang et al. 1989). Feeding of ALA to etiolated seedlings bypasses the suppression of the C5 pathway and turns the plants greenish due to Pchlide accumulation. (2) In angiosperms Pchlide can only be reduced to Chlide in light. In dark-grown seedlings Pchlide accumulates to a certain degree before repression of ALA synthesis prevents excessive Pchlide accumulation. Light-dependent conversion of Pchlide parallels enhanced ALA synthesis and structural changes in thylakoid membranes (see below) although the mechanism of this coordinated control is not understood. (3) It has been demonstrated that heme inhibits ALA synthesis in plants. Collectively, most of the recent results indicate that heme functions both directly and indirectly as a transcriptional corepressor and as a feedback inhibitor of glutamyl tRNA reductase (Rieble and Beale 1991; Pontoppidan and Kannangara 1994).

Several groups have studied gene expression and enzyme activities of the C5 pathway and contributed to the current view that ALA synthesis is rate limiting for Chl synthesis. Because of the unique activation of glutamate by ligation to a tRNA, it was proposed that modulation of the structure and the amount of the cofactor is responsible for diverting glutamate to the tetrapyrrole pathway (see above). However, alterations of tRNA levels could not be detected in tissue isolated in the light and the dark of photoperiodically grown organisms (Mayer and Beale 1990; Jahn 1992). Levels of mRNA encoding glutamyl-tRNA reductase were elevated in response to light (Ilag et al. 1994). Expression analysis of the small gene family revealed that the mRNA species accumulate differently in light and in plant organs (Bougri and Grimm 1996, Tanaka et al. 1996). Glutamyl-tRNA reductase activity was increased in greening cucumber cotyledons in comparison to etiolated tissues (Masuda et al. 1996). Increased mRNA and activity levels for GSA aminotransferase were determined in response to light in *Chlamydomonas* cultures and *Arabidopsis* (Kannangara and Gough 1978a; Ilag et al. 1994; Matters and Beale 1994), while its RNA levels did not significantly alter in greening barley seedlings (Grimm 1990). Calcium and calmodulin are involved in mediating blue light signal perception for the activation of the GSA aminotransferase gene in *Chlamydomonas* (Im et al. 1996). Collectively, at present it is difficult to judge whether a single enzymatic step is rate limiting or the control is shared between all enzymes involved.

Recently, tremendous progress has been made in the understanding of regulated expression of genes encoding the isoenzymes A and B of Pchlide oxidoreductase. RNA and protein levels of POR A are highly abundant in dark-grown seedlings whereas light-grown seedlings contain only traces of the isoform (Apel 1981). POR B is more continuously expressed in etiolated and green tissue (Holtorf et al. 1995). Additionally, the POR A precursor is imported by plastids only in the presence of Pchlide. Fully developed chloroplasts have reduced POR A import because the concentration of Pchlide is reduced (Reinbothe et al. 1995).

A change in the need for Chl and heme is assumed to result in varying metabolic activities. The author's group analysed expression and activity of many tetrapyrrolic enzymes during dark/light growth or under continuous conditions and focused on the mechanisms controlling the channelling of Proto IX to the heme or Chl-synthesising branches of the pathway (Papenbrock et al., submitted). Under cycling light conditions, a transient maximum in ALA synthesis and Mg chelatase activity parallel maximal RNA levels for glutamyl-tRNA reductase and the CHL H Mg chelatase subunit in the first hours of illumination. Such regulatory changes allow a massive flow of substrate into the Chl synthesising branch of the pathway. Fe chelatase activity and the level of the corresponding RNA oscillate synchronously in a rhythm opposite to that of Mg chelatase. Under constant conditions, ALA synthesis and Fe chelatase activity cycle, while Mg chelatase shows two activity peaks in a 24h period. The control of ALA synthesis and the allocation of Proto IX to Mg or Fe chelatase reflect the functional coordination of tetrapyrrole biosynthesis in response to daily fluctuations in tetrapyrrole requirements (Papenbrock et al. 1998). This is in general agreement with previous findings (Hudson et al. 1993; Bougri and Grimm 1996; Jensen et al 1996; Kruse et al. 1997). Moreover, the photoperiodic rhythm of ALA synthesis and Mg chelatase activity matches the synthesis of LHC II apoproteins, reflecting a coordinated biosynthesis of Chl and apoprotein for their assembly into functional complexes (Nagy et al. 1988; Plumley and Schmidt 1995).

Expression and activity profiles for other single enzymatic steps are well documented. The analysis of mRNA steady state levels encoding a single enzyme will perhaps indicate consistencies or differences in the expression pattern observed in several plant species grown under comparable conditions. These analyses do not necessarily contribute to a better understanding of the potential mechanisms of metabolic regulation. Comparative analyses covering gene expression and activity of many enzymatic steps are more useful for understanding regulatory mechanisms.

It is worth mentioning that tetrapyrrole biosynthesis could be involved in the concerted interplay between the nuclear and plastid genome. Plastid biogenesis requires coordinated information from nuclear and plastidal genes. Nuclear control affects plastidal gene expression, but also some factors which originate in the plastids are required for the transcriptional activity of nuclear genes (Taylor 1989; Susek et al. 1993). Impairment of the integrity of the plastids by application of photobleaching inhibitors or mutations leading to photooxidation of the plastid compartment can cause modified expression of nuclear photosynthetic genes (Taylor 1989). This indicates that the actual status of the plastids can be conveyed to the nucleus and the existence of plastid-derived signals was proposed from these experiments. However, the nature of the plastid signal has remained elusive. Johannigmeier and Howell (1984) suggested Mg porphyrins or Proto IX as signals that inhibit mRNA formation for LHC II apoproteins in dark-grown *Chlamydomonas*. The authors exclude a signalling function in the light because of the high consumption of these metabolic intermediates.

8.4 Mutants and Transgenic Plants with Deregulated Tetrapyrrole Biosynthesis

Genetic and biochemical characterisation of tetrapyrrole biosynthesis has progressed by using approaches to genetically dissect the tetrapyrrole biosynthetic pathway and to examine either the regulatory mechanism of the complete pathway or the impact of single enzymatic steps on the regulation of metabolic flux. The use of effectors in tetrapyrrole biosynthesis, such as inhibitors, activators and derepressors, are not considered and the reader is referred to other chapters in this volume.

8.4.1 Pigment Mutants

Pigment-deficient mutants have proved to be useful for examining the mechanisms of metabolic control or analysing biochemically the enzymatic steps which are affected by the mutation. However it must be kept in mind that mutant characterisation suffers from the difficulty in assigning a mutation to a particular function. The pleotropic nature of pigment mutants on chloroplast structure limits the analysis of these mutants. Because of the mutual dependence of Chl synthesis and chloroplast development, it remains difficult to distinguish between a primary malfunction of pigment synthesis and an impairment of other cellular activities unless the mutant gene was identified.

Mutants of the *xantha* and *albino* phenotype were used to investigate specific blocks in the entire biosynthetic pathway (von Wettstein et al. 1971; Henningsen et al. 1993; Runge et al. 1995). Mutants which accumulate more Proto IX upon feeding ALA in darkness than wild-type plants have been suggested to contain genetic lesions for the oxidation of Protogen IX or the chelation of Proto IX. The maize mutants *blandy 4, l13* or *oy-1040* also accumulate Proto IX after feeding with ALA in the dark (Mascia 1978). *Xantha* mutants of *Arabidopsis* with genetic lesions for the entire pathway were recently obtained by screening for Chl deficiency under defined low light conditions (Runge et al. 1995).

Mutants of *Chlamydomonas*, designated *br_s-1* and *br_c-1*, have been isolated which block Chl synthesis at the level of Proto IX (Wang et al. 1974). Two other mutants, *chl1* and *lts3*, which are most likely allelic, have been described with the same phenotype (Chekunova et al. 1993). The mutant *br_c-1(lts3)* controls a step in light-independent Chl formation, while the *br_s-1* mutation prevents the Mg chelation of Proto IX. A regulatory function has been assigned to the genetic lesions of the barley *tigrina* mutants which also exhibit a loss of repression of ALA synthesis or the formation of the ALA synthesising enzymes. The barley *tigrina* mutants accumulate up to 15- to 20-fold the wild-type level of Pchlide in dark-grown barley seedlings (Nielsen 1974).

Phytochrome chromophore-deficient mutants have genetic lesions in the biosynthesis of phytochromobilin (Terry 1997). This survey cannot cover mutations in genes encoding the phytochrome apoprotein. Several mutants of different plants have been shown to be effected at two enzymatic steps in the heme degradation pathway. Biochemical analyses indicate that the *yellow-green-2 (yg-2)* mutant of tomato (Terry and Kendrick 1996) and the *pcd1* mutant of pea (Weller et al. 1996) are deficient in heme oxygenase, while the *aurea (au)* mutant of tomato (Terry and Kendrick 1996) and the *pcd2* mutant of pea (Weller et al. 1997) lack phytochromobilin synthase activity. However these mutants are viable and their seedlings have yellow-green to green wild-type-like phenotypes.

8.4.2
Genetic Manipulation of Tetrapyrrole Biosynthesis

The approach of insertion mutagenesis for identification of genes involved in tetrapyrrole biosynthesis has been introduced earlier in context with the identification of coding sequences for Mg chelatase subunits. For studies on the physiological consequences of deregulated tetrapyrrole synthesis the use of mutants is limited due to their pleiotropic phenotype, particularly if the mutant genes are not identified. A physiological aberration can more easily be interpreted if it directly correlates with a genetically engineered enzymatic step. In the author's loboratory, antisense RNA techniques have been used to gradually reduce expression of many enzymes in tetrapyrrole biosynthesis, allowing the analysis of the consequences of these enzyme deficiencies. The transgenic approach was undertaken to examine the regulatory and regulating competence of single enzymatic steps. Modification of gene expression and protein activity can affect the substrate flow and the expression or stability of other constituents in the pathway. These analyses reflect the interactive relationship between gene expression of single enzymes and chloroplast biogenesis. Moreover, transgenic plants expressing antisense RNA for proteins involved in the metabolic pathway of pigment synthesis can be used to evaluate new target sites for herbicides. The reduction of protein content mimics the inhibitory effect of herbicides. In the following section a few examples of deregulated transgenic plants are outlined. The detailed analysis of these plants was aimed at gaining insights into the regulatory mechanism of tetrapyrrole biosynthesis and its relevance for photosynthesis or antioxidative stress defence.

8.4.2.1
Transgenic Plants with Reduced GSA Aminotransferase Activity

A full-length cDNA sequence encoding a tobacco GSA aminotransferase was inserted in reverse orientation behind the constitutive CaMV 35S promoter on a binary vector. This vector construct was introduced into the tobacco genome by an *Agrobacterium tumefaciens*-mediated leaf disc transformation technique (Höfgen et al. 1994). The transformants selected for further analysis were characterised by a gradually reduced Chl content and growth rate (Fig. 3). The chlorosis was visible from the earliest stage of development. Lower pigment contents corresponded to reduced activity of GSA aminotransferase. Analysis of the steady state levels of RNAs and proteins for enzymes in the pathway indicated that the impaired RNA synthesis due to antisense RNA expression and the reduced GSAT activity did not, apparently, affect gene

Fig. 3. Tobacco transformants expressing antisense RNA for GSA aminotransferase and wild-type plants (*right back*). The three transgenic lines 57 (*left back*), 25 (*left front*) and 42 (*right front*) represent the broad range of Chl deficiency in the transformants. GSA aminotransferase activity of these transformants correlates with reduced Chl content in analysed leaves

expression of other tetrapyrrolic enzymes (Härtel et al. 1997). Therefore, the reduced Chl content is clearly due to the reduced GSA aminotransferase activity.

Other physiological processes are affected as a consequence of the reduced Chl content and/or the antisense inhibition of GSA aminotransferase. The transformants respond differently to various light intensities compared to wild-type. The preferential assembly of complete antenna and core proteins accompanied by a drastic increase in mRNA for LHC apoproteins has been discussed above (Härtel et al. 1997; see Sect. 8.3.).

Regardless of the extent to which Chl synthesis was reduced, the ratios of total Chl to carotenoids remained constant under low light intensities. In contrast, plants grown in high light had an elevated carotenoid content relative to Chl. The carotenoid levels progressively increased with decreasing capacity for Chl synthesis. The large pool size of xanthophyll-cycle pigments upon suppression of Chl was found to be correlated with a decrease in the quantum yield of photosystem II (Härtel and Grimm 1998). Therefore, regulation of carotenoid synthesis conclusively depends on the Chl-synthesising capacity and on the efficiency of photosynthesis. Moreover, the Chl-deficient transgenic plants reflect also the strong regulatory interdependence of Chl and carotenoid synthesis as previously proposed. The special role of carotenoids for light energy transfer and dissipation in response to adaptation to environmental conditions was illustrated in the transgenic plants with gradually reduced Chl content.

Fig. 4. Eight-week-old primary transformants containing antisense RNA for Urogen decarboxylase (*right*) and a wild-type plant (*left*)

8.4.2.2
Transgenic Plants with an Alternative ALA Synthesising Pathway

Synthesis of the yeast ALA synthase and its translocation to plastids leads to additional synthesis of ALA molecules and complements chlorphyll deficiency of the GSA aminotransferase antisense plants (Zavgorodnyaya et al. 1997). Due to the bypass of the endogenous C5 pathway the transgenic plants show an increased tolerance towards the gabaculine, a powerful inhibitor of GSA aminotransferase.

8.4.2.3
Transgenic Plants Expressing Antisense RNA for Urogen Decarboxylase and Coprogen Oxidase

Gene expression of Urogen decarboxylase and Coprogen oxidase was modified by antisense RNA synthesis in transgenic tobacco (Kruse et al. 1995b; Mock and Grimm 1997). In contrast to the chlorotic phenotype of transformants with reduced activities for GSA aminotransferase, these plants possess characteristic leaf necrosis (Figs. 4 and 5). Leaves of Urogen decarboxylase antisense plants have spacious lesions on the leaves, while a fine net of necrotic lesions is formed mainly along the leaf veins of Coprogen oxidase antisense plants. The premature leaves are less injured, but the photodynamic damage occurs earlier or later in the expanding leaves depending on the effectiveness of the antisense transformation.

Both types of transformants contain only slightly reduced levels of Chl and heme. The transgenic plants appear not to suffer from a deficiency in Chl, but rather from the accumulating substrate of each target protein. The accumula-

Fig. 5. Coprogen oxidase antisense plant PL1-15 (*right*) and wild-type plant (*left*) grown for 8 weeks under greenhouse conditions

tion of the non-metabolised porphyrins [either Uro(ogen) or Copro(ogen)] correlates with the reduced enzyme activity of the corresponding enzyme. Gene expression and activity of other enzymes in the pathway are not initially deregulated by the reduced protein activity, but are later diminished as a consequence of photodynamic damage. Deregulated gene expression of other nuclear and plastid encoded photosynthetic proteins indicates an interference in the signalling between plastids and nucleus (see above). It was suggested that gene expression is adapted in response to photosensitisation in plastids (Kruse et al. 1995b). The photoreactive process of accumulating porphyrins (porphogenesis) resembles other oxidative stress reactions in the cell, in which reactive oxygen species are involved, acting as signalling or damaging agents (Alscher et al. 1997). The photodynamic action of accumulating porphyrins is strictly dependent on light intensity and/or length of the light period. A 16-h light period of medium intensity (300 μM photons $m^{-2}s^{-1}$) stimulates the formation of leaf necrosis compared to a 16-h low intensity light period or a 6-h high light treatment (Mock and Grimm 1997).

The subsequent cellular responses were investigated in these tetrapyrrole-sensitised transgenic plants (Mock et al. 1998). In comparison to control plants, the transformants had increased levels of antioxidant mRNA, particularly those encoding superoxide dismutase, catalases and glutathione peroxidase. These elevated transcript levels correlate with increased activities of cytosolic Cu-/Zn-dependent and mitochondrial Mn-dependent superoxide dismutase. Most of the enzymes of the Halliwell-Asada pathway also displayed increased activities. However, despite the elevated enzyme activities the limited capacity of the antioxidative system was apparent from decreased levels of ascorbate and glutathione, which led to the formation of damaged leaf tissue.

The cellular antioxidative response triggered by accumulating porphyrins resembles that obtained by treatment with photodynamic herbicides (Knörzer et al. 1996). The minor differences in the cellular antioxidative reaction between the transgenic plants with reduced activity of Urogen decarboxylase or Coprogen oxidase and the herbicide-treated plants can, most likely, be explained by the dose dependency of the herbicide treatment, the difference in the photosensitivity of uroporphyrin (Uro), coproporphyrin (Copro and Proto IX) which have an ascending photoreactivity, and the difference between the short time exposure of the herbicide in soybean cell cultures and the permanent photooxidative stress from accumulating porphyrins in the transgenic plants.

Acknowledgements. I thank Dr. Matthew Terry for critically reading the manuscript and making valuable suggestions and the members of my research group for helpful discussion during preparation of this chapter.

References

Akhtar M (1994) The modification of acetate and propionate side chains during the biogenesis of haem and chlorophylls: mechanistic and stereochemical studies. In: Chadwick DJ, Ackrill K (eds) The biosynthesis of the tetrapyrrole pigments. Ciba Foundation Symposium 180. Wiley, Chichester, pp 209–220

Alscher RG, Donahue JL, Cramer CL (1997) Reactive oxygen species and antioxidants: relationships in green cells. Physiol Plant 100:224–233

Anandan S, Morishige DT, Thornber JP (1993) Light-induced biogenesis of light-harvesting complex I (LHCI) during chloroplast development in barley. Plant Physiol 101:227–236

Andersen RV (1992) Characterization of a barley tRNA synthetase involved in chlorophyll biosynthesis. In: Murata N (ed) Research in photosynthesis, vol III. Kluwer Academic Publishers, Amsterdam, pp 27–30

Apel K (1981) The protochlorophyllide holochrome of barley (*Hordeum vulgare* L.): phytochrome induced decrease of translatable mRNA coding for the NADPH:protochlorophyllide oxidoreductase. Eur J Biochem 120:89–93

Armstrong GA, Runge S, Frick G, Sperling U, Apel K (1995) Identification of NADPH:protochlorophyllide oxidoreductases A and B: a branched pathway for light-dependent chlorophyll biosynthesis in *Arabidopsis thaliana*. Plant Physiol 108:1505–1517

Avissar YJ, Beale SI (1989a) Identification of the enzymatic basis for δ-aminolevulinic acid auxotrophy in a *hemA* mutant of *Escherichia coli*. J Bacteriol 171:2919–2924

Avissar YJ, Beale SI (1989b) Biosynthesis of tetrapyrrole pigment precursors: pyridoxal requirement of the aminotransferase step in the formation of δ-aminolevulinate from glutamate in extracts of *Chlorella vulgaris*. Plant Physiol 89:852–859

Avissar YJ, Beale SI (1990) Cloning and expression of a structural gene from *Chlorobium vibrioforme* that complements the *hemA* mutation in *Escherichia coli*. J Bacteriol 172:1656–1659

Avissar YJ, Ormerod JG, Beale SI (1989) Distribution of δ-aminolevulinic acid biosynthetic pathways among phototrophic bacterial groups. Arch Microbiol 151:513–519

Beale SI (1993) Biosynthesis of cyanobacterial tetrapyrrole pigments: hemes, chlorophylls and phycobilins. In: Bryant (ed) The molecular biology of cyanobacteria. Kluwer Academic Publishers, Amsterdam, pp 520–558

Beale SI, Gough SP, Granick S (1975) The biosynthesis of δ-aminolevulinic acid from the intact carbon skeleton of glutamic acid in greening barley. Proc Natl Acid Sci USA 72:2719–2723

Becerril JM, Duke SO (1989) Protoporphyrin IX content correlates with activity of photobleaching herbicides. Plant Physiol 90:1175–1181

Berry-Lowe S (1987) The chloroplast glutamate tRNA gene required for δ-aminolevulinate synthesis. Carlsberg Res Commun 52:197–210

Boese QF, Spano AJ, Li J, Timko MP (1991) Aminolevulinic acid dehydratase in pea (*Pisum sativum* L.). Identification of an unusual metal-binding domain in the plant enzyme. J Biol Chem 266:17060–17066

Bogorad L (1950) Factors associated with the synthesis of chlorophyll in the dark in seedlings of *Pinus jeffreyi.* Bot Gaz 111:221–241

Bollivar DW, Beale SI (1995) Formation of the isocyclic ring of chlorophyll by isolated *Chlamydomonas reinhardtii* chloroplasts. Photosyn Res 43:113–124

Bollivar DW, Jiang Z., Bauer CE, Beale SI (1994a) Heterologous expression of the *bchM* gene product from *Rhodobacter capsulatus* and demonstration that it encodes *S*-adenosyl-L-methionine:Mg-protoporphyrin IX methyltransferase. J Bacteriol 176:5290–5296

Bollivar DW, Suzuki JK, Beatty JT, Dobrowolski JM, Bauer CE (1994b) Directed mutational analysis of bacteriochlorophyll *a* biosynthesis in *Rhodobacter capsulatus.* J Mol Biol 237:622–640

Bollivar DW, Wang S, Allen J, Bauer CE (1994c) Molecular genetic analysis of terminal steps in bacteriochlorophyll *a* biosynthesis: characteriaztion of a *Rhodobacter capsulatus* strain that synthesizes geranylgeraniol-esterified bacteriochlorophyll *a.* Biochemistry 33:12763–12768

Bougri O, Grimm B (1996) Members of a low-copy number gene family encoding glutamyl-tRNA reductase are differentially expressed in barley. Plant J 9:867–878

Bruyant P, Kannangara CG (1987) Biosynthesis of δ-aminolevulinate in greening barley leaves, VIII: purification and characterization of the glutamate-tRNA ligase. Carlsberg Res Commun 52:99–109

Burke DH, Alberti M, Hearst JE (1993) *BchFNBH* bacteriochlorophyll synthesis genes of *Rhodobacter capsulatus* and identification of the third subunit of light-independent protochlorophyllide reductase in bacteria and plants. J Bacteriol 175:2414–2422

Camadro J, Chambon H, Jolles J, Labbe P (1986) Purification and properties of coproporphyrinogen oxidase from the yeast *Saccharomyces cerevisiae.* Eur J Biochem 156:579–587

Castelfranco PA, Weinstein JD, Schwarcz S, Pardo AD, Wezelman BE (1979) The Mg insertion step in chlorophyll biosynthesis. Arch Biochem Biophys 192:592–598

Chadwick DJ, Ackrill K (1994) The biosynthesis of the tetrapyrrole pigments. Wiley, Chichester (Ciba Foundation symposium 180)

Chauhan S, O'Brian MR (1995) A mutant *Bradyrhizobium japonicum* δ-aminolevulinic acid dehydratase with an altered metal requirement in situ for tetrapyrrole biosynthesis in soybean root nodules. J Biol Chem 270:19823–19827

Chekunova EM, ShalygoNV, Yaronskaya EB, Averina NG, Chunaev AS (1993) Regulation of biosynthesis of chlorophyll precursors in mutants of the green algae *Chlamydomonas reinhardtii.* Biockhimiya (Russ) 58:1430–1436

Chen MW, Jahn D, Schön A, O'Neill GP, Söll D (1990) Purification of the glutamyl-tRNA reductase from *Chlamydomonas reinhardtii* involved in δ-aminolevulinic acid formation during chlorophyll biosynthesis. J Biol Chem 265:4058–4063

Chen TC, Miller GW (1974) Purification and characterization of uroporphyrinogen decarboxylase from tobacco leaves. Plant Cell Physiol 15:993–1005

Chereskin BA, Wong YS, Castelfranco PA (1982) In vitro synthesis of the chlorophyll isocyclic ring: transformation of the magnesium-protoporphyrin IX and magnesium protoporphyrin IX monomethyl ester into magnesium-2,4-divinyl pheoporphyrin a_5. Plant Physiol 70:987–993

Choquet Y, Rahire M, Girad-Bascou J, Erikson J, Rochaix JD (1992) A chloroplast is required for the light-independent accumulation of chlorophyll in *Chlamydomonas reinhardtii.* EMBO J 11:1697–1704

Chow KS, Singh DP, Roper JM, Smith AG (1998) A single precursor protein for ferrochelatase I from *Arabidopsis* is imported in vitro into both chloroplasts and mitochondria. J Biol Chem 272:27565–27571

Crocket N, Alefounder PR, Battersby AR, Abell C (1991) Uroporphyrinogen III synthase: studies on its mechanism of action, molecular biology and biochemistry. Tetrahedron 47:6003–6014

Darrah PM, Kay SA, Teakle GR, Griffiths WT (1990) Cloning and sequencing of protochlorophyllide reductase. Biochem J 265:789–798

Dreyfuss BW, Thornber JP (1994) Organization of the light-harvesting complex of photosystem I and its assembly during plastid development. Plant Physiol 106:841–848

Douglas SE, Reith M (1993) A *bchI* homolog, encoding a subunit of Mg chelatase, is located on the plastid genomes of red and cryptomonad algae. J Mar Biotechnol 1:135–141

Drechsler-Thielmann B, Dörnemann D, Senger H (1993) Synthesis of protoheme via both the C5 pathway and the Shemin pathway in the pigment mutant C-2A′ of *Scenedesmus obliquus*. Z Naturforsch C 48:584–589

Ebbon JG, Tait GH (1969) Studies on S-adenosyl-methionine-magnesium-protoporphyrin-methyl-transferase in *Euglena gracilis* strain Z. Biochem J 111:573–582

Elder GH, Evans JO (1978) Evidence that coproporphyrinogen oxidase activity of rat liver is situated in the intermembrane space of mitochondria. Biochem J 172:345–347

Ellsworth RK, Dullaghan JP, Pierre MES (1974) The reaction mechanism of S-adenosyl-L-methionine:Mg protoporphyrin methyltransferase of wheat. Photosynthetica 8:375–383

Fereirra GC, Andrew TL, Karr SW, Dailey HA (1988) Organization of the terminal two enzymes of the heme biosynthetic pathway. J Biol Chem 263:3855–3839

Fuesler TP, Hanamoto CM, Castelfranco PA (1982) Separation of Mg-protoporphyrin IX and Mg-protoporphyrin IX monomethyl ester synthesized de novo by developing cucumber etioplasts. Plant Physiol 69:421–423

Fuesler TP, Wong YS, Castelfranco PA (1984) Localization of Mg-chelatase and Mg-protoporphyrin IX monomethyl ester (oxidative) cyclase activities within isolated, developing cucumber chloroplasts. Plant Physiol 75:662–664

Fujita Y, Takahashi Y, Chuganji M, Matsubara M (1992) The *nifH*-like (*frxC*) gene is involved in the biosynthesis of chlorophyll in the filamentous cyanobacterium *Plectonema boryanum*. Plant Cell Physiol 33:81–92

Gaubier P, Wu HJ, Laudie M, Delseny M, Grellet F (1995) A chlorophyll synthase gene from *Arabidopsis thaliana*. Mol Gen Genet 249:58–64

Gibson KD, Laver WG, Neuberger A (1958) Initial stages in the biosynthesis of porphyrins. The formation of δ-aminolevulinic acid from glycine and succinyl-coenzyme A by particles from chicken erythrocytes. Biochem J 70:71–81

Gibson LCD, Hunter CN (1994) The bacteriochlorophyll biosynthesis gene, *bchM*, of *Rhodobacter sphaeroides* encodes S-adenosyl-L-methionine:Mg-protoporphyrin IX methyltransferase. FEBS Lett 352:127–130

Gibson LCD, Willows RD, Kannangara CG, von Wettstein D, Hunter CN (1995) Magnesium-protoporphyrin chelatase of *Rhodobacter sphaeroides*: reconstitution of activity by combining the products of the *bchH, -I,* and *-D* genes expressed in *Escherichia coli*. Proc Natl Acad Sci USA 92:1941–1944

Gibson LCD, Marrison JL, Leech RM, Jensen PE, Bassham DC, Gibson M, Hunter CN (1996) A putative Mg-chelatase subunit from *Arabidopsis thaliana* cv C24. Plant Physiol 111:61–71

Gorchein A (1972) Magnesium protoporphyrin chelatase activity in *Rhodopseudomonas spheroides* studies with whole cells. Biochem J 127:97–106

Gough SP, Kannangara CG (1977) Synthesis of δ-aminolevulinate by a chloroplast stroma preparation from greening barley leaves. Carlsberg Res Commun 42:459–464

Griffiths WT (1978) Reconstitution of chlorophyll formation by isolated etioplast membranes. Biochem J 174:681–692

Grimm B (1990) Primary structure of a key enzyme in plant tetrapyrrole synthesis: glutamate-1-semialdehyde aminotransferase. Proc Natl Acad Sci USA 87:4169–4173

Grimm B, Bull A, Welinder KG, Gough SP, Kannangara CG (1989) Purification and partial amino acid sequence of the glutamate 1-semialdehyde aminotransferase of barley and *Synechococcus*. Carlsberg Res Commun 54:67–79

Grimm B, Smith AJ, Kannangara GC, Smith M (1991) Gabaculine-resistant glutamate 1-semialdehyde aminotransferase of *Synechococcus*. J Biol Chem 266:12495–12501

Grimm B, Smith MA, von Wettstein D (1992) The role of Lys272 in the pyridoxal 5-phosphate active site of *Synechococcus* glutamate 1-semialdehyde aminotransferase. Eur J Biochem 206:579–585

Hansson M, Hederstedt L (1992) Cloning and characterization of the *Bacillus subtilis hemEYH* gene cluster which encodes protoheme IX biosynthetic enzymes. J Bacteriol 174:8081–8093

Hart GJ, Battersby AR (1985) Purification and properties of uroporphyrinogen III synthase (co-synthase) from *Euglena gracilis*. Biochem J 232:151–160

Hart GJ, Miller AD, Leeper FJ, Battersby AR (1987) Biosynthesis of the natural porphyrins: proof that hydroxymethylbilane synthase (porphobilinogen deaminase) uses a novel binding group in ist catalytic action. J Chem Soc Chem Commun 1987:1762–1765

Härtel H, Grimm B (1998) Consequences of chlorophyll deficiency for leaf carotenoid composition in tobacco expressing glutamate 1-semialdehyde aminotransferase antisense RNA: dependence on developmental age and growth light. Exp Botany 49:535–546

Härtel H, Kruse E, Grimm B (1997) Restriction of chlorophyll synthesis due to expression of glutamate-1-semialdehyde aminotransferase antisense RNA does not reduce the light-harvesting antenna size in tobacco. Plant Physiol 113:1113–1124

Hennig M, Grimm B, Contestabile R, John RA, Jansonius JN (1997) Crystal structure of glutamate 1-semialdehyde aminomutase. An $alpha_2$-dimeric Vitamin B_6 dependent enzyme with asymmetry in structure and active site reactivity. Proc Natl Acad Sci USA 94:4866–4871

Henningsen KW, Boynton JE, von Wettstein D (1993) Mutants at *xantha* and *albina* loci in relation to chloroplast biogenesis in barley (*Hordeum vulgare* L.). Biologiske Skrifter 42: Munksgaard, Copenhagen

Herrin DL, Battey JF, Greer K, Schmidt GW (1992) Regulation of chlorophyll apoprotein expression and accumulation. J Biol Chem 267:8260–8269

Higuchi M, Bogorad L (1975) The purification and properties of uroporphyrinogen I synthases and uroporphyrinogen II cosynthase: interactions between the enzymes. Ann N Y Acad Sci 244:401–418

Hill KL, Merchant S (1995) Coordinate expression of coproporphyrinogen oxidase and cytochrome c6 in the green alga *Chlamydomonas reinhardtii* in response to changes in copper availability. EMBO J 14:857–865

Höfgen R, Axelsen KB, Kannangara CG, Schüttke I, Pohlenz HD, Willmitzer L, Grimm B, von Wettstein D (1994) A visible marker for antisense mRNA expression in plants: inhibition of chlorophyll synthesis with a glutamate-1-semialdehyde aminotransferase antisense gene. Proc Natl Acad Sci USA 91:1726–1730

Holtorf H, Reinbothe S, Reinbothe C, Bereza B, Apel K (1995) Two routes of chlorophyllide synthesis that are differentially regulated by light in barley (*Hordeum vulgare* L.). Proc Natl Acad Sci USA 92:3254–3258

Hsu WP, Miller GW (1970) Coproporphyrinogen oxidase in tobacco. Biochem J 117:215–220

Huang L, Bonner BA, Castelfranco PA (1989) Regulation of 5-aminolevulinic acid (ALA) synthesis in developing chloroplasts. II. Regulation of ALA-synthesizing capacity by phytochrome. Plant Physiol 90:1003–1008

Hudson A, Carpenter R, Doyle R, Coen ES (1993) *Olive*: a key gene required for chlorophyll biosynthesis in *Antirrhinum majus*. EMBO J 12:3711–3719

Ilag LL, Kumar AM, Söll D (1994) Light regulation of chlorophyll biosynthesis at the level of 5-aminolevulinate formation in *Arabidopsis*. Plant Cell 6:265–275

Im CZ, Matters GL, Beale SI (1996) Calcium and calmodulin are involved in blue light induction of the *GSA* gene for an early chlorophyll biosynthetic step in *Chlamydomonas*. Plant Cell 8:2245–2253

Ito H, Tanaka Y, Tsuji H, Tanaka A (1993) Conversion of chlorophyll b to chlorophyll a in isolated cucumber etioplasts. Arch Biochem Biophys 306:148–151

Jacobs JM, Jacobs NJ (1987) Oxidation of protoporphyrinogen to protoporphyrin, a step in chlorophyll and haem biosynthesis: purification and partial characterization of the enzyme from barley organelles. Biochem J 244:219–244

Jaffe EK (1995) Porphobilinogen synthase, the first source of heme's asymmetry. J Bioenerg Biomembr 27:169–179

Jahn D (1992) Complex formation between glutamyl-tRNA synthetase and glutamyl-tRNA reductase during tRNA-dependent synthesis of 5-aminolevulinic acid in *Chlamydomonas.* FEBS Lett 314:77–80

Jahn D, Michelsen U, Söll D (1991) Two glutamyl-tRNA reductase activities in *Escherichia coli.* J Biol Chem 266:2542–2548

Jensen PE, Willows RD, Petersen BL, Vothknecht UC, Stummann BM, Kannangara CG, von Wettstein D, Henningsen KW (1996a) Structural genes for Mg-chelatase subunits in barley: *Xantha-f, -g* and *-h.* Mol Gen Genet 250:383–394

Jensen PE, Gibson LCD, Henningsen KW, Hunter CN (1996b) Expression of the *chlI, chlD,* and *chlH* genes from the cyanobacterium *Synechocystis* PCC6803 in *Escherichia coli* and demonstration that the three cognate proteins are required for magnesium-protoporphyrin chelatase activity. J Biol Chem 271:16662–16667

Johanningmeier U, Howell SH (1984) Regulation of light-harvesting chlorophyll-binding protein mRNA accumulation in *Chlamydomonas reinhardtii.* J Biol Chem 259:12541–13549

Jordan PM (1994) Highlights in haem biosynthesis. *Curr.* Opin Struct Biol 4:902–911

Jordan PM, Gibbs PNB (1985) Mechanism of action of 5-aminolevulinate dehydratase from human erythrocytes. Biochem J 227:1015–1020

Jordan PM, Mgbeje IAB, Thomas SD, Alwan AF (1988) Nucleotide sequence for the *hemD* gene of *Escherichia coli* encoding uroporphyrinogen III synthase and initial evidence for *hem* operon. Biochem J 249:613–616

Juknat AA, Seubert A, Seubert S, Ippen H (1989) Studies on uroporphyrinogen decarboxylase of etiolated *Euglena gracilis* Z. Eur J Biochem 179:423–428

Juknat AA, Dörnemann D, Senger H (1994) Purification and kinetic studies on a porphobilinogen deaminase from the unicellular green alga *Scenedesmus obliquus.* Planta 193:123–130

Kannangara CG, Gough SP (1978a) Biosynthesis of δ-aminolevulinate in greening barley leaves:glutamate-1-semialdehyde aminotransferase. Carlsberg Res Commun 43:185–194

Kannangara CG, Gough SP (1978b) Biosynthesis of δ-aminolevulinate in greening barley leaves: induction of enzyme synthesis in the light. Carlsberg Res Commun 44:11–20

Kannangara CG, Gough SP, Oliver RP, Rasmussen SK (1984) Biosynthesis of δ-aminolevulinate in greening barley leaves. VI: activation of glutamte by ligation to RNA. Carlsberg Res Commun 49:417–437

Kannangara CG, Vothknecht UC, Hansson M, von Wettstein D (1997) Magnesium chelatase: association with ribosomes and mutant complementation studies identify subunit Xantha-G as a functional counterpart of the *Rhodobacter* subunit BchD. Mol Gen Genet 254:85–92

Kaczor C, Smith MW, Sangwan I, O'Brian MR (1994) Plant δ-aminolevulinic acid dehydratase. Expression in soybean root nodules and evidence for a bacterial lineage of the *Alad* gene. Plant Physiol 104:1411–1417

Kiel JAKW, Ten Berge AM, Venema G (1992) Nucleotide sequence of the *Synechococcus* sp. PCC7942 *hemE* gene encoding the homologue of mammalian uroporphyrinogen decarboxylase. J DNA Sequenc Mapping 2:415–418

Klemm DJ, Barton LL (1987) Purification and properties of protoporphyrinogen oxidase from an anaerobic bacterium *Desulfovibrio gigas.* J Bacteriol 169:5209–5215

Koncz C, Meyerhofer R, Koncz-Kalman Z, Nawrath C, Reiss B, Redei GP, Schell J (1990) Isolation of a gene encoding a novel chloroplast protein by T-DNA tagging in *Arabidopsis thaliana.* EMBO J 9:1337–1346

Knörzer OC, Durner J, Böger P (1996) Alteration in the antioxidative system of the suspension-cultured soybean cells (*Glycine max*) induced by oxidative stress. Physiol Plant 97:388–396

Kruse E, Mock HP, Grimm B (1995a) Coproporphyrinogen III oxidase from barley and tobacco – sequence analysis and initial expression studies. Planta 196:796–803

Kruse E, Mock HP, Grimm B (1995b) Reduction of coproporphyrinogen oxidase level by antisense RNA synthesis leads to deregulated gene expression of plastid proteins and affects the oxidative defense system. EMBO J 14:3712–3720

Kruse E, Grimm B, Beator J, Kloppstech K (1997a) Developmental and circadian control of the capacity for δ-aminolevulinic acid synthesis in green barley. Planta 202:235–241

Kruse E, Mock HP, Grimm B (1997b) Isolation and characterisation of tobacco (*Nicotiana tabacum*) cDNA clones encoding proteins involved in magnesium chelation into protoporphyrin IX. Plant Mol Biol 35:1053–1056

Lee HJ, Ball MD, Parham R, Rebeiz CA (1992) Chloroplast biogenesis 65: enzymic conversion of protoporphyrin IX to Mg-protoporphyrin IX in a subplastidic membrane fraction of cucumber etiochloroplasts. Plant Physiol 99:1134–1140

Lehnen LP, Sherman TD, Becerril JM, Duke SO (1990) Tissue and cellular localization of acifluorfen-induced porphyrins in cucumber cotyledons. Pestic Biochem Physiol 37:329–248

Lermontova I, Kruse E, Mock HP, Grimm B (1997) Cloning and characterization of plastidal and mitochondrial isoform of tobacco protoporphyrinogen IX oxidase. Proc Natl Acad Sci USA 94:8895–8900

Li JM, Brathwaite O, Cosloy SD, Russel CS (1989) 5-Aminolevulinic acid synthesis in *Escherichia coli*. J Bacteriol 171:2547–2552

Li J, Goldschmidt-Clermont M, Timko MP (1993) Chloroplast-encoded *chlB* is required for light-independent protochlorophyllide reductase activity in *Chlamydomonas reinhardtii*. Plant Cell 5:1817–1829

Liedgens W, Grützmann R, Schneider HAW (1980) Highly efficient purification of the labile plant enzyme 5-aminolevuliante dehydratase by means of monoclonal antibodies. Z Naturforsch 35c:958–962

Little HN, Jones OTG (1976) The subcellular localization and properties of the ferrochelatase of etiolated barley. Biochem J 156:309–314

Lopez J, Ryan S, Blankenship RE (1996) Sequence of the *bchG* gene from *Chloroflexus aurantiacus*: relationship between chlorophyll synthase and other polyprenyltransferase. J Bacteriol. 178:3369–3373

Louie GV, Brownlie PD, Lambert R, Coooper JB, Blundell TL, Wood SP, Warren MJ, Woodcock SC, Jordan PM (1992) Structure of porphobilinogen deaminase reveals flexible multidomain polymerase with a single catalytic site. Nature 359:33–39

Luo J, Lim K (1993) Order of uroporphyrinogen III decarboxylation on incubation of porphobilinogen and uroporphyrinogen III with erythrocyte uroporphyrinogen decarboxylase. Biochem J 289:529–532

Madsen O, Sandal L, Sandal NN, Marcker KA (1993) A soybean coproporphyrinogen oxidase gene is highly expressed in root nodules. Plant Mol Biol 23:35–43

Mascia P (1978) An analysis of precusors accumulated by several chlorophyll biosynthetic mutants of maize. Mol Gen Genet 161:237–244

Masuda T, Ohta H, Shoi Y, Takamiya KI (1996) Light regulation of 5-aminolevulinic acid-synthesis in *Cucumis sativus*: light stimulates activity of glutamyl-tRNA reductase during greening. Plant Physiol Biochem 34:11–16

Matters GL, Beale SI (1994) Structure and light-regulated expression of the *gsa* gene encoding the chlorophyll biosynthetic enzyme glutamate-1-semialdehyde aminotransferase in *Chlamydomonas reinhardtii*. Plant Mol Biol 24:617–629

Matters GL, Beale SI (1995) Structure and expression of the *Chlamydomonas reonhardtii alad* gene encoding the chlorophyll biosynthetic enzyme, δ-aminolevulinic acid dehydratase (porphobilinogen synthase). Plant Mol Biol 27:607–617

Matringe M, Camadro JM, Labbe P, Scalla R (1989) Protoporphyrinogen oxidase as a molecular target for diphenyl ether herbicides. Biochem J 260:231–235

Matringe M, Camadro JM, Block MA, Joyard J, Scalla R, Labbe P, Douce R (1992) Localization within the chloroplasts of protoporphyrinogen oxidase the target enzyme for diphenylether-like herbicides. J Biol Chem 267:4646–4651

Matters GL, Beale SI (1995) Blue-light regulated expression of genes for two early steps of chlorophyll biosynthesis in *Chlamydomonas reinhardtii*. Plant Physiol 109:471–479

Mayer SM, Beale SI (1990) Light regulation of δ-aminolevulinic acid biosynthetic enzymes and tRNA in *Euglena gracilis*. Plant Physiol 94:1365–1375

Miyamoto K, Tanaka R, Teramoto H, Masuda T, Tsuji H, Inokuchi H (1994) Nucleotide sequences of the cDNA clones encoding ferrochelatase from barley and cucumber. Plant Physiol 105:769–770

Mock HP, Grimm B (1997) Reduction of uroporphyrinogen decarboxylase by antisense RNA expression affects activities of other enzymes involved in tetrapyrrole biosynthesis and leads to light-dependent necrosis. Plant Physiol 113:1101–1112

Mock HP, Trainotti L, Kruse E, Grimm B (1995) Isolation, sequencing and expression of cDNA sequences encoding uroporphyrinogen decarboxylase from tobacco and barley. Plant Mol Biol 28:245–256

Mock HP, Keetman U, Kruse E, Rank B, Grimm B (1998) Defense responses to tetrapyrrole-induced oxidative stress in transgenic plants with reduced uroporphyrinogen decarboxylase or coproporphyrinogen oxidase activity. Plant Physiol 116:107–116

Nagy F, Kay SA, Chua NH (1988) A circadian clock regulates transcription of the wheat *Cab-1* gene. Genes Dev 2:376–382

Nair SP, Harwood JL, John RA (1991) Direct identification and quantification of the cofactor in glutamate semialdehyde aminotransferase from pea leaves. FEBS Lett 283:4–6

Nakayama M, Masuda T, Sato N, Yamagata H, Bowler C, Ohta H, Shioi Y, Takamiya K (1995) Cloning, subcellular localization and expression of *ChlI*, a subunit of Mg-chelatase in soybean. Biochem Biophys Res Commun 215:422–428

Narita S, Tanaka R, Ito T, Okada K, Taketani S, Inokuchi H (1996) Molecular cloning and characterization of a cDNA that encodes protoporphyrinogen oxidase of *Arabidopsis thaliana*. Gene 182:169–175

Nasrulhaq-Boyce A, Griffiths WT, Jones OTG (1987) The use of continuous assays to characterize the oxidative cyclase that synthesizes the chlorophyll isocyclic ring. Biochem J 243:23–29

Nielsen OF (1974) Macromolecular physiology of plastids. Hereditas 76:269–304

Ohtsuka T, Ito H, Tanaka A (1997) Conversion of chlorophyll *b* to chlorophyll *a* and the assembly of chlorophyll with apoproteins by isolated chloroplasts. Plant Physiol 113:137–147

Oliver RP, Griffiths WT (1982) Pigment-protein complexes of illuminated etiolated leaves. Plant Physiol 70:1019–1025

Orsat B, Monfort A, Chatellard P, Stutz E (1992) Mapping and sequencing of an actively transcribed *Euglena gracilis* chloroplast gene (*ccs*A) homologous to the *Arabidopsis thaliana* nuclear gene *cs* (*ch*-42). FEBS 303:181–184

Oster U, Bauer CE, Rüdiger W (1997) Characterization of chlorophyll *a* and bacteriochlorophyll *a* synthases by heterologous expression in *Escherichia coli*. J Biol Chem 272:9671–9676

Papenbrock J, Gräfe S, Kruse E, Hänel F, Grimm B (1997) Mg-chelatase of tobacco: identification of a *Chl D* cDNA sequence encoding a third subunit, analysis of the interaction of the three subunits with the yeast two-hybrid system and reconstitution of the enzyme activity by co-expression of recombinant CHL D, CHL H and CHL I. Plant J 12:981–990

Plumley FG, Schmidt GW (1995) Light-harvesting chlorophyll *a*/*b* complexes: interdependent pigment synthesis and protein assembly. Plant Cell 7:689–704

Polking GF, Hannapel DJ, Gladon RJ (1995) Characterization of a cDNA encoding 5-aminolevulinic acid dehydratase in tomato (*Lycopersicon esculentum* Mill.). Plant Cell Rep 14:366–369

Pontoppidan B, Kannangara CG (1994) Purification and partial characterisation of barley glutamyl $tRNA^{Glu}$ reductase. Eur J Biochem 225:529–537

Porra RJ (1997) Recent progress in porphyrin and chlorophyll biosynthesis. Photochem Photobiol 65:492–516

Porra RJ, Lascelles J (1968) Studies on ferrochelatase: the enzymic formation of haem in proplastids, chloroplasts and plant mitochondria. Biochem J 108:343–348

Porra RJ, Schäfer W, Cmiel E, Katheder I, Scheer H (1993) Derivation of the formyl group oxygen of chlorophyll *b* from molecular oxygen in greening leaves of a higher plant *(Zea mays)*. FEBS Lett 371:21–24

Radmer RJ, Bogorad L (1967) S-adenosyl-L-methionine magnesium protoporphyrin methyltransferase, an enzyme in the biosynthetic pathway of chlorophyll in *Zea mays*. Plant Physiol 42:463–465

Rebeiz CA, Lascelles J (1982) Biosynthesis of pigments in plants and bacteria. In: Govindjee (ed) Photosynthesis: energy conversion by plants and bacteria. vol 1. pp 699–780

Rebeiz CA, Wu SM, Kuhadja M, Daniell H, Perkins EJ (1983) Chlorophyll *a* biosynthetic routes and chlorophyll *a* chemical heterogeneity in plants. Mol Cell Biochem 57:97–125

Reinbothe S, Reinbothe C (1996) The regulation of enzymes involved in chlorophyll biosynthesis. Eur J Biochem 237:323–342

Reinbothe S, Runge S, Reinbothe C, van Cleve B, Apel K (1995) Substrate dependent transport of the NADPH:protochlorophyllide oxidoreductase into isolated plastids. Plant Cell 7:161–172

Reith M, Munholland J (1993) A high gene map of the chloroplast genome of the red alga *Porphyra purpurea*. Plant Cell 5:465–475

Rieble S, Beale SI (1991) Purification of glutamyl tRNA reductase from *Synechocystis* sp. PCC6803. J Biol Chem 266:9740–9745

Roitgrund C, Mets LJ (1990) Localization of two novel chloroplast functions: *trans*-splicing of RNA and protochlorophyllide reduction. Curr Genet 17:147–153

Rüdiger W, Benz J, Guthoff C (1980) Detection and partial characterozation of activity of chlorophyll synthetase in etioplast membranes. Eur J Biochem 109:193–200

Runge S, van Cleve B, Lebedev N, Armstrong G, Apel K (1995) Isolation and classification of chlorophyll deficient *xantha* mutants of *Arabidopsis thaliana*. Planta 197:490–500

Sangwan I, O´Brian MR (1993) Expression of the soybean glutamate 1-semialdehyde aminotransferase gene in symbiotic root nodules. Plant Physiol 102:829–834

Schaumburg A, Schneider-Poetsch HAW, Eckerskorn C (1992) Characterization of plastid 5-aminolevulinate dehydratase (ALAD; EC 4.2.1.24) from spinach (*Spinacia oleracea* L.) by sequencing and comparison with non-plant ALAD enzymes. Z Naturforsch 47c:77–84

Schneegurt MA, Beale SI (1992) Origin of the chlorophyll b formyl oxygen in *Chlorella vulgaris*. Biochemistry 31:11677–11683

Schoch S, Lempert U, Rüdiger W (1977) Über die letzten Stufen der Chlorophyll-Biosynthese. Zwischenprodukte zwischen Chlorophyllid und phytolhaltigem Chlorophyll. Z Pflanzenphysiol 83:427–436

Schön A, Krupp G, Gough S, Berry-Lowe S, Kannangara CG, Söll D (1986) The RNA required in the first step of chlorophyll biosynthesis is a chloroplast glutamate tRNA. Nature 322:281–284

Schultz R, Steinmüller K, Klaas M, Forreiter C, Rasmussen S, Hiller C, Apel K (1989) Nucleotide sequence of a cDNA coding for the NADPH-protochlorophyllide oxidoreductase (PCR) of barley (*Hordeum vulgare* L.) and its expression in *Escherichia coli*. Mol Gen Genet 217:335–361

Sharif AL, Smith AG, Abell C (1989) Isolation and characterization of a cDNA clone for a chlorophyll synthesis enzyme from *Euglena gracilis*: the chloroplast enzyme hydroxymethylbilane synthase (porphobilinogen deaminase) is synthesized with a very long transit peptide in *Euglena*. Eur J Biochem 184:353–359

Smith AG (1988) Subcellular localization of two porphyrin-synthesis enzymes in *Pisum sativum* (pea) and *Arum* (cuckoo-pint) species. Biochem J 249:423–428

Smith AG, Griffiths WT (1993) Enzymes of chlorophyll and heme biosynthesis. In: Dey PM, Harborne JB (eds), Methods in plant biochemistry, vol 9. Academic Press, London, pp 299–343

Smith AG, Marsh O, Elder GH (1993) Investigation of the subcellular location of the tetrapyrrole-biosynthesis enzyme coproporphyrinogen oxidase in higher plants. Biochem J 292:503–508

Smith AG, Santana MA, Wallace-Cook ADM, Ropert JM, Labbe-Bois R (1994) Isolation of a cDNA encoding chloroplast ferrochelatase from *Arabidopsis thaliana* by functional complementation of a yeast mutant. J Biol Chem 269:13405–13413

Smith MA, Grimm B, Kannangara CG, von Wettstein D (1991a) Spectral kinetics of glutamate 1-semialdehyde aminomutase of *Synechococcus.* Proc Natl Acad Sci USA 88:9775–9779

Smith MA, Kannangara CG, Grimm B, von Wettstein D (1991b) Characterization of glutamate-1-semialdehyde aminotransferase of *Synechococcus.* Eur J Biochem 202:49–757

Soll J, Schultz G, Rüdiger W, Benz J (1983) Hydrogenation of geranylgeraniol. Two pathways exist in spinach chloroplasts. Plant Physiol 71:849–854

Spano AJ, Timko MP (1991) Isolation, characterization and partial amino acid sequence of a chloroplast-localized porphobilinogendeaminase from pea (*Pisum sativum* L.). Biochim Biophys Acta 1076:29–36

Spano AJ, He ZH, Michel H, Hunt DF, Timko MP (1992a) Molecular cloning nuclear gene structure and developmental expression of NADPH-protochlorophyllide oxidoreductase in pea (*Pisum sativum* L.). Plant Mol Biol 18:967–972

Spano AJ, He ZH, Timko MP (1992b) NADPH-protochlorophyllide oxidoreductase in white pine (*Pinus strobus*) and loblolly pine (*P. taeda*). Mol Gen Genet 236:86–95

Stange-Thomann N, Thomann HU, Lloyd AJ, Lyman H, Söll D (1994) A point mutation in *Euglena gracilis* chloroplast $tRNA^{Glu}$ uncouples protein and chlorophyll biosynthesis. Proc Natl Acad Sci USA 91:7947–7951

Susek RE, Ausubel FM, Chory J (1993) Signal transduction mutants of *Arabidopsis* uncouple nuclear CAB and RBCS gene expression from chloroplast development. Cell 74:787–799

Suzuki JY, Bauer CE (1992) Light-independent chlorophyll biosynthesis: involvement of the chloroplast gene *chlL (frxC).* Plant Cell 4:929–940

Tait GH (1972) Coproporphyrinogenase activities in extracts *of Rhodopseudomonas shaeroides* and *Chromaticum* strain D. Biochem J 128:1159–1169

Tanaka R, Yoshida K, Nakayashiki T, Masuda T, Tsuji H, Inokuchi H, Tanaka A (1996) Differential expression of two *hemA* mRNAs encoding glutamyl-tRNA reductase proteins in greening cucumber seedlings. Plant Physiol 110:1223–1230

Taylor WC (1989) Regulatory interaction between nuclear and plastid genomes. Annu Rev Plant Physiol Plant Mol Biol 40:211–233

Teakle JR, Griffiths WT (1993) Cloning, characterization and import on protochlorophyllide reductase from wheat (*Triticum aestivum*). Biochem J 296:225–230

Terry MJ (1997) Phytochrome chromophore-deficient mutants. Plant Cell Environm 20:740–745

Terry MJ, Kendrick RE (1996) The *aurea* and *yellow-green-2* mutants of tomato are deficient in phytochrome chromophore synthesis. J Biol Chem 271:21681–21688

van Tuinen A, Hanhart CJ, Kerckhoffs LHJ, Nagatani A, Boylan MT, Quail PH, Kendrick RE, Koorneef M (1996) Analysis of the phytochrome deficient *yellow-green-2* and *aurea* mutants of tomato. Plant J 9:173–182

von Wettstein D, Henningsen KW, Boynton JE, Kannangara GC, Nielsen OF (1971) The genetic control of chloroplast development in barley. In: Boardman NH, Linnane AW, Smillie RM (eds). Autonomy and biogenesis of mitochondria and chloroplastsNorth Holland, Amsterdam, pp 205–223

von Wettstein D, Gough S, Kannangara CG (1995) Chlorophyll biosynthesis. Plant Cell 7:1039–1057

Vothknecht UC, Kannangara CG, von Wettstein D (1996) Expression of catalytically active barley glutamyl $tRNA^{Glu}$ reductase in *Escherichia coli* as a fusion protein with glutathione S-transferase. Proc Natl Acad Sci USA 93:9287–9291

Walker CJ, Weinstein JD (1991a) In vitro assay of the chlorophyll biosynthetic enzyme Mg-chelatase: resolution of the activity into soluble and membrane-bound fractions. Proc Natl Acad Sci USA 88:5789–5793

Walker CJ, Weinstein JD (1991b) Further characterization of the magnesium chelatase in isolated developing cucumber chloroplasts. Plant Physiol 95:1189–1196

Walker CJ, Weinstein JD (1994) The magnesium-insertion step of chlorophyll biosynthesis is a two-stage reaction. Biochem J 299:277–284

Walker CJ, Weinstein JD (1995) Re-examination of the localization of Mg-chelatase within the chloroplast. Physiol Plant 94:419–424

Walker CJ, Castelfranco PA, Whyte BJ (1991) Synthesis of the divinyl protochlorophyllide. Enzymological properties of the Mg-protoporphyrin IX monomethyl ester oxidative cyclase system. Biochem J 276:691–697

Ward ER, Volrath S (1995) Manipulation of protoporphyrinogen oxidase enzyme activity in eucaryotic organisms. PCT patent WO 95/34659

Wang WY, Wang WL, Boynton JE, Gillham NW (1974) Genetic control of chlorophyll biosynthesis in *Chlamydomonas.* J Cell Biol 63:806–823

Weinstein JD, Beale SI (1983) Separate physiological roles and subcellular compartments for two tetrapyrrole biosynthetic pathways in *Eiglena gracilis.* J Biol Chem 258:6799–6807

Weller JL, Terry MJ, Rameau C, Reid JB, Kendrick RE (1996) The phytochrome-deficient pcd1 mutant is unable to convert heme to biliverdin IXα. Plant Cell 8:55–67

Weller JL, Terry MJ, Reid JB, Kendrick RE (1997) The phytochrome-deficient pcd2 mutant is unable to convert biliverdin IXα to 3Z-phytochromobilin. Plant J 11:1177–1186

Willows RD, Kannangara CG, Pontoppidan B (1995) Nucleotides of $tRNA^{glu}$ involved in recognition by barley chloroplast glutamyl-tRNA synthetase and glutamyl-tRNA reductase. Biochim Biophys Acta 1263:228–234

Willows RD, Gibson LCD, Kannangara CG, Hunter CN, von Wettstein D (1996) Three separate proteins constitute the magnesium chelatase of *Rhodobacter sphaeroides.* Eur J Biochem 235:438–443

Witty M, Wallace-Cook ADM, Albrecht H, Spano AJ, Michel H, Shabanowitz J, Hunt DF, Timko MP, Smith AG (1993) Structure and expression of the chloroplast-localized porphyobilinogen deaminase of pea (*Pisum sativum* L.) isolated by redundant polymerase chain reaction. Plant Physiol 103:139–147

Witty M, Jones RM, Robb MS, Shoolingin Jordan PM, Smith AG (1996) Subcellular location of the tetrapyrrole synthesis enzyme porphyobilinogen deaminase in higher plants: an immunological investigation. Planta 199:557–664

Wong YS, Castelfranco (1984) Resolution and reconstitution of the Mg-protoporphyrin IX monomethyl ester (oxidative) cyclase, the enzyme system responsible for the formation of the chlorophyll isocyclic ring. Plant Physiol 75:658–661

Wong YS, Castelfranco (1985) Properties of the Mg-protoporphyrin IX monomethyl ester (oxidative) cyclase system. Plant Physiol 79:730–733

Yang Z, Bauer CE (1990) *Rhodobacter capsulatus* genes involved in early steps of the bacteriochlorophyll biosynthetic pathway. J Bacteriol 172:5001–5010

Zavgorodnyaya A, Papenbrock J, Grimm B (1997) Yeast 5-aminolevulinate synthase provides additional chlorophyll precursor in transgenic plants. Plant J 12:169–178

CHAPTER 9

Characteristics of Protoporphyrinogen Oxidase

Jean-Michel Camadro[1], Sylvain Arnould[1], Laurence Le Guen[1], Renata Santos[2], Michel Matringe[3], and René Mornet[4]

Contents

9.1 Introduction

Protoporphyrinogen oxidase (EC 1.3.3.4) catalyzes the oxidative O_2-dependent aromatization of the colorless protoporphyrinogen IX to the highly conjugated protoporphyrin IX, the precursor of both hemes and chlorophylls (Fig. 1). It is the final enzyme in the common branch of the heme and chlorophyll biosynthetic pathways in plants (Fig. 2).

The protoporphyrinogen oxidase reaction was for a long time the least well known of the eight enzyme reactions involved in heme biosynthesis. Studies on protoporphyrinogen oxidase were mostly dedicated to elucidating the molecular basis of the human disease variegate porphyria, an inherited autosomal dominant disease characterized by a 50% deficit in protoporphyrinogen oxidase activity in lymphocytes from heterozygous patients (Brenner and Bloomer 1980a; Deybach et al. 1981). Almost no protoporphyrinogen oxidase activity was detected in lymphocytes or cultured fibroblasts from homozygous patients (Korda et al. 1984; Murphy et al. 1986; Mustajoki et al. 1987). Variegate

[1]Laboratoire de Biochimie des Porphyrines, Département de Microbiologie, Institut Jacques-Monod, UMR 7592 CNRS-Université Paris, 7-Université Paris 6, 2 Place Jussieu, F-75251 Paris Cedex 05, France
[2]Laboratoire de Génétique Moléculaire des Réponses Adaptatives, Département de Microbiologie, Institut Jacques-Monod, UMR 7592 CNRS -Université Paris, 7-Université Paris 6, 2 Place Jussieu, F-75251 Paris Cedex 05, France
[3]Laboratoire Mixte, UMR 41 CNRS-Rhône-Poulenc, BP 9163, F-69263 Lyon Cedex 09, France
[4]Laboratoire d'Ingénierie Moléculaire et Matériaux Organiques, UMR 6501 CNRS-Université d'Angers, Faculté des Sciences, 2 Boulevard Lavoisier, F-49045 Angers Cedex, France

Peter Böger, Ko Wakabayashi
Peroxidizing Herbicides

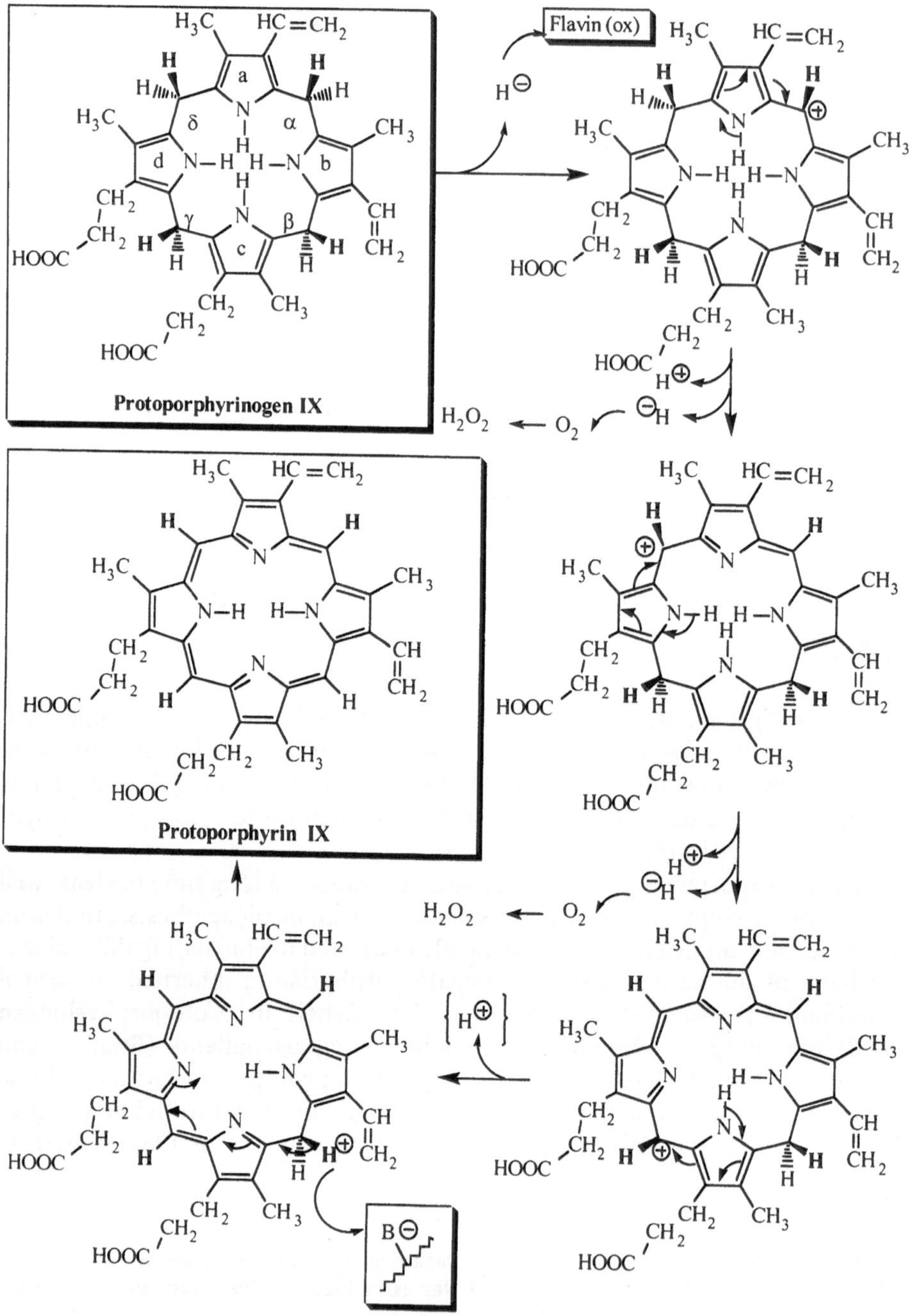

Fig. 1. Structures of protoporphyrinogen IX and protoporphyrin IX and some of the intermediates postulated in the enzyme reaction with yeast protoporphyrinogen oxidase

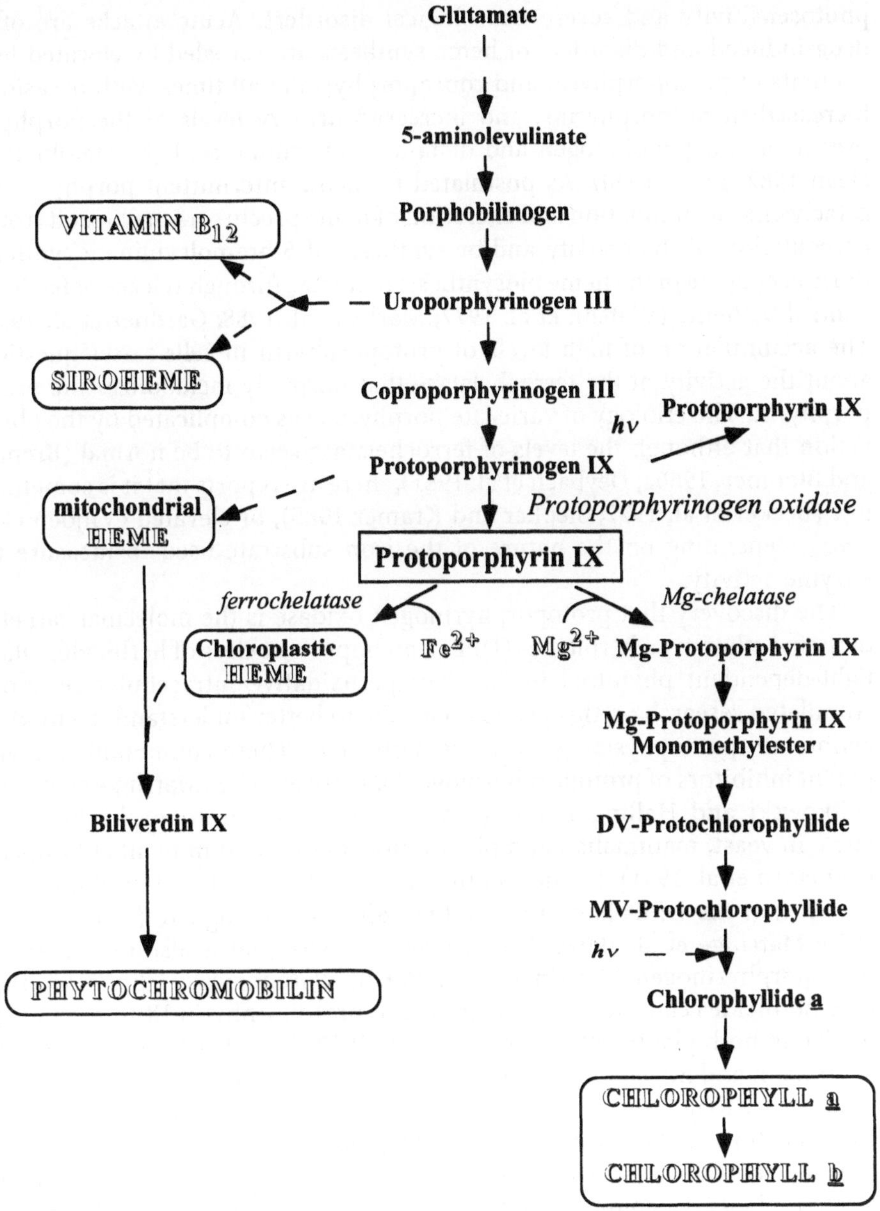

Fig. 2. General organization of heme and chlorophyll biosynthesis pathways

porphyria has been particularly studied in South Africa, where the incidence of the disease in the Afrikaner populations is high because the gene defect occurred in the original Dutch settlers in the Cape (Dean 1982; Jenkins 1996; Hift et al. 1997). The clinical symptoms in affected adults are variable with marked photosensitivity and severe neurological disorders. Acute attacks are often drug-induced and disorders of heme synthesis are revealed by elevated fecal contents of protoporphyrin and coproporphyrin at all times, with occasional increased urine porphyrins, and increased urinary levels of the porphyrin precursors porphobilinogen and delta-aminolevulinic acid (Mustajoki 1978; Dean 1982; Elder 1990). As postulated for acute intermittent porphyria, the cataclysmic accumulation and excretion of heme precursors may result from a deregulation of the activity and/or synthesis of 5-aminolevulinate synthase, the first enzyme in the heme biosynthesis pathway, through release of feedback control by heme (Watson et al. 1977; Marks et al. 1988; Gardner et al. 1991). The accumulation of high levels of protoporphyrin in cells raises questions about the activity of the ferrochelatase that normally metabolizes the protoporphyrin. The etiology of variegate porphyria was complicated by the observation that although the levels of ferrochelatase seem to be normal (Brenner and Bloomer, 1980a; Deybach et al. 1981), there are reports that it is sometimes low (Becker et al. 1977; Siepker and Kramer 1985), or elevated (Viljoen et al. 1983), depending on the nature of the iron substrate used to measure the enzyme activity.

The discovery that protoporphyrinogen oxidase is the molecular target of diphenyl ether-type herbicides (DPEs), an important class of herbicides whose light-dependent phytotoxicity involves peroxidative intracellular reactions, stimulated research on this peculiar enzyme to better understand the mode of action and pathophysiology of these herbicides. These compounds are very potent inhibitors of protoporphyrinogen oxidase activity (Matringe et al. 1989; Witkowski and Halling 1989). They compete with protoporphyrinogen in vitro, in yeast, mammalian and plant mitochondria and in plant chloroplasts (Camadro et al. 1991). Studies on the binding of radiolabeled inhibitors have shown that the DPEs have a single high-affinity binding site (Varsano et al. 1990; Matringe et al. 1992b; Nicolaus et al. 1995), that is also recognized by protoporphyrinogen (Matringe et al. 1992b). Paradoxically, diphenyl ether-type herbicides can cause large amounts of protoporphyrin IX to accumulate in vivo in both plants (Matringe and Scalla 1987, 1988; Lydon and Duke 1988; Sandmann and Böger 1988; Witkowski and Halling 1988) and mammals (Krijt et al. 1992; Krijt et al. 1993). The accumulated protoporphyrin IX is not a substrate for the chelatases of heme or magnesium-protoporphyrin synthesis. The mechanism by which protoporphyrin accumulates in diphenyl ether-treated plants is complex and may involve non-specific porphyrinogen-oxidizing enzymes, such as peroxidases (Lee et al. 1993; Lee and Duke 1994; Yamato et al. 1994), oxidizing protoporphyrinogen to protoporphyrin, and also coproporphyrinogen or uroporphyrinogen to their corresponding porphyrins. These activities are not inhibited by diphenyl ether (Lee et al. 1993). Similarity of the symptoms developed in diphenyl ether-treated plants to those of patients with variegate porphyria suggested that protoporphyrinogen oxidase occupies a key position in the control of metallo-porphyrins synthesis. This raises several fundamental problems of cell biology, one of which is the critical role of the subcellular location of enzymes and metabolites in the control of metabolic fluxes.

This chapter examines several aspects of the structure and function of protoporphyrinogen oxidase, from early biochemical studies to the analysis of mutants and molecular genetics of protoporphyrinogen oxidase.

9.2 Biochemistry

The protoporphyrinogen oxidase reaction is classically described as the removal of six hydrogen atoms from protoporphyrinogen IX via six electron-transfer reactions. In terms of molecular evolution, this reaction appears to be the least well conserved of the eight enzyme reactions involved in the heme biosynthetic pathway, with considerable differences in the catalytic strategies developed in prokaryotes (Gram negative or Gram positive bacteria) and eukaryotes. Since the pioneering work of Porra and Falk (1961) and Sano and Granick (1961), who first studied the metabolism of coproporphyrinogen III to protoporphyrin IX in mammalian mitochondria, there has been much debate as to whether this oxidative step is enzyme-catalyzed, because porphyrinogens oxidize rapidly to their corresponding porphyrins in the presence of air via a light-sensitive autocatalytic reaction (Falk 1964).

This property of porphyrinogens, including protoporphyrinogen, makes it difficult to assay protoporphyrinogen oxidase activity due to the many possible sources of artifacts. Protoporphyrinogen oxidase activity should be quite simple to assay, since the substrate of the reaction, protoporphyrinogen IX, does not absorb visible light or fluoresce, whereas the product of the reaction, protoporphyrin IX, has a characteristic absorption spectrum ($\lambda max = 410\,nm$) and fluoresces strongly ($\lambda exc = 410\,nm$; $\lambda em = 632\,nm$). The most straightforward way to measure protoporphyrinogen oxidase activity is thus to measure the rate of appearance of the product using spectrophotometric or spectrofluorimetric detection (Poulson and Polglase 1975; Brenner and Bloomer 1980b; Camadro et al. 1982, 1993a).

However, measuring the activity accurately is a matter of compromise between contradictory requirements. Firstly, protoporphyrinogen IX is obtained by the chemical reduction of protoporphyrin IX with sodium amalgam (Falk 1964; Poulson and Polglase 1975), and the quality of the substrate is critical. Obtaining a satisfactory sample of protoporphyrinogen is not only a matter of training; certain "environmental" factors, such as the relative humidity, may significantly interfere with the preparation of the amalgam. The non-enzymatic formation of protoporphyrin must be kept as low as possible because of the propency of protoporphyrinogen to auto-oxidize in the presence of light, oxygen and protoporphyrin. This is usually done by preparing a stock solution of protoporphyrinogen containing reducing agents (dithiothreitol or ascorbate) and by adding these agents to the assay medium. This, of course, may interfere with some of the enzyme catalytic properties.

Secondly, protoporphyrinogen oxidase activity can be measured in crude extracts or the membrane fractions of plants, mammals or microorganisms. However, since protoporphyrinogen oxidase uses dioxygen as a substrate, care must be taken to avoid anaerobiosis during the assay by flushing the reaction buffer with air and by washing the membrane fraction thoroughly to remove endogeneous respiratory substrates.

Thirdly, care must be taken when assaying protoporphyrinogen oxidase in such crude extracts, that the protoporphyrin generated during the reaction is not further metabolized by the chelatases that incorporate a divalent ion (usually iron or zinc) into the tetrapyrrole nucleus to give a metalloporphyrin with spectral properties very different from those of protoporphyrin. The simple addition of a chelator of divalent ions such as EDTA to the assay medium is usually sufficient to inhibit the chelatases. However, high concentrations of iron-EDTA complexes may promote the non-enzymatic oxidation of protoporphyrinogen by initiating Fenton-type reactions of activation of dioxygen. Fourthly, protoporphyrin IX is a highly hydrophobic compound that tends to aggregate or stack in aqueous media (Brown and Shillcock 1976; Nishide et al. 1977; Margalit et al. 1983). The spectral and fluorescence properties of aggregated protoporphyrin are very different from those of "monomeric" protoporphyrin (Margalit et al. 1983) and care should be taken to ensure maximum monomerization of protoporphyrin. This is usually achieved by adding small quantities of a neutral detergent with a low critical micellar concentration, such as Tween-80.

Finally, assaying protoporphyrinogen oxidase in plant extracts is slightly more complex, since protoporphyrin IX fluorescence is quenched by the chlorophylls or chlorophyll precursors present in the enzyme fractions. This problem can be overcome by measuring the quenching of known concentrations of standard protoporphyrin IX by the enzyme fraction (Camadro et al. 1993a). When these possible pitfalls are taken into account, the protoporphyrinogen oxidase assay is very sensitive and reproducible.

The first conclusive description of the enzymatic basis of protoporphyrinogen oxidation was provided by Poulson and Polglase who characterized and partially purified protoporphyrinogen oxidase from the mitochondrial membrane fraction of yeast (Poulson and Polglase 1975) and rat liver (Poulson 1976). Although the two enzymes appeared to be quite different in terms of apparent molecular masses (180 and 35 kDa) and catalytic properties (pH for the activity optimum of 7.5 and 8.7), they both required dioxygen for their activity, and no other electron acceptor supported the enzyme reaction under anaerobic conditions. This may not be surprising for a mammalian enzyme, but the yeast *Saccharomyces cerevisiae* is able to grow and synthesize some heme under anaerobic conditions where its energy metabolism is based upon fermentation. This raised the question of how the enzyme can function without dioxygen. There is no definitive answer to this problem, but it has been shown later that purified yeast protoporphyrinogen oxidase has a very high affinity for dioxygen (Camadro et al. 1994) that may allow protoporphyrin synthesis under the micro-aerophylic growth conditions usually referred to as anaerobic growth conditions.

The oxidation of protoporphyrinogen in facultative aerobic organisms was examined further in early studies of protoporphyrinogen oxidation in *Escherichia coli.* Jacobs and Jacobs (1975, 1976) showed that nitrate and fumarate may serve as electron acceptors for the anaerobic oxidation of protoporphyrinogen to protoporphyrin in cell-free extracts of *E. coli* grown anaerobically in the presence of nitrate or fumarate respectively. Both anaerobic protoporphyrin formation from protoporphyrinogen and protoheme formation from protoporphyrinogen are markedly stimulated by nitrate. In contrast, anaerobic formation of protoheme from protoporphyrin does not depend upon adding nitrate or fumarate. Neither nitrate nor fumarate increases the aerobic formation of protoporphyrin from protoporphyrinogen by cell-free extracts of *E. coli* when oxygen is the electron acceptor in the reaction. The aerobic oxidation of protoporphyrinogen is not inhibited by diphenyl ether-type herbicides (Jacobs et al. 1990), but the coupling of nitrate or fumarate reduction is strongly inhibited by 2-heptyl-4-hydroxy quinoline-N-oxide (HQNO), an inhibitor of the electron transport chain, suggesting that anaerobic oxidation of protoporphyrinogen in *E. coli* involves coupling into the anaerobic electron transport system (Jacobs and Jacobs 1977). Menaquinone may act as a hydrogen carrier for the anaerobic protoprophyrinogen oxidation, with fumarate as electron acceptor, as shown by studies in which extracts and particles of *E. coli* mutants unable to synthesize menaquinone did not couple protoporphyrinogen oxidation to fumarate reduction, whereas mutants containing menaquinone but lacking ubiquinone or cytochromes had this activity (Jacobs and Jacobs 1978). The *E. coli* anaerobic protoporphyrinogen oxidation system appears to be relatively specific, since coproporphyrinogen III cannot replace protoporphyrinogen IX as substrate. However, methylene blue, triphenyl tetrazolium and nitrate (but not nitrite) can all replace fumarate as anaerobic hydrogen acceptor.

There is the same diversity of electron acceptor in the obligate anaerobe *Desulfovibrio gigas*, where the anaerobic oxidation of protoporphyrinogen to protoporphyrin is catalyzed by a membrane-bound enzyme system (Klemm and Barton 1985). Protoporphyrin is formed in the presence of nitrite, hydroxylamine, sulfite, thiosulfate, ATP plus sulfate, NAD+, NADP+, flavin adenine dinucleotide, flavin mononucleotide, fumarate, 2,6-dichlorophenol-indophenol, methyl viologen, and 3-(4,5-dimethylthiazol-2-yl)-2,5-diphenyltetrazolium bromide. Dialyzed cell extracts were most active with sulfite, NAD+, and NADP+ as electron acceptors. Protoporphyrinogen oxidase has been purified to apparent homogeneity from the plasma membrane of *Desulfovibrio gigas* (Klemm and Barton 1987). The enzyme has a molecular weight of 148 kDa and was found to have three dissimilar subunits (12, 18.5, and 57 kDa) held together by disulfide bonds. Unlike other protoporphyrinogen oxidases, which use molecular oxygen as an electron acceptor, this enzyme does not couple to oxygen. In contrast to the membrane-bound enzyme, protoporphyrinogen oxidase donates electrons to 2,6-dichlorophenol-indophenol, but not to NAD+, NADP+, flavin adenine dinucleotide, or flavin mononucleotide.

Several protoporphyrinogen oxidases have been purified from eukaryotic cells. Like the bacterial enzymes, the molecular properties of these enzymes seem to vary considerably. The protoporphyrinogen oxidase from bovine liver mitochondria was purified with a 68% yield and 870-fold purification (Siepker et al. 1987). The enzyme contained FAD as a cofactor, had an apparent Mr of 57 kDa, and the Km for protoporphyrinogen IX was 16.6 μM. The activity of the isolated enzyme was markedly stimulated by fatty acids such as oleic acid. Surprisingly, antibodies raised against purified protoporphyrinogen oxidase cross-reacted with ferrochelatase (and anti ferrochelatase IgG cross-reacted with protoporphyrinogen oxidase). In addition, radiolabeled peptides of both enzymes, generated by chymotrypsin, demonstrated common peptides when analyzed by two-dimensional chromatography.

Protoporphyrinogen oxidase has been purified to apparent homogeneity from mouse liver mitochondria (Dailey and Karr 1987). The purified protein has a molecular weight of approximately 65 kDa under both native (gel filtration chromatography on Sepharose CL-6B in the presence of 0.5% sodium cholate) and denaturating (sodium dodecyl sulfate-polyacrylamide gel electrophoresis) conditions, suggesting that the enzyme is a monomer. The absorption spectrum of the enzyme as purified showed no evidence of a chromophoric cofactor. Purified protoporphyrinogen oxidase has a Km for protoporphyrinogen IX of 5.6 μM with a Vmax of 2300 $nmol.h^{-1}.mg^{-1}$. It uses meso- and hematoporphyrinogen at about 10% of the level of protoporphyrinogen. The pH optimum is broad with a maximum at 7.1. Divalent cations do not stimulate or inhibit it, and sulfhydryl reagents do not inhibit the purified enzyme. The kinetic parameters for the two-substrate (protoporphyrinogen and oxygen) reaction were determined (Ferreira and Dailey 1988) and the Km for oxygen with saturating concentrations of protoporphyrinogen was 125 μM. The stoichiometry of the reaction was 1 mol protoporphyrin formed for 3 mol dioxygen consumed. Low concentrations (less than 15 μM) of potential alternative electron acceptors, such as ubiquinone-6, ubiquinone-10 and dicoumarol, stimulated protoporphyrinogen oxidase activity whereas coenzyme Q0 and menadione did not activate at these concentrations. Above 30 μM, all five quinones inhibited the enzyme activity. FAD did not significantly affect the activity of the enzyme, but bilirubin, a product of heme catabolism, was a competitive inhibitor of protoporphyrinogen oxidase, with a Ki of 25 μM. An improved purification of mouse protoporphyrinogen oxidase provided new information and revealed the presence of a non-covalently bound flavin moiety (Proulx and Dailey 1992). This flavin is present at approximately stoichiometric amounts in the purified enzyme and was identified by its fluorescence spectrum and high performance liquid chromatography as flavin mononucleotide (FMN). No other cofactor was detected in the purified enzyme, neither protein-bound metal ions nor protein-associated pyrroloquinoline quinone which may be involved in the activation of dioxygen. Circular dichroism studies predicted that the native protein had 30.5% alpha helix, 40.5% beta sheet, 13.7% turn, and 15.3% random coil. Denaturation of protoporphyrinogen oxidase with urea resulted in a biphasic curve when ellipticity was plotted against urea concentration, typical of amphipathic proteins.

Almost 20 years after the initial studies of Poulson and Polglase (1975) protoporphyrinogen oxidase was purified to homogeneity from yeast mitochondrial membranes and found to be a 55-kDa polypeptide with a pI of 8.5 and a specific activity of 40000 nmol of protoporphyrin. h^{-1}.mg $protein^{-1}$ at 30 °C (Camadro et al. 1994). The Km for protoporphyrinogen IX is 0.1 μM and, unlike the mouse enzyme, the yeast enzyme has a high affinity for dioxygen and the Km for oxygen is 0.5–1.5 μM. The purified enzyme contains stoichiometric amounts of FAD as flavin cofactor. It is inhibited by diphenyl ether-type herbicides with kinetic constants for the inhibition similar to those reported for the membrane-bound enzyme. Studies with rabbit antibodies to yeast protoporphyrinogen oxidase indicate that the enzyme is synthesized as a high molecular weight precursor (58 kDa) that is rapidly converted in vivo to the mature (55-kDa) membrane-bound form. Protoporphyrinogen oxidase activity was found only in purified yeast mitochondrial inner membrane (not in the outer membrane) and, while remaining fully active, appears to be extremely sensitive to proteolysis.

There is protoporphyrinogen oxidase in both the mitochondria and chloroplasts of plants (Jacobs and Jacobs 1987; Matringe et al. 1989; Smith et al. 1993) and the two enzymes have similar properties (Jacobs and Jacobs 1987; Matringe et al. 1989; Camadro et al. 1991). However, protoporphyrinogen oxidase is an integral protein of both the thylakoid and envelope membranes of spinach chloroplasts (Matringe et al. 1992a). This dual distribution is probably due to the subsequent enzymes of the heme and chlorophyll biosynthesis pathways lying within the chloroplast. They chelate a divalent metal ion (magnesium or iron) to the protoporphyrin released by the protoporphyrinogen oxidase. The chloroplast ferrochelatase is associated with the thylakoid membrane (Matringe et al. 1994), whereas the magnesium-chelating complex is associated with the plastid envelope (Walker and Weinstein 1994; Walker et al. 1997). Thus, within the plastids, the enzymes involved in the biosynthesis of heme and chlorophyll precursors are physically segregated at the level of the first membrane-bound enzyme of these pathways, protoporphyrinogen oxidase. The control of the flux of protoporphyrin through these pathways is unknown, but understanding this control would greatly increase our understanding of the physiology of plant cells.

The biochemical characterization of protoporphyrinogen oxidase activity in plant cells was made difficult by the presence of high levels of peroxidase type protoporphyrinogen oxidizing activity. Jacobs and Jacobs (1984) first demonstrated that protoporphyrinogen oxidation was slower in chloroplasts from older barley or mature spinach, suggesting that it was involved in chlorophyll synthesis. The activity in spinach chloroplasts has a pH optimum of 7. The activity is inhibited by glutathione or excess detergent, and is readily lost at room temperature. The plant activity is less specific for porphyrinogen substrates, oxidizing mesoporphyrinogen, but not coproporphyrinogen or uroporphyrinogen, as rapidly as protoporphyrinogen. Partial purification and characterization of the enzyme from plant organelles (Jacobs and Jacobs 1987) indicated no readily detectable differences between the enzyme isolated from mitochondrial or etioplast fractions. The enzyme from both organelle fractions had a Km of 5 μM and was labile to mild heat and acidification. The purest fractions showed a polypeptide of 36 kDa on SDS-PAGE. Purified barley mitochondrial protoporphyrinogen oxidase contained a variety of lipids, including phosphatidyl ethanolamine and free fatty acids in molar ratio to the protein of 30/1 and 60/1 respectively (Jacobs et al. 1989b). Iron, but no flavins or cytochromes, was detected. Enzymatic oxidation, in the presence of glutathione, was inhibited by the iron chelator o-phenanthroline but was stimulated by iron-EDTA. The purified enzyme was also inhibited by reducing agents such as glutathione, ascorbate, NADH and NADPH; this type of inhibition is characteristic of protoporphyrinogen oxidizing enzymes found in the microsomal and plasma membrane fractions prepared from 7-day-old, etiolated barley leaves where the activities were almost as high as in crude or purified etioplasts (Lee et al. 1993). Uroporphyrinogen I and coproporphyrinogen I oxidizing activities were found in all fractions; however, the etioplast fractions were significantly more specificic for protoporphyrinogen IX than the other fractions. The plasma membrane-associated protoporphyrinogen oxidizing activity was strongly inhibited by dithiothreitol and was stimulated by quinones like duroquinone, juglone, or pyrroloquinoline-quinone. These quinones had little or no effect in etioplasts (Lee and Duke 1994). The plasma-membrane enzyme had a lower affinity for protoporphyrinogen IX (172 μM) than did the etioplastic enzyme (26 μM).

The reactivities of the plasma-membrane enzyme toward various effectors (diethyldithiocarbamate, a copper chelator, hydrogen peroxide, cyanide and catalase) led Lee and Duke (1994) to suggest that the plasma membrane-associated protoporphyrinogen oxidizing activity had characteristics similar to those of a peroxidase. However, peroxidases able to oxidize porphyrinogens to porphyrins were not restricted to the plasma membrane and Yamato et al. (1994) purified a protoporphyrinogen oxidizing enzyme from the soluble fraction of tobacco cell lines. Approximately 90% of the total activity was in the soluble fraction of the SL cells. The purified enzyme had a molecular weight of approximately 48 kDa by SDS-PAGE, an apparent Km of 78.9 μM and Vmax of 1.3 μmol.min^{-1}.mg protein^{-1}. It used uroporphyrinogen I and coproporphyrinogen I as substrates, contained a heme, and showed peroxidase activity toward guaiacol and pyrogallol. Amino acid sequences from this soluble protoporphyrinogen oxidizing enzyme corresponded to the acid/base catalysis and heme binding regions of plant peroxidases (Yamato et al. 1995).

It has been shown that horseradish peroxidase rapidly oxidizes uroporphyrinogen and some synthetic porphyrinogens to porphyrins in vitro (Jacobs et al. 1996) by a catalytic process that is inhibited by ascorbic acid and may lead to the formation of green compounds with the spectral characteristics of a chlorin with a large peak at 638 nm. The latter reaction required a sulfhydryl reducing agent such as glutathione and was inhibited by ascorbic acid. Glutathione and ascorbic acid are two major components of plant cell defenses against reactive species of oxygen through the Haliwell-Asada cycle, and are therefore likely to be involved in their protection against photodamage caused by the accumulation of protoporphyrin in cells treated with diphenyl ethers.

Since many protoporphyrinogen oxidizing enzymes exist in plant cells, we undertook the purification of protoporphyrinogen oxidase from lettuce etioplasts, with restrictive criteria for the selection of the active fractions during chromatographic processes (Camadro et al. 1993b). We further purified only those fractions containing protoporphyrinogen oxidase activity not inhibited by reducing agents and strongly inhibited by diphenyl ether-type herbicides. Lettuce protoporphyrinogen oxidase was found to be a 55-kDa polypeptide with a pI of 6.5 and a specific activity of 9300 nmol protoporphyrinogen IX oxidized. h^{-1}. mg $protein^{-1}$ at 30 °C. The Km for protoporphyrinogen was 0.3 μM. Fluorescence spectra of the purified enzyme revealed the presence of a flavin. The intensity of fluorescence was identical at neutral and acidic pH, suggesting either the presence of FMN or the presence of a flavin covalently bound to the polypeptide chain (Decker 1993). The purified enzyme activity was inhibited by diphenyl ether-type herbicides. The IC_{50} were found to be identical to those measured on the membrane-bound enzyme.

9.3 Molecular Genetics

Extensive mutagenesis of *Escherichia coli* K12 (Sasarman et al. 1968) and *Salmonella typhimurium* (Sasarman et al. 1970) by neomycin and the selection of dwarf colonies have provided a useful way of isolating heme-deficient mutants. One of the *E. coli* mutants, designated SASX38, accumulates uroporphyrin, coproporphyrin and protoporphyrin (Sasarman et al. 1979). This mutant grows very poorly, even on rich medium, and is supplemented with heme in Brain-Heart Infusion-based growth media. Since it has normal ferrochelatase activity, it was assumed to be deficient in protoporphyrinogen oxidase activity. The gene affected in the mutant was designated *hemG*. Mapping of the *hemG* gene by phage P1-mediated transduction showed that it was located very close to the *chlB* gene (frequency of cotransduction 78.7%), between the *metE* and *rha* markers. This location, at 86′ on the *E. coli* genetic linkage map (Bachmann 1983), is distinct from the other known *hem* loci in *E. coli* K12. Xu et al. (1992) analyzed a large collection of heme-deficient mutants of *S. typhimurium*, a close relative of *E. coli*, obtained by inserting Mu-derived transposons and diethyl sulfate chemical mutagenesis. There were surprisingly few *hemG* mutations (1/93 Mud-J insertion mutants, and 1/48 mutants induced by diethyl sulfate). Backcross analysis of a *hemG*::Mud-J insertion indicated that the mutation probably does not require a second-site suppressor for viability. An other *hemG* mutant in *E. coli* was obtained by Nishimura et al. (1995a) by selection of photoresistant revertants of a Δ*visA* (Δ*hemH*) strain of *Escherichia coli* K12. This approach could be interesting for studies of the photo-damage that occurs in other cell types treated with inhibitors of protoporphyrinogen oxidase. Mutations in the *visA* gene of *E. coli* caused the mutant bacteria to die when exposed to visible light (Nakahigashi et al. 1991). The *visA* gene was found to be the structural gene for ferrochelatase (*hemH*) (Miyamoto et al. 1991, 1992; Frustaci and O'Brian 1993).

Mutations in the early genes involved in the biosynthesis of heme can cure the photosensitivity and the light-induced cell death appears to be due to the accumulation of protoporphyrin IX, one of the substrates of ferrochelatase. As for the human disease protoporphyria, accumulation of protoporphyrin IX due to a defect of ferrochelatase promotes the production of active species of oxygen (probably singlet oxygen and/or superoxide anions) when cells are illuminated with visible light, leading to peroxidation of cellular components and cell death. One of the photoresistant revertants of the Δ*visA* (Δ*hemH*) strain of *E. coli* K12, the VSR751 strain, accumulated uroporphyrin, coproporphyrin and protoporphyrin IX, but did not accumulate as much protoporphyrin as cells of the parent strain (Δ*hemH*). The pattern of porphyrin accumulation indicates that strain VSR751 is defective in protoporphyrinogen oxidase, and genetic analysis showed that the mutation mapped at 86′ and was therefore likely to be in the *hemG* gene. The mutant had similar growth defects on rich media as the SASX38 mutant. Both *visA* and *hemG* mutants are 100-fold more sensitive in tumbling to blue light than the wild-type parent (Yang et al. 1996). When heme is present, as in the wild type, the non-iron (non-heme) porphyrins are at a relatively low concentration and tumbling to blue light at an intensity effective for *hemG* or *hemH* strains did not occur, indicating that the function of tumbling to light is most likely to allow escape from the lethal effect of intense light.

The photoresistant revertants of the light-sensitive strain Δ*visA* of *E. coli* included a double mutant (H103) with mutations in *hemA* [a gene involved in the biosynthesis of 5-aminolevulinic acid (ALA), the committed precursor of heme] and in another gene downstream of *hemA* (Nakayashiki et al. 1995). This gene, designated *hemK*, was located at 27 min on the linkage map of the *E. coli* chromosome. The mutant strain H103 formed small colonies and had no catalase activity even in the presence of ALA, indicating its inability to catalyze a step in the biosynthesis of heme from ALA. However, extracts of H103 cells had readily detectable ALA dehydratase and porphobilinogen deaminase activities. H103 cells carrying a plasmid that included only *hemA* as an insert accumulated coproporphyrin and protoporphyrin, but were not sensitive to light, a phenotype similar to that of a *hemG* mutant. This suggested that it also lacked protoporphyrinogen oxidase activity.

The *hemG* gene of *E. coli* K12 was first isolated by a mini-Mu in vivo cloning procedure (Sasarman et al. 1993). The *hemG* gene restored normal growth to the *hemG* mutant, and the transformed cells displayed higher protoporphyrinogen oxidase activity that a wild-type control strain, ranging from 2 times (when expressed under its own promoter) to 25 times (when cloned into an expression vector such as pBluescript or pTrc99A). The enzyme activity was recovered in the membrane fraction of transformed cells. Sequencing of the *hemG* gene identified an open reading frame of 546 nucleotides (181 amino acids), within the minimal fragment able to complement the mutant. The molecular mass of the HemG protein was 21 kDa in a DNA-directed coupled transcription-translation system in agreement with values found by SDS-PAGE. The initiation codon was GTG and not ATG and the identity of the first 18 amino acids at the amino-terminal end of the protein was confirmed by microsequencing. *hemG* was independently cloned by complementation of the gene defect in the VSR751 strain (Nishimura et al. 1995a) and subcloned by PCR from Kohara's DNA library in Lambda phage. The *hemG* gene lies between *trkH* and the *rrnA* operon, and is transcribed clockwise in the same direction as the *rrnA* operon. Sequence analysis revealed the presence of flavodoxin motif [LIV]-[LIVFY]-[FY]-x-[ST]-x(2)-[AGC]-x-T-x(3)-A-x(2)-[LIV] (Bairoch 1992) in the N terminus of the protein that may be involved in the binding of the phosphate group of a flavin mononucleotide cofactor. Flavodoxins are electron-transfer proteins that function in various electron transport systems (Wakabayashi et al. 1989) where the FMN molecule serves as a redox-active prosthetic group. Flavodoxins are functionally interchangeable with ferredoxins. In the recent releases of the PROSITE database (Bairoch 1992), HemG is referred to as a false positive protein containing the flavodoxin motif. However, the flavodoxin consensus is conserved not only in sequence but also in position in the sequence of HemG and it is conceivable that the *E. coli* protoporphyrinogen oxidase is a true flavodoxin that donates electrons from protoporphyrinogen to some component of the aerobic (and perhaps anaerobic) electron transport chain. The biochemistry of the purified HemG is still incomplete and its function unknown.

A potential partner for HemG may be the product of the *hemK* gene. This gene has been cloned and sequenced (Nakayashiki et al. 1995). HemK forms part of the *hemA-prfA-hemK* operon and encodes a 225 amino acid polypeptide. Sequence analysis has revealed no specific features of the protein (binding of flavinic cofactor or metal ion, etc.) and there is no biochemical evidence to support the direct involvement of HemK in the oxidation of protoporphyrinogen to protoporphyrin. A striking difference between HemG and HemK is that whereas HemG appears to have no similar sequence, HemK has significant similarity with many bacterial genes. Since HemK might be involved in protoporphyrinogen oxidation, databases contain many entries to it as "possible protoporphyrinogen oxidase", or even "protoporphyrinogen oxidase". Some bacteria such as *Haemophilus influenzae* or *Synecocystis* sp. have several homologues to *hemK*. HemK is also similar to yeast or human putative adenine-specific methylases. Thus the role of this protein in the aerobic or aenerobic oxidation of protoporphyrinogen is still a matter of speculation.

The involvement of electron-transfer protein in protoporphyrinogen oxidation is indicated by analysis of mutants of *Bradyrhizobium japonicum* unable to generate c-type cytochromes, perhaps due to a lack of protoporphyrinogen oxidase (O'Brian et al. 1987b; Ramseier et al. 1989). Rhizobiacae are an important group of bacteria involved in fixing dinitrogen in plant-bacterial symbiotic nodules. The reduction of dinitrogen is extremely energy consuming (16 ATP per dinitrogen reduced) and the bacteria synthesize large amounts of cytochromes involved in oxidative phosphorylation. The nitrogenase enzyme complex catalyzing the reduction of dinitrogen is very susceptible to inactivation by dioxygen, and a specific hemoprotein, leghemoglobin, is synthesized in nodules to trap dissolved oxygen. The apoleghemoglobin is synthesized by the plant, and the heme moiety is provided by the bacteria (O'Brian et al. 1987a,b). Two mutant strains with pleiotropic respiratory deficiency have been found. The *B. japonicum* strain LO505 is a transposon Tn5-induced cytochrome-deficient mutant; it excretes the oxidized heme precursor coproporphyrin III into the growth medium. Cell-free extracts from the mutant strain LO505 contain very little protoporphyrinogen oxidase activity. Soybean root nodules formed with this mutant do not contain leghemoglobin, but the apoprotein is synthesized nevertheless. The other mutant strain, 2606::Tn5-20 (Ramseier et al. 1989), has a phenotype similar to that of strain LO505. Molecular cloning of the genes complementing the two mutations revealed unsuspected features of the proteins involved in protoporphyrinogen oxidation. The Tn5 insertions are very close together but in different open reading frames (ORFs) of a cluster of *Cyc* genes (*Hel* in *Rhodobacter capsulatus*, *Ccm* in *E. coli*) encoding components of a heme-ABC-transporter, *CycV* (*HelA*, *CcmA*), the ATP-binding subunit, *CycW* (*HelB*, *CcmB*) and *CycZ* (*HelC*, *CcmC*) (for a review see Thony-Meyer 1997). Tn5 lies in a small open reading frame, *CycX* (*HelD*) encoding a small (61 amino acids) hydrophobic protein in the LO505 strain but Tn5 is inserted at the 3′ end of the coding region of a larger open reading frame, *CycY* (*HelX*) in the 2606::Tn5-20 strain (Ramseier et al. 1991). This ORF encodes a 162-amino acid protein that is very similar to the bacterial thioredoxins.

Thioredoxins (Holmgren 1985; Gleason and Holmgren 1988) are small proteins (about 100 residues) which take part in various redox reactions via the reversible oxidation of an active center disulfide bond. They can exist in reduced or oxidized forms; in the oxidized form the two cysteine residues form an intramolecular disulfide bond. Thioredoxins are present in prokaryotes and eukaryotes and the sequence around the redox-active disulfide bond is well conserved. A number of eukaryotic proteins contain domains related to thioredoxin. All of them seem to be protein disulphide isomerases (EC 5.3.4.1), an endoplasmic reticulum enzyme that catalyzes the rearrangement of disulfide bonds in various proteins. Bacterial proteins that act as thiol:disulfide interchange proteins for disulfide bond formation in some periplasmic proteins also contain a thioredoxin domain. CycY protein is a peculiar thioredoxin in that it is a membrane-anchored periplasmic protein. Its role in the biogenesis of c-type cytochromes may be to activate the cysteine residues of the apocytochrome-*c* which are involved in establishing the thioether bond with the vinyl side chains of protoheme (Fabianek et al. 1997).

The relationship between the roles of CycX and CycY in the control of protoporphyrinogen oxidation is unclear, but these results may be examined with reference to the earliest work on protoporphyrinogen oxidation, which was carried out over 30 years ago. Sano and Granick (1961) and Sano et al. (1964a,b), in their studies on cytochrome *c* biogenesis, found that porphyrin-*c*, the desferri-form of heme-*c*, was readily formed from protoporphyrinogen rather than from protoporphyrin. Little is known of protoporphyrinogen oxidation in rhizobiacae (Keithly and Nadler 1983; Jacobs et al. 1989a), but recent advances in the molecular biology of the system indicate that it needs to be further developed.

In their analysis of hemin-auxotroph mutants of *Bacillus subtilis*, Hanson and Hederstedt (1992) reported the cloning and sequencing of two operons encoding enzymes of the heme biosynthesis pathway, the *hemAXCDBL* operon located at 244 degrees on the genetic map of *B. subtilis* encoding the enzymes involved in the synthesis of 5-aminolevulinic acid to that of uroporphyrinogen III and the *HemEHY* operon located at 94 degrees on the genetic map of *B. subtilis* encoding the enzymes uroporphyrinogen III decarboxylase (*hemE*) and ferrochelatase (*hemH*). The HemY is a 53-kDa protein with coproporphyrinogen oxidase activity, or protoporphyrinogen oxidase activity, or both. Mutations in the *hemY* gene, or deletion of the gene caused accumulation of coproporphyrinogen III or coproporphyrin III in the growth medium and accumulation of trace amounts of other porphyrinogens or porphyrins in the cells. HemY is a peripheral membrane-bound protein (Hansson and Hederstedt 1994a). Most of the recombinant protein is a soluble enzyme in *E. coli* (Dailey et al. 1994) and oxidizes coproporphyrinogen III to coproporphyrin and protoporphyrinogen IX to protoporphyrin, but not uroporphyrinogen III to uroporphyrin III (Dailey et al. 1994; Hansson et al. 1997). The apparent specificity constant, kcat/Km, for HemY is about 12-fold greater with coproporphyrinogen III as a substrate than with protoporphyrinogen IX as a substrate (Hansson et al. 1997). The protoporphyrinogen IX oxidase activity is thus consistent with the function of HemY in a late step of protoheme IX biosynthesis, but the control of the enzyme activity in vivo is not understood and the efficient coproporphyrinogen III to coproporphyrin oxidase activity is not explained by the current view of protoheme IX biosynthesis (for further discussion of this point *vide infra*). A databases search for bacterial homologues of the *B. subtilis* HemY identified open reading frames with sequences similar to that of HemY in microorganisms such as *Mycobacterium leprae* and *Myxococcus xanthus*. Dailey and Dailey (1996) expressed the *Myxococcus xanthus hemY* gene in *E. coli* and found that the recombinant protein oxidized protoporphyrinogen to protoporphyrin but not coproporphyrinogen to coproporphyrin in vitro. The *hemY* gene of *Propionibacterium freudenreichii* has been cloned recently and sequenced as part of a *hemYHBXRL* operon (Hashimoto et al. 1997). The *hemY* gene complemented the gene defect in a *hemG* mutant of *E. coli* and the HemY protein is very similar to that of *B. subtilis*.

The fact that the yeast *Saccharomyces cerevisiae* is a facultative aerobe made this eukaryotic organism a powerful tool for isolating respiratory-deficient mutants defective in heme synthesis. Mutants blocked in one of each of the eight steps of heme biosynthesis were isolated using different methods of selection and/or enrichment in heme mutants and characterized (Sugimura et al. 1966; Gollub et al. 1977; Urban-Grimal and Labbe-Bois 1981; Amillet and Labbe-Bois 1995). A *hem14* mutant has been isolated which lacks cytochromes, accumulates protoporphyrin, is devoid of protoporphyrinogen oxidase (Urban-Grimal and Labbe-Bois 1981; Camadro et al. 1982) and makes normal amounts of immunodetectable protein (Camadro et al. 1994). The structural gene for protoporphyrinogen oxidase, HEM14, was isolated by functional complementation of the *hem14-1* mutant (Camadro and Labbe 1996). The *hem14-1* mutation was genetically linked to URA3, a marker on chromosome V, and HEM14 was physically mapped on the right arm of this chromosome, between PRP22 and FAA2. Disruption of the HEM14 gene leads to protoporphyrinogen oxidase deficiency in vivo (heme deficiency and accumulation of heme precursors) and in vitro (lack of immunodetectable protein or enzyme activity). Smith et al. (1996) analyzed the function of the genes of yeast chromosome V by genetic footprinting. The only phenotype found with the Ty1 insertion in the HEM14 ORF (referred to as YER044w) was an increased sensitivity for growth on high salt medium. The Ty1 insertion was not precisely mapped and may lie in some part of the gene that allows, nevertheless, some production of protoporphyrinogen oxidase, or part of it, that complements the usual glycerol-minus phenotype of heme mutants.

The HEM14 gene encodes a 539-amino acid protein (59, 665 Da; pI 9.3) somewhat similar to the *hemY* gene product of *Bacillus subtilis*. Studies on protoporphyrinogen oxidase overexpressed in yeast and purified as wild-type enzyme showed that the NH_2-terminal mitochondrial targeting sequence of protoporphyrinogen oxidase is not cleaved during importation. The enzyme was strongly inhibited by diphenyl ether-type herbicides and readily photolabeled by a diazoketone derivative of tritiated acifluorfen. The mutant allele *hem14-1* contains two mutations, L422P and K424E, responsible for the inactive enzyme. Both mutations introduced independently in the wild-type HEM14 gene completely inactivated the protein when analyzed in an *Escherichia coli* expression system. The HEM14 gene was isolated independently by functional complementation of a respiratory mutant that accumulated protoporphyrin and was rescued by exogenous heme (Glerum et al.

1996). An open reading frame on chromosome I of the yeast *Schizosaccharomyces pombe* was described in the yeast genome sequencing project; it is very similar to HEM14 and, as in *S. cerevisiae*, lies between the genes homologous to PRP22 and FAA2, and is therefore the protoporphyrinogen oxidase gene in this organism.

The strategy of cloning eukaryotic cDNAs encoding protoporphyrinogen oxidase by functional complementation of the *hemG* mutant of *E. coli* appears to be extremely productive for the characterization of protoporphyrinogen oxidase from mammals, humans (Nishimura et al. 1995b; Dailey and Dailey 1996) and mice (Dailey et al. 1995; Taketani et al. 1995b), and also those from plants, *Arabidopsis thaliana* (Ward and Volrath 1995; Narita et al. 1996) and two isoforms from *Nicotiana tabacum* (Lermontova et al. 1997).

The cDNA has been used as a molecular probe to locate the human protoporphyrinogen oxidase gene. Taketani et al. (1995a) showed by Southern blotting of human genomic DNA that there is a single copy of the protoporphyrinogen oxidase gene, and they mapped the gene to 1q22 by fluorescence in situ hybridization. This contradicts previous results showing a genetic linkage between the gene locus for variegate porphyria and the alpha-1-antitrypsin gene and the immunoglobulin heavy chain cluster on chromosome 14 (Bissbort et al. 1988). The reason for this is unclear and it was suggested that an additional gene product on chromosome 14 may interact with protoporphyrinogen oxidase protein to maintain the normal enzyme activity or to affect expression of the protoporphyrinogen oxidase gene (Taketani et al. 1995a). However, Roberts et al. (1995) showed that the variegate porphyria gene defect is linked to microsatellite and other markers in the region 1q21–q23. They conclude that locus heterogeneity is unlikely in variegate porphyria. The protoporphyrinogen oxidase gene has 13 exons and spans about 8 kb on 1q22 (Taketani et al. 1995a). The exon/intron boundary sequences conform to consensus acceptor (GT) and donor (AG) sequences, and exons in the gene appear to encode functional protein domains. Primer extension analysis revealed two major transcriptional initiation sites in a region with sequence motifs characteristic of a promoter. The human protoporphyrinogen oxidase cDNA encodes a 477-amino acid protein. Northern blotting detected a 1.8-kb mRNA in human K562 and HepG2 cells. The human protoporphyrinogen oxidase cDNA expressed in COS-1 monkey cells produces much higher protoporphyrinogen oxidase activity, which was inhibited by acifluorfen. The human protein can also be detected in the mitochondria of the transfected cells, despite the fact that the protein lacks the apparent transport-specific leader sequence found in a number of other mitochondrial proteins. The expression of the protoporphyrinogen oxidase gene is only slightly modulated in cells undergoing erythroid differentiation and producing large amounts of heme (Taketani et al. 1995b).

Two different, full-length tobacco cDNA sequences have been cloned by complementation of the *hemG* mutant of *E. coli* (Lermontova et al. 1997). The first is very similar to a cDNA from *A. thaliana* isolated using the same screening procedure (Narita et al. 1996). It encodes a 548-amino acid protein, with a putative transit sequence of 50 residues. The second encodes a 504-residue protein. The deduced protein sequences have 28% identical amino acid residues (Lermontova et al. 1997). The first protoporphyrinogen IX oxidase translated in vitro could be translocated to plastids, and the approximately 53-kDa mature protein was detected in the stroma and membrane fraction. The second enzyme was targeted to mitochondria without any detectable reduction in size. The location of both enzymes in subcellular fractions was confirmed immunologically. Steady-state RNA analysis indicates an almost synchronous expression of both genes during tobacco plant development, greening of young seedlings, and diurnal and circadian growth. The mature plastid and mitochondrial isoenzymes were overproduced in *E. coli.* Extracts of bacteria containing the recombinant mitochondrial enzyme have more protoporphyrinogen IX oxidase activity than the control strains, whereas the plastid enzyme was found only as an inactive peptide. Since protoporphyrinogen oxidase is present in three independent membrane fractions of plant cells, there may be a third cDNA encoding one of the plastid isoenzymes or a single gene may encode both forms of the chloroplast protoporphyrinogen oxidase. This would involve targeting the protein to the envelope and the thylakoidal membrane by selective processing of the targeting sequence. Alternatively, the genomic organization of the chloroplast-specific protoporphyrinogen oxidase gene may give rise to the synthesis of two different mRNAs with identical coding sequences but different targeting sequences, by differential splicing of a common pre-mRNA or through two transcription start points on the gene. These problems remain to be studied.

9.4 Structure and Function

All the eukaryotic protoporphyrinogen oxidases appear to be structurally related to the HemY gene product of *Bacillus subtilis.* Within the family of HemY homologues (see Fig. 3), sequence identities may be as low as 6% (*P. freudenreichii* vs *S. cerevisiae*) to moderate (26% between *M. xanthus* and *A. thaliana* POX1); the average level of identity is 20–21%. The sequence similarities within orders are much higher, 88% (human vs mouse) and 72% (*A. thaliana* POX1 vs *N. tabaccum* POX1) but the two cDNAs known from plants (POX1 and POX2) have only 28% similar sequences. The aligment of the sequences, using the Clustalw software (Fig. 4) shows that they all have a highly conserved domain in their N terminus that contains a consensus sequence GxGxxG which is part of a β-α-β ADP binding fold found in many flavoproteins. The glycine residues are believed to interact with the phosphoryl groups of the adenosine moeity of FAD.

One remarkable feature of all protoporphyrinogen oxidases is that, although they are biochemically membrane-bound proteins (and may even have

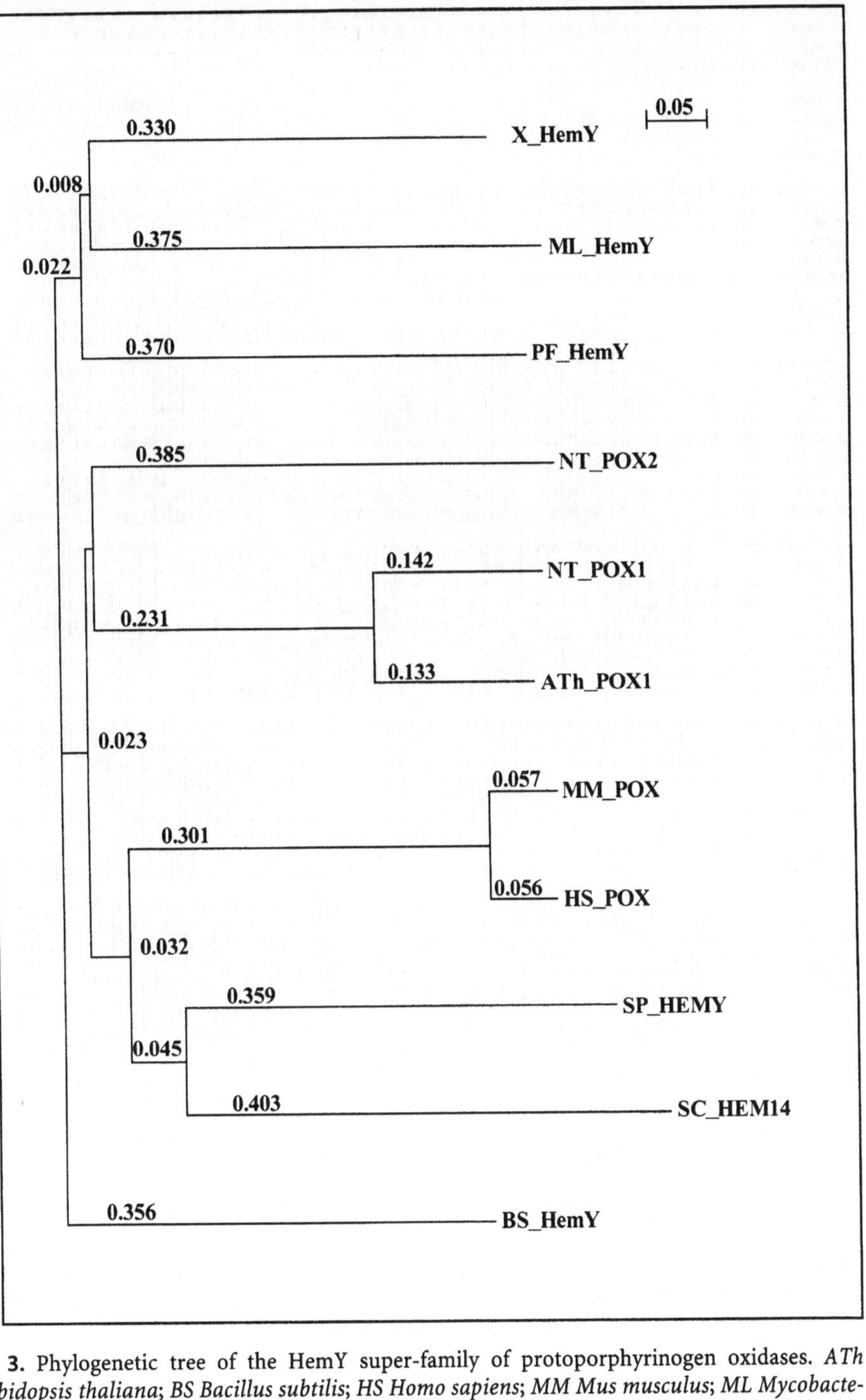

Fig. 3. Phylogenetic tree of the HemY super-family of protoporphyrinogen oxidases. *ATh Arabidopsis thaliana; BS Bacillus subtilis; HS Homo sapiens; MM Mus musculus; ML Mycobacterium leprae; MX Myxococcus xanthus; NT Nicotiana tabaccum; PF Propionibacterium freudenreichii; SC Saccharomyces cerevisiae; SP Schizosaccharomyces pombe*

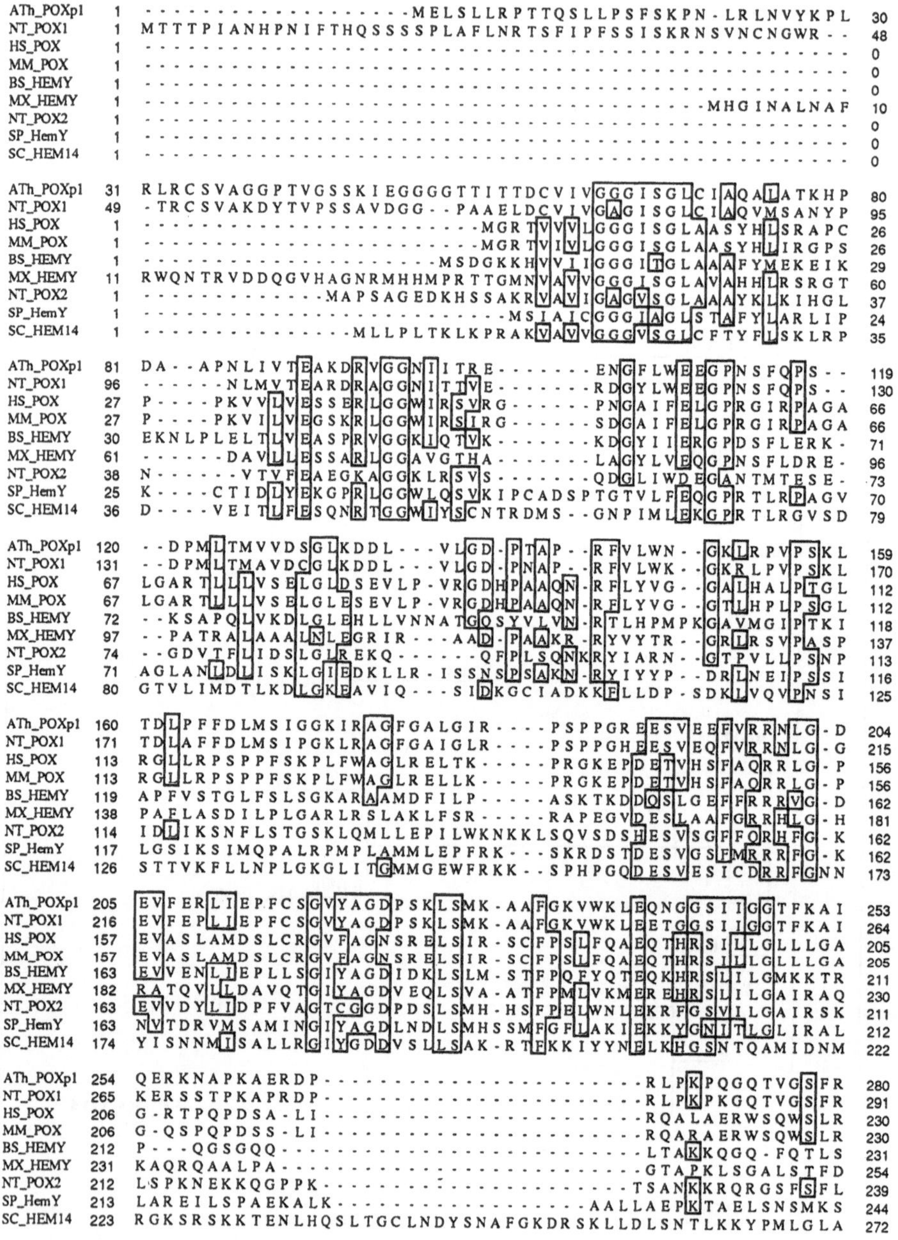

Fig. 4. Alignment of HemY sequences (Clustalw). *ATh Arabidopsis thaliana*; *BS Bacillus subtilis*; *HS Homo sapiens*; *MM Mus musculus*; *MX Myxococcus xanthus*; *NT Nicotiana tabaccum*; *SC Saccharomyces cerevisiae*; *SP Schizosaccharomyces pombe*

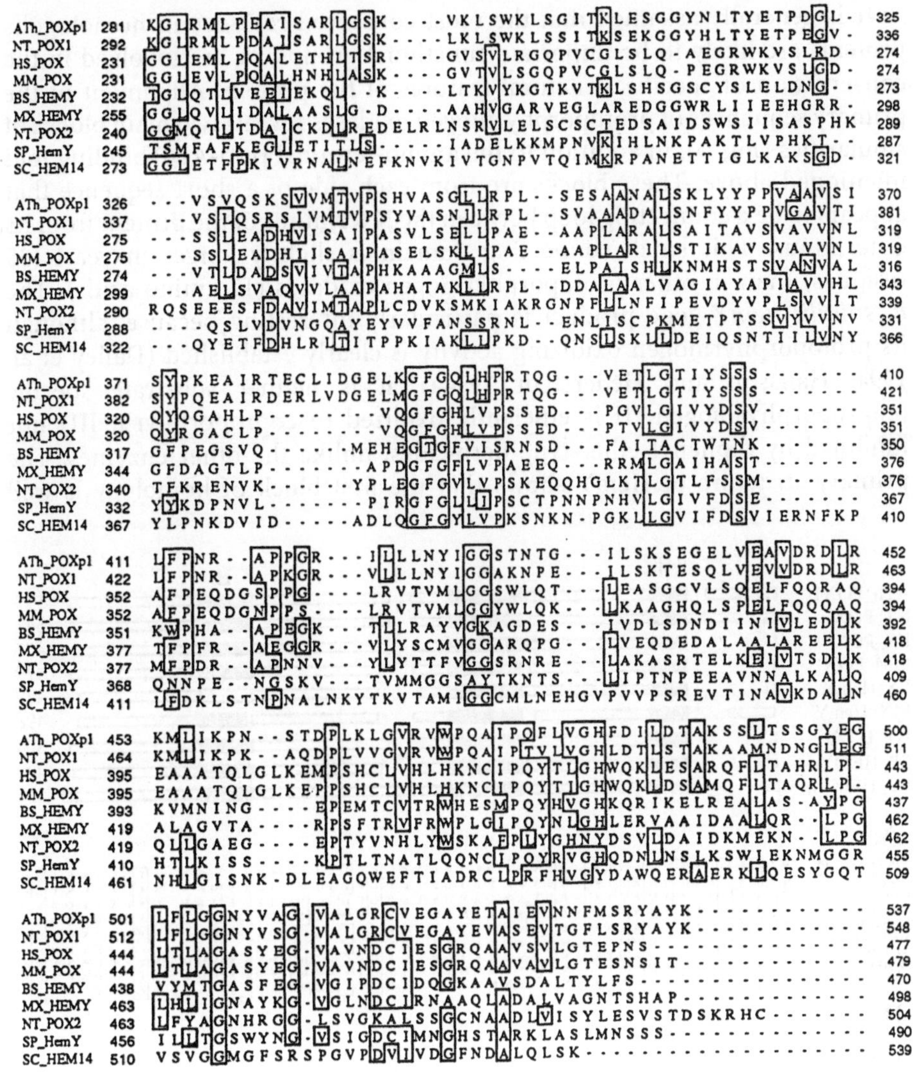

Fig. 4. (*Continued*)

the characteristics of integral membrane proteins), they all lack the typical transmembrane domains in the hydropathy plots derived from their sequences. This has been confirmed by analysis of the aligments of the sequences using the TMAP software (EMBL computational services), which detected no transmembrane segments. This has very important implications in terms of the topology of protoporphyrinogen oxidase, since it implies that the enzyme must be on one side of the membrane where it is located, and thus the coupling of the enzyme activity with the chelatases may be very different when the two enzymes are on the same surface of a membrane, or on two opposite surfaces.

It is generally assumed that the most conserved residues in homologous proteins are likely to be important functional groups, or are involved in the conservation of specific structural features of the protein. Alignment of the primary sequences of protoporphyrinogen oxidases reveals two main blocks of similarities (Fig. 5), in addition to the amino-terminal βαβ-ADP binding fold mentioned above. These blocks are from each side of a short sequence that appears to be a connecting loop between the N terminus of the protein from its C terminus. This small connecting domain is slightly longer in yeast (35 residues) than in other protoporphyrinogen oxidases (12 amino acids). The *B. subtilis hemY* gene product is particularly interesting because, although its protoporphyrinogen oxidizing activity is clearly established (Dailey et al. 1994; Hansson and Hederstedt 1994a), it is neither substrate specific (coproporphyrinogen III is efficiently oxidized to coproporphyrin III) nor inhibited by diphenyl ether-type herbicides, unlike the other enzymes. The primary structure of HemY shows that the first block of homology is well

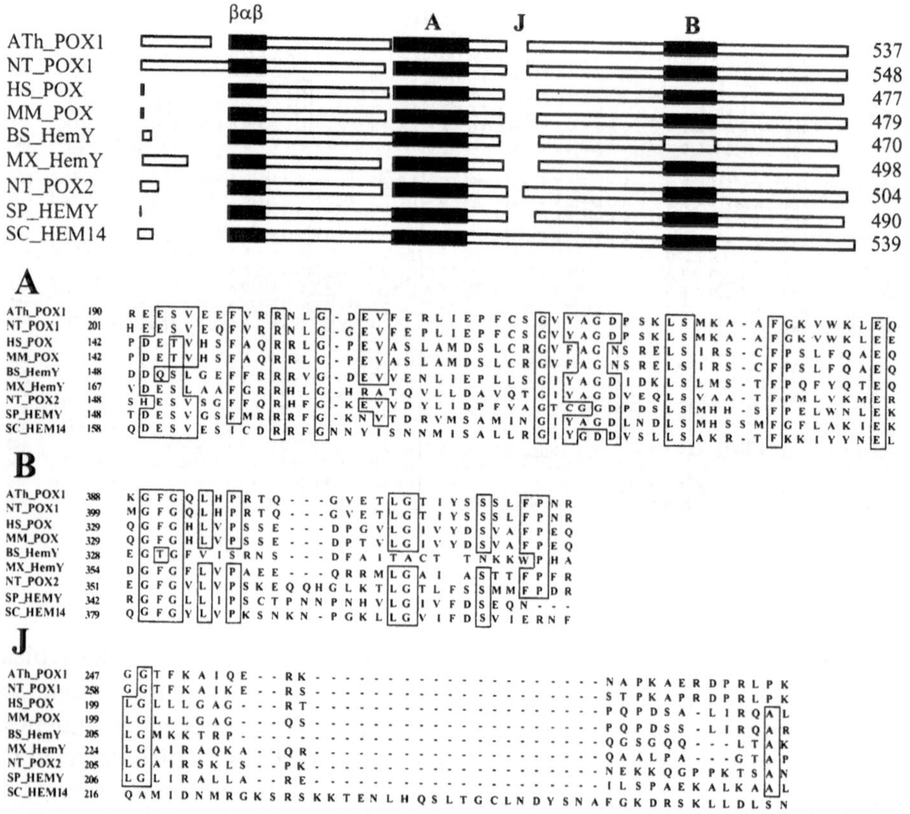

Fig. 5. Domain organization of protoporphyrinogen oxidases, and conserved blocks of similarities. *ATh Arabidopsis thaliana*; *BS Bacillus subtilis*; *HS Homo sapiens*; *MM Mus musculus*; *MX Myxococcus xanthus*; *NT Nicotiana tabaccum*; *SC Saccharomyces cerevisiae*; *SP Schizosaccharomyces pombe*

conserved and similar to that of the other protoporphyrinogen oxidases, but it lacks the second block of homology. The enzyme from *M. xanthus* that has the same substrate specificity and inhibition characteristics as eukaryotic enzymes has both conserved sequences. Two missense mutations in the yeast gene (L422P and K424E) that abolished the enzyme activity (Camadro and Labbe 1996) show the importance of some residues in this block. This suggests that some of the determinants of the specificity of protoporphyrinogen oxidases lie within this structure, and it is therefore a privileged target for investigating the molecular basis of the interaction of diphenyl ether-type herbicides with protoporphyrinogen oxidase.

Several mutations have been found in the human protoporphyrinogen oxidase gene that make the altered protein inactive or partially active. The first is a missense mutation leading to substitution of arginine for glycine (G232R) (Deybach et al. 1996). This mutation co-segregated in patients from a family with variegate porphyria. The glycine-232 is a conserved residue located at the begining of the C-terminal domain. An arginine to tryptophane mutation (R59W) is present in most of the patients with variegate porphyria in South African families (Meissner et al. 1996; Warnich et al. 1996), and this defect may be the founder gene defect causing variegate porphyria in South Africa. Rarer mutations, H20P and R168C (Warnich et al. 1996), have each been found in one patient. Dailey and Dailey (1997) studied three mutations, R59W, R168C and A433P, in heterologous expression system, and showed that they result in decreased enzyme activity by causing a decrease in kcat without any significant change in Km for the substrate protoporphyrinogen IX. The purified R59W protoporphyrinogen oxidase apparently lacks the FAD cofactor. This mutation is at the edge of the β-α-β ADP binding fold, a domain that may be involved in the correct assembly of the apo-protein with the cofactor during the biogenesis of protoporphyrinogen oxidase and maps very close to the position of a small deletion in the *B. subtilis* gene (ΔRK79; Hansson and Hederstedt 1994a) that also leads to the synthesis of an inactive enzyme. Only one in the coding region of human protoporphyrinogen gene has been described to date that converts an arginine to a histidine (R304H) without affecting the enzyme activity.

The presence of a flavin at the active site of protoporphyrinogen oxidase may explain some catalytic properties of the enzyme. Jones et al. (1984) studied the oxidation of protoporphyrinogen stereospecifically tritiated at the methylene bridges. They found that three hydrogen atoms from the methylene bridges are removed from one side of the protoporphyrinogen nucleus, while the fourth is removed from the other side of the cyclic tetrapyrrole. The protons from bridges α and β (Fig. 1) also appear to be removed by two distinct mechanisms. Jones et al. have produced a model of protoporphyrinogen oxidation that involves the removal of three hydrides and one proton. Thus, while the flavin may be involved in removal of the hydrides, another functional group on the protein, possibly a basic amino-acid residue, may be involved in the removal of the fourth proton. This model is supported by calculations of molecular dynamics based on the protoporphyrinogen oxidase inhibitors QSAR (Nandihalli et al. 1992; Akagi and Sakashita 1993), which also indicate that charge transfer and electrostatic interactions may be important in protoporphyrinogen oxidase functioning. Little is known about the reactivity of the flavin in protoporphyrinogen oxidase. Yeast and mammalian protoporphyrinogen oxidases have been shown to be FAD-containing enzymes (Camadro et al. 1994; Dailey et al. 1995; Camadro and Labbe 1996; Dailey and Dailey 1996). The spectra of the purified yeast enzyme show that the flavin is stable in a semi-quinone form (Camadro and Labbe 1996). It is highly probable that all the enzymes of the HemY family are flavoproteins as they all have the β-α-β ADP binding fold but this has not been formally demonstrated for other protoporphyrinogen oxidases.

Arnould et al. (1997) investigated the role of the flavin in catalysis by studying the reactivity of yeast protoporphyrinogen oxidase toward a potential inhibitor of flavoproteins, the diphenyleneiodonium cation. Diphenyleneiodonium and related bis(aryliodonium) species appear to be time-dependent, mechanism-based inhibitors of several flavoproteins such as mitochondrial NADH ubiquinone oxidoreductase (Gatley and Sherratt 1976; Ragan and Bloxham 1977), neutrophil NADPH oxidase (Doussiere and Vignais 1992; O'Donnell et al. 1993), xanthine oxidase (Doussiere and Vignais 1992), nitric oxide synthase (Stuehr et al. 1991) and cytochrome P450 reductase (Tew 1993). These compounds seem to act on the reduced flavins generated during enzyme turnover that could act as electron donors to the inhibitor, allowing the generation of phenyl radicals that would then covalently modify the flavin or some amino acid side chain important for catalysis (O'Donnell et al. 1994). Alkynyl and aryl mono- and diiodonium salts are potent inhibitors of pyrroloquinolin quinone related redox processes in addition to inhibiting flavoproteins (Gallop et al. 1993; Bishop et al. 1994). The typical slow-binding kinetics observed with the yeast protoporphyrinogen oxidase suggest that the enzyme with a reduced flavin rapidly combines with the inhibitor to form an initial complex which then slowly isomerizes to a modified enzyme-inhibitor complex according to a bi-bi ping-pong mechanistic model for the enzyme reaction (Fig. 6) where diphenyleneiodonium competes with molecular oxygen in the reoxidation step of the reduced flavin.

This model raises an important question regarding the mode of action of protoporphyrinogen oxidase. What is the number of redox reactions of the flavin during the catalysis? The mechanism of reaction implies that protoporphyrinogen oxidase must be with an oxidized flavin when the protoporphyrinogen binds to the enzyme and that the reduced flavin generated during protoporphyrin formation must be reoxidized by molecular oxygen (Fig. 1). The reaction must thus be of ordered type. However, the sequence of the molecular events occurring during the reaction is not clearly established and two main mechanisms are possible. In the first one, protoporphyrinogen binds to the enzyme with an oxidized flavin, and is oxidized by three consecutive steps of reduction/oxidation of the flavin with rotation of the

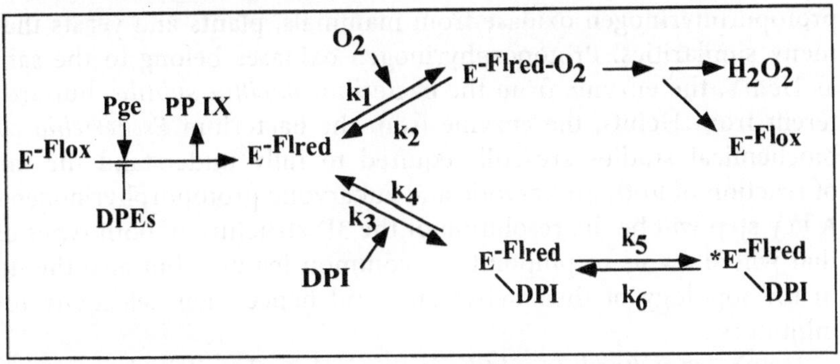

Fig. 6. Postulated mechanism of action of yeast protoporphyrinogen oxidase. *DPEs* diphenyl ethers; *DPI* diphenyleneiodonium; *E* protoporphyrinogen oxidase; *Flox* oxidized flavin; *Flred* reduced flavin; *Pge* protoporphyrinogen; *PP IX* protoporphyrin IX

partially oxidized tetrapyrrole intermediates within the active site. A second possible mechanism of protoporphyrinogen oxidase reaction is that protoporphyrinogen binds to the enzyme with an oxidized flavin, and is oxidized through a single reduction-oxidation cycle of the flavin, corresponding to the removal of two hydrogen atoms. Then, the conjugated system obtained might undergo spontaneous and fast oxidation by O_2 with removal of two hydrides from two of the remaining methylene bridges, through a different process, uncatalyzed or involving an unknown cofactor. In both mechanisms, the full aromatization to protoporphyrin IX involves the abstraction of a proton from the last methylene bridge. Whatever the mechanism, the order of the reaction must be one with respect to oxygen, assuming the first step of the reaction is the slowest one.

Substituted diphenyleneiodoniums have been synthesized and shown to be even better inhibitors than 2,2′-diphenyleneiodonium. The 4-nitro-2,2′-diphenyleneiodonium is a quasi-irreversible inhibitor ($k5/k6 > 5000$). Diphenyleneiodoniums appear to be a new class of protoporphyrinogen oxidase inhibitors that act very differently from diphenyl ether-type herbicides. They could be promising tools for studies on the structure-function relationships of protoporphyrinogen oxidase.

9.5 Conclusions and Perspectives

Diphenyl ether-type herbicides proved to be extremely powerful tools to study protoporphyrinogen oxidase. They showed that, despite some discrepancies in the description of their molecular properties, protoporphyrinogen oxidases from eukaryotic cells were all inhibited the same way and were therefore likely to share some structural features in their catalytic sites. This was largely confirmed by the analysis of the sequences of the cDNAs and genes encoding protoporphyrinogen oxidase from mammals, plants and yeasts that showed many similarities. Protoporphyrinogen oxidases belong to the same family as HemY, the enzyme from the bacterium *Bacillus subtilis*, but are very different from HemG, the enzyme from the bacterium *Escherichia coli*. Many biochemical studies are still required to fully understand the mechanism of reaction of both prokaryotic and eukaryotic protoporphyrinogen oxidases. A key step will be the resolution of the 3D structure of both types of enzyme that will allow us to pinpoint the common features but also the differences in the topology of their active site, and hence their selectivity toward the inhibitors.

HemY from *Bacillus subtilis* is atypical in that it is the prototype of the eukaryotic protoporphyrinogen oxidases, but it differs considerably from these enzymes since it is not inhibited by diphenyl ether-type herbicides and is more active in oxidizing coproporphyrinogen III to coproporphyrin III than protoporphyrinogen IX to protoporphyrin IX. Under normal conditions, protoporphyrinogen oxidase activity must be tightly coupled to the activity of coproporphyrinogen oxidase, which provides protoporphyrinogen, the substrate for protoporphyrinogen oxidase, and to the activity of the chelatases that further metabolize protoporphyrin IX. Two different genes, *hemF* and *hemN*, encode coproporphyrinogen oxidase in most bacteria (Xu et al. 1992); *hemF* encodes an oxygen-dependent enzyme that is very similar to the eukaryotic coproporphyrinogen oxidases (Xu and Elliott 1993), while *hemN* encodes an oxygen-independent enzyme (Troup et al. 1995), which is not found in eukaryotes. Surprisingly, only the *hemN* gene has been found in *B. subtilis*, although this bacterium is usually described as an obligate aerobe microorganism. The biochemical characterization of purified HemN is still very poor and it would be interesting to test experimentally the hypothesis that, in *B. subtilis*, HemY catalyzes the production of coproporphyrin III, while HemN catalyzes the decarboxylation of coproporphyrin III to protoporphyrin IX in the cytosol of the cells, channeling its substrate to ferrochelatase, which is a soluble enzyme in *B. subtilis* (Hansson and Hederstedt 1994b). This would prevent the sequestration of the liposoluble protoporphyrin IX into the cell membranes making it unavailable for the chelatase. The understanding of the relationship between coproporphyrinogen oxidase, protoporphyrinogen oxidase and the chelatases (ferrochelatase or magnesium-chelatase) and their role in the control of the flux of heme and chlorophyll precursors, especially in plant cells where many subcellular compartments are involved, is the new frontier in tetrapyrroles synthesis studies.

Acknowledgements. We thank Prof. Pierre Labbe for many helpful discussions, Dr. René Scalla for his interest in the early stages of our work on protoporphyrinogen oxidase, and our coworkers, past and present and Dr. O. Parkes for editing part of the English text. The work reported from J.-M. C.'s lab was supported by grants from CNRS, Université Paris 7, Ministère de l'Enseignement Supérieur et de la Recherche (ACC-SV5) and Association de Recherche sur le Cancer. R. S. was a recipient of a research grant from the Fundação para Ciência e a Tecnologia (Portugal).

References

Akagi T, Sakashita N (1993) A quantum chemical study of "light-dependent herbicides". Z Naturforsch 48c:345–349

Amillet JM, Labbe-Bois R (1995) Isolation of the gene HEM4 encoding uroporphyrinogen III synthase in *Saccharomyces cerevisiae*. Yeast 11:419–424

Arnould S, Berthon JL, Hubert C, Dias M, Cibert C, Mornet R, Camadro JM (1997) Kinetics of protoporphyrinogen oxidase inhibition by diphenyleneiodonium derivatives. Biochemistry 36:10178–10184

Bachmann BJ (1983) Linkage map of *Escherichia coli* K12, 7th edn. Microbiol Rev 47:180–230

Bairoch A (1992) PROSITE: a dictionary of sites and patterns in proteins. Nucleic Acids Res 20 [Suppl]:2013–2018

Becker DM, Viljoen JD, Katz J, Kramer S (1977) Reduced ferrochelatase activity: a defect common to porphyria variegata and protoporphyria. Br J Haematol 36:171–179

Bishop A, Paz MA, Gallop PM, Karnovsky ML (1994) Methoxatin (PQQ) in guinea-pig neutrophils. Free Radic Biol Med 17:311–20

Bissbort S, Hitzeroth HW, du Wentzel DP, Van den Berg CW, Senff H, Wienker TF, Bender K (1988) Linkage between the variegate porphyria (VP) and the alpha-1-antitrypsin (PI) genes on human chromosome 14. Hum Genet 79:289–290

Brenner DA, Bloomer JR (1980a) The enzymatic defect in variegate prophyria Studies with human cultured skin fibroblasts. N Engl J Med 302:765–769

Brenner DA, Bloomer JR (1980b) A fluorometric assay for measurement of protoporphyrinogen oxidase activity in mammalian tissue. Clin Chim Acta 100:259–266

Brown SB, Shillcock M (1976) Equilibrium and kinetic studies on the aggregation of porphyrins in aqueous solution. Biochem J 153:279–285

Camadro JM, Labbe P (1996) Cloning and characterization of the yeast HEM14 gene coding for protoporphyrinogen oxidase, the molecular target of diphenyl ether-type herbicides. J Biol Chem 271:9120–9128

Camadro JM, Urban-Grimal D, Labbe P (1982) A new assay for protoporphyrinogen oxidase – evidence for a total deficiency in that activity in a heme-less mutant of *Saccharomyces cerevisiae*. Biochem Biophys Res Commun 106:724–730

Camadro JM, Matringe M, Scalla R, Labbe P (1991) Kinetic studies on protoporphyrinogen oxidase inhibition by diphenyl ether herbicides. Biochem J 277:17–21

Camadro JM, Matringe M, Scalla R, Labbe P (1993a) Fluorimetric assay of protoporphyrinogen oxidase in chloroplast and in plant, yeast and mammalian mitochondria. In: Böger P, Sandmann G (eds) Target assays for modern herbicides and related phytotoxic compounds,: CRC Press Lewis Publishers, Boca Raton, pp 29–34

Camadro JM, Matringe M, Thome F, Brouillet N, Labbe P (1993b) Molecular properties of yeast and lettuce protoporphyrinogen oxidases. In: Duke SO, Rebeiz CA (eds) Porphyric pesticides: Design, mechanism of action, toxicology and relationships to pharmaceuticals ACS Books

Camadro JM, Thome F, Brouillet N, Labbe P (1994) Purification and properties of protoporphyrinogen oxidase from the yeast *Saccharomyces cerevisiae* Mitochondrial location and evidence for a precursor form of the protein. J Biol Chem 269:32085–32091

Dailey HA, Dailey TA (1996) Protoporphyrinogen oxidase of *Myxococcus xanthus* Expression, purification, and characterization of the cloned enzyme. J Biol Chem 271:8714–8718

Dailey HA, Dailey TA (1997) Characteristics of human protoporphyrinogen oxidase in controls and variegate porphyrias Cell Mol Biol (Noisy-le-grand) 43:67–73

Dailey HA, Karr SW (1987) Purification and characterization of murine protoporphyrinogen oxidase. Biochemistry 26:2697–2701

Dailey TA, Dailey HA (1996) Human protoporphyrinogen oxidase: expression, purificmation, and characterization of the cloned enzyme. Protein Sci 5:98–105

Dailey TA, Meissner P, Dailey HA (1994) Expression of a cloned protoporphyrinogen oxidase. J Biol Chem 269:813–815

Dailey TA, Dailey HA, Meissner P, Prasad AR (1995) Cloning, sequence, and expression of mouse protoporphyrinogen oxidase. Arch Biochem Biophys 324:379–384

Dean G (1982) Porphyria variegata. Acta Derm Venereol Suppl (Stockh) 100:81–85

Decker KF (1993) Biosynthesis and function of enzymes with covalently bound flavin. Annu Rev Nutr 13:17–41

Deybach JC, de Verneuil H, Nordmann Y (1981) The inherited enzymatic defect in porphyria variegata. Hum Genet 58:425–428

Deybach JC, Puy H, Robreau AM, Lamoril J, Da Silva V, Grandchamp B, Nordmann Y (1996) Mutations in the protoporphyrinogen oxidase gene in patients with variegate porphyria. Hum Mol Genet 5:407–410

Doussiere J, Vignais PV (1992) Diphenylene iodonium as an inhibitor of the NADPH oxidase complex of bovine neutrophils. Factors controlling the inhibitory potency of diphenylene iodonium in a cell-free system of oxidase activation. Eur J Biochem 208:61–71

Elder GH (1990) The cutaneous porphyrias. Semin Dermatol 9:63–69

Fabianek RA, Huber-Wunderlich M, Glockshuber R, Kunzler P, Hennecke H, Thony-Meyer L (1997) Characterization of the *Bradyrhizobium japonicum* CycY protein, a membrane-anchored periplasmic thioredoxin that may play a role as a reductant in the biogenesis of c-type cytochromes. J Biol Chem 272:4467–4473

Falk JE (1964) Porphyrins and metalloporphyrins, vol. 2. Elsevier, Amsterdam

Ferreira GC, Dailey HA (1988) Mouse protoporphyrinogen oxidase Kinetic parameters and demonstration of inhibition by bilirubin. Biochem J 250:597–603

Frustaci JM, O'Brian MR (1993) The *Escherichia coli visA* gene encodes ferrochelatase, the final enzyme of the heme biosynthetic pathway. J Bacteriol 175:2154–2156

Gallop PM, Paz MA, Flückiger R, Stang PJ, Zhdankin VV, Tykwinski RR (1993) Highly effective inhibition by alkynyl and aryl mono and diiodonium salts. J Am Chem Soc 115:11702–11704

Gardner LC, Smith SJ, Cox TM (1991) Biosynthesis of delta-aminolevulinic acid and the regulation of heme formation by immature erythroid cells in man. J Biol Chem 266:22010–22018

Gatley SJ, Sherratt SA (1976) The effects of diphenylene iodonium on mitochondrial reactions. Biochem J 158:307–315

Gleason FK, Holmgren A (1988) Thioredoxin and related proteins in procaryotes. FEMS Microbiol Rev 4:271–297

Glerum DM, Shtanko A, Tzagoloff A, Gorman N, Sinclair PR (1996) Cloning and identification of HEM14, the yeast gene for mitochondrial protoporphyrinogen oxidase. Yeast 12:1421–1425

Gollub EG, Liu KP, Dayan J, Adlersberg M, Sprinson DB (1977) Yeast mutants deficient in heme biosynthesis and a heme mutant additionally blocked in cyclization of 2,3-oxidosqualene. J Biol Chem 252:2846–2854

Hansson M, Hederstedt L (1992) Cloning and characterization of the *Bacillus subtilis* hemEHY gene cluster, which encodes protoheme IX biosynthetic enzymes. J Bacteriol 174:8081–8093

Hansson M, Hederstedt L (1994a) *Bacillus subtilis* HemY is a peripheral membrane protein essential for protoheme IX synthesis which can oxidize coproporphyrinogen III and protoporphyrinogen IX. J Bacteriol 176:5962–5970

Hansson M, Hederstedt L (1994b) Purification and characterisation of a water-soluble ferrochelatase from *Bacillus subtilis*. Eur J Biochem 220:201–208

Hansson M, Gustafsson MC, Kannangara CG, Hederstedt L (1997) Isolated *Bacillus subtilis* HemY has coproporphyrinogen III to coproporphyrin III oxidase activity. Biochim Biophys Acta 1340:97–104

Hashimoto Y, Yamashita M, Murooka Y (1997) The *Propionibacterium freudenreichii hemYHBXRL* gene cluster, which encodes enzymes and a regulator involved in the biosynthetic pathway from glutamate to protoheme. Appl Microbiol Biotechnol 47:385–392

Hift RJ, Meissner PN, Corrigall AV, Ziman MR, Petersen LA, Meissner DM, Davidson BP, Sutherland J, Dailey HA, Kirsch RE (1997) Variegate porphyria in South Africa, 1688–1996–new developments in an old disease. S Afr Med J 87:722–731

Holmgren A (1985) Thioredoxin. Annu Rev Biochem 54:237–271

Jacobs NJ, Jacobs JM (1975) Fumarate as alternate electron acceptor for the late steps of anaerobic heme synthesis in *Escherichia coli.* Biochem Biophys Res Commun 65:435–441

Jacobs NJ, Jacobs JM (1976) Nitrate, fumarate, and oxygen as electron acceptors for a late step in microbial heme synthesis. Biochim Biophys Acta 449:1–9

Jacobs NJ, Jacobs JM (1977) Evidence for involvement of the electron transport system at a late step of anaerobic microbial heme synthesis. Biochim Biophys Acta 459:141–144

Jacobs NJ, Jacobs JM (1978) Quinones as hydrogen carriers for a late step in anaerobic heme biosynthesis in *Escherichia coli.* Biochim Biophys Acta 544:540–546

Jacobs JM, Jacobs NJ (1984) Protoporphyrinogen oxidation, an enzymatic step in heme and chlorophyll synthesis: partial characterization of the reaction in plant organelles and comparison with mammalian and bacterial systems. Arch Biochem Biophys *229*, 312–319

Jacobs JM, Jacobs NJ (1987) Oxidation of protoporphyrinogen to protoporphyrin, a step in chlorophyll and haem biosynthesis Purification and partial characterization of the enzyme from barley organelles. Biochem J 244:219–224

Jacobs NJ, Borotz SE, Guerinot ML (1989a) Protoporphyrinogen oxidation, a step in heme synthesis in soybean root nodules and free-living rhizobia. J Bacteriol 171:573–576

Jacobs NJ, Borotz SE, Jacobs JM (1989b) Characteristics of purified protoporphyrinogen oxidase from barley. Biochem Biophys Res Commun 161:790–796

Jacobs JM, Jacobs NJ, Borotz SE, Guerinot ML (1990) Effects of the photobleaching herbicide, acifluorfen-methyl, on protoporphyrinogen oxidation in barley organelles, soybean root mitochondria, soybean root nodules, and bacteria. Arch Biochem Biophys 280:369–375

Jacobs JM, Jacobs NJ, Duke SO (1996) Protoporphyrinogen destruction by plant extracts and correlation with tolerance to protoporphyrinogen oxidase-inhibiting herbicides. Pestic Biochem Physiol 55:77–83

Jenkins T (1996) The South African malady (news; comment) Nat Genet 13:7–9

Jones C, Jordan PM, Akhtar M (1984) Mechanism and stereochemistry of the porphobilinogen deaminase and protoporphyrinogen IX oxidase reactions: stereospecific manipulation of hydrogen atoms at the four methylene bridges during the biosynthesis of haem. J Chem Soc Perkin Trans I, 2625–2633

Keithly JH, Nadler KD (1983) Protoporphyrin formation in *Rhizobium japonicum.* J Bacteriol 154:838–845

Klemm DJ, Barton LL (1985) Oxidation of protoporphyrinogen in the obligate anaerobe *Desulfovibrio gigas.* J Bacteriol 164:316–320

Klemm DJ, Barton LL (1987) Purification and properties of protoporphyrinogen oxidase from an anaerobic bacterium, *Desulfovibrio gigas.* J Bacteriol 169:5209–5215

Korda V, Deybach JC, Martasek P, Zeman J, da Silva V, Nordmann Y, Houstkova H, Rubin A, Holub J (1984) Homozygous variegate porphyria (letter). Lancet 1:851

Krijt J, Pleskot R, Sanistrak J, Janousek V (1992) Experimental porphyria induced by oxadiazon in male mice and rats. Pestic Biochem Physiol 42:180–187

Krijt J, Van HI, Hassing I, Vokurka M, Blaauboer BJ (1993) Effect of diphenyl ether herbicides and oxadiazon on porphyrin biosynthesis in mouse liver rat primary hepatocyte culture and hepg2 cells. Arch Toxicol 67:255–261

Lee HJ, Duke SO (1994) Protoporphyrinogen IX-oxidizing activities involved in the mode of action of peroxidizing herbicides. J Agric Food Chem 42:2610–2618

Lee HJ, Duke MV, Duke SO (1993) Cellular localization of protoporphyrinogen-oxidizing activities of etiolated barley *Hordeum vulgare* L leaves: relationship to mechanism of action of protoporphyrinogen oxidase-inhibiting herbicides. Plant Physiol (Rockv) 102:881–889

Lermontova I, Kruse E, Mock HP, Grimm B (1997) Cloning and characterization of a plastidal and a mitochondrial isoform of tobacco protoporphyrinogen IX oxidase. Proc Natl Acad Sci USA 94:8895–8900

Lydon J, Duke SO (1988) Porphyrin synthesis is required for photobleaching activity of the p-nitrosubstituted diphenyl ether herbicides. Pest Biochem Physiol 31:74–83

Margalit R, Shaklai N, Cohen S (1983) Fluorimetric studies on the dimerization equilibrium of protoporphyrin IX and its haemato derivative. Biochem J 209:547–552

Marks GS, McCluskey SA, Mackie JE, Riddick DS, James CA (1988) Disruption of hepatic heme biosynthesis after interaction of xenobiotics with cytochrome P-450. Faseb J 2:2774–2783
Matringe M, Scalla R (1987) Photoreceptors and respiratory electron flow involvement in the activity of acifluorfen-methyl and LS 82556 on non chlorophyllous soybean cells. Pestic Biochem Physiol 27:267–274
Matringe M, Scalla R (1988) Studies on the mode of action of Acifluorfen-Methyl in non-chlorophyllous cells Accumulation of tetrapyrroles. Plant Physiol 86:619–622
Matringe M, Camadro JM, Labbe P, Scalla R (1989) Protoporphyrinogen oxidase as a molecular target for diphenyl ether herbicides. Biochem J 260:231–235
Matringe M, Camadro JM, Block MA, Joyard J, Scalla R, Labbe P, Douce R (1992a) Localization within chloroplasts of protoporphyrinogen oxidase, the target enzyme for diphenylether-like herbicides. J Biol Chem 267:4646–4651
Matringe M, Mornet R, Scalla R (1992b) Characterization of [3H]acifluorfen binding to purified pea etioplasts, and evidence that protoporphyrinogen oxidase specifically binds acifluorfen. Eur J Biochem 209:861–868
Matringe M, Camadro JM, Joyard J, Douce R (1994) Localization of ferrochelatase activity within mature pea chloroplasts. J Biol Chem 269:15010–15015
Meissner PN, Dailey TA, Hift RJ, Ziman M, Corrigall AV, Roberts AG, Meissner DM, Kirsch RE, Dailey HA (1996) A R59W mutation in human protoporphyrinogen oxidase results in decreased enzyme activity and is prevalent in South Africans with variegate porphyria. Nat Genet 13:95–97
Miyamoto K, Nakahigashi K, Nishimura K, Inokuchi H (1991) Isolation and characterization of visible light-sensitive mutants of *Escherichia coli* K12. J Mol Biol 219:393–398
Miyamoto K, Nishimura K, Masuda T, Tsuji H, Inokuchi H (1992) Accumulation of protoporphyrin IX in light-sensitive mutants of *Escherichia coli.* FEBS Lett 310:246–248
Murphy GM, Hawk JL, Magnus IA, Barrett DF, Elder GH, Smith SG (1986) Homozygous variegate porphyria: two similar cases in unrelated families. J R Soc Med 79:361–363
Mustajoki P (1978) Variegate porphyria. Ann Intern Med 89:238–244
Mustajoki P, Tenhunen R, Niemi KM, Nordmann Y, Kaariainen H, Norio R (1987) Homozygous variegate porphyria A severe skin disease of infancy. Clin Genet 32:300–305
Nakahigashi K, Nishimura K, Miyamoto K, Inokuchi H (1991) Photosensitivity of a protoporphyrin-accumulating, light-sensitive mutant (*visA*) of *Escherichia coli* K-12. Proc Natl Acad Sci USA 88:10520–10524
Nakayashiki T, Nishimura K, Inokuchi H (1995) Cloning and sequencing of a previously unidentified gene that is involved in the biosynthesis of heme in *Escherichia coli.* Gene 153:67–70
Nandihalli U, Duke M, Duke S (1992) Quantitative structure-activity relationships of protoporphyrinogen oxidase-inhibiting diphenyl ether herbicides. Pest Biochem Biophys 43:193–211
Narita S, Tanaka R, Ito T, Okada K, Taketani S, Inokuchi H (1996) Molecular cloning and characterization of a cDNA that encodes protoporphyrinogen oxidase of *Arabidopsis thaliana.* Gene 182:169–175
Nicolaus B, Johansen JN, Boger P (1995) Binding affinities of peroxidizing herbicides to protoporphyrinogen oxidase from corn. Pestic Biochem Biophys 51:20–29
Nishide H, Mihayashi K, Tsuchida E (1977) Dissociation of aggregated ferroheme complexes and protoporphyrin IX by water-soluble polymers. Biochim Biophys Acta 498:208–214
Nishimura K, Nakayashiki T, Inokuchi H (1995a) Cloning and identification of the *hemG* gene encoding protoporphyrinogen oxidase (PPO) of *Escherichia coli* K12. DNA Res 2:1–8
Nishimura K, Taketani S, Inokuchi H (1995b) Cloning of a human cDNA for protoporphyrinogen oxidase by complementation in vivo of a *hemG* mutant of *Escherichia coli.* J Biol Chem 270:8076–8080
O'Brian MR, Kirshbom PM, Maier RJ (1987a) Bacterial heme synthesis is required for expression of the leghemoglobin holoprotein but not the apoprotein in soybean root nodules. Proc Natl Acad Sci USA 84:8390–8393

O'Brian MR, Kirshbom PM, Maier RJ (1987b) Tn5-induced cytochrome mutants of *Bradyrhizobium japonicum*: effects of the mutations on cells grown symbiotically and in culture. J Bacteriol 169:1089–1094

O'Donnell V, Tew D, Jones O, England P (1993) Studies on the inhibitory mechanism of iodonium compounds with special reference to neutrophil NADPH oxidase. Biochem J 290:41–49

O'Donnell VB, Smith GC, Jones OT (1994) Involvement of phenyl radicals in iodonium inhibition of flavoenzymes. Mol Pharmacol 46:778–785

Porra RJ, Falk JE (1961) Protein-bound porphyrins associated with protoporphyrin biosynthesis. Biochem Biophys Res Commun 5:179–184

Poulson R (1976) The enzymic conversion of protoporphyrinogen IX to protoporphyrin IX in mammalian mitochondria. J Biol Chem 251:3730–3733

Poulson R, Polglase WJ (1975) The enzymic conversion of protoporphyrinogen IX to protoporphyrin IX. Protoporphyrinogen oxidase activity in mitochondrial extracts of *Saccharomyces cerevisiae*. J Biol Chem 250:1269–1274

Proulx KL, Dailey HA (1992) Characteristics of murine protoporphyrinogen oxidase. Prot Sci 6:801–809

Ragan CI, Bloxham DP (1977) Specific labelling of a constituent polypeptide of bovine heart mitochondrial reduced nicotinamide-adenine nucleotide-ubiquinone reductase by the inhibitor diphenylene iodonium. Biochem J 163:605–615

Ramseier TM, Kaluza B, Studer D, Gloudemans T, Bisseling T, Jordan PM, Jones RM, Zuber M, Hennecke H (1989) Cloning of a DNA region from *Bradyrhizobium japonicum* encoding pleiotropic functions in heme metabolism and respiration. Arch Microbiol 151:203–212

Ramseier TM, Winteler HV, Hennecke H (1991) Discovery and sequence analysis of bacterial genes involved in the biogenesis of *c*-type cytochromes. J Biol Chem 266:7793–7803

Roberts AG, Whatley SD, Daniels J, Holmans P, Fenton I, Owen MJ, Thompson P, Long C, Elder GH (1995) Partial characterization and assignment of the gene for protoporphyrinogen oxidase and variegate porphyria to human chromosome 1q23. Hum Mol Genet 4:2387–2390

Sandmann G, Böger P (1988) Accumulation of protoporphyrin IX in the presence of peroxidizing herbicides. Z Naturforsch 43c:699–704

Sano S, Granick S (1961) Mitochondrial coproporphyrinogen oxidase and protoporphyrin formation. J Biol Chem 236:1173–1180

Sano S, Ikeda K, Sakakibara S (1964a) Synthesis of cytochrome *c*-type core from protoporphyrinogen. Biochem Biophys Res Commun 15:284–289

Sano S, Nanzyo N, Rimington C (1964b) Synthesis of porphyrin *c*-type compounds from protoporphyrinogen. Biochem J 93:270–280

Sasarman A, Surdeanu M, Szegli G, Horodniceanu T, Greceanu V, Dumitrescu A (1968) Hemin-deficient mutants of *Escherichia coli* K12. J Bacteriol 96:570–572

Sasarman A, Sanderson KE, Surdeanu M, Sonea S (1970) Hemin-deficient mutants of *Salmonella typhimurium*. J Bacteriol 102:531–536

Sasarman A, Chartrand P, Lavoie M, Tardif D, Proschek R, Lapointe C (1979) Mapping of a new hem gene in *Escherichia coli* K12. J Gen Microbiol 113:297–303

Sasarman A, Letowski J, Czaika G, Ramirez V, Nead MA, Jacobs JM, Morais R (1993) Nucleotide sequence of the *hemG* gene involved in the protoporphyrinogen oxidase activity of *Escherichia coli* K12. Can J Microbiol 39:1155–1161

Siepker LJ, Kramer S (1985) Protoporphyrin accumulation by mitogen stimulated lymphocytes and protoporphyrinogen oxidase activity in patients with porphyria variegata and erythropoietic protoporphyria: evidence for deficiency of protoporphyrinogen oxidase and ferrochelatase in both diseases. Br J Haematol 60:65–74

Siepker LJ, Ford M, de Kock R, Kramer S (1987) Purification of bovine protoporphyrinogen oxidase: immunological cross-reactivity and structural relationship to ferrochelatase. Biochim Biophys Acta 913:349–358

Smith AG, Marsh O, Elder GH (1993) Investigation of the subcellular location of the tetrapyrrole-biosynthesis enzyme coproporphyrinogen oxidase in higher plants. Biochem J 292:503–508

Smith V, Chou KN, Lashkari D, Botstein D, Brown PO (1996) Functional analysis of the genes of the yeast chromosome V by genetic footprinting. Science 274:2069–2074

Stuehr DJ, Fasehun OA, Kwon NS, Gross SS, Gonzalez JA, Levi R, Nathan CF (1991) Inhibition of macrophage and endothelial cell nitric oxide synthase by diphenyleneiodonium and its analogs. Faseb J 5:98–103

Sugimura T, Okabe K, Nagao M, Gunge N (1966) A respiration-deficient mutant of *Saccharomyces cerevisiae* which accumulates porphyrins and lacks cytochromes. Biochem Biophys Res Commun 115:267–275

Taketani S, Inazawa J, Abe T, Furukawa T, Kohno H, Tokunaga R, Nishimura K, Inokuchi H (1995a) The human protoporphyrinogen oxidase gene (PPOX): organization and location to chromosome 1. Genomics 29:698–703

Taketani S, Yoshinaga T, Furukawa T, Kohno H, Tokunaga R, Nishimura K, Inokuchi H (1995b) Induction of terminal enzymes for heme biosynthesis during differentiation of mouse erythroleukemia cells. Eur J Biochem 230:760–765

Tew DG (1993) Inhibition of cytochrome P450 reductase by diphenyliodonium cation. Kinetic analysis and covalent modifications. Biochemistry (USA) 32:10209–10215

Thony-Meyer L (1997) Biogenesis of respiratory cytochromes in bacteria. Microbiol Mol Biol Rev 61:337–376

Troup B, Hungerer C, Jahn D (1995) Cloning and characterization of the *Escherichia coli hemN* gene encoding the oxygen-independent coproporphyrinogen III oxidase. J Bacteriol 177:3326–3331

Urban-Grimal D, Labbe-Bois R (1981) Genetic and biochemical characterization of mutants of *Saccharomyces cerevisiae* blocked in six different steps of heme biosynthesis. Mol Gen Genet 183:85–92

Varsano R, Matringe M, Magnin N, Mornet R, Scalla R (1990) Competitive interaction of three peroxidizing herbicides with the binding of [3H]acifluorfen to corn etioplast membranes. FEBS Lett 272:106–108

Viljoen DJ, Cummins R, Alexopoulos J, Kramer S (1983) Protoporphyrinogen oxidase and ferrochelatase in porphyria variegata. Eur J Clin Invest 13:283–287

Wakabayashi S, Kimura T, Fukuyama K, Matsubara H, Rogers LJ (1989) The amino acid sequence of a flavodoxin from the eukaryotic red alga *Chondrus crispus*. Biochem J 263:981–984

Walker CJ, Weinstein JD (1994) The magnesium-insertion step of chlorophyll biosynthesis is a two-stage reaction. Biochem J 299:277–284

Walker CJ, Yu GH, Weinstein JD (1997) Comparative study of heme and Mg-protoporphyrin (monomethyl ester) biosynthesis in isolated pea chloroplasts: effects of ATP and metal ions. Plant Physiol Biochem (Paris) 35:213–221

Ward ER, Volrath S (1995) Manipulation of protoporphyrinogen oxidase enzyme activity in eucaryotic organisms. International patent WO95/34659

Warnich L, Kotze MJ, Groenewald IM, Groenewald JZ, van Brakel MG, van Heerden CJ, de Villiers JN, van de Ven WJ, Schoenmakers EF, Taketani S, Retief AE (1996) Identification of three mutations and associated haplotypes in the protoporphyrinogen oxidase gene in South African families with variegate porphyria. Hum Mol Genet 5:981–984

Watson CJ, Pierach CA, Bossenmaier I, Cardinal R (1977) Postulated deficiency of hepatic heme and repair by hematin infusions in the "inducible" hepatic porphyrias. Proc Natl Acad Sci USA 74:2118–2120

Witkowski DA, Halling BP (1988) Accumulation of photodynamic tetrapyrroles induced by acifluorfen-methyl. Plant Physiol 87:632–637

Witkowski DA, Halling BP (1989) Inhibition of plant protoporphyrinogen oxidase by the herbicide acifluorfen-methyl. Plant Physiol 90:1239–1242

Xu K, and Elliott, T (1993) An oxygen-dependent coproporphyrinogen oxidase encoded by the hemF gene of *Salmonella typhimurium*. J Bacteriol 175:4990–4999

Xu K, Delling J, Elliott T (1992) The genes required for heme synthesis in *Salmonella typhimurium* include those encoding alternative functions for aerobic and anaerobic coproporphyrinogen oxidation. J Bacteriol 174: 3953–3963

Yamato S, Katagiri M, Ohkawa H (1994) Purification and characterization of a protoporphyrinogen-oxidizing enzyme with peroxidase activity and light-dependent herbicide resistance in tobacco cultured cells. Pestic Biochem Biophys 50:72–82

Yamato S, Ida T, Katagiri M, Ohkawa H (1995) A tobacco soluble protoporphyrinogen-oxidizing enzyme similar to plant peroxidases in their amino acid sequences and immunochemical reactivity. Biosci Biotechnol Biochem (1995) 59:558–559

Yang H, Sasarman A, Inokuchi H, Adler J (1996) Non-iron porphyrins cause tumbling to blue light by an *Escherichia coli* mutant defective in *hemG*. Proc Natl Acad Sci USA 93:2459–2463

CHAPTER 10

Assay Systems for Peroxidizing Herbicides

Hiroyuki Watanabe[1] and Tsutomu Shimizu[2]

Contents

10.1 Introduction

As shown in previous chapters, peroxidizing herbicides inhibit chlorophyll biosynthesis and induce lipid peroxidation, disruption of membranes, chlorophyll degradation, and desiccation of plants in the light. They primarily inhibit protoporphyrinogen oxidase (Protox), which leads to rapid accumulation of protoporphyrin IX (Proto IX), an intermediate of chlorophyll biosynthesis. Abnormal accumulation of Proto IX in plant tissue is assumed to cause light-induced formation of active oxygen which initiates the peroxidative chain reaction described previously (Böger and Sandmann 1990; Chapter 6).

In this chapter we present some examples of assay procedures to evaluate the effects of peroxidizing herbicides on Protox activity, Proto IX accumulation, and phytotoxic activity using some higher plants. In addition, we propose a simple assay system to screen active peroxidizers using unicellular green algae, *Scenedesmus*, for very short time in vitro culture.

[1] Yokohama Research Center, Mitsubishi Chemical Corporation, 1000 Kamoshida, Aoba-ku, Yokohama, Kanagawa 227-8502, Japan
[2] Life Science Research Institute, Kumiai Chemical Industry Co. Ltd., 3360 Kamo, Kikukawa-cho, Ogasa-gun, Shizuoka 439-0031, Japan

Peter Böger, Ko Wakabayashi
Peroxidizing Herbicides

10.2 Assay Systems Using Higher Plants

10.2.1 Materials and Methods

Plant Material. We selected five species of higher plants for this experiment: cotton (*Gossypium hirsutum* L. var. Cocher), corn (*Zea mays* L. var. Pioneer 3352), soybean [*Glycine max* (L.) Merr. var. Deltapine 506], velvetleaf (*Abutilon theophrasti* Medic) and cocklebur (*Xanthium strumarium* L.). These plants were germinated and grown on vermiculite in darkness for 5–7 days at 27 °C. Corn and soybean are tolerant to one of the peroxidizers, fluthiacet-methyl (Fig. 1), while cotton, velvetleaf and cocklebur are sensitive in field applications (Miyazawa et al. 1993).

Chemicals. Fluthiacet-methyl and the derivatives of this herbicide (Fig. 1) were used in this experiment. Other compounds were purchased from commercial sources as follows: oxadiazon [5-tert-butyl-3-(2,4-dichloro-5-isopropoxyphenyl)-1,3,4-oxadiazolin-2-one] and nitrofen (2,4-dichlorophenyl 4-nitrophenyl ether) from Wako Pure Chemical (Osaka, Japan), Proto IX from Sigma Chemical Co. (St. Louis, Missouri, USA); and a porphyrin acids chromatographic marker kit, which contained mesoporphyrin IX, coproporphyrin I, 5 carboxyl porphyrin, 6 carboxyl porphyrin, 7 carboxyl porphyrin and uroporphyrin I from Porphyrin Products (Logan, Utah, USA).

Measurement of Leakage of Electrolytes. Etiolated cotton seedlings, which had been grown as described above, were allowed to stand for 1 h in room light

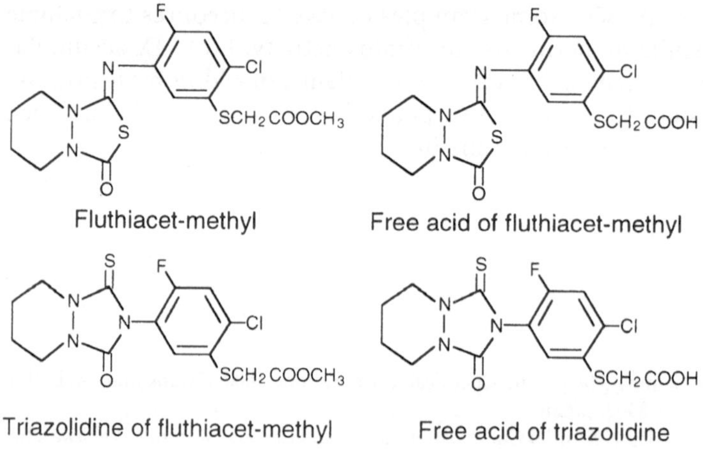

Fig. 1. Fluthiacet-methyl and its derivatives

(approx. 1500 lx) and then cotyledons were excised. Etiolated velvetleaf seedlings, also grown as described above, were placed in darkness for 8 h after removal of seed coats. Then they were placed for 1 h in room light and finally cotyledons were excised. Cotyledons (0.5–1.0 g fr. wt.) of these plants were incubated in 25 ml distilled water that contained a compound to be tested at 27 °C for appropriate periods of time under illumination from fluorescent lamps (15000–18200 lx). The conductivity of the solution was then measured with a conductivity meter (DS-14; Horiba, Kyoto, Japan).

Measurement of Chlorophyll Biosynthetic Activity. The biosynthesis of chlorophyll was monitored in terms of the greening of etiolated cotyledons of cotton and velvetleaf. Chlorophyll was extracted with 20 ml of 80% acetone from cotyledons (0.5–1.0 g fr. wt.) that had been incubated as described above for the measurement of electrolyte leakage. Total chlorophyll contents were determined by measuring the absorbance at 665 and 649 nm with a spectrophotometer (Model 220A, Hitachi, Tokyo, Japan) and fitting the data to the following equation (Comer and Zscheile 1942):

$$\text{Total chlorophyll (mg/ml)} = 6.45\ A_{665} + 17.72\ A_{649}\,.$$

Extraction and Analysis of Porphyrins. Porphyrins were extracted with basic acetone as described by Sandmann and Böger (1988) from etiolated cotton cotyledons that had been treated with fluthiacet-methyl. Etiolated cotton seedlings were sprayed with fluthiacet-methyl and allowed to stand for 48 h in darkness. Then, the cotyledons (20 g fr. wt.) were homogenized with 100 ml of a mixture of acetone/1 N NH_4OH/50 mM 2-hydroxyethylpiperazine-N′-2-ethanesulfonic acid (HEPES) buffer, pH 8 (15:2:2, v/v/v). The homogenate was centrifuged at 15000 × *g* for 20 min. After washing the supernatant three times with hexane, the acetone phase was diluted with water, brought to pH 2 by HCl and extracted three times with 100 ml diethyl ether. The diethyl ether phase was dried with anhydrous Na_2SO_4, evaporated to dryness in vacuo, and then dissolved in 5 ml dimethylsulfoxide (DMSO). The DMSO solution was filtered before porphyrin analysis. To prepare the porphyrin standard mixtures, one tube of porphyrin acid chromatographic marker kit was dissolved in 1 ml DMSO to yield a final concentration of 10 μM. Proto IX was dissolved in DMSO to a final concentration of 17.8 μM.

Porphyrins were analyzed as reported by Bonkovsky et al. (1986) by HPLC (Maxima 820 Chromatography Workstation equipped with Model 510 HPLC Pump, WISP710B Sample Processor, Model 420 Fluorescence Detector and Model 441 Absorbance Detector; Waters, Milford, Massachusetts, USA) on a NOVA-PAK C_{18} column (3.9 × 150 mm, Waters) using a gradient system. Solvent A was prepared as follows: 0.1 M ammonium phosphate solution was adjusted to pH 3.5 using concentrated phosphoric acid and filtered. Methanol was added such that the final volume ratio was 56 parts ammonium phosphate solution to 44 parts methanol. The pH of the final mixture was adjusted to 3.4 with phosphoric acid. Solvent B was pure methanol. A 16-min gradient was employed to increase solvent B from 30% to 100%. Solvent B was maintained at 30% for 2 min before gradient and at 100% for 7 min after gradient. A flow rate of 1 ml was maintained throughout. Sample injection volume was 30 μl. Porphyrins were detected by fluorescence above 530 nm (with excitation at 410–440 nm) and they were identified by co-chromatography with the authentic standards.

Assay of Protox Inhibition. Protoporphyrinogen IX (Protogen IX), the substrate of Protox reaction, was prepared by reducing Proto IX with sodium amalgam by the method reported by Jacobs and Jacobs (1982) with slight modification. Two kinds of reducing agent were used in order to avoid autooxidation of Protogen IX. One was dithiothreitol (DTT), as in the standard method, and the other was sodium isoascorbate. Thus, 9 mg Proto IX was dissolved in 10 mM KOH that contained 20% (v/v) ethanol and 8 ml this solution was brought to 20 ml with 10 mM KOH. The solution was bubbled with N_2 gas to remove oxygen and reduced by addition of 25 g of freshly ground sodium amalgam. The solution was filtered through glass wool under a stream of N_2 gas, and the filtrate was supplemented with 18 ml tris(hydroxymethyl) aminomethane (Tris-HCl) buffer, pH 8.7, that contained 1 mM ethylenediaminetetraacetic acid (EDTA) and 4 mM DTT [or 1.4% (w/v) sodium isoascorbate]. The solution was passed through a membrane filter (Millex-GS; Millipore, Bedford, Massachusetts, USA), and then adjusted to pH 8.5–8.8 with 40% H_3PO_4, which had been bubbled with N_2 gas. This solution was stable for at least 1 week at −80 °C. All operations were conducted under a dim light.

Etioplasts were prepared from the etiolated cotyledons or leaves of corn, cotton, soybean, velvetleaf and cocklebur, which turned slightly green upon standing for approximately 3 h in room light. Corn leaves and soybean cotyledons were ground with five volumes of 30 mM HEPES-KOH buffer (pH 7.8) containing 0.5 M sucrose, 1 mM DTT, 1 mM $MgCl_2$, 1 mM EDTA and 0.2% (w/v) bovine serum albumin (BSA) in an ice-cold mortar. Cotyledons of cotton, velvetleaf and cocklebur were homogenized with five volumes of 30 mM HEPES-KOH buffer (pH 7.8) containing 0.5 M sucrose, 5 mM DTT, 1 mM $MgCl_2$, 1 mM EDTA, 0.2% (w/v) BSA, 0.5 mM KCN and 25% (w/fr. wt. of tissues) polyvinylpolypyrrolidone. The homogenate was squeezed through one layer of nylon gauze and centrifuged at 300 × *g* for 5 min. The supernatant was centrifuged at 10000 × *g* for 1 min and the precipitate was suspended in 30 mM HEPES-KOH buffer (pH 7.8) containing 0.5 M sucrose, 1 mM DTT, 1 mM $MgCl_2$, 1 mM EDTA and 0.2% (w/v) BSA. The suspension was centrifuged at 150 × *g* for 5 min and then the supernatant at 2000 × *g* for 5 min. The resultant precipitate was suspended in the above buffer and the suspension was centrifuged at 500 × *g* for 20 min. The precipitate thus obtained was suspended in the same buffer and used for the experiments as etioplasts. The etioplasts were sonicated with 0.5% (v/v) Tween 20, just before assays of enzymatic activity.

The assay of Protox activity was carried out basically as described by Jacobs and Jacobs (1982), but two reducing agents, DTT and sodium isoascorbate, were also used in this enzyme reaction. The reaction mixture (1 ml) contained 100 mM Tris-HCl buffer (pH 8.0), 1 mM EDTA, 4 mM DTT or 5 mM sodium isoascorbate, 60 μM Protogen IX and etioplasts. Chemical compounds were added to the mixtures as acetone solutions. The final concentrations of acetone were always kept below 1% (v/v). The reaction was initiated by the addition of etioplasts. After incubation for 60 min at 30 °C in darkness, a portion of the mixture (25–100 μl) was transferred to 2.9–2.975 ml 100 mM Tris-HCl buffer (pH 8.7) that contained 1 mM EDTA, 5 mM DTT and 1% (v/v) Tween 80, and then fluorescence at 630 nm (with excitation at 410 nm) was immediately measured with a fluorescence spectrophotometer (Model 650–60, Hitachi, Tokyo, Japan). Heat-denatured etioplasts were used in the control experiments.

10.2.2
Results

Leakage of Electrolytes. The leakage of electrolytes from both etiolated and green cotyledons of cotton that had been treated with fluthiacet-methyl at 27 °C and 15 000 lx occurred after a delay of approximately 12 h (Fig. 2). The extent of the leakage increased as the concentration of the compound was raised. The leakage from etiolated velvetleaf cotyledons treated with fluthiacet-methyl was also increased with increasing concentrations of the compound

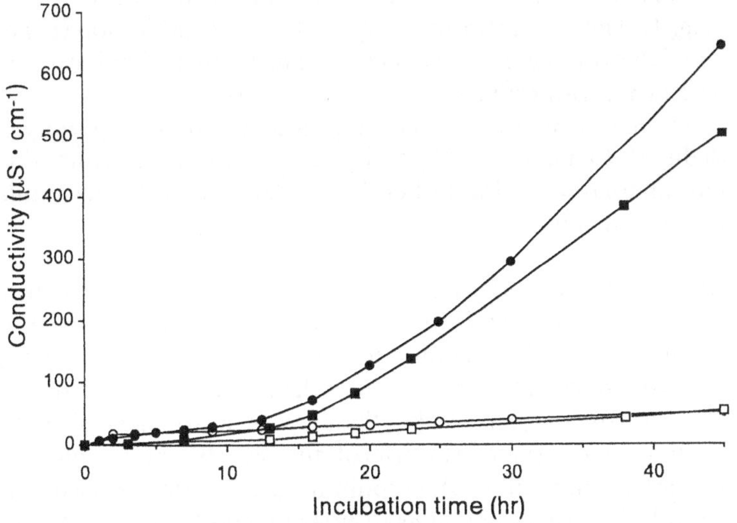

Fig. 2. Leakage of electrolytes from cotton cotyledons treated with fluthiacet-methyl (15 000 lx, 27 °C). ● Etiolated cotyledons treated with fluthiacet-methyl; ○ etiolated cotyledons without herbicide; ■ green cotyledons treated with fluthiacet-methyl; □ green cotyledons without herbicide fluthiacet-methyl (15 000 lx, 27 °C). ● Etiolated cotyledons treated with fluthiacet-methyl; ○ etiolated cotyledons without herbicide; ■ green cotyledons treated with fluthiacet-methyl; □ green cotyledons without herbicide

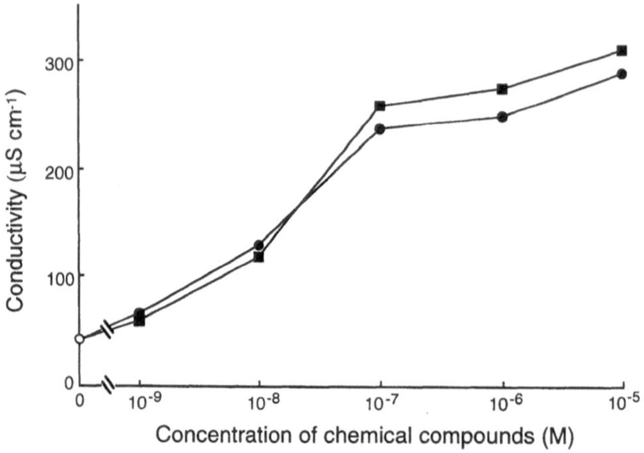

Fig. 3. Leakage of electrolytes from etiolated velvetleaf cotyledons treated with fluthiacet-methyl and its triazolidine (18 200 lx, 27 °C, 18-h incubation) ● fluthiacet-methyl; ■ triazolidine

(Fig. 3), with 17 nM being the concentration required for 50% of the maximum leakage (L_{50}). Nearly the same value of L_{50} (20 nM) was obtained with the triazolidine of fluthiacet-methyl.

Inhibition of Chlorophyll Biosynthesis and Accumulation of Porphyrins. The chlorophyll contents of etiolated cotyledons of cotton and velvetleaf were significantly reduced by treatment with fluthiacet-methyl or its triazolidine (Fig. 4). The concentrations required for 50% inhibition (I_{50}) of the biosynthesis of chlorophyll in velvetleaf by fluthiacet-methyl and its triazolidine were 10 and 12 nM, respectively.

Fluthiacet-methyl induced the accumulation of porphyrins, which were deduced to be Proto IX and mesoporphyrin IX from the results of co-chromatography with authentic standards, and the accumulation was dose-dependent (Fig. 5).

Inhibition of Protox Activity. The Protox activity of corn was inhibited by fluthiacet-methyl and its derivatives. The strength of the inhibition by fluthiacet-methyl was similar to that by oxadiazon and rather greater than that by nitrofen when DTT was used as the reducing agent. However, inhibition by fluthiacet-methyl was weaker than that by nitrofen when DTT was replaced by sodium isoascorbate. The triazolidine was the most potent inhibitor among the tested compounds when sodium isoascorbate was used. The free acid of the triazolidine was as potent as oxadiazon, and the free acid of fluthiacet-methyl was almost as potent as fluthiacet-methyl when sodium isoascorbate was used as the reducing agent (Table 1).

The inhibitory potencies of the triazolidine and the free acid of the triazolidine were unchanged when we tested enzymes from different sources,

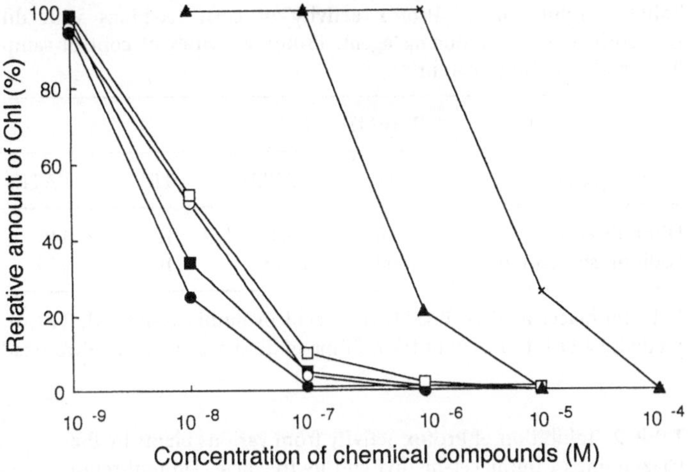

Fig. 4. Chlorophyll (*Chl*) content of cotyledons treated with fluthiacet-methyl and its triazolidine. Chlorophyll was extracted after a 24-h incubation at 27 °C and 18 200 lx. Total chlorophyll content of cotyledons of the control samples of cotton and velvetleaf was 380 and 330 µg per g tissue, respectively. ● Cotton treated with fluthiacet-methyl; ■ cotton treated with triazolidine; ▲ cotton treated with oxadiazon; × cotton treated with nitrofen; ○ velvetleaf treated with fluthiacet-methyl; □ velvetleaf treated with triazolidine

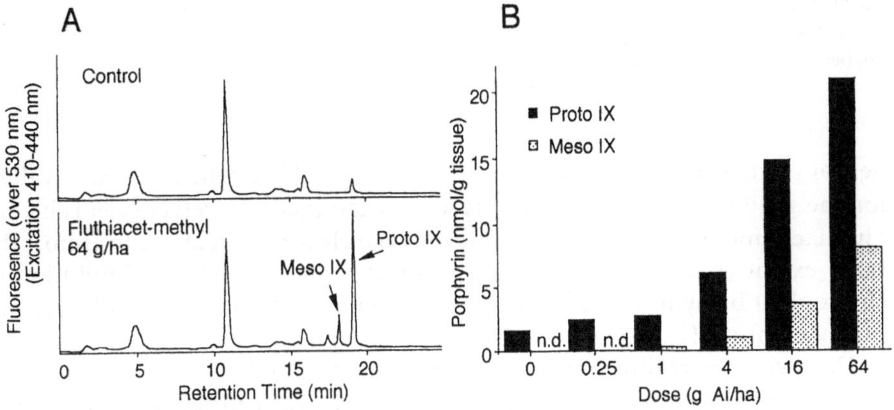

Fig. 5. Dose-response relationship for the accumulation of porphyrins in etiolated cotton cotyledons treated with fluthiacet-methyl. Etiolated cotton seedlings were sprayed with fluthiacet-methyl at doses described in the figure and allowed to stand for 48 h in darkness. Then, porphyrins were extracted from cotyledons as described in the text. *Black bars* Proto IX; *gray bars* mesoporphyrin IX; *n.d.* not detected

indicating that there was no difference in the sensitivity of the Protox from different plants to these two inhibitors (Table 2).

Correlations Among the Leakage of Electrolytes, Inhibition of Chlorophyll Biosynthesis and Inhibition of Protox Activity. Correlations among L_{50} val-

Table 1. Inhibition of Protox activity of corn seedlings with dithiothreitol or sodium isoascorbate as the reducing agent. Protox activities of control samples ranged from 0.3 to 0.49 nmol $min^{-1}mg$ $protein^{-1}$

	I_{50}(nM)					
Reducing agent	FLU	FAFLU	TRI	FATRI	OXA	NIP
Dithiothreitol	110	350	12	110	130	1000
Sodium isoascorbate	1500	2000	10	120	120	840

FLU, fluthiacet-methyl; FAFLU, free acid of fluthiacet-methyl; TRI, triazolidine of fluthiacet-methyl; FATRI, free acid of triazolidine of fluthiacet-methyl; OXA, oxadiazon; NIP, nitrofen.

Table 2. Inhibition of Protox activity from various plants by the triazolidine of fluthiacet-methyl and its free acid. Dithiothreitol was used as the reducing agent

	I_{50}(nM)	
Enzyme source	Triazolidine	Free acid
Cotton	11	66
Velvetleaf	5.4	100
Cocklebur	5.1	100
Corn	9.1	110
Soybean	8.1	66

ues for electrolyte leakage, I_{50} values for chlorophyll biosynthesis and I_{50} values for the inhibition of Protox activity were examined for velvetleaf (Table 3). Fluthiacet-methyl and its triazolidine induced electrolyte leakage to almost the same extent (L_{50}: 17 and 20 nM, respectively). The potency for inhibition of chlorophyll biosynthesis of fluthiacet-methyl and that of its triazolidine was nearly identical (I_{50}: 10 and 12 nM, respectively).

However, the abilities of these compounds to inhibit Protox activity were very different. The I_{50} value of fluthiacet-methyl was 830 nM, whereas that of the triazolidine was 5.4 nM, which was close to the values of L_{50} and I_{50} for the inhibition of chlorophyll biosynthesis. In contrast, the values of L_{50} and I_{50} of the free acid of the triazolidine for leakage, inhibition of chlorophyll biosynthesis and inhibition of Protox activity were 310, 220 and 100 nM, respectively. These results indicated that the effects in vivo (electrolyte leakage and the inhibition of chlorophyll biosynthesis) of fluthiacet-methyl were correlated with the effects in vitro (the inhibitions of Protox activity) of neither fluthiacet-methyl itself nor the free acid of the triazolidine, but they were correlated with the effects in vitro of the triazolidine. This result led to the finding that fluthiacet-methyl inhibits Protox activity after conversion to the corresponding triazolidine by glutathione S-transferase (Shimizu et al. 1995).

Table 3. I_{50} values of fluthiacet-methyl and its derivatives for inhibition of Protox activity and of chlorophyll biosynthesis, and L_{50} values for electrolyte leakage from velvetleaf

	I_{50} (nM)		
Compound	Protox	Chlorophyll	L_{50} (nM)
Fluthiacet-methyl	830	10	17
Triazolidine of fluthiacet-methyl	5.4	12	20
Free acid of triazolidine of fluthiacet-methyl	100	220	310

10.3 Assay Systems Using Green Algae

10.3.1 Materials and Methods

Algal Cultivation. The strain of unicellular green algae, *Scenedesmus acutus*, was no. 276-3a of the Algae Collection, University of Göttingen, Germany. The mineral medium, as given by Böger and Nicolaus (1993), was sterilized and inoculated with a cell density of 2 µl packed cell volume (pcv) ml^{-1} culture suspension. The cells started the exponential growth just after inoculation and attained a cell density of 5 µl pcv ml^{-1} for about 24 h. The sterile cultivation was carried out in a thermostated waterbath at 22 °C in closed glass culture vessels (4 × 40 cm), which could be gassed with sterile CO_2-enriched air (4% v/v). Gassing was through an inlet reaching to the bottom of the vessels, and stirred the culture suspension and prevented cell sedimentation. Light was supplied with white fluorescent lamps at the intensity of approximately 80 µmol $m^{-2}s^{-1}$, equivalent to about 6000 lx.

Chemicals. Chlorophthalim, N-(4-chlorophenyl)-3,4,5,6-tetrahydrophthalimide (shown in Fig. 6), and other 16 cyclic imide compounds (see Watanabe et al. 1993) were prepared by the method reported by Ohta et al. (1980). Oxyfluorfen, 2-chloro-1-(3-ethoxy-4-nitrophenoxy)-4-(trifluoromethyl)benzene, was purchased from Rohm and Haas, Philadelphia, Pennsylvania.

Determination of Short-Term Accumulation of Proto IX. After 24-h cultivation, chlorophthalim or an other peroxidizing compound was added in *Scenedesmus* cell suspensions at their exponential growth stage, and Proto IX in the cells was then determined after a 1-h treatment or at the time indicated. The compound was dissolved in ethanol, keeping the final ethanol content below 0.1% (v/v) in the culture medium. A 50-ml volume of the cell suspension was quickly centrifuged at 2500 × *g* for 5 min at about 6 °C in the dark, and

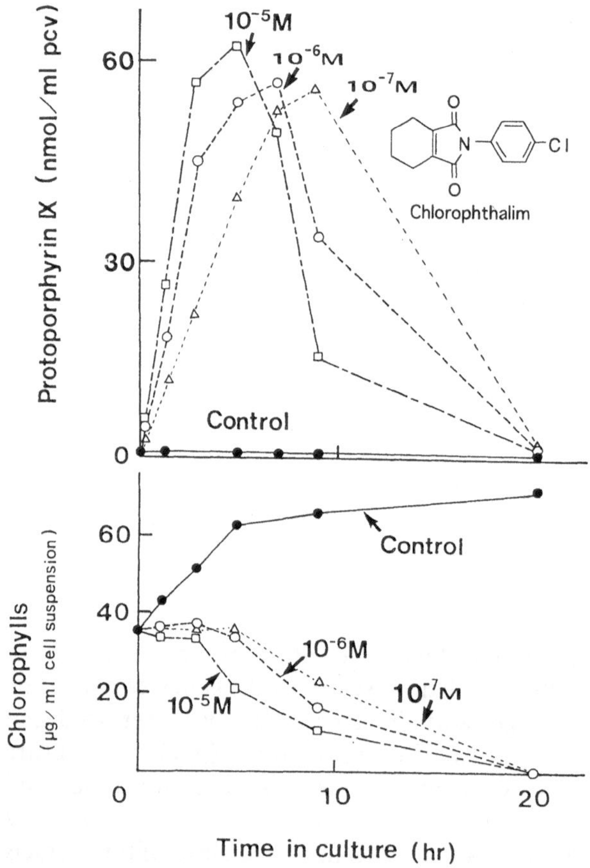

Fig. 6. Time-dependent changes in Proto IX and chlorophyll content in *Scenedesmus* cells. Chlorophthalim with the concentrations indicated was administered to the cell suspension 24 h after inoculation. Note that after approximately 5–6 h Proto IX decreases, as does chlorophyll, apparently due to photooxidation. Lower figure also demonstrates that peroxidizing herbicides cause an immediate halt of chlorophyll biosynthesis

approximately 250 μl pcv of *Scenedesmus* cells was used for Proto IX determination.

Extraction and determination of Proto IX was performed according to the method of Ho et al. (1987) with some modifications. After washing twice with 25 ml ice-cold distilled water the cells were extracted twice with 2 ml of the extraction mixture containing methanol, tetrahydrofuran, and a 5-mM aqueous solution of trifluoroacetic acid (30:16:5, v/v/v) at 55 °C for 10 min with shaking in the dark. After removal of the precipitate by centrifugation, Proto IX in the dark-green extract is determined by HPLC System (Model D-6000, Hitachi, Tokyo, Japan) which is composed of an F-1050 Fluorescent Spectrophotometer, L-4200 UV-VIS Detector, L-6200 Intelligent Pump, 655A-40 Autosampler and D-6000 HPLC Manager. The column used was Senshu Pak C_6H_5-1252-N (250 × 4.5 mm inner diameter; Senshu Scientific Co., Tokyo, Japan) with a Guard Pak CN pre-column (Waters Associate, Milford, Connecticut, USA). The eluent was a mixture of methanol, tetrahydrofuran, and a 5-mM aqueous solution of trifluoroacetic acid (24:13:13, v/v/v) with a flow rate of 1 ml min^{-1}. Proto IX concentration was measured by fluorescence using the excitation wavelength of 405 nm and the emission wavelength of 633 nm. For calibration, an authentic inner standard of Proto IX from Sigma Chemical Co., St. Louis, Missouri, USA, was used.

Determination of Chlorophyll Content and Cell Growth. Total chlorophyll content was determined by the measurement of optical density of the methanol extract at 650 and 665 nm. A 2-ml volume of the cell suspension was centrifuged at 2500 × *g* for 10 min and washed once with ice-cold distilled water. Chlorophyll was extracted by incubation of the pellet with 5 ml of methanol at 65 °C for 10 min. The total amount of chlorophyll was calculated by the following equation:

$$\text{Total chlorophyll } (\mu\text{g/ml algal suspension}) = (25.5 \times A_{650} + 4.03 A_{665}) \times C$$

where C is the dilution constant (volume of methanol/algal cell suspension); C = 2.5 in this experiment.

Cell growth was estimated by the changes of the packed cell volume of *Scenedesmus.* Packed cell volume was measured by centrifugation of 2 ml cell suspension at 500 × *g* for 5 min with a special graduated microcentrifuge tube, which was purchased from Kummer Co., Freiburg, Germany.

The pI_{50} values (log 1/molar I_{50} values) were used in this experiment to quantitate and discuss the influence of compounds on chlorophyll decrease or cell growth.

Herbicidal Activity of Peroxidizing Compounds. The root growth inhibition of peroxidizing compounds against sawa millet (*Echinochloa utilis* L.) was determined using the petri dish test described previously (Ohta et al. 1976). Each dish with 30 seeds was kept in a humid glass chamber at a conditioning temperature (20–25 °C) and 60% relative humidity for 7 days: for the first 48 h in the dark (20 °C), followed by a 5-days' light-dark regime, namely daylight for 12 h (25 °C, the light intensity of 30000 lx) and darkness for 12 h (20 °C). The molar I_{50} values of the compounds, tested against the growth of the root relative to the control, were estimated from the dose-response relationship by means of a probit analysis. A good correlation was reported between the pI_{50} values and the herbicidal activity of peroxidizing compounds against various kinds of weeds (Wakabayashi et al. 1979).

10.3.2 Results

Time and Dose Dependency of Proto IX Accumulation. Time-dependent changes in Proto IX and chlorophyll content were shown in Fig. 6. Proto IX accumulated in the cells after chlorophthalim application (0.1–10 µM) and reached the maximum level after 5–10 h. The initial velocity of Proto IX accumulation clearly depended on chlorophthalim concentration, but apparently the maximum level attained eventually was not dependent on the concentration. After the maximum accumulation, Proto IX rapidly decreased concurrently with chlorophyll degradation.

Exact 1-h application of chlorophthalim dose-dependently accumulated Proto IX in *Scenedesmus* cells (data were not shown). The 1-h incubation with *Scenedesmus* cell suspension is the minimum period in which to take samples and to reliably determine the Proto IX content. It was suggested that the short-term accumulation of Proto IX just after herbicide application is related to herbicide activity of chlorophthalim.

Relationship Between Short-Term Accumulation of Proto IX and Herbicidal Activity. We determined short-term accumulation of Proto IX in *Scenedesmus* cells exactly 1 h after incubation with 17 cyclic imide compounds and the diphenyl ether oxyflurfen at 1 μM concentration, and also chlorophyll decrease (pI_{50}) after 20-h application of the compounds, and the root growth inhibition (pI_{50}) of sawa millet against the compounds (see Watanabe et al. 1992 for the results in detail). The quantitative correlations between each parameter of determinations were expressed as the regression equations given in Table 4. The decreases in chlorophyll (bleaching effects) and root growth inhibition of sawa millet by these compounds were well correlated with the 1-h accumulation of Proto IX. No quantitative correlation was found with long-term Proto IX accumulation.

We concluded that the measurements of short-term accumulation of Proto IX in *Scenedesmus* cells represented reliable and quantitative indicators for phytotoxicity of peroxidizing herbicides, and also could be used for simple and convenient screenings for active compounds with small handling for short time.

Table 4. Quantitative correlation between short-term Proto IX accumulation and other phytotoxic parameters (from Watanabe et al. 1992)

(1) pI_{50} (chlorophyll) = 0.040 Proto IX + 4.918
(±0.008) (±0.421)
n = 18, r = 0.936, F = 113.7, s = 0.495

(2) pI_{50} (growth) = 0.039 Proto IX + 4.884
(±0.009) (±0.456)
n = 18, r = 0.920, F = 88.1, s = 0.536

(3) pI_{50} (*Echinochloa*) = 0.032 Proto IX + 3.830
(±0.013) (±0.701)
n = 18, r = 0.784, F = 25.5, s = 0.824

(1) and (2) are from *Scenedesmus acutus* cultures; (3) is root growth inhibition of sawa millet (*Echinochloa utilis*). Data in parentheses are the 95% confidence intervals.

References

Böger P, Nicolaus B (1993) Ethane formation by peroxidizing herbicides. In: Böger P, Sandmann G (eds) Target assays for modern herbicides and related phytotoxic compounds, chap 8. Lewis Publishers/CRC Press, Boca Raton, pp 51–60

Böger P, Sandmann G (1990) Modern herbicides affecting typical plant processes. In: Bowers WS, Ebing W, Martin D, Wegler R (eds) Chemistry of plant protection, vol 6. Springer, Berlin Heidelberg New York, pp 173

Bonkovsky HL, Wood SG, Howell SK, Sinclair PR, Lincoln B, Healey JF, Sinclair JF (1986) High-performance liquid chromatographic separation and quantitation of tetrapyrroles from biological materials. Anal Biochem 155:56–64

Comer CL, Zscheile FP (1942) Analysis of plant extracts for chlorophylls *a* and *b* by photoelectric spectrophotometric method. Plant Physiol 17:198–209

Ho J, Guthrie R, Tieckelmann H (1987) Quantitative determination of porphyrins, their precursors and zinc protoporphyrin in whole blood and dried blood by high-performance liquid chromatography with fluorometric detection. J Chromatogr (Biomed Appl) 417:269–275

Jacobs NJ, Jacobs JM (1982) Assay for enzymatic protoporphyrinogen oxidation, a late step in heme synthesis. Enzyme 28:206–219

Miyazawa T, Kawano K, Shigematsu S, Yamaguchi M, Matsunari K (1993) KIH-9201, a new low-rate post-emergence herbicide for maize (*Zea mays*) and soybean (*Glycine max*). In: Proceedings of Brighton Crop Protection Conference, Weeds, The British Crop Protection Council, pp 23–28

Ohta H, Suzuki S, Watanabe H, Jikihara T, Matsuya K, Wakabayashi K (1976) Structure-activity relationship of cyclic imide herbicides. Agric Biol Chem 40:745–752

Ohta H, Jikihara T, Wakabayashi K, Fujita T (1980) Quantitative structure-activity study of herbicidal N-acryl-3,4,5,6- tetrahydrophthalimides and related cyclic imides. Pestic Biochem Physiol 14:153–159

Sandmann G, Böger P (1988) Accumulation of porphyrin IX in the presence of peroxidizing herbicides. Z Naturforsch 43c:699–704

Shimizu T, Hashimoto N, Nakayama I, Nakao T, Mizutani H, Unai T, Yamaguchi M, Abe H (1995) A novel isotriazolidine herbicide, fluthiacet-methyl, is a potent inhibitor of protoporphyrinogen oxidase after isomerization by glutathione S-transferase. Plant Cell Physiol 36:625–632

Wakabayashi K, Matsuya K, Ohta H, Jikihara T (1979) Structure-activity relationship of cyclic imide herbicides. In: Geissbuehler H (eds) Advances in pesticide science, part 2. Pergamon Press, Oxford, pp 256–260

Watanabe H, Ohori Y, Sandmann G, Wakabayashi K, Böger P (1992) Quantitative correlation between short-term accumulation of protoporphyrin IX and peroxidative activity of cyclic imides. Pestic Biochem Physiol 42:99–109

Watanabe H, Sandmann G, Wakabayashi K, Böger P (1993) Short-term formation of protoporphyrin IX in *Scenedesmus* using cyclic imide analogues. In: Böger P, Sandmann G (eds) Target assays for modern herbicides and related phytotoxic compounds, chap 7. Lewis Publishers/CRC Press, Boca Raton, pp 43–49

CHAPTER 11

Herbicidal Efficacy of Protoporphyrinogen Oxidase Inhibitors

EIKI NAGANO[1]

Contents

11.1 Introduction

Compounds which inhibit protoporphyrinogen oxidase (Protox) were known as "photobleaching herbicides" before their site of action was discovered. Photobleaching herbicides cause very strong bleaching of the treated part of higher plants. It was known that a photobleaching herbicide requires oxygen and light to express its herbicidal activity. After intensive investigations on the mode of action of photobleaching herbicide, protoporphyrinogen oxidase has been identified as the molecular target. It is generally accepted that the inhibition of Protox and oxidation of protoporphyrinogen leads to the accumulation of a strong photosensitizer, namely protoporphyrin IX (PP IX), followed by activation of oxygen and lipid peroxidation (Duke et al. 1993). This mechanism explains the necessity of light and oxygen. After discovery of the molecular target of photobleaching herbicides, several chemical structures were known to inhibit Protox. It is possible to classify Protox inhibitors into several groups according to their chemical structure.

The typical symptom of their activity by foliar treatment is shown as follows. Bleaching in the treated part is detected first within 24h, and the bleached area often changes to a brown color in a few days after treatment. In soil treatment, inhibition of germination is observed at a very high dosage. At lower dosages usually seedling growth inhibition and seedling browning are typical symptoms.

[1] Agricultural Chemicals Research Laboratory, Sumitomo Chemical Co. Ltd., 4-2-1 Takatukasa, Takarazuka, Hygo 665-0051, Japan

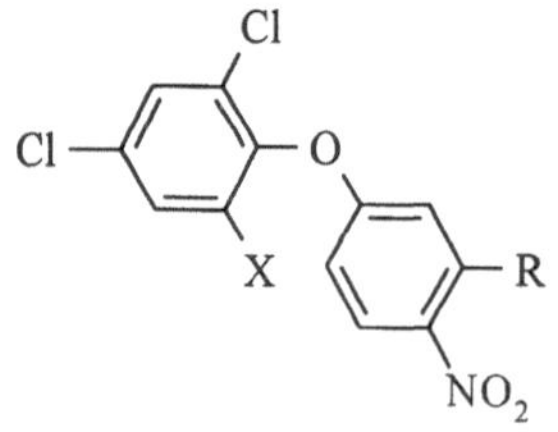

R = H, X = H (nitrofen)
R = H, X = Cl (chloronitrofen)
R = OCH_3, X = H (chloromethoxyfen)
R = $COOCH_3$, X = H (bifenox)

Fig. 1. Diphenyl ether rice herbicides

Generally, annual broad-leaf weeds are sensitive to Protox inhibitors. Perennial broad-leaf weeds and grasses are less sensitive in both pre-emergence and post-emergence treatment.

Since the chemical structure is the important factor to determine the biological characteristics, significant differences in the biological activity such as weed control spectrum have been observed among Protox inhibitors belonging to different chemical classes. In this chapter, we will discuss the difference in herbicidal activity among Protox inhibitors with emphasis on their chemical structures.

11.2 Diphenyl Ether Herbicides

The introduction of several diphenyl ethers, shown in Fig. 1, in transplanted rice was the first success of Protox inhibitors from a commercial standpoint. Nitrofen was introduced first for controlling barnyardgrass (*Echinochloa oryzicola*) and broad-leaf weeds, followed by chloronitrofen, chloromethoxyfen and bifenox with some improved efficacy and rice selectivity.

Oxyfluorfen was the first one introduced in upland fields. The efficacy was non-selective in foliar treatment, and selective to soybean and cotton in soil treatment. Many field tests in various crops have been conducted, which are not discussed in this chapter.

Acifluorfen was the first commercial Protox inhibitor showing crop selectivity in post-emergence treatment. Although the strong bleaching was detected sometimes in treated soybean leaves in the greenhouse test, this type of phytotoxicity has become acceptable to farmers. Many field tests showed that the recovery of treated crops is very fast, and the phytotoxicity of treated leaves does not affect the yield. I guess the phytotoxicity caused by acifluorfen indicated the acceptable level in the development of the following Protox inhibitors (see Fig. 2). The success of acifluorfen was followed by analogous herbicides, fomesafen and lactofen.

Although the chemical structures of the three compounds shown in Fig. 2 are similar, significant difference in the weed control spectrum was reported. Excellent control against jimsonweed (*Datura stramonium*), redroot pigweed (*Amaranthus retroflexus*), hemp sesbania (*Sesbania exaltata*) and black night-

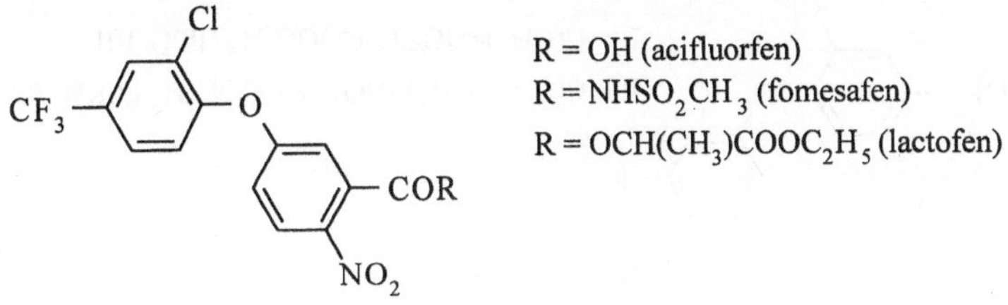

Fig. 2. Phenoxybenzoic acid analogues

shade (*Solanum nigrum*) was observed with a common use rate of acifluorfen, fomesafen and lactofen, and excellent control against pitted morningglory (*Ipomoea lacuanos*) and good control against ivyleaf morningglory (*Ipomoea hederacea*) with acifluorfen and fomesafen, but lactofen provided only good control against pitted morningglory (*Ipomoea lacuanos*) (Mathis and Oliver 1980; Higgins et al. 1988). Good velvetleaf (*Abutilon theophrasti*) and cocklebur (*Xanthium pensylvanicum*) control was obtained with acifluorfen when applied to weeds below the three leaf stage (Glenn et al. 1985). Excellent common sunflower (*Helianthus annus*) control was provided with lactofen, in contrast to fomesafen which provided poor control (Engelken et al. 1987). Lactofen is a more lipophilic compound than acifluorfen and fomesafen, because lactofen does not have an acidic proton. The difference of the weed control spectrum of lactofen is thought to be strongly affected by the relatively lipophilic and neutral property. Fomesafen showed better selectivity to soybean than the other two herbicides (Higgins et al. 1988b). A mechanism of soybean selectivity was reported; the rapid nucleophilic migration of diphenyl ether by homoglutathione and slower uptake were the important factors (Frear et al. 1983; Evans et al. 1987). The mechanism of the difference of weed control spectrum is thought to be complicated. Some reports showed that the different leaf surface causes the difference in uptake in less lipophilic acifluorfen and fomesafen (Higgins et al. 1988a,b).

In contrast to post-emergence treatment, these selective herbicides provide practically insufficient weed control in pre-emergence treatment at the same dosage. More than a several-fold higher dosage of acifluorfen or fomesafen is required in soil treatment for effective weed control even in the greenhouse test, and controllable weeds are limited to small seed broad-leaf weeds with soybean and corn selectivity.

Two diphenyl ether families were reported as the soybean selective post-emergence herbicide. One is the phenoxybenzoic acid family including acifluorfen and fomesafen mentioned above and the other is the phenoxyacetophenonoxime family. The later chemical class includes PPG-1013 and AKH-7088 (Fig. 3). As reported by Hayashi (1990), AKH-7088 provided excellent velvetleaf (*Abutilon theophrasti*) and redroot pigweed (*Amaranthus retroflexus*) control with acceptable soybean and corn injury. In the greenhouse, PPG-1013 showed better control of many broad-leaf weeds than AKH-

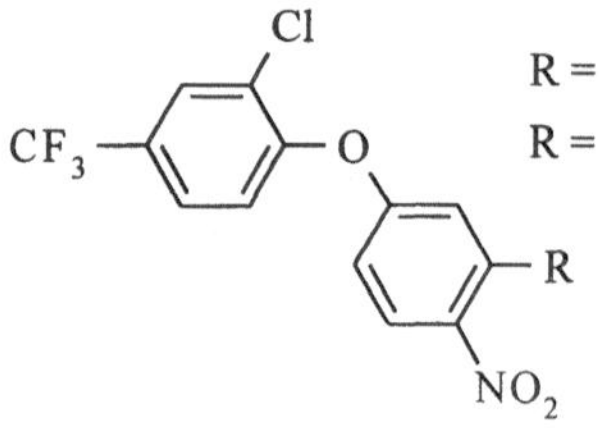

R = $C(CH_3)$=$NOCH_2COOC_2H_5$ (PPG-1013)
R = $C(CH_2OCH_3)$=$NOCH_2COOC_2H_5$ (AKH-7088)

Fig. 3. Phenoxybenzoxime herbicides

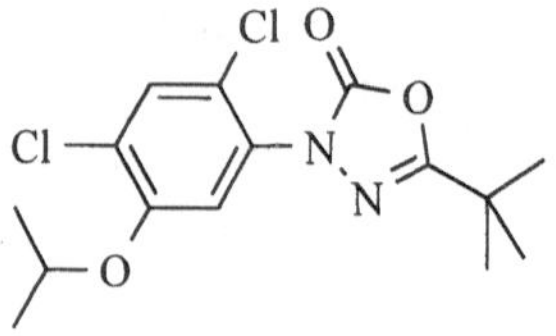

Fig. 4. Structure of oxadiazon

7088 but with stronger phytotoxicity (Hayashi 1987). The lipophilicity of these two herbicide is close to that of flumiclorac pentyl mentioned below in Section 11.3 and they resemble each other in the weed control spectrum with one clear exception, namely smartweed (*Polygonum spp.*). The mechanism of soybean selectivity of AKH-7088 has been discussed by Kouji et al. (1990). Resistant soybean absorbed the compound faster than sensitive velvetleaf, and metabolized slower. While soybean metabolized into inactive compounds, velvetleaf produced metabolite-active acid analogues. These explanations cannot explain the selective difference between soybean and velvetleaf sufficiently, because the acid is not stronger in phytotoxicity than AKH-7088, and larger amounts of AKH-7088 remained in soybean leaves. Like the phenoxybenzoic acid family, the herbicidal control spectrum of the phenoxyacetophenonoxime family is limited to small seed broad-leaf weeds in pre-emergence treatment.

11.3 N-Phenyl Heterocyclic Compounds

Oxadiazon (Fig. 4) was the first commercial N-phenyl heterocyclic compound in this chemical class. In our greenhouse test with upland field conditions, oxadiazon at 500 g a.i./ha provided good control of velvetleaf (*Abutilon theophrasti*), smartweed (*Polygonum spp.*), lambsquarter (*Chenopodium album*), and redroot pigweed (*Amaranthus retroflexus*) with marginal effects on soybean, rice, and cotton phytotoxicity in pre-emergence treatment.

Discovery of phenyl heterocycles with the phenyl group substituted by a fluorine atom in 2-position, a chlorine in 4-position and alkoxy group in 5-position opened the way to the screening of a new generation of Protox inhibitors. Flumiclorac pentyl discovered in our laboratory is the first one in

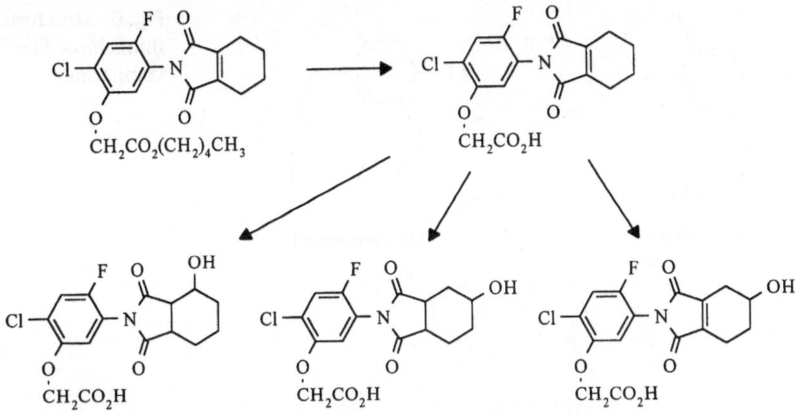

Fig. 5. Metabolic pathway of flumiclorac pentyl in soybean

this area. Excellent velvetleaf (*Abutilon theophrasti*), black nightshade (*Solanum nigrum*), pigweeds (*Amaranthus spp.*), ragweed (*Ambrosia artemisiifolia*) and prickly sida (*Sida spinosa*) control was obtained at 30 g a.i./ha with flumiclorac pentyl (Kamoshita et al. 1992; Hashimoto et al. 1995). Velvetleaf is very sensitive to flumiclorac pentyl and controlled until 10-leaf stage without phytotoxicity in soybean. *In vitro* study suggested that flumiclorac pentyl showed selective inhibition at the enzyme level. Protox found in the plastid fraction isolated from velvetleaf was 50-fold more sensitive to flumiclorac pentyl than that of soybean (Hashimoto et al. 1995). In contrast to the diphenyl ether family, flumiclorac pentyl was characterized by fast biological degradation.

The degradation study in plants, animals, soil and water revealed very fast degradation (Hashimoto et al. 1995); thus, the instability of flumiclorac pentyl was thought to be one reason for the relatively weaker herbicidal activity against some weeds. Uptake and metabolism in soybean and velvetleaf are shown in Fig. 5. About 10% of treated flumiclorac pentyl was taken up by soybean leaf within 3 h after treatment and 7% by the velvetleaf. About 10% of the parent compound remained in soybean leaf and 60% in velvetleaf. However, the same metabolites were detected both in soybean and velvetleaf. These results suggest that the tolerance of soybean to flumiclorac pentyl is primarily due to a relatively high rate of metabolism of flumiclorac pentyl (Hashimoto et al. 1995). The major metabolic reactions in soybean are ester hydrolysis and oxidation of the cyclohexene ring followed by conjugation.

From this hypothesis, the chemical structure which is less reactive than α,β-unsaturated imide structure is thought to be a promising factor for stronger herbicidal activity and wider weed control spectrum. The reactive electrophilic imide structure of flumiclorac pentyl inspired many chemists to synthesize compounds analogous to flumiclorac pentyl.

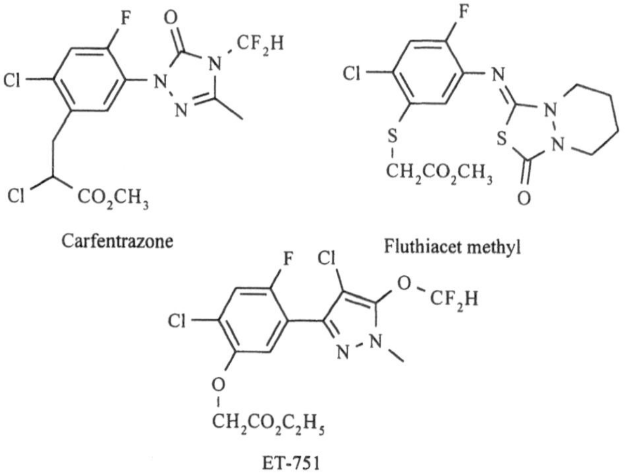

Fig. 6. Structure of the flumiclorac family compounds

Fluthiacet methyl, ET-751 and carfentrazone are newcomers of the flumiclorac family (Fig. 6). Fluthiacet methyl as soybean and corn herbicide (Miyazawa et al. 1993), carfentrazone as wheat and corn herbicide (Van Saun et al. 1993) and ET-751 as wheat herbicide (Miura et al. 1993) are under development. These compounds have a relatively more stable heterocyclic ring against nucleophile than the imide structure of flumiclorac. In most cases velvetleaf (*Abutilon theophrasti*) and black nightshade (*Solanum nigrum*) are most sensitive to the compounds of the flumiclorac family, and grasses, smartweeds (*Polygonum spp.*) and sicklepod (*Cassia obtusifolia*) generally are often resistant. Enhanced control of morningglory (*Ipomoea spp.*), redroot pigweed (*Amaranthus retroflexus*) and lambsquarter (*Chenopodium album*) was obtained with fluthiacet methyl and ET-751 in soybeans or with fluthiacet methyl, carfentrazone and ET-751 in corn. *Galium spp.* and *Veronica spp.* control became possible in wheat and barley at 8 g a.i. of fluthiacet methyl/ha or 8 g a.i. of carfentrazone/ha or 4 g a.i. of ET-751/ha. Although a metabolic pathway was reported, the investigation of the mechanism of selectivity requires more research.

The compounds discussed above have poor pre-emergence activity. Two compounds, flumioxazin (Yoshida et al. 1991) and sulfentrazone (Van Saun et al. 1991), are commercially the first Protox inhibitors in this chemical class to show significant pre-emergence activity in soybeans and peanuts (Fig. 7).

Excellent control of redroot pigweed (*Amaranthus retroflexus*), lambsquarter (*Chenopodium album*), jimsonweed (*Datura stramonium*), morningglory (*Ipomoea spp.*) and nutsedge (*Cyprerus spp.*) was obtained with sulfentrazone in the range 125–500 g a.i./ha without phytotoxicity to soybean. The methanesulfonamide structure characterizes both the chemical property and the herbicidal activity. As sulfentrazone is a weak acid (pKa = 6.56), the soil pH influences the efficacy. It was more active at pH 5–6 than pH 7 for most weed species (Wehtje et al. 1995). Nutsedge activity may be an attractive

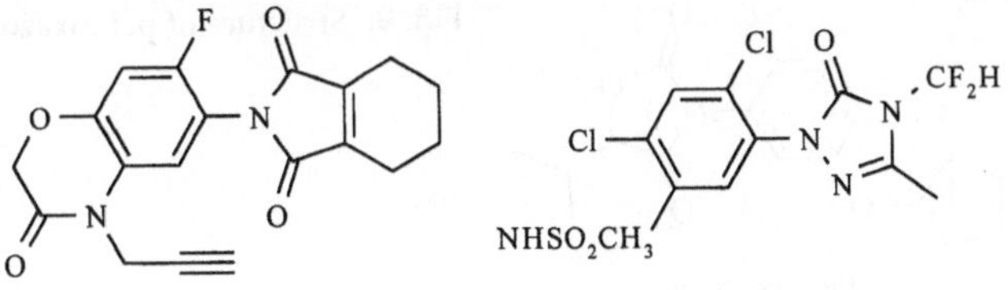

Flumioxazin Sulfentrazone

Fig. 7. Structure of flumioxazin and sulfentrazone

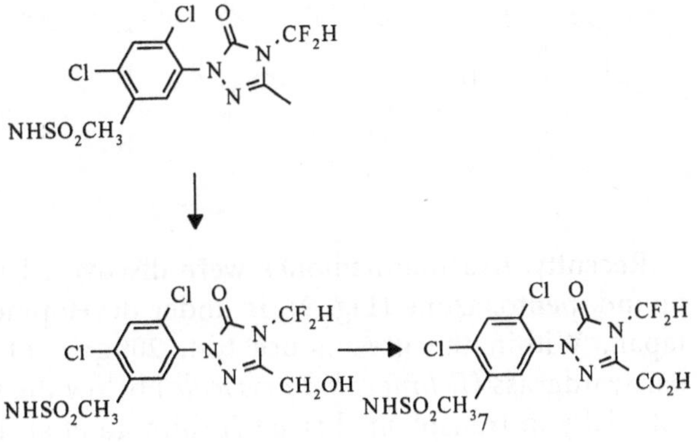

Fig. 8. Metabolic pathway of sulfentrazone in soybean

feature of sulfentrazone. Flumioxazin is very effective in controling black nightshade (*Solanum nigrum*), veletleaf (*Abutilon theophrasti*), smartweed (*Polygonum spp.*), jimsonweed (*Datura stramonium*) and lambsquarter (*Chenopodium album*) in the range 63–100 g a.i./ha without phytotoxicity to soybean and peanut. Additionally, selective Florida beggarweed (*Desmodium tortuosum*) control is possible in peanut fields. In addition to pre-emergence treatment, flumioxazin applied at more than 8 g a.i./ha post-emergence shows strong nonselective activity against broad-leaf weeds. Therefore, flumioxazin also provides long-term nonselective weed control when applied at the relatively higher dosage, due to the strong nonselective post-emergence activity with the residual activity.

The selective action of sulfentrazone was explained by selective oxidative metabolism in soybean as shown in Fig. 8 (Theodoridis et al. 1992). The tolerant weed sicklepod (*Cassia obtusifolia*) also has the same metabolic pathway (Dayan et al. 1996).

The selective metabolism is also expected to be an important factor for the selective action of flumioxazin. As described above, selective action was observed mainly by soil treatment. The mechanism of the selective action between soybean and weeds is under investigation. Preliminary studies suggest that metabolism will be the important factor for the selective action (M. Sakaki 1992, unpubl. res.).

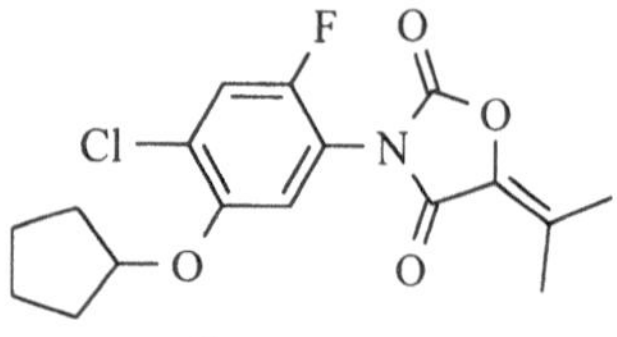

Fig. 9. Structure of pentoxazone

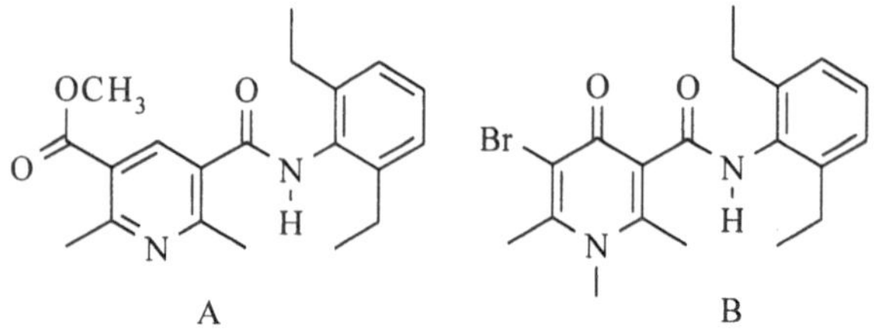

Fig. 10. Nicotinamides tested in the greenhouse

Recently, oxazolidinediones were discovered in this class, and one compound, pentoxazone (Fig. 9), is under development for transplanted rice in Japan. Within the greenhouse test, 200 g a.i. of pentoxazone/ha controled barnyardgrass (*Echinochloa oryzicola*) below the two-leaf-stage without phytotoxicity in transplanted rice (Yoshimura et al. 1992; Ueda et al. 1995). The mechanism of selective action has not been reported yet.

11.4 Nicotinanilide Analogues

Several nicotinanilide and many 4-oxo-1,4-dihydronicotinanilide compounds were reported as Protox inhibitor. The relationship between herbicidal activity against barnyardgrass (*Echinochloa oryzicola*) and the structure of 4-oxo-1,4-dihydronicotinanilide was analyzed with the Hansch-Fujita approach. Finally, DLH-1777 was found to be the most active compound. It controls giant foxtail (*Stetaria viridis*), barnyardgrass (*Echinochloa oryzicola*), crabgrass (*Digitaria* spp.), lambsquarter (*Chenopodium album*) and redroot pigweed (*Amaranthus retroflexus*) in corn fields with either pre- or post-emergence treatment from 250 g to 1 kg a.i./ha (Morishima et al. 1990b).

To compare the two nicotinanilide families, the structurally typical two compounds A and B shown in Fig. 10 were tested in our laboratory. The hydrophilic character of nicotinanilide substructure may determine their physical properties. Compound A controlled velvetleaf (*Abutilon theophrasti*) and black nightshade (*Solanum nigrum*) at 250 g a.i./ha in foliar treatment and at 500 g a.i./ha in pre-emergence treatment. Compound B also controlled ivyleaf morningglory (*Ipomoea hederacea*), veletleaf (*Abutilon theophrasti*) and black nightshade (*Solanum nigrum*) at the same dosage in each treatment. In the quantitative structure–activity relationship (QSAR) report, β-phenethyl analogue had very high herbicidal potential like oxyfluorfen against barnyardgrass (*Echinochloa oryzicola*) (Osabe 1991).

11.5 Concluding Remarks

As described in this chapter, Protox inhibitors are highly variable in their chemical structure. This structural variability led to the commercialization of several compounds with different weed control activity and crop selectivity, and the mechanistic aspect of the different activities of different Protox inhibitors has been discussed. Protox inhibitors are potentially very active, and some of them show excellent weed control at dosages as low as 5 g a.i./ha. Up until now, they have been used without any serious resistant weeds problem and it is thought that Protox inhibitors are an effective tool in resistant weed management. Further detailed study of the mode of action and structure-activity relationship of Protox inhibitors will lead to a new generation of herbicides.

References

Dayan FE, Weete JD, Hancock HG (1996) Physiological basis for differential sensitivity to sulfentrazone by sicklepod (*Cassia obtusifolia*) and coffee senna (*Senna occidentalis*). Weed Sci 44:12–17

Duke SO, Rebeiz CA (1993) Porphyric pesticides. ACS Symp Seri 559, Am Chem Soc, Washington, DC

Engelken LK, Brenneman LG, Creswel JLI, Czapar GF, Owen MDK (1987) Common sunflower control in soybeans. Proc North Central Weed Sci Soci 42:42

Evans JDHL, Cavell BD, Hignett RR (1987) Fomesafen-metabolism as a basis for its selestivity in soya. 1987 Bri Crop Protect Conf Weeds, pp 345–352

Fenyes JG, Steffens JJ (eds) Synthesis and chemistry of agrochemicals III. ACS Symp Seri 504. Am Chem Soc, Washington, DC, pp 135–146

Frear DS, Swanson HR, Mansager ER (1983) Acifluorfen metabolism in soybean: diphenylether bond cleavage and the formation of homoglutathione, cysteine, and glucose conjugates. Pestic Biochem Physiol 20:299–310

Glenn S, Hook BJ, Peregoy RS, Wiepke T (1985) Control of velvetleaf (*Abutilon theophrasti*) and common cocklebur (*Xanthium pensylvanicum*) in soybeans (*Glycine max*) with sequential applications of mefluidide and acifluorfen. Weed Sci 33:244–249

Hashimoto S, Nagano E, Otsubo T, Nambu K, Hosokawa S, Takemoto I (1995) Resource a new herbicide. Sumitomokagaku 1995-I:4–18

Hayashi Y (1987) United States patent 4708734

Hayashi Y (1990) Synthesis and selective herbicidal activity of methyl (E,Z)-[[[1-[5-[2-chloro-4-(trifluoromethyl)phenoxy]-2-nitrophenyl]-2-methoxyethylidene]-amino]oxy]acetate and analogous compounds. J Agric Food Chem 38:839–844

Higgins JM, Whitewell T, Corbin FT, Carter JR. GE, Hill JR. HS (1988a) Absorption, translocation, and metabolism of acifluorfen and lactofen in pitted morningglory (*Ipomoea lacunosa*) and ivyleaf morningglory (*Ipomoea bederacea*). Weed Sci 36:141–145

Higgins J, Whitewell T, Murdock ED, Toler JE (1988b) Recoverry of pitted morningglory (*Ipomoea lacunosa*) and ivyleaf morningglory (*Ipomoea hederacea*) following application of acifluorfen, fomesafen, and lactofen. Weed Sci 36:345–353

Kamoshita K, Nagano E, Saito K, Sakaki M, Yoshida R, Sato R, Oshio H (1992) S-23031 – a new post-emergence herbicide for soybeans. In: Copping LG, Green MB, Rees RT (eds) Pest management in soybean. SCI Elsevier Applied Science, London, pp 31–325

Kouji H, Masuda T, Matsunaka S (1990) Mechanism of herbicidal action and soybean selectivity of AKH-7088, a novel diphenyl ether herbicide. Pestic Biochem Physiol 37:219–226

Mathis WD, Oliver LR (1980) Control of six morningglory (*Ipomoea*) species in soybean (*Glycine max*). Weed Sci 28:409–415

Miura Y, Onishi M, Mabuti T, Yanai I (1993) A new herbicide for use in cereals. 1993 Bri Crop Protect Conf, pp 35–40

Miyazawa T, Kawano K, Shigematsu S, Yamaguchi M, Matsunari K, Porpiglia P, Gutbrod KG (1993) KIH-9201, a new low-rate post-emergence herbicide for maize (*Zea mays*) and soybean (*Glycine max*). 1993 Bri Crop Protect Conf, pp 23–28

Morishima Y, Osabe H, Goto Y, Masamoto K, Yagihara H (1990a) Studies on herbicide 4-pyridone-3-carboxamide derivatives 1: herbicidal activity under paddy conditions and quantitative structure–activity relationships. Weed Res Jpn 35:273–281

Morishima Y, Osabe H, Goto Y, Masamoto K, Yagihara H (1990b) Studies on herbicide 4-pyridone-3-carboxamide derivatives 2: herbicidal activity under upland condition and physiological action of 2′,6′-disubstituted anilides. Weed Res Jpn 35:282–289

Osabe H (1991) Quantitative structure–activity studies of light-dependent herbicidal pyridone-carboxyanilides. paper, Kyoto University, Kyoto, Japan

Theodoridis G, Baum JS, Hotzman FW, Manfredi MC, Maravetz LL, Lyga JW, Tymonko JM, Poss KM, Wyle MJ (1992) Synthesis and herbicidal properties of aryltriazolinones: a new class of pre- and post-emergence herbicides. In: Baker DR, Fenyes JG, Steffens JJ (eds) Synthesis and chemistry of agrochemicals III. ACS Symp Ser 504:134–146. Am Chem Soc, Washington, DC

Ueda T, Ugai S, Hori M, Hirai K (1995) Action mechanism of a new herbicide, KPP-314. II. Effect on the porphyrin pathway. Proceedings of the 43th congress of weed science of Japan, Weed Science of Japan, Tokyo, pp 10–11

Van Saun WA, Bahr JT, Crosby GA, Fore ZQ, Guscar HL, Harnish WN, Hooten RS, Marquez MS, Parrish DS, Theodoridis G, Tymonko JM, Wilson KR, Wyle MJ (1991) F-6285 – a new herbicide for the pre-emergence selective control of broadleaved and grass weeds in soybeans. 1991 Bri Crop Protect Conf, pp 77–82

Van Saun WA, Bahr JT, Bourdouxhe LJ, Gargantiel FJ, Hotzman FW, Shires SW, Sladen NA, Tutt SF, Wilson KR (1993) F-8426 – a new, rapidly acting, low rate herbicide for the post-emergence selective control of broad-leaved weeds in cereals. 1993 Bri Crop Protect Conf, pp 19–22

Wehtje G, Walker RH, Grey TL, Spratlin CE (1995) Soil effects of sulfentrazone. Proc Southern Weed Sci Soc 48:224

Yoshida R, Sakaki M, Sato R, Haga T, Nagano E, Oshio H, Kamoshita K (1991) S-53482 – a new N-phenylphthalimide herbicide. 1991 Bri Crop Protect Conf, pp 69–75

Yoshimura T, Ugai S, Nagato S, Hori M, Hirai K, Yano T, Ejiri E (1992) Proceedings of the 17th congress of the Pesticide Science of Japan, 48

CHAPTER 12

Antagonizing Peroxidizing Herbicides

OLIVER C. KNÖRZER and PETER BÖGER[1]

Contents

12.1 Introduction

Peroxidizing herbicides such as *p*-nitrodiphenyl ethers, cyclic imides, oxadiazoles or pyrazole derivatives have a common mode of action, i.e. inhibition of chlorophyll biosynthesis. Binding of these herbicides to their target enzyme, protoporphyrinogen oxidase, results in the rapid accumulation of protoporphyrin IX. This chlorophyll intermediate causes light-induced formation of reactive oxygen species (ROS) by energy transfer to oxygen, resulting in peroxidative destruction of pigments, proteins and nucleic acids as well as disintegration of cellular membrane systems by lipid peroxidation (reviewed by Böger and Sandmann, 1990; Böger and Wakabayashi 1995; Wakabayashi and Böger, this vol.).

Several diphenyl ether compounds, e.g. oxyfluorfen or acifluorfen (see Fig. 2 and Table 3 for chemical name), have been developed to commercial herbicides to control weeds in wheat, barley, rice or soybean cultures. Selectivity is

[1]Lehrstuhl für Physiologie und Biochemie der Pflanzen, Universität Konstanz, D-78457 Konstanz, Germany

Peter Böger, Ko Wakabayashi
Peroxidizing Herbicides

mainly due to differences in detoxification pathways of the active compounds or to an efficient antioxidative defense to remove the ROS formed during herbicide action.

This chapter will focus on antagonists of peroxidizing herbicides, either physiological adaptations within the antioxidative pathway leading to increased ROS scavenging or induction of defense pathways by safeners, salicylic acid and its functional analogue BTH [benzo(1,2,3)thiadiazole-7-carbothioic acid *S*-methylester]. Possibilities for future applications are discussed.

12.2 Antioxidative Defense

Elevated levels of ROS, such as singlet oxygen, hydroxyl radicals, superoxide anions and hydrogen peroxide, are formed in plant cells in many different situations under aerobic conditions. Environmental stress, like elevated levels of air pollutants such as ozone and SO_2, high UV doses, salinity and cellular senescence as well as pathogen attack and herbicide application can induce oxidative stress in plants (Foyer and Mullineaux 1994; Inzé and van Montagu 1995; Streb and Feierabend 1996; Alscher et al. 1997). To cope with these ROS, higher plants have developed complex protection mechanisms. The main pathway of detoxification of ROS is outlined in Fig. 1a. Superoxide anions, either arising from photoreduction of oxygen by photosystem I (Mehler reaction) or by chemical reduction of oxygen, are scavenged by superoxide dismutase (SOD), yielding H_2O_2 and O_2 by dismutation of two molecules of $O_2^{\cdot -}$. Hydrogen peroxide, is a toxic metabolite itself, and is scavenged in higher plants by the ascorbate - glutathione cycle, also referred to as the Halliwell - Asada pathway (Foyer and Halliwell 1976; Nakano and Asada 1980), in the chloroplast where catalase is absent. Ascorbate peroxidase (APX) catalyzes the reduction of H_2O_2 to water using ascorbate (AsA) as electron donor; oxidation of ascorbate by ascorbate peroxidase proceeds in a one electron step forming the monodehydroascorbate radical (Nakano and Asada 1981). It either spontaneously disproportionates to give ascorbate and dehydroascorbate (DAsA), or is reduced to ascorbate by NAD(P)H (Winkler et al. 1994). The latter reduction is catalyzed by the flavoenzyme monodehydroascorbate reductase (MDAR) (Hossain et al. 1984). For reduction of dehydroascorbate to ascorbate the

→

Fig. 1. Pathways of antioxidative defense and glutathione-mediated detoxification in higher plants (according to Elstner 1990; Kreuz et al. 1996; Knörzer 1997, modified). *APX* ascorbate peroxidase; *DHAR* dehydroascorbate reductase; *GR* glutathione reductase; *GPX* glutathione peroxidase activity; *GST* glutathione *S*-transferase; *MDAR* monodehydroascorbate reductase; *L-OO·* peroxyl radical; *L-OOH* lipid hydroperoxide; *L-OH* lipid hydroxyalkenal; *α-tocopherol·* chromanoxyl radical; *XZ* xenobiotic or toxic metabolite (note that hydroxyalkenals can also be subject to glutathione conjugation); *Z* nucleophile moiety displaced in the GST-catalyzed reaction; *X-SG*, glutathione conjugate

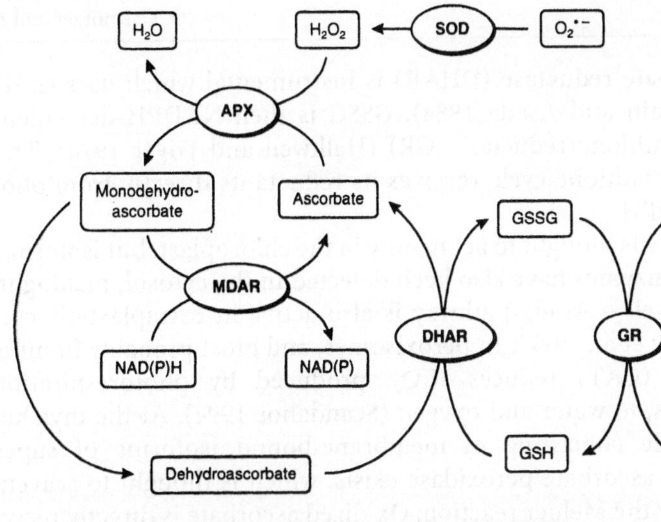

(a) Ascorbate-glutathione pathway (Halliwell-Asada pathway)

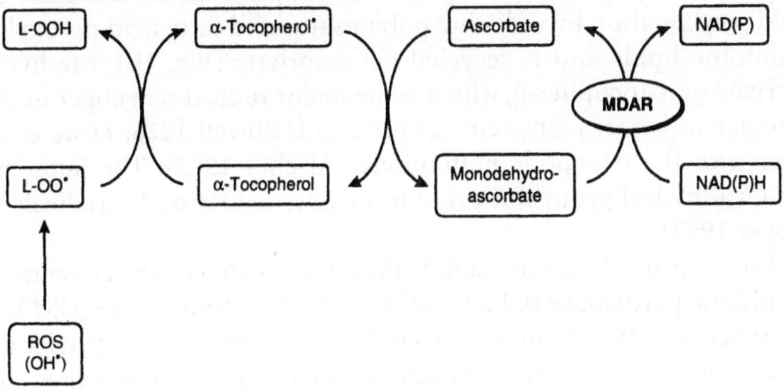

(b) Ascorbate/α-tocopherol pathway

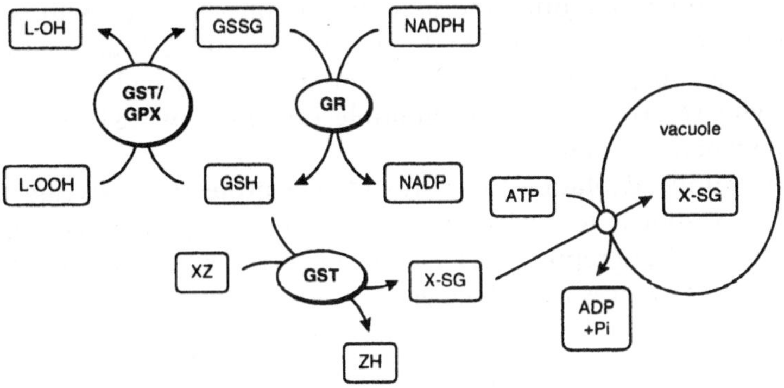

(c) Glutathione pathway

dehydroascorbate reductase (DHAR) is instrumental which uses GSH as reductant (Hossain and Asada 1984). GSSG is then NADPH-dependently reduced by glutathione reductase (GR) (Halliwell and Foyer 1978). Thus, the ascorbate – glutathione cycle receives its reductants directly from photosynthesis via NADPH.

This pathway is thought to act mainly in the chloroplast, but isoforms of the participating enzymes have also been detected in the cytosol, making it likely that the Halliwell – Asada pathway is also active in extraplastidic compartments (Jiménez et al. 1997). In peroxisomes, and most probably in mitochondria, catalase (CAT) reduces H_2O_2, produced by photorespiration and oxidative stress, to water and oxygen (Scandalios 1994). At the thylakoids an analogous cycle consisting of membrane-bound isoforms of superoxide dismutase and ascorbate peroxidase exists, which is thought to scavenge the ROS formed by the Mehler reaction. Oxidized ascorbate is directly recycled by ferredoxin (Asada 1994; Polle 1996).

Several low molecular weight antioxidants in higher plants act as radical scavengers. The lipophilic α-tocopherol interrupts radical chain reactions during lipid peroxidation by reducing polyunsaturated fatty acid peroxyl radicals of membrane lipids and is recycled via ascorbate (Fig. 1b), the hydrophilic counterpart of α-tocopherol, which is the major radical scavenger in the aqueous compartments of plant cells (Foyer and Halliwell 1976; Luwe et al. 1993) together with the tripeptide glutathione (Alscher 1989). The latter protects, e.g., free sulfhydryl groups of proteins against oxidation by radicals (Kunert and Foyer 1993).

Additional protection against ROS may be provided by the enzymatic action of glutathione peroxidase (GPX) and glutathione S-transferase (GST). Though the existence of GPX in higher plants is discussed controversely, GPX activities have been detected in various species of higher plants (Drotar et al. 1985). Eshdat et al. (1997) report that GPX in plants acts as a protecting enzyme during lipid peroxidation by scavenging lipid peroxides. A similar role has been proposed for GST, an enzyme which normally functions as a detoxification enzyme for xenobiotics and toxic metabolites, such as hydroxyalkenals resulting from lipid peroxidation. Some isoforms of GST reportedly catalyze the glutathione-dependent reduction of lipid hydroperoxides, a peroxidative side activity (Marrs, 1996). The multiple functions of GSTs give rise to a glutathione pathway (Fig. 1c).

Changes and adaptations in antioxidative protection systems of higher plants as a response to environmental stresses and peroxidizing herbicides have frequently been reported. The following sections will give a brief overview emphasizing responses to herbicides.

12.2.1
Antioxidative Responses to Peroxidizing Herbicides

12.2.1.1
Antioxidants

Ascorbate and α-tocopherol play a key role in cellular protection against oxidative damage by ROS (Sies 1993). Finckh and Kunert (1985) measured the contents of these antioxidants in various plant species. When these plants were treated with oxyfluorfen it became apparent that cellular damage induced by the herbicide depended on the ratio of ascorbate to α-tocopherol. Plants with a ratio of 10 to 15:1 were most tolerant against the phytotoxic effects of oxyfluorfen (Fig. 2). Higher or lower ratios resulted in stronger susceptibility against oxyfluorfen. These results indicate, that the balance of antioxidants in plants is crucial for effective protection against oxidative damage.

Plants can adapt their antioxidant concentrations as a response to environmental stress and herbicide treatment. High light intensities, UV radiation, ozone and SO_2 have been shown to increase the cellular levels of ascorbate and glutathione (Mehlhorn et al. 1986; Mishra et al. 1995; Takeuchi et al. 1996). Schmidt and Kunert (1986) investigated the synthesis of the two antioxidants in bean leaves treated with acifluorfen and found that ascorbate was more than twofold increased and the total glutathione content over threefold. GST was also markedly increased. In sensitive tobacco leaves glutathione and ascorbate

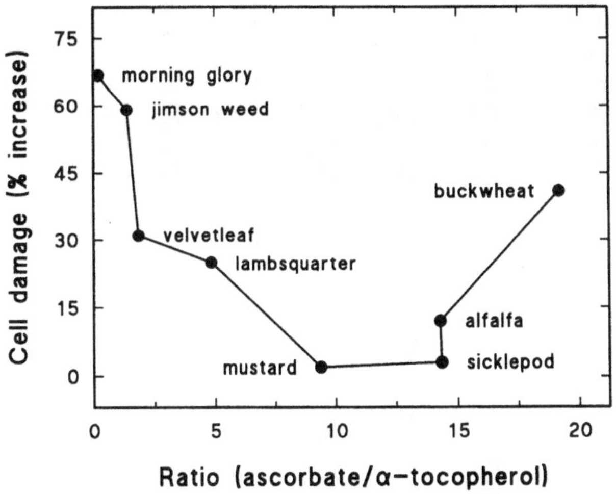

Fig. 2. Ratio of ascorbate content to α-tocopherol content (both in milligram per gram fresh weight) and percent increase of cell damage in different plant species. Seedlings were treated for 6 days with 1 kg oxyfluorfen ha^{-1}. Cell damage was determined as increase of the ratio of dry weight to fresh weight in percent. Oxyfluorfen is 2-chloro-1-(3-ethoxy-4-nitrophenoxy)-4-(trifluoromethyl)benzene. (Modified after Finckh and Kunert 1985)

were strongly elevated by treatment of the plants with acifluorfen (Gullner et al. 1991b). In this study tolerance against peroxidizing herbicides was attributed to the enhanced inducibility of the antioxidative system in a tolerant tobacco cultivar compared to a sensitive one.

Our group studied the effects of oxyfluorfen on the cellular concentration of ascorbate and glutathione in a cell culture system (Knörzer et al. 1996). The soybean cell suspensions investigated were able to modulate their antioxidative system as a response to the herbicide. Whereas the cellular ascorbate content was not increased by herbicide treatment, the total glutathione pool (reduced + oxidized form) was raised about threefold (Fig. 3). Herbicidal action of oxyfluorfen resulted in a shift of the reduction state of these two antioxidants. The ratio of oxidized to reduced partner strongly increased during treatment. For both ascorbate and glutathione this ratio increased twofold, indicative of a decreased reduction state of the aqueous antioxidant pool. In the case of ascorbate the shifted ratio was due to a decreased level of reduced ascorbate, as dehydroascorbate concentrations remained constant; in the case of glutathione this effect was due to a stronger increase of GSSG compared to GSH. This result reflects an oxidative stress induced by oxyfluorfen. This redox-shift might be necessary to activate (or regulate) the transcription of genes of the antioxidative system. In plants, GSH and other thiols have been

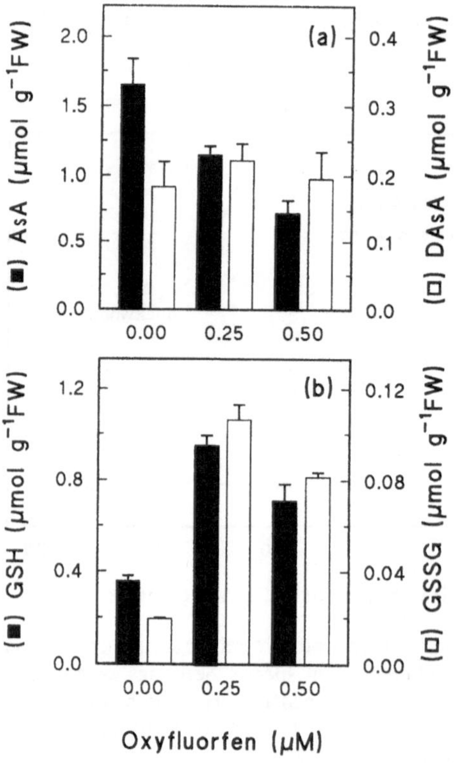

Fig. 3. Influence of oxyfluorfen on antioxidants ascorbate (a) and glutathione (b) in soybean cell suspension cultures. Cellular contents of ascorbic acid (*AsA*), dehydroascorbic acid (*DAsA*), reduced glutathione (*GSH*) and oxidized glutathione (*GSSG*) were determined after a 48-h incubation with oxyfluorfen. All values are means of four independent experiments ±SD. (Modified after Knörzer et al. 1996)

shown to induce the genes of various proteins involved in plant defense or in scavenging active oxygen species (Mehdy 1994; Nemat Alla 1995).

12.2.1.2
Antioxidative Enzymes

High levels of oxyfluorfen lead to oxidation of antioxidants (Fig. 3). Obviously, maintaining the antioxidant pool reduced is crucial for the efficacy of the antioxidative defense. As summarized in Table 1, activities of the recycling enzymes of the ascorbate – glutathione pathway, i.e. glutathione reductase and monodehydroascorbate reductase, are upregulated as a response of soybean cells to oxyfluorfen, indicative of a direct response to an oxidative attack. As GSH is a major antioxidant and the most important cellular reductant, obviously glutathione reductase, which reduces GSSG to the physiologically active form GSH, has to be modulated in order to maintain the stress tolerance of the soybean cells. Apparently monodehydroascorbate reductase plays the key role as the ascorbate-reducing enzyme during oxidative stress in plant cells. Dehydroascorbate is very unstable at physiological pH (Winkler et al. 1994) and therefore rapidly removed from the ascorbate pool. Conceivably, dehydroascorbate reductase activity was not increased during oxyfluorfen treatment, but showed a decreasing activity. As the monodehydroascorbate radical is the intermediate formed by reducing H_2O_2 by ascorbate peroxidase, recycling of ascorbate proceeds mainly by action of monodehydroascorbate

Table 1. Influence of oxyfluorfen and salicylic acid on the specific activities of enzymes participating in antioxidative defense. (Modified after Knörzer et al. 1996; Knörzer 1997)

	Oxyfluorfen treatment			Salicylate treatment		
	Specific activity[a] (μmol min^{-1} mg^{-1})		Percent of control	Specific activity[a] (μmol min^{-1} mg^{-1})		Percent of control
Enzyme	Control	Oxyfluorfen[b]	(%)	Control	Salicylate[c]	(%)
APX	1.76	2.72	154	2.44	2.80	115
GR	0.19	0.28	150	0.13	0.26	204
MDAR	0.86	1.26	147	0.66	1.33	200
DHAR	0.32	0.23	72	0.22	0.17	77
GST	0.20	1.16	569	0.14	1.05	773
CAT	33.77	57.32	170	41.32	79.76	193

APX, ascorbate peroxidase; GR, glutathione reductase; MDAR, monodehydroascorbate reductase; DHAR, dehydroascorbate reductase; GST, glutathione *S*-transferase; CAT, catalase.

[a] All values are means of four independent experiments.

[b] Soybean cell suspensions were treated for 48 h with 0 (control) and 500 nM oxyfluorfen, respectively, prior to protein extraction and enzyme determination.

[c] Soybean cell suspensions were treated for 48 h with 0 (control) and 1 mM Na-salicylate, respectively, prior to protein extraction and enzyme determination.

reductase. Directly ROS-scavenging enzyme activities (ascorbate peroxidase and catalase) increased in soybean cells treated with oxyfluorfen (Table 1), which can be regarded as a direct response to enhanced cellular radical concentrations.

For glutathione reductase a corresponding induction of enzyme activity has been reported by Schmidt and Kunert (1986) for bean leaves treated with acifluorfen. Gullner et al. (1991a) measured increasing activities of glutathione reductase and ascorbate peroxidase by treatment of tobacco plants with acifluorfen. Resistance against paraquat in *Conyza bonariensis* was attributed to enhanced levels of superoxide dismutase, glutathione reductase and ascorbate peroxidase by Amsellem et al. (1993) and Ye and Gressel (1994). The latter study showed, that comparably weak increases of enzyme activities by 30 to 60% result in over 300-fold resistance of the plants against paraquat. The inductions of antioxidative enzyme activities present in the soybean cells are in the same range and we agree with Ye and Gressel (1994) that small shifts of enzyme activities can contribute strongly to an enhanced defense if the enzymes are present in small compartments. This holds particularly true for the ascorbate – glutathione cycle, which is mainly localized in the plastids.

Adaptations of the antioxidative system in transgenic tobacco plants accumulating protoporphyrin IX have been shown to occur at the gene level by transcriptional analyses (H.P. Mock, pers. comm.). In accordance, recent studies of our own revealed elevated transcript levels of ascorbate peroxidase and catalase in oxyfluorfen-treated tobacco plants (unpubl. results). Immunoblotting of glutathione reductase showed about twofold higher enzyme levels in treated cells (Knörzer 1997). So, inductions of antioxidative enzymes seem to be due to amplified gene expression. The signalling pathways, however, have yet to be investigated.

12.2.2 Antioxidative Response in Transgenic Plants

Gene transfer experiments directed towards increased resistance of plants to oxidative stress have contributed much to our present knowledge of stress adaptations of defense pathways and the importance of enzymatic components. Several detailed reviews have been published during recent years (Hérouart et al. 1993a; Foyer and Mullineaux 1994; Allen et al. 1997). When ascorbate peroxidase was transferred into tobacco, cytosolic and chloroplastic overexpression resulted in enhanced tolerance against paraquat and oxidative stress (Pitcher et al. 1994; Webb and Allen 1995, 1996). Enhanced CuZn–SOD levels in the plastids of tobacco were not able to protect the plants against ozone and paraquat (Teppermann and Dunsmuir 1990; Pitcher et al. 1991). Contrasting results have been published by Sen Gupta et al. (1993): a twofold overexpression of SOD resulted in enhanced tolerance against ozone. Cytosolic or chloroplastic overexpression of a bacterial GR in transgenic tobacco resulted in enhanced tolerance against SO_2 and paraquat but not against ozone

(Aono et al. 1991, 1993). Again, contradictory results can be found in the literature: Foyer et al. (1991) reported that transgenic plants exhibiting two- to tenfold enhanced GR activities were not protected against paraquat. Size and redox state of the glutathione pool did not change.

These different results reflect the complex regulation of the antioxidative response in plants. Compartmentation of defense pathways as well as formation of ROS at different sites in plants, e.g. chloroplastic formation of ROS during paraquat inhibition of photosystem I and primary ozone attack through the apoplast, require distinct stress sensors and signal molecules to trigger the response of specific antioxidative isozymes and antioxidants.

Elucidation of this complex regulation network will therefore be crucial for future approaches to enhance tolerance of plants against oxidative stress produced by peroxidizing herbicides. Especially overproduction of signalling molecules in transgenic plants may be a promising mechanism to increase tolerance by a general upregulation of the cellular defense. Glutathione and other thiols have been shown to induce genes of various proteins involved in plant defense or in scavenging ROS (Mehdy 1994; Nemat Alla 1995). Hydrogen peroxide or other ROS may themselves be involved in regulation of defensive genes (Hérouart et al. 1994), although it seems more likely that ROS trigger the response via second messengers like redox-shifts in cellular thiols or the plastoquinone pool (Escoubas et al. 1995). Metabolites like nicotinamides or salicylic acid have also been described as signal transducers of cellular redox shifts (Berglund 1994; Dempsey and Klessig 1994).

12.3 Detoxification of Xenobiotics in Plants

Metabolism of pesticides in higher plants largely contributes to the specificity of these compounds for different plant species. Several enzymatic mechanisms act together to detoxify herbicides in plants. Cytochrome P450 monoxygenases generally carry out the first step in metabolism of hydrophobic compounds, also referred to as phase 1. Hydroxylation of aromatic rings or alkyl groups as well as *O,N*-dealkylation yields oxidized substrates for further conjugation reactions (Cole 1994). Most important is *O*- and *N*-glucosylation by UDP-glucosyl transferases (phase 2). The glucosylated products are water soluble and can either be translocated into the vacuole for storage or excreted to the apoplast (Kreuz et al. 1996). Herbicides of the phenoxyphenoxypropionate type and sulfonylureas are detoxified by this pathway (Kreuz et al. 1996).

An important role in conjugating toxic xenobiotics to less harmful metabolites is played by glutathione *S*-transferases, which transfer the major cellular thiol glutathione to electrophilic centers (Marrs, 1996). Several herbicides like thiocarbamate sulfoxides, chloroacetamides, triazines and *p*-nitrodiphenyl ethers are detoxified via glutathione conjugation (Lamoureux and Rusness 1986; Farago et al. 1994; Neuefeind et al. 1997). The conjugated products are translocated to the vacuole by an ATP-dependent glutathione pump and sequestered (Martinoia et al. 1993). The various GST enzymes found in plants are homo- or heterodimers of subunits with molecular masses of 23 to 29 kDa and differ in their substrate specifity, inducibility and regulation (Marrs, 1996). The physiological functions of GSTs are thought to involve phenylpropanoid secondary metabolism, conjugation of toxic lipid metabolites (hydroxyalkenals) as well as scavenging of lipid hydroperoxides (Kreuz et al. 1996; Marrs 1996). The latter function implies a role for plant GSTs as protection enzymes during oxidative stress situations. Human as well as plant GSTs have been characterized as lipid hydroperoxide glutathione peroxidases (Berhane et al. 1994; Edwards 1996). In higher plants this function may substitute for lack of Se-dependent glutathione peroxidases.

12.3.1 Metabolism of Peroxidizing Herbicides: the Role of Glutathione S-Transferase

Peroxidizing herbicides of the *p*-nitrodiphenyl ether type are rapidly metabolized by glutathione conjugation. A nucleophilic attack on the diphenyl ether bond by glutathione yields a phenolic intermediate as well as a glutathione S-nitrophenyl conjugate. Both metabolites are subject to glucosylation and acylation, or the glutathione conjugate may be hydrolyzed. This type of metabolism has been shown for fluorodifen (2,4′-dinitro-4-trifluoromethyl diphenyl ether) in *Pisum sativum* and *Picea spec.* as well as for acifluorfen in *Glycine max* (Frear and Swanson 1973; Frear et al. 1983; Lamoureux et al. 1991). So GSTs play a major role in metabolizing peroxidizing herbicides.

As GSTs exhibit also antioxidative properties by scavenging lipid hydroperoxides, lipid peroxidation initiated by light-induced radical formation could be interrupted by GST action. Several GST isozymes, like GST II, show glutathione peroxidase activity; other isozymes do not (Edwards 1996). More detailed studies on differential inductions of GST isoforms during the impact of peroxidizing herbicides are necessary to elucidate the role of plant GSTs as oxidative protectants.

In soybean cell suspensions the specific activity of GST dramatically increased by treatment of the soybean cells with the diphenyl ether oxyfluorfen (Fig. 4). An approximately sixfold increase above the control level implies a role for GST in oxyfluorfen detoxification. As mentioned above, the closely related *p*-nitrodiphenyl ether herbicide acifluorfen is split in soybean via conjugation with glutathione (Frear et al. 1983). Possibly oxyfluorfen is metabolized by a similar process. In addition, GST may have a general role in the plant defense response. GSTs are induced by a wide variety of stress signals in plants, including xenobiotic action, pathogen attack, oxidative stress and H_2O_2 (reviewed by Marrs 1996). An elevated level of hydrogen peroxide induces a strong increase of GST mRNA in soybean cells (Levine et al. 1994). Ulmasov et al. (1995) report that the promotor of the soybean *GH2/4* gene, which encodes a GST, is activated by several chemical agents, including the plant signalling

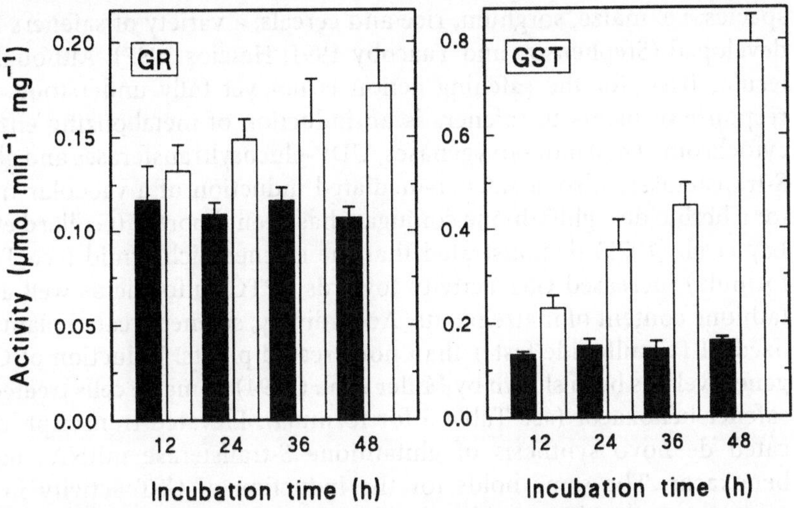

Fig. 4. Time course of induction of glutathione reductase (*GR*) and glutathione *S*-transferase (*GST*) activities. Soybean cell suspension cultures were incubated without (control, *solid bars*) and with 0.5 µM oxyfluorfen (*open bars*), respectively, for periods indicated prior to determination of enzyme activities. All values are means of two independent experiments ±SD

compounds salicylic acid and jasmonate, as well as glutathione. In addition, GST is induced by ozone fumigation of *Arabidopsis thaliana* (Sharma et al. 1996). In wheat the induction of two GST subunits by paraquat and atrazine exhibited light-dependency and the authors conclude that the observed induction was likely to occur as a response to a light-induced oxidative impact (Mauch and Dudler 1993).

Besides GST, also glutathione reductase (GR) was strongly activated by oxyfluorfen treatment of cultured soybean cells (Fig. 4). As shown in Section 12.2.1.1 the glutathione content was increased too as a response to oxyfluorfen (Fig. 3). These results indicate a dominant role of the glutathione pathway (Fig. 1c) related to the action of peroxidizing herbicides. Herbicide conjugation and antioxidative counteraction by GST require delivery of reduced glutathione to protect plant cells against severe damage. This is reflected by increased glutathione content and GR activity.

12.3.2 Antidotes for Peroxidizing Herbicides

The "*safening principle*" is a means to increase tolerance of crop species against herbicidal compounds by selectively increasing their ability to metabolize the applied substances without affecting susceptibility of weeds. Herbicide safeners are synthetic compounds of structural diversity, which are applied to the seed before planting (seed dressing) or sprayed together with the herbicide (reviewed by Farago et al. 1994; Hatzios 1997). For monocotyledonous crop species, i.e. maize, sorghum, rice and cereals, a variety of safeners have been developed (Stephenson and Yaacoby 1991; Hatzios 1997). Although the molecular basis for the safening action is not yet fully understood, a general response of plants to safeners is an induction of metabolizing enzymes like cytochrome P450 monooxygenases, UDP-glucosyltransferases and glutathione *S*-transferases. Also, a safener-mediated induction of a vacuolar transporter for a herbicide – glutathione conjugate has been reported (Gaillard et al. 1994). Lay et al. (1975) demonstrated that the safener dichlormid (see Table 3 for formula) increased GST activity towards EPTC-sulfoxide as well as the glutathione content of maize plants. Accordingly, safener-treated plants metabolized EPTC-sulfoxide faster than non-treated plants. Induction of GST at the gene level has been shown by Miller et al. (1994) in maize cells treated with the safener benoxacor (see Table 3 for formula). Elevated transcript levels indicated de novo synthesis of glutathione *S*-transferase mRNAs induced by benoxacor. The same holds for the induction of GST activity in *Sorghum* against metolachlor (2-chloro-*N*-[2-ethyl-6-methylphenyl]-*N*-[2-methoxy-1-methylethyl]-acetamide) by safeners (Dean et al. 1990).

At present no safeners for peroxidizing herbicides are offered. Induction of glutathione *S*-transferase activity by safeners could be a useful means to increase selectivity of diphenyl ether herbicides. Matsunaka and Wakabayashi (1989) demonstrated that treatment of maize with 1,8-naphthalic anhydride (NA) increased tolerance against chlorophthalim (Table 2). The mechanism is not yet understood. Possibly induction of GST by NA contributes to the observed tolerance despite the fact that chlorophthalim is not metabolized. As

Table 2. Safening of chlorophthalim phytotoxicity by 1,8-naphthalic anhydride (NA). (Modified after Matsunaka and Wakabayashi 1989)

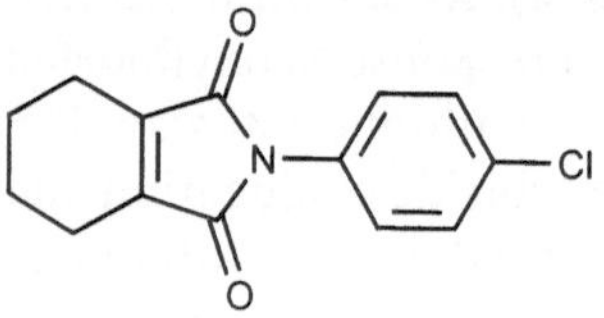

	Injury[a]			
Chlorophthalim (kg a.i. ha^{-1})	Maize	Crabgrass	Lambs-quarter	Velvet-leaf
1	3	10	10	7
2	5	10	10	9
1 + Na[b]	0	10	10	7
2 + Na[b]	0	10	10	9

a.i., Active ingredient.
[a] Score: 0 = no injury, 10 = lethal.
[b] NA: 0.5% seed dressing.

safeners have been shown to increase plant glutathione levels (Farago and Brunold 1990), the activation of the glutathione pathway (Fig. 1c) by safeners could protect plants from herbicide-induced oxidative attack. Devlin and Zbiec (1993) reported that preemergent treatment of corn with dichlormid and NA diminished shoot growth inhibition by the cyclic imide compound flumioxazin (Fig. 5).

Treatment of cress seedlings with different safeners strongly repressed protoporphyrin IX accumulation induced by the diphenyl ether herbicide acifluorfen methyl (Table 3; Böger and Miller 1994). This decrease was not due to destruction of protoporphyrin IX by mixed function oxidases. Furthermore, it seems unlikely that the herbicide was metabolized by the seedlings during the 16-h incubation with the safeners and herbicide. This conclusion is supported by further results presented in that study, showing that the same safener-induced decrease of protoporphyrin-IX accumulation was found with chlorophthalim (Böger and Miller 1994). Probably the safeners applied induce enzyme activities which decompose tetrapyrroles. Protoporphyrinogen-IX destruction by barley leaf extracts has been reported by Jacobs et al. (1996). A possible role of GST related to porphyrin metabolites has not yet been investigated. The findings indicated above may be used to start development

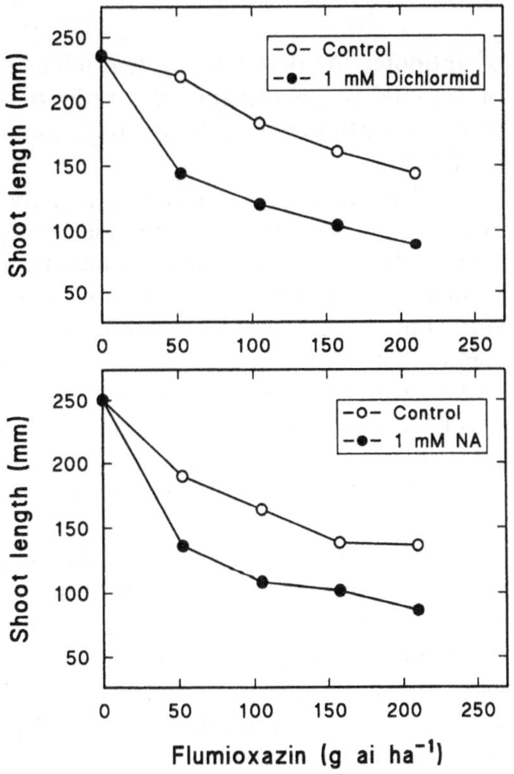

Fig. 5. Effects of safeners dichlormid and 1,8-naphthalic anhydride (*NA*) against herbicidal inhibition of shoot growth of corn by the cyclic imide herbicide flumioxazin. *Open circles*: Effects of herbicide alone; *solid circles* effects of herbicide and safener applied together. (Modified after Devlin and Zbiec 1993)

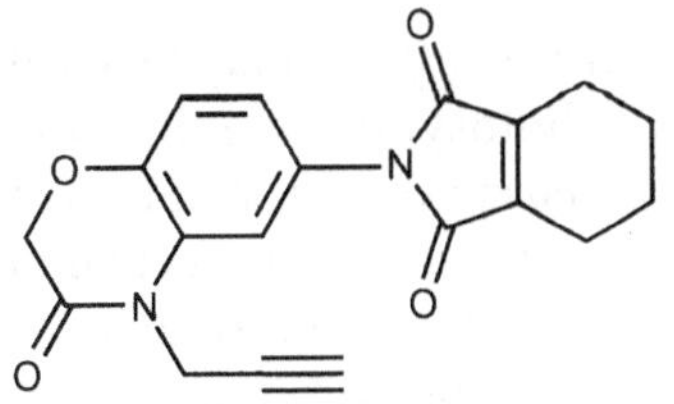

Flumioxazin

Table 3. Influence of different safeners on herbicide-induced protoporphyrin-IX formation (nmol g^{-1} FW) in cress seedlings. (Modified after Böger and Miller 1994)

	Acifluorfen methyl	
Safener (1 mM)	Control	50 nM[2]
No safener	1.4	28
Fenclorim	1.7	14
Benoxacor	1.3	10
Dichlormid	1.6	9
Oxabetrinil	1.8	8
BAS 145 138	0.8	8
NA	1.4	4

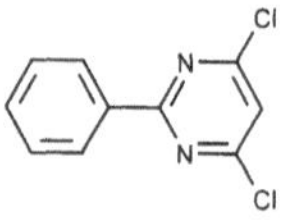
Fenclorim

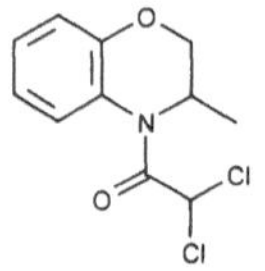
Benoxacor

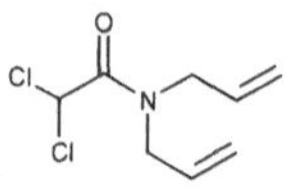
Dichlormid

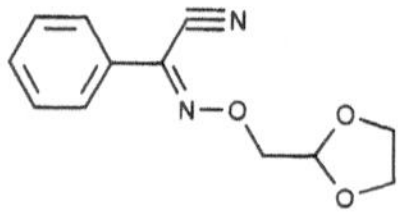
Oxabetrinil

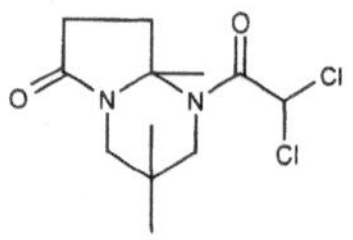
BAS 145 138

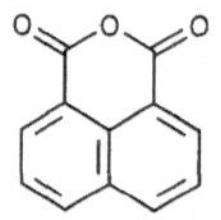
Naphthalic anhydride (NA)

[a] Cress seedlings were incubated for 16 h in dim light with acifluorfen methyl or 1% (v/v) acetone (control), respectively, and different safeners prior to protoporphyrin-IX extraction and quantification. Acifluorfen methyl is 5-[2-chloro-4-(trifluoromethyl)phenoxy]-2-nitrobenzoic acid methyl ester.

of antidotes for peroxidizing herbicides since their major weakness is lack of selectivity. Conceivably the ratio of safener to herbicide used in the studies mentioned is still too high and not yet appropriate for practical application.

As safeners increase both GST activity and the glutathione level (Farago and Brunold 1990) the antioxidative system is involved to counteract peroxidation through radical quenching. We have reported that shifts in the content of reduced glutathione in the cell, either decrease by buthionine sulfoximine or increase by application of 2-oxothiazolidine-4-carboxylate, lead to increased or decreased peroxidative ethane evolution (Sandmann and Böger 1990). Cross tolerance of several plant species against the non-selective herbicide paraquat or diphenyl ethers is mediated by enhanced antioxidative mechanisms, either enzymatic or non-enzymatic ones (Kömives et al. 1997). Possibly GSTs contribute to the observed tolerances due to their peroxidase activity.

In closing it should be mentioned that also the "*diuron effect*" may be used to decrease the activity of peroxidizing herbicides. Photosynthesis inhibitors applied together with the peroxidizer to autotrophic cells prevent formation of excess protoporphyrin IX (see Böger and Sandmann 1990 for reference; Ohki et al. 1997; Fig. 3 of Chap. 6, this Vol.) and subsequent phytotoxic peroxidation. In case the photosynthesis inhibitor is degraded by the crop within a reasonable time after application, the peroxidizing herbicide will start the accumulation of protoporphyrin IX, but slowly and with a time lag. This leads to delayed and slow radical formation which will allow the (crop) plant to increase its antioxidative capacity and to develop tolerance. Bentazone or triazines may be appropriate photosynthesis inhibitors.

12.4 The Salicylic Acid Defense Pathway

The activation of elaborate defense mechanisms in plants to both biotic and abiotic stress situations requires the recognition of a stimulus by a receptor and the subsequent use of second messengers and effector proteins to trigger the appropriate response. Salicylic acid (SA) is a key signal molecule in many plant processes (Enyedi et al. 1992; Bowler and Chua 1994; Ryals et al. 1996). Recently, a great deal of attention has been directed towards elucidating its role in plant disease resistance against pathogens, and it has emerged as a key signalling molecule in both local defense reactions at infection sites of pathogens as well as in the induction of "*systemic acquired resistance*" (SAR), characterized by the induction of defense genes such as glucanases or peroxidases (Dangl et al. 1996; Hammond-Kosack and Jones 1996; Ryals et al. 1996).

12.4.1 Salicylic Acid as an Activator of the Antioxidative System

The mechanism(s) by which SA induces pathogen resistance are still under discussion. One mode of action appears to be inhibition of catalase, thereby elevating endogenous levels of H_2O_2 (Chen et al. 1993). According to this hypothesis, the elevated H_2O_2, or other ROS derived from it, would serve as a second messenger and activate plant defense-related genes. ROS participate in activation of plant genes encoding glutathione *S*-transferase, glutathione peroxidase, polyubiquitin and other genes associated with oxidative stress or chilling tolerance (Levine et al. 1994; Prasad et al. 1994). On the contrary, H_2O_2 or H_2O_2-inducing chemicals, as well as UV treatment and ozone stimulate SA biosynthesis (León et al. 1995; Neuenschwander et al. 1995; Summermatter et al. 1995), suggesting that SA may act downstream of H_2O_2. Nevertheless, SA and H_2O_2 or other ROS are mechanistically connected in signal transduction during pathogen response. Consequently, a connection between the antioxidative pathway and SA-mediated signalling has been proposed (Fodor et al. 1997; Knörzer 1997; Rao et al. 1997).

In tobacco, either infection with tobacco mosaic virus or treatment with SA leads to increases of antioxidative enzymes, which is thought to suppress necrotic symptoms in leaves expressing systemic acquired resistance (Fodor et al. 1997). Treatment of *Arabidopsis thaliana* with 1 to 5 mM SA resulted in slight increases of superoxide dismutase and ascorbate peroxidase (Rao et al. 1997). It should be noted that endogenous concentrations reach only levels of about 150 μM (Enyedi et al. 1992). Therefore effects elicited with SA concentrations of up to 5 mM can hardly be considered as physiological. Consequently, the latter study reports SA-induced lipid peroxidation in plants treated with high doses of SA, which therefore may act prooxidatively.

Using our well characterized soybean cell culture system, we have investigated the influence of SA on antioxidant levels and antioxidative enzyme activities (Knörzer et al. 1998). Incubation of soybean cell suspension cultures with SA increased the endogenous levels of the antioxidants ascorbate and glutathione (Fig. 6). The total ascorbate content (ascorbate + dehydroascorbate) reached a maximum 52% increase vs. control level. The redox state of the ascorbate pool, calculated as oxidized (dehydroascorbate) to reduced ascorbate was shifted towards the oxidized component. This redox relation increased 3.1-fold in cells treated with 1 mM SA, due to a stronger increase of dehydroascorbate compared to ascorbate (Fig. 6a). As documented in Fig. 6b, the same holds true for the pool of cellular glutathione. Total glutathione of the cell cultures (GSH + GSSG) increased strongly by SA treatment, reaching 3.6-fold the control level. The somewhat stronger increase of GSSG, compared to GSH, shifted the glutathione redox state (calculated as GSSG to GSH) slightly towards the oxidized partner.

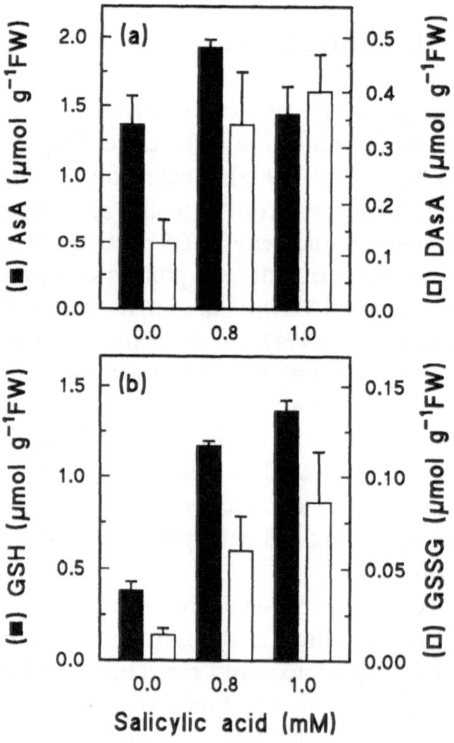

Fig. 6. Influence of salicylic acid on antioxidants ascorbate (a) and glutathione (b) in soybean cell suspension cultures. Cellular contents of ascorbic acid (*AsA*), dehydroascorbic acid (*DAsA*), reduced glutathione (*GSH*) and oxidized glutathione (*GSSG*) were determined after a 48-h incubation with salicylic acid. All values are means of four independent experiments ±SD. (Modified after Knörzer et al. 1998)

These results support the view that SA is a signal molecule for stress responses. The stronger induction of glutathione in relation to ascorbate may be explained by the possible function of GSH or GSSG as a gene regulator. Glutathione and other thiols have been shown to induce GR, GST and superoxide dismutase in plants (Hérouart et al. 1993b; Mehdy 1994; Nemat-Alla 1995). According to this hypothesis glutathione may act downstream of SA in activation of defense genes.

As documented in Table 1, several enzymes related to oxidative stress in plants are activated by SA. Glutathione reductase, catalase (CAT) and monodehydroascorbate reductase (MDAR) exhibited about twofold higher specific activities than in untreated controls. Glutathione reductase and MDAR maintain the reducing power of the Halliwell – Asada pathway by regenerating the two major reductants glutathione and ascorbate. Thus the stimulation of these two enzymes is indicative of an activation of this pathway. The other two enzymes participating in this metabolic cycle, ascorbate peroxidase and dehydroascorbate reductase, were not activated (Table 1). CAT, the major extraplastidic H_2O_2-scavenging enzyme, was twofold induced, although CAT was shown to be strongly inhibited by SA in vitro (Durner and Klessig, 1996). These results are in accordance with the findings of Guan and Scandalios (1995), reporting a regulatory function of SA on CAT genes. Again, GST exhibited the strongest response to the SA treatment. Its specific activity was induced more than eightfold. Recently, Ulmasov et al. (1995) found that the GST promotor is regulated by SA.

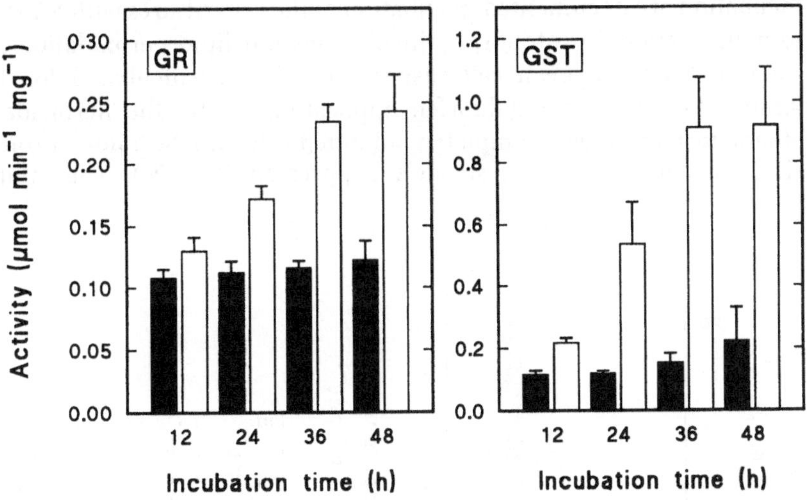

Fig. 7. Induction time of glutathione reductase (*GR*) and glutathione *S*-transferase (*GST*) activities. Soybean cell suspension cultures were incubated without (control, *solid bars*) and with 1 mM salicylic acid (*open bars*), respectively, for periods indicated prior to determination of enzyme activities. All values are means of two independent experiments ±SD. (Modified after Knörzer et al. 1998)

Compared with the results obtained by oxyfluorfen treatment, the kinetics of the GST and GR inductions to SA revealed a faster response (Figs. 4 and 7). The maximum induction of enzyme activities in SA-treated cells was attained after 36h, compared to 48h by oxyfluorfen treatment. These time courses indicate inductions at transcriptional or translational level. Accordingly, elevated enzyme levels of GR in cells treated with either of the two chemicals could be detected by immunoblotting (Knörzer 1997). On the other hand, the somewhat faster response of these two enzymes may be indicative of SA being a downstream signal of the oxidative attack caused by oxyfluorfen.

12.4.2 Salicylic Acid and BTH as Antagonists of Oxyfluorfen

The inductions of antioxidants and antioxidative enzymes by either salicylic acid or oxyfluorfen exhibit remarkable similarity. Consequently, we investigated whether SA as a putative signal transducer of the antioxidative response was able to protect plant cells against herbicide-induced lipid peroxidation (Knörzer et al. 1998). The primary target of oxyfluorfen and related nitrodiphenyl ethers is protoporphyrinogen oxidase. Its inhibition leads to accumulation of protoporphyrin IX and in the light sensitized protoporphyrin IX induces lipid peroxidation (Böger and Wakabayashi 1995; Chapter 6, this Vol.). The latter can be measured by short-chain hydrocarbons, i.e. ethane, produced during the herbicide treatment. In Fig. 8 oxyfluorfen-induced accumulation of protoporphyrin IX, as well as ethane formation during a 24-h incubation are documented. Incubation of the cell cultures with SA alone leads to neither altered protoporphyrin-IX levels nor lipid peroxidation (data not shown). When soybean cell suspensions were preincubated for 16h with 0.5mM SA prior to oxyfluorfen application, firstly the herbicide-induced ethane formation was completely inhibited (Fig. 9a). Secondly, protoporphyrin-IX accumulation was strongly suppressed (Fig. 9c). The inhibition of

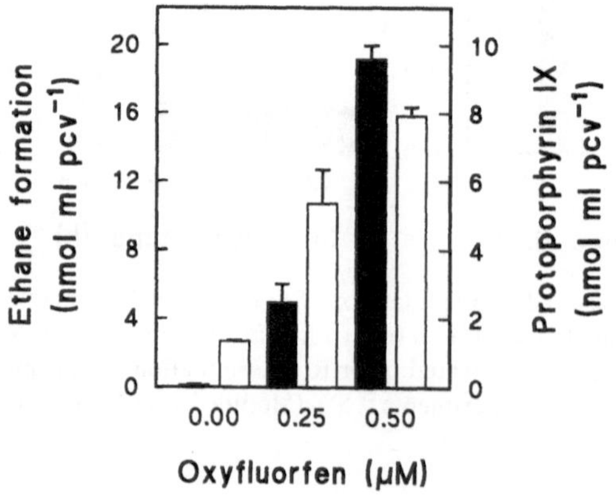

Fig. 8. Oxyfluorfen-induced lipid peroxidation (*solid bars*) and protoporphyrin IX accumulation (*open bars*) in soybean cells. Suspension cultures were incubated for 24h in light with the oxyfluorfen concentrations indicated prior to protoporphyrin IX extraction and quantification. Lipid peroxidation was measured by GC-determination of ethane formation (Knörzer 1997)

isolated protoporphyrinogen oxidase by peroxidizing compounds was not influenced by SA present.

Apparently, SA protects plant cells against oxidative damage by peroxidizing herbicides by two mechanisms: (1) it induces the antioxidative system, leading to increased protection against oxidative damage like lipid peroxidation; and (2) SA suppresses oxyfluorfen-induced protoporphyrin-IX accumulation by a yet unknown mechanism. Possibly the latter effect is due to the induction of a tetrapyrrole-degrading enzyme activity. Referring to the study of Böger and Miller (1994), inhibition of protoporphyrin accumulation by SA treatment strongly resembles the mode of action of different safeners applied to cress seedlings (Table 3).

The recently developed plant protection compound Bion® [benzo-(1,2,3)thiadiazole-7-carbothioic acid *S*-methylester; BTH] acts as an SA analogue and induces the systemic acquired resistance pathway, leading to

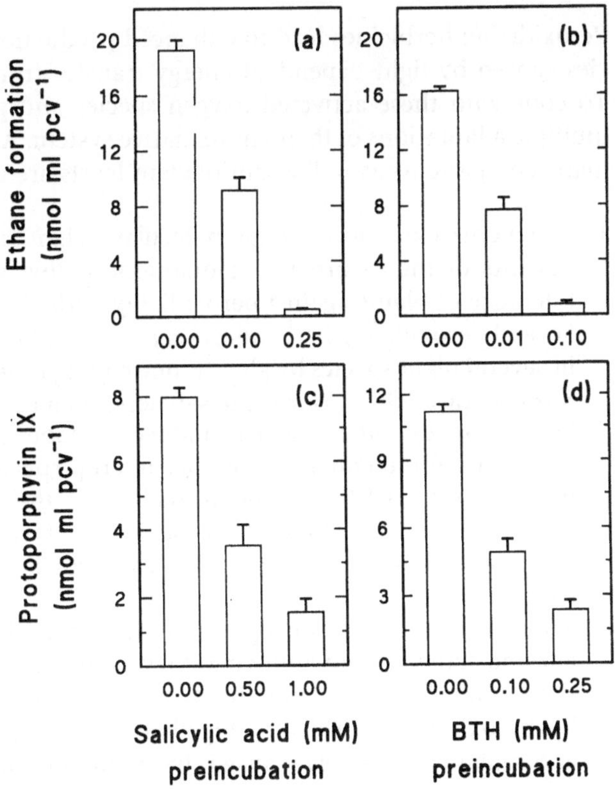

Fig. 9. Inhibition of oxyfluorfen-induced lipid peroxidation and accumulation of protoporphyrin IX by salicylic acid (a, c) and BTH [benzo(1,2,3)thiadiazole-7-carbothioic acid *S*-methylester] (b, d). Soybean cell suspension cultures were preincubated for 16 h with the concentrations indicated prior to application of 0.5 μM oxyfluorfen and were further incubated for 24 h with herbicide. All values are means of three independent experiments ±SD

improved tolerance against fungal infections (Friedrich et al. 1996; Lawton et al. 1996). As demonstrated in Fig. 9b, d, BTH exhibited similar effects compared to SA. Oxyfluorfen-induced lipid peroxidation as well as protoporphyrin-IX accumulation could be counteracted by preincubating the soybean cells with BTH (Knörzer et al., in preparation). Although the influence of BTH on antioxidants and antioxidative enzymes has yet to be investigated, it seems likely that BTH acts in a similar way to SA. These findings suggest that the SA-mediated defense pathway is connected to the antioxidative response of higher plants. Possibly it acts as a signal transducer during activation of the antioxidative system. Experiments with whole plants on the improvement of the capacity of the oxidative stress response have not been conducted as yet. It is conceivable that SA analogues may represent antidote leads to develop safeners against peroxidizing herbicides.

12.5 Summary

Peroxidizing herbicides lead to enhanced production of reactive oxygen species caused by light-dependent energy transfer from porphyrin metabolites. To cope with these activated oxygen species and peroxides plants react by multiple adaptations of their antioxidative system. As a response several enzymatic components as well as antioxidant levels are upregulated.

1. The increase of low and high molecular weight antioxidative defense components of the ascorbate – glutathione pathway can be used to achieve tolerance of plants against peroxidizing herbicides.
2. Several peroxidizing herbicides, i.e. *p*-nitrodiphenyl ethers are metabolized in several plant species by glutathione conjugation catalyzed by glutathione *S*-transferase (GST). Herbicide safeners, such as 1,8-naphthalic anhydride, benoxacor or others which induce metabolizing enzymes like GST, may be useful to enhance tolerance of crop species in the field. Additional protection by GST may be provided by the antioxidative activity of GST itself against lipid hydroperoxides formed during peroxidative attack against cellular membranes by herbicide-induced reactive oxygen species.
3. Salicylic acid and its analogue BTH were found to increase antioxidative enzyme activities and antioxidant levels in plants.
4. Both salicylic acid and BTH alleviate lipid peroxidation induced by peroxidizing herbicides together with decreased protoporphyrin-IX accumulation possibly caused by induction of tetrapyrrole-degrading enzyme(s).

Especially the protoporphyrin-IX degrading mode of action appears to be a promising objective for future studies. This finding may represent a novel mechanism to protect plants against the impact of peroxidizing herbicides.

Acknowledgements. The author's work was supported by the Deutsche Forschungsgemeinschaft (grant no. Bo 310/116) and by the Fonds der Chemischen Industrie. We would like to thank Dr. J. Durner at the Waksman Institute, Piscataway, USA, for helpful discussions.

References

Allen RD, Webb RP, Schake SA (1997) Use of transgenic plants to study antioxidant defenses. Free Radic Biol Med 23:473–479

Alscher RG 1989. Biosynthesis and antioxidant function of glutathione in plants. Physiol Plant 77:457–464

Alscher RG, Donahue JL, Cramer CL (1997) Reactive oxygen species and antioxidants: relationships in green cells. Physiol Plant 100:224–233

Amsellem Z, Jansen MAK, Driesenaar ARJ, Gressel J (1993) Developmental variability of photooxidative stress tolerance in paraquat-resistant *Conyza*. Plant Physiol 103:1097–1106

Aono M, Kubo A, Saji H, Natori T, Tanaka K, Kondo N (1991) Resistance to active oxygen toxicity of transgenic *Nicotiana tabacum* that expresses the gene for glutathione reductase from *Escherichia coli*. Plant Cell Physiol 32:691–697

Aono M, Kubo A, Saij H, Tanaka K, Kondo N (1993) Enhanced tolerance to photo-oxidative stress of transgenic *Nicotiana tabacum* with high chloroplastic glutathione reductase activity. Plant Cell Physiol 34:129–135

Asada K (1994) Production and action of active oxygen species in photosynthetic tissues. In: Foyer CH, Mullineaux PM (eds) Causes of photooxidative stress and amelioration of defence systems in plants. CRC Press, Boca Raton, pp 77–104

Berglund T (1994) Nicotinamide, a missing link in the early stress response in eukaryotic cells: a hypothesis with special reference to oxidative stress in plants. FEBS Lett 351:145–149

Berhane K, Widersten M, Engstrom A, Kozarich JW, Mannervick B (1994) Detoxification of base propenals and other alpha, beta-unsaturated aldehyde products of radical reactions and lipid peroxidation by human glutathione transferases. Proc Natl Acad Sci USA 91:1480–1484

Böger P, Miller R (1994) Protoporphyrin accumulation induced by peroxidizing herbicides is counteracted by safeners. Z Naturforsch 49c:775–780

Böger P, Sandmann G (1990) Modern herbicides affecting typical plant processes. In: Bowers WS, Ebing, W, Martin D, Wegler S (eds) Chemistry of plant protection, vol 6, Springer, Berlin Heidelberg New York, pp 173–216

Böger P, Wakabayashi K (1995) Peroxidizing herbicides I: mechanism of action. Z Naturforsch 50c:159–166

Bowler C, Chua NH (1994) Emerging themes of plant signal transduction. Plant Cell 6:1529–1541

Chen Z, Ricigliano JW, Klessig DF (1993) Purification and characterization of a soluble salicylic acid-binding protein from tobacco. Proc Natl Acad Sci, USA 90:9533–9537

Cole DJ (1994) Detoxification and activation of agrochemicals in plants. Pestic Sci 42:209–222

Dangl JL, Dietrich RA, Richberg MH (1996) Death don't have no mercy: cell death programs in plant – microbe interactions. Plant Cell 8:1793–1807

Dean JV, Gronwald JW, Eberlein CV (1990) Induction of glutathione S-transferase isoenzymes in *Sorghum* by herbicide antidotes. Plant Physiol 92:467–473

Dempsey DA, Klessig DF (1994) Salicylic acid, active oxygen species and systemic acquired resistance in plants. Trends Cell Biol 4:334–338

Devlin RM, Zbiec II (1993) Effect of four safeners against the herbicidal activity of V-53482 in corn (*Zea mays*). Plant Growth Regul Soc Am Q 21(4):190–197

Drotar A, Phelbs P, Fall R (1985) Evidence for glutathione peroxidase activities in cultured plant cells. Plant Sci 42:35–40

Durner J, Klessig DF (1996) Salicylic acid is a modulator of tobacco and mammalian catalases. J Biol Chem 271:28492–28501

Edwards R (1996) Characterisation of glutathione transferases and glutathione peroxidases in pea (*Pisum sativum*). Physiol Plant 98:594–604

Elstner EF (1990) Der Sauerstoff: Biochemie, Biologie, Medizin. BI-Wissenschaftsverlag, Mannheim
Enyedi AJ, Yalpani N, Silverman P, Raskin I (1992) Signal molecules in systemic plant resistance to pathogens and pests. Cell 70:879–886
Escoubas JM, Lomas M, LaRoche J, Falkowsky PG (1995) Light intensity regulation of *cab* gene transcription is signaled by the redox state of the plastoquinone pool. Proc Natl Acad Sci USA 92:10237–10241
Eshdat Y, Holland D, Faltin Z, Ben-Hayim G (1997) Plant glutathione peroxidases. Physiol Plant 100:234–240
Farago S, Brunold C (1990) Regulation of assimilatory sulfate reduction by herbicide antidotes. Plant Physiol 94:1808–1812
Farago S, Brunold C, Kreuz K (1994) Herbicide safeners and glutathione metabolism. Plant Physiol 91:537–542
Finckh BF, Kunert KJ (1985) Vitamins C and E: an antioxidative system against herbicide-induced lipid peroxidation in higher plants. J Agric Food Chem 33:574–577
Fodor J, Gullner G, Adam AL, Barna B, Kömives T, Kiraly Z (1997) Local and systemic responses of antioxidants to tobacco mosaic virus infection and to salicylic acid in tobacco: role in systemic acquired resistance. Plant Physiol 114:1443–1451
Foyer C, Lelandais M, Galap C, Kunert KJ (1991) Effects of elevated cytosolic glutathione reductase activity on the cellular glutathione pool and photosynthesis in leaves under normal and stress conditions. Plant Physiol 97:863–872
Foyer CH, Halliwell B (1976) The presence of glutathione and glutathione reductase in chloroplasts: a proposed role in ascorbate metabolism. Planta 133:21–25
Foyer CH, Mullineaux PM (1994) Causes of photooxidative stress and amelioration of defence systems in plants. CRC Press, Boca Raton
Frear DS, Swanson HR (1973) Metabolism of substituted diphenylether herbicides in plants. I Enzymatic cleavage of fluorodifen in peas (*Pisum sativum* L.). Pestic Biochem Physiol 3:473–482
Frear DS, Swanson HR, Mansager ER (1983) Acifluorfen metabolism in soybean: diphenylether bond cleavage and the formation of homoglutathione, cysteine, and glucose conjugates. Pestic Biochem Physiol 20:299–310
Friedrich L, Lawton K, Ruess W, Masner P, Specker N, Gut Rella M, Meier B, Dincher S, Staub T, Uknes S, Métraux JP, Kessmann H, Ryals J (1996) A benzothiadiazole derivative induces systemic acquired resistance in tobacco. Plant J 10:61–70
Gaillard C, Dufaud A, Tommasini R, Kreuz K, Amrhein N, Martinoia E (1994) A herbicide antidote (safener) induces the activity of both the herbicide detoxifying enzyme and of a vacuolar transporter for the detoxified herbicide. FEBS Lett 352:219–221
Guan L, Scandalios JG (1995) Developmentally related responses of maize catalase genes to salicylic acid. Proc Natl Acad Sci USA 92:5930–5934
Gullner G, Kiraly L, Kömives T (1991a) Nitrodiphenyl ether and phenylimide resistance of a tobacco biotype is due to enhanced inducibility of its antioxidant systems. In: Brighton Crop Protection Conference, Weeds, pp 1111–1118
Gullner G, Kömives T, Király L (1991b) Enhanced inducibility of antioxidant systems in a *Nicotiana tabacum* L. biotype results in acifluorfen resistance. Z Naturforsch 46c:875–881
Halliwell B, Foyer CH (1978) Properties and physiological function of a glutathione reductase purified from spinach leaves by affinity chromatography. Planta 139:9–17
Hammond-Kosack KE, Jones JDG (1996) Resistance gene-dependent plant defense responses. Plant Cell 8:1773–1791
Hatzios KK (1997) Regulation of enzymatic systems detoxifying xenobiotics in plants: a brief overview and directions for future research. In: Hatzios KK (ed) Regulation of enzymatic systems detoxifying xenobiotics in plants. Kluwer, Dordrecht, pp 1–5 (NATO ASI series, vol 37)
Hérouart D, Bowler C, Willekens H, Van Camp W, Slooten L, Van Montagu M, Inzé D (1993a) Genetic engineering of oxidative stress resistance in higher plants. Philos Trans R Soc Lond B 342:235–240

Hérouart D, Van Montagu M, Inzé D (1993b) Redox-activated expression of the cytosolic copper/zinc superoxide dismutase gene in *Nicotiana*. Proc Natl Acad Sci USA 90:3108–3112

Hérouart D, Van Montagu M, Inzé D (1994) Developmental and environmental regulation of the *Nicotiana plumbaginifolia* cytosolic Cu/Zn-superoxide dismutase promoter in transgenic tobacco. Plant Physiol 104:873–880

Hossain MA, Asada K (1984) Purification of dehydroascorbate reductase from spinach and its characterization as a thiol enzyme. Plant Cell Physiol 25:85–92

Hossain MA, Nakano Y, Asada K (1984) Monodehydroascorbate reductase in spinach chloroplasts and its participation in regeneration of ascorbate for scavenging hydrogen peroxide. Plant Cell Physiol 25:385–395

Inzé D, Van Montagu M (1995) Oxidative stress in plants. Curr Opin Biotechnol 6:153–158

Jacobs JM, Jacobs NJ, Duke SO (1996) Protoporphyrinogen destruction by plant extracts and correlation with tolerance to protoporphyrinogen oxidase-inhibiting herbicides. Pestic Biochem Physiol 55:77–83

Jiménez A, Hernández JA, del Río LA, Sevilla F (1997) Evidence for the presence of the ascorbate – glutathione cycle in mitochondria and peroxisomes of pea leaves. Plant Physiol 114:275–284

Knörzer OC (1997) Oxidativer Stress in Pflanzen: Charakterisierung des antioxidativen Systems in Zellkulturen der Sojabohne (*Glycine max*). PhD thesis, University of Konstanz

Knörzer OC, Durner J, Böger P (1996) Alterations in the antioxidative system of suspension-cultured soybean cells (*Glycine max*) induced by oxidative stress. Physiol Plant 97:338–396

Knörzer OC, Lederer B, Durner J, Böger P (1998) Salicylic acid activates the antioxidative defense in suspension-cultured soybean cells. Physiol Plant (submitted)

Kömives T, Gullner G, Kiraly Z (1997) The ascorbate – glutathione cycle and oxidative stresses in plants In: Hatzios KK (ed) Regulation of enzymatic systems detoxifying xenobiotics in plants. Kluwer, Dordrecht, pp 85–96 (NATO ASI series, vol 37)

Kreuz K, Tommasini R, Martinoia E (1996) Old enzymes for a new job. Plant Physiol 111:349–353

Kunert KJ, Foyer C (1993) Thiol/disulfide exchange in plants. In: De Kok LJ (ed) Sulfur nutrition and assimilation in higher plants. SPB Academic, The Hague, pp 139–151

Lamoureux GL, Rusness DG (1986) Tridiphane[2-(3,5-dichlorophenyl)-2-(2,2,2-trichlorethyl) oxirane], an antrazine synergist: enzymatic conversion to a potent glutathione S-transferase inhibitor. Pestic Biochem Physiol 26:323–342

Lamoureux GL, Rusness DG, Schröder P, Rennenberg H (1991) Diphenylether herbicide metabolism in a spruce cell suspension culture: the identification of two novel metabolites derived from a glutathione conjugate. Pestic Biochem Physiol 39:291–301

Lawton KA, Friedrich L, Hunt M, Weymann K, Delaney T, Kessmann H, Staub T, Ryals J (1996) Benzothiadiazole induces disease resistance in *Arabidopsis* by activation of the systemic acquired resistance signal transduction pathway. Plant J 10:71–82

Lay MM, Hubbell JP, Casida JE (1975) Dichloroacetamide antidotes for thiocarbamate herbicides: mode of action. Science 189:287–289

León J, Lawton MA, Raskin I (1995) Hydrogen peroxide stimulates salicylic acid biosynthesis in tobacco. Plant Physiol 108:1673–1678

Levine A, Tenhaken R, Dixon R, Lamb C (1994) H_2O_2 from the oxidative burst orchestrates the plant hypersensitive disease resistance response. Cell 79:583–593

Luwe MWF, Takahama U, Heber U (1993) Role of ascorbate in detoxifying ozone in the apoplast of spinach (*Spinacia oleracea* L.) leaves. Plant Physiol 101:969–976

Martinoia E, Grill E, Tommasini R, Kreuz K, Amrhein N (1993) ATP-dependent glutathione *S*-conjugate export pump in the vacuolar membrane of plants. Nature 364:237–249

Marrs KA (1996) The functions and regulation of glutathione *S*-transferases in plants. Annu Rev Plant Physiol Plant Mol Biol 47:127–158

Matsunaka S, Wakabayashi K (1989) Crop safening against herbicides in Japan. In: Hatzios KK, Hoagland RE (ed) Crop safeners for herbicides, Academic Press, San Diego, pp 47–62

Mauch F, Dudler R (1993) Differential induction of distinct glutathione S-transferases of wheat by xenobiotics and by pathogen attack. Plant Physiol 102:1193–1201

Mehdy MC (1994) Active oxygen species in plant defence against pathogens. Plant Physiol 106:467–472

Mehlhorn M, Seufert G, Schmidt A, Kunert KJ (1986) Effect of SO_2 and O_3 on production of antioxidants in conifers. Plant Physiol 82:336–338

Miller KD, Irzyk GP, Fuerst EP (1994) Benoxacor treatment increases glutathione S-transferase activity in suspension cultures of *Zea mays*. Pestic Biochem Physiol 48:123–134

Mishra NP, Fatma T, Singhal GS (1995) Development of antioxidative defense system of wheat seedlings in response to high light. Physiol. Plant 95:77–82

Mock H-P, Keetmann U, Kruse E, Rank B, Grimm B (1998) Defense responses to tetrapyrrole-induced oxidative stress in transgenic plants with reduced uroporphyrinogen decarboxylase or coproporphyrinogen oxidase activity. Plant Physiol 116:107–116

Nakano Y, Asada K (1980) Spinach chloroplasts scavenge hydrogen peroxide on illumination. Plant Cell Physiol 21:1295–1307

Nakano Y, Asada K (1981) Hydrogen peroxide is scavenged by ascorbate-specific peroxidase in spinach chloroplasts. Plant Cell Physiol 22:867–880

Nemat Alla MM (1995) Glutathione regulation of glutathione-S-transferase and peroxidase activity in herbicide-treated *Zea mays*. Plant Physiol Biochem 33:185–192

Neuefeind T, Reinemer P, Bieseler B (1997) Plant glutathione *S*-transferase and herbicide detoxification. J Biol Chem 378:199–205

Neuenschwander U, Vernooij B, Friedrich L, Uknes S, Kessmann, H, Ryals J (1995) Is hydrogen peroxide a second messenger of salicylic acid in systemic acquired resistance? Plant J 8:27–233

Ohki A, Ohki S, Koizumi K, Sato Y, Kohno H, Böger P, Wakabayashi K (1997) Phytotoxicity caused by peroxidizing herbicides is alleviated by 2-substituted-4,6-bis(ethylamino)-1,3,5-triazines. J Pestic Sci Japan 22:309–313

Pitcher LH, Brennan A, Hurley P, Dunsmuir M, Tepperman J, Zilinskas BA (1991) Overproduction of *Petunia* chloroplastic copper/zinc superoxide dismutase does not confer ozone tolerance in transgenic tobacco. Plant Physiol 97:452–455

Pitcher LH, Repetti P, Zilinskas BA (1994) Overproduction of ascorbate peroxidase protects transgenic tobacco against oxidative stress, abstract no 623, Plant Physiol 105:116

Polle A (1996) Mehler reaction: friend or foe in photosynthesis? Bot Acta 109:84–89

Prasad TK, Anderson MD, Martin BA, Stewart CR (1994) Evidence for chilling-induced oxidative stress in maize seedlings and a regulatory role for hydrogen peroxide. Plant Cell 6:65–74

Rao MV, Paliyath G, Ormrod DP, Murr DP, Watkins CB (1997) Influence of salicylic acid on H_2O_2 production, oxidative stress, and H_2O_2-metabolizing enzymes. Plant Physiol 115:137–149

Ryals JA, Neuenschwander UH, Willits MG, Molina A, Steiner HY, Hunt MD (1996) Systemic acquired resistance. Plant Cell 8:1809–1819

Sandmann G, Böger P (1990) Peroxidizing herbicides: some aspects of tolerance. In: Green MB, LeBaron HM, Moberg WK, (eds) Managing resistance to agrochemicals. Am Chem Soc, Washington DC, pp 407 (ACS symposium series 421)

Scandalios JG (1994) Regulation and properties of plant catalases. In: Foyer CH, Mullineaux (eds) Causes of photooxidative stress and amelioration of defence systems in plants. CRC Press, Boca Raton, pp 275–314

Schmidt A, Kunert KJ (1986) Lipid peroxidation in higher plants. The role of glutathione reductase. Plant Physiol 82:700–702

Sen Gupta A, Heinen JL, Holaday AS, Burke J, Allen RD (1993) Increased resistance to oxidative stress in transgenic plants that overexpress chloroplastic Cu/Zn superoxide dismutase. Proc Natl Acad Sci USA 90:1629–1633

Sharma YK, León J, Raskin I, Davis KR (1996) Ozone-induced responses in *Arabidopsis thaliana*: the role of salicylic acid in the accumulation of defense-related transcripts and induced resistance. Proc Natl Acad Sci USA 93:5099–5104

Sies H (1993) Strategies of antioxidant defense. Eur J Biochem 215:213–219

Stephenson GR, Yaacoby T (1991) Milestones in the development of herbicide safeners. Z Naturforsch 46c:794–797

Streb P, Feierabend J (1996) Oxidative stress responses accompanying photoinactivation of catalase in NaCl-treated rye leaves. Bot Acta 109:125–132

Summermatter K, Sticher L, Métreaux JP (1995) Systemic responses in *Arabidopsis thaliana* infected and challenged with *Pseudomonas syringae* pv *syringae*. Plant Physiol 108:1379–1385

Takeuchi Y, Kubo H, Kasahara H, Sakaki T (1996) Adaptive alterations in the activities of scavengers of active oxygen in cucumber cotyledons irradiated with UV-B. J Plant Physiol 147:589–592

Tepperman JM, Dunsmuir P (1990) Transformed plants with elevated levels of chloroplastic SOD are not more resistant to superoxide toxicity. Plant Mol Biol 14:501–511

Ulmasov T, Ohmiya A, Hagen G, Guilfoyle T (1995) The soybean GH2/4 gene that encodes a glutathione *S*-transferase has a promoter that is activated by a wide range of chemical agents. Plant Physiol 108:919–927

Webb RP, Allen RD (1995) Overexpression of pea cytosolic ascorbate peroxidase in *Nicotiana tabacum* confers protection against the effects of paraquat. Plant Physiol [Suppl] 108:64

Webb RP, Allen RD (1996) Overexpression of pea cytosolic ascorbate peroxidase confers protection against oxidative stress in transgenic *Nicotiana tabacum*. Plant Physiol [Suppl] 111:48

Winkler BS, Orselli SM, Rex TS (1994) The redox couple between glutathione and ascorbic acid: a chemical and physiological perspective. Free Radic Biol Med 17:333–349

Ye B, Gressel J (1994) Constitutive variation of ascorbate peroxidase activity during development parallels that of superoxide dismutase and glutathione reductase in paraquat- resistant *Conyza*. Plant Sci 102:147–151

CHAPTER 13

Strategy for Peroxidizing Herbicide-Resistant Crops

TADAO ASAMI and SHIGEO YOSHIDA[1]

Contents

13.1 Introduction

The widespread and persistent use of herbicides has increased the efficiency of modern crop production, and chemicals are now available for the control of most weeds in most crops. However, it is inevitable that there would be a biological result from dependence on a single control method. One result has been the selection and enrichment of genes which resulted in herbicide resistance in weed populations. A breakthrough in weed science was the finding of resistance to triazine herbicides (Ryan 1970), and the triazine-resistant weed biotypes then provided superb experimental materials for researchers. The dramatic progress in plant molecular biology made possible the isolation of the gene encoding D1 protein which is the target for triazine. Resistance was found to be due to the exchange of one or more amino acid(s) of the D1 protein. These findings led to the use of genetic engineering techniques in order to create herbicide-resistant crops. For example, a gene conferring resistance to glyphosate has been inserted into the genome of a number of crop

[1] Plant Function Laboratory, The Institute of Physical and Chemical Research (RIKEN) 2-1 Hirosawa, Wako-shi, Saitama 351-01, Japan

Peter Böger, Ko Wakabayashi
Peroxidizing Herbicides

species where it possesses glyphosate resistance (Dyer 1994). The resulting possible application of glyphosate to crops has received increasing attention and transgenic plants have been developed into commercial products.

Peroxidizing herbicides are generally applied as foliar sprays which induce rapid accumulation of photosensitizers such as protoporphyrin IX due to inhibition of chlorophyll biosynthesis in chloroplasts. Protoporphyrinogen IX oxidase is the target enzyme in the porphyrin pathway for the most powerful peroxidizing herbicides, which show neither an oxidative force nor a strong phytotoxicity by themselves. Since some of these herbicides are effective at a very low dosage, below a few grams per hectare or less, this class of herbicides (especially inhibitors of protoporphyrinogen oxidase) has become an attractive subject of modern agrochemistry as to how they provide an excellent method of weed control. The amount of information available for elucidating the mode of action of peroxidizing herbicides has increased rapidly in the last decade. This progress has been supported by plant molecular biology, which opens an exciting area of research for making transgenic crops resistant to herbicides.

Some plant species such as rice and soybean are known to be naturally resistant against protoporphyrinogen oxidase inhibitors (Lee et al. 1991; Pornprom et al. 1994). However, there is no evidence that any species has evolved resistance under the selection pressure of peroxidizing herbicides. Recently, a few techniques have become available to mutate a plant for herbicide resistance by selection of cultured cells with protoporphyrinogen oxidase inhibitors (Sato et al. 1994; Ichinose et al. 1995; Prasad and Dailey 1995) or by expression of a resistant gene transferred from other organisms (Choi et al. 1998). These new techniques hopefully will bring about an extensive use of peroxidizing herbicides, which will allow a broader weed control spectrum with environmental security. This chapter discusses possible mechanisms to exhibit herbicide resistance in the peroxidizing process which is triggered by inhibition of chlorophyll biosynthesis.

13.2 Mechanisms of Herbicide Resistance

Peroxidizing herbicides are believed to require a very unique process to exert their herbicidal action, which is started with protoporphyrinogen oxidase inhibition. However, little is known about resistance mechanisms of plants to this class of herbicides. Recently, a new type of photobleaching compounds was found to induce the photoperoxidizing effect on green cultured cells without either protoporphyrinogen oxidase inhibition or protoporphyrin IX accumulation (Wang et al. 1997). This result indicates that "peroxidizing herbicides" may involve not only protoporphyrinogen oxidase inhibitors but also those attacking other sites, suggesting that characterization of resistance to these herbicides may be difficult. To avoid confusion, "peroxidizing herbicides" refers to protoporphyrinogen oxidase inhibitors only in this chapter. Stalker

(1989) classified mechanisms of herbicide resistance into four types (Stalker 1989). However, the complicated mode of action of peroxidizing herbicides suggests that more mechanisms are involved, as shown in Fig. 1.

13.2.1 Structural Modification of Herbicide Target Enzymes

13.2.1.1 *Alteration by Gene Selection*

Alteration of a herbicide target enzyme comprises structural modification of its herbicide binding domain that prevents stable binding of specific herbicides. This category of mutation is often obtained by gene selection under herbicide stress. This technique has been employed for production of sulfonylurea-resistant (Haughn et al. 1988; McHughen 1989; Wiersma et al. 1989) and photosystem II inhibitor-resistant biotypes (Darmency and Pernes 1985; Barsby et al. 1987; Von Wettstein and Chua 1987; Ayotte et al. 1989; Mets and Thiel 1989; Trebst 1991). These mutants occurred by replacement of amino acid(s) in acetolactate synthase or in D1 proteins reducing sensitivity to the respective herbicides. However, this type of gene selection is not always easy, as shown in the unsuccessful attempt to select *Arabidopsis thaliana* mutants resistant to glyphosate (Fig. 2) (Padgette et al. 1996). This negative result suggested that the glyphosate target enzyme in the plant system requires multiple mutations in order to obtain glyphosate resistance. Although peroxidizing herbicides have been widely used for more than 20 years, there is no report on naturally evolved plants resistant to peroxidizing herbicides. Recently, some efforts have been made to produce peroxidizing-resistant plants (Abe et al. 1997); however, there is no information on the herbicide-binding domain of protoporphyrinogen oxidase and/or herbicide-resistant protoporphyrinogen oxidase yet. Since peroxidizing herbicides were implied to be competitive inhibitors of protoporphyrinogen oxidase because of their bicyclic structures fulfilling the complementary space at the binding site of protoporphyrinogen IX (Nandihalli et al. 1992), alteration of this domain may cause serious problems to normal protoporphyrinogen oxidase activity. In other words, mutation by structural modification of the herbicide-binding domain according to amino acid(s) substitution is presumably incapable of creating crops resistant to a protoporphyrinogen oxidase inhibitor since many mutants may be lethal. This may be the reason why there is no report on natural mutation of protoporphyrinogen oxidase.[2] As genes of protoporphyrinogen oxidase have recently been cloned (Narita et al. 1996;

[2] Recently, Horikoshi et al. (Nihon Noyaku Co., Ltd.) reported the tobacco cell lines resistant to ET-751, a potent photobleaching herbicide, in which the amino acid sequences of protoporphyrinogen oxidase were altered by point mutations (Horikoshi et al. 1998).

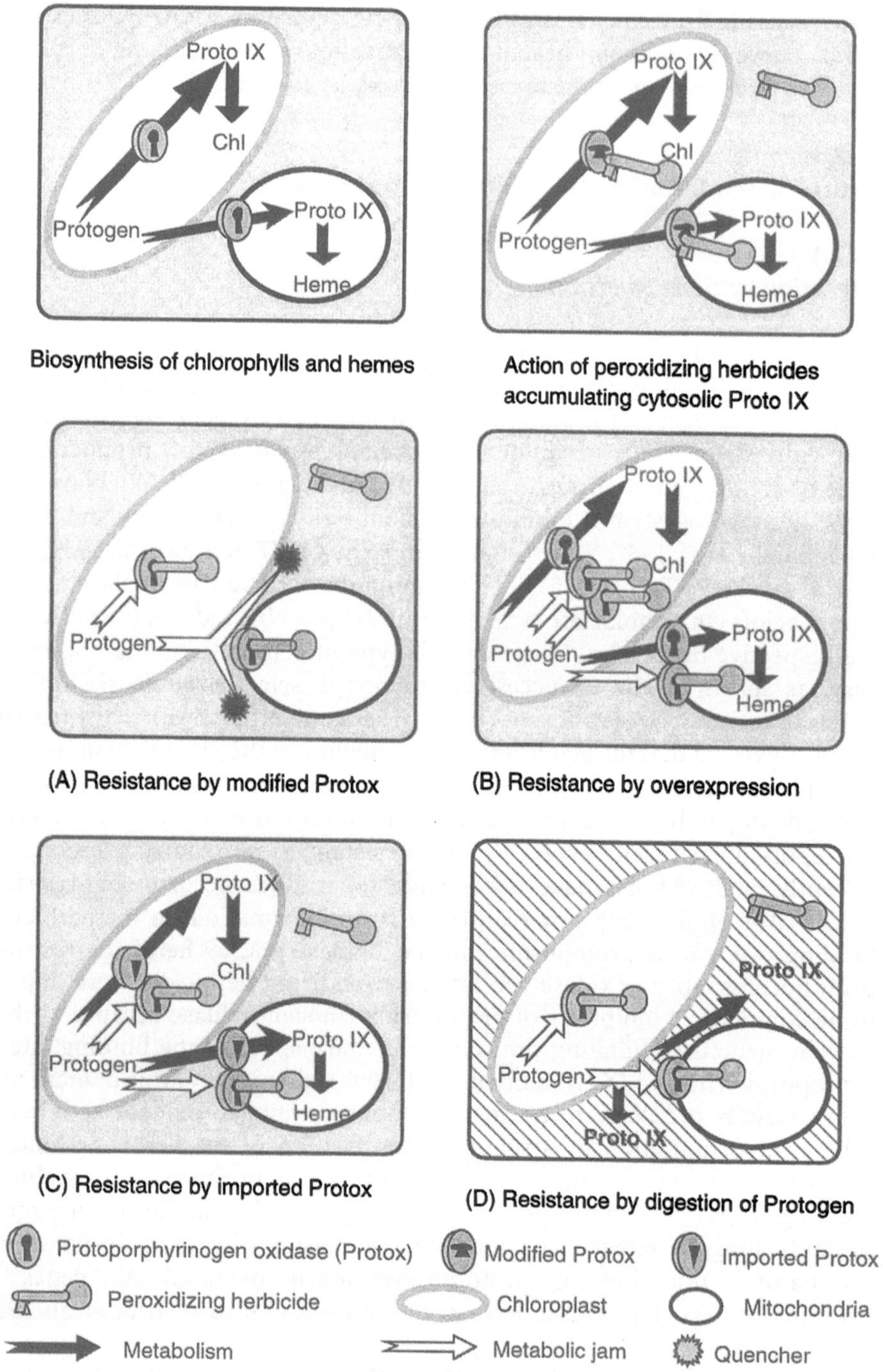

Fig. 1. Protoporphyrin IX biosynthesis and its inhibition, with four possible mechanisms for obtaining resistance in plants under peroxidizing herbicide stress

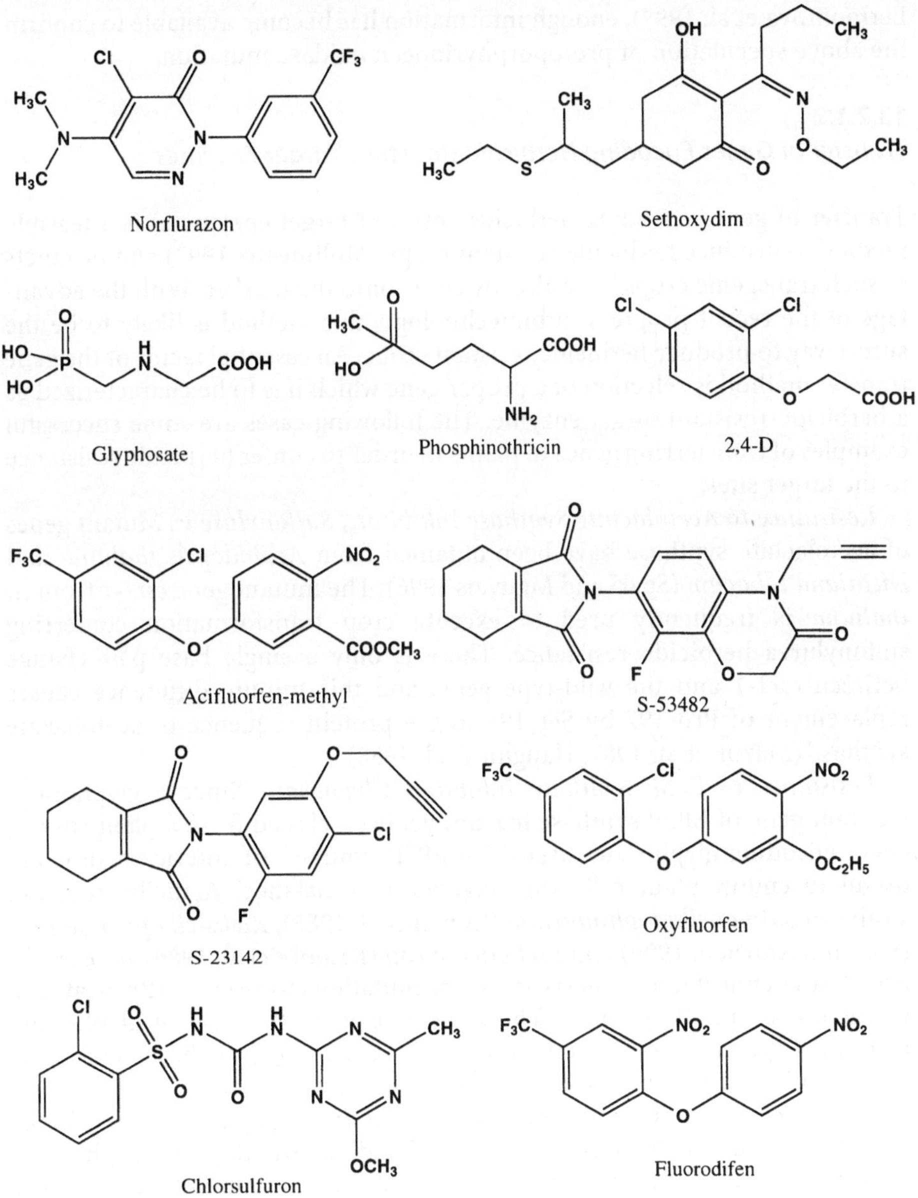

Fig. 2. Structures of herbicides cited in the text

Lermontova et al. 1997), enough information has become available to confirm the above speculation of protoporphyrinogen oxidase mutation.

13.2.1.2
Transfer of Genes Encoding Herbicide-Resistant Target Enzymes

Transfer of genes encoding herbicide-resistant target enzyme(s) is a feasible method to produce herbicide-resistant crops (Mullineaux 1992) and products of such transgenic crops have already come onto the market. With the advantage of the recent progress in biotechnology, this method is likely to be the surest way to produce herbicide resistant crops. An essential factor of the gene transfer method is selection of a proper gene which has to be characterized as a herbicide-resistant target enzyme. The following cases are some successful examples of transferring genes to plants in order to confer herbicide resistance to the target sites.

Resistance to Acetolactate Synthase Inhibitors, Sulfonylureas. Mutant genes of acetolactate synthase have been obtained from *Arabidopsis thaliana* and *Nicotiana tabacum* (Saari and Mauvais 1996). The mutant gene *csr1-1* from *A. thaliana* is frequently used to execute crop transformation conferring sulfonylurea herbicide resistance. There is only a single base pair change between *csr1-1* and the wild-type gene, and this minute difference causes replacement of Pro 197 by Ser 197 in the protein sequence of acetolactate synthase (Mazur et al. 1987; Haughn et al. 1988).

Resistance to EPSP Synthase Inhibitor, Glyphosate. Since a glyphosate-resistant gene of EPSP synthase has not yet been cloned from a plant source, genes encoding glyphosate-insensitive EPSP synthase of enteric bacteria are useful to endow plant cells with glyphosate resistance. Actually, resistant strains of *Salmonella typhimurium* (Comai et al. 1983), *Klebsiella pneumoniae* (Sost and Amrhein 1990) and *Escherichia coli* (Kishore et al. 1986) were established by chemical mutagenesis and these mutations were evidently located in the genes for EPSP synthase. The same is true in the case of glyphosate-resistant *Euglena gracilis* (Reinbothe et al. 1991). Some of these genes have now been introduced into several crop species (Dyer 1994).

Resistance to Phytoene Desaturase Inhibitors, Bleaching Herbicides. Sandmann et al. (1996) reported that herbicide-resistant genes of phytoene desaturase were hardly selected in their tests. A resistant mutant was produced in cyanobacteria, *Synechococcus* PCC 7942, and the gene characterized as responsible for the resistance was cloned to be a phytoene desaturase containing a single point mutation (Chamovitz et al. 1991). On the basis of the sequence similarity between cyanobacteria and some crops, it is suggested that the mutant isolated in cyanobacteria may be useful in planning site-directed mutagenesis in crop species to obtain norflurazon (Fig. 2) resistance (Sherman et al. 1996).

Alternatively, insertion of a gene encoding herbicide-resistant phytoene desaturase into cyanobacteria was carried out using a naturally norflurazon-resistant foreign gene, *crtI*, coding for phytoene desaturase of the bacterium *Erwinia uredovora* (Misawa et al. 1990). Norflurazon resistance of the cyanobacteria transformant was estimated to be 1000-fold higher than that of the control strain. This strategy was applied for a herbicide-resistant tobacco (Misawa et al. 1993). The 5′ region of *crtI* was attached to the sequence of the transit peptide for a pea Rubisco small subunit due to the location of phytoene desaturase on the thylakoid membranes of chloroplasts (Linden et al. 1993), and the vector was put into tobacco plants under the control of the 35S promoter of cauliflower mosaic virus by means of *Agrobacterium*. Immunogold localization then detected the *crtI* gene product at the thylakoid membrane in chloroplasts (Misawa et al. 1993). The bacterial phytoene desaturase in the transgenic tobacco plants enhanced conversion of phytoene into β-carotene and conferred strong resistance against norflurazon. Thus, the norflurazon-resistant plant was obtained by a phytoene desaturase gene which had naturally evolved in bacterial systems insensitive to norflurazon, and the function of the gene was well conserved in the transformant in coexistence with the endogenous plant-type phytoene desaturase. This strategy is presumably applicable for a wide range of herbicides.

13.2.2
Overproduction of the Herbicide-Sensitive Target Enzyme

Overproduction of a herbicide-sensitive target enzyme is an established idea to introduce herbicide resistance into plant cells, as shown by a sethoxydim (Fig. 2)-resistant cell line of maize which exhibited a 90-fold increase in resistance due to a 2.6-fold increase of acetyl coenzyme A carboxylase activity (Parker et al. 1989). Similar results were reported for an L-phosphinothricin (Fig. 2)-resistant cell line of alfalfa with a 100-fold increase in resistance due to a seven fold overproduction of glutamin synthase (Donn et al. 1984), and for a glyphosate-resistant carrot cell line due to a two fold increase of 5-enolpyruvylshikimate-3-phosphate synthase (Nafziger et al. 1984). The above examples strongly imply a realistic way for making up a plant cell line resistant to peroxidizing herbicides provided that it is possible to enhance protoporphyrinogen oxidase activity in chloroplasts.

Overproduction of an enzyme in a plant cell can be generated by either amplification of the structural gene sequence encoding the protein or mutation in the promoter region of the gene, leading to an enhanced rate of transcription. This technique may overcome herbicide damage to the plant cells due to overproduction of a herbicide target enzyme (Stark and Wahl 1984). Gene amplification for overproduction of target protein was first applied to animal cells (O'Hare et al. 1981) and insects (Mouches et al. 1986), and then used in plant biotechnology in order to obtain herbicide resistance. As glyphosate-resistant photoautotrophic cells of *E. gracilis* were selected in the presence of 6 μM glyphosate, it was found that these cells had up to 40-fold higher levels of EPSP synthase compared to the herbicide-susceptible wild-type (Reinbothe et al. 1991). This overproduction correlated with a selective amplification of two out of five genomic sequences that hybridized with the *Saccharomyces cerevisiae* AOR1 gene probe in Southern blots. One of the amplified genomic fragments is assumed to encode the previously identified monofunctional EPSP synthase (Reinbothe et al. 1993). Cell cultures of *Corydalis sempervirens*, tolerant to the herbicide glyphosate, have a 30- to 40-fold increased level of EPSP synthase, and a 10-fold enhanced level of the corresponding mRNA, due to a higher rate of transcription of the gene (Holleander-Czytko et al. 1992). While chlorsulfuron (Fig. 2)-resistant *Daucus carota* cell lines were isolated in a stepwise selection procedure, the specific activity of acetolactate synthase increases along with the herbicide resistance. Southern hybridization analysis has revealed that this resistance is due to gene amplification (Caretto et al. 1994). Also, a glyphosate-tolerant Petunia hybrid cell line overproducing EPSP synthase was shown to overproduce mRNA for the target enzyme as a result of a 20-fold amplification of the corresponding gene (Shah et al. 1986). Mutation in the promoter region of the gene encoding the target protein was found in the case of a norflurazon-resistant mutant of *Synechococcus* which overproduced the herbicide target protein, phytoene desaturase. The amount of phytoene desaturase in resistant *Synechococcus* was estimated to be 20-fold higher than that in the wild-type (Chamovitz et al. 1993). All of the resistant lines described above were obtained by selection for herbicide-resistant mutants in the presence or absence of mutagens.

Transfer of a gene encoding a herbicide-resistant target enzyme was often carried out under control of a constitutive promoter, e.g. 35-S promoter with a duplicated enhancer like p70 (Tourneur et al. 1993), which generally results in high levels of gene expression and, in turn, leads to overproduction of the target enzyme in plants.

13.2.3
Detoxification of Herbicides

Metabolic quenching of herbicide effects in plant systems is a possible strategy for producing herbicide-resistant crops. Plants possess a versatile metabolism for degradation of various compounds so as to detoxify phytotoxic factors including herbicides. A typical detoxification mechanism for herbicides has been explained by structural modification due to glutathione S-transferase, glucosyltransferase or cytochrome P-450 (Coleman et al. 1997). In order to apply this mechanism for peroxidizing herbicide resistance, a new concept of herbicide research is necessary in order to find a suitable herbicide with a well-balanced structure in fragility and phytotoxicity.

A microbial source of 2,4-D (Fig. 2) degradation genes is a suitable example to explain this method (Llewellyn and Last 1996). The breakdown of 2,4-D in the soil is performed by a variety of microorganisms including bacteria, yeasts and fungi from several taxonomic groups. This process was well characterized in *Alcaligenes eutrophus*, and the *tfdA* gene encoding the enzyme which catalyzes the first step in the 2,4-D degradation pathway was used to confer resistance in transgenic plants. By expressing *tfdA* in tobacco under the control of CaMV35S promoter, transgenic tobacco showed a significant degree of resistance which was at least ten fold that of non-transgenic tobacco plants (Lyon et al. 1989; Streber and Willmitzer 1989; Bayley et al. 1992). The enzyme encoded by the *tfdA* gene is specific for 2,4-D and does not degrade natural plant auxins, such as indole acetic acid.

13.3 Methods for Making Crops Resistant to Peroxidizing Herbicides

As for a creation of cops resistant to peroxidizing herbicides, searching for mutant genes of protoporphyrinogen oxidase has many problems similar to the case for phytoene desaturase inhibitors (see Sect. 13.2.1). Microorganisms, animals or even plants occasionally show resistance to peroxidizing herbicides at the cellular level due to different types of protoporphyrinogen oxidase. However, there is no information about mutations of herbicide-sensitive protoporphyrinogen oxidase at all.

13.3.1 Resistant Enzymes and Cells

Early studies (Matringe et al. 1989a,b) showed that diphenyl ether-type herbicides inhibit protoporphyrinogen oxidases of plants and of mouse liver mitochondria. The enzymes from plants, yeast and mouse varied in their response to the herbicides. However, this variation of herbicide-sensitivity between individual enzymes, protoporphyrinogen oxidase, cannnot be explained satisfactory at present. Protoporphyrinogen oxidase of *E. coli* was reported to be resistant against peroxidizing herbicides (Jacobs et al. 1990) and Sasarman et al. (1993) cloned the *hemG* gene from *E. coli* which may encode protoporphyrinogen oxidase. Although *hemG* has been listed as a key gene in the tetrapyrrole biosynthesis, protoporphyrinogen oxidase activity of its gene product has not yet been demonstrated beyond doubt due to poor biochemical data on this matter. Prasad and Dailey reported that murine erythroleukemia cells were acclimated with 5 μM acifluorfen by long term maintenance so as to grow normally even in 100 μM acifluorfen (Fig. 2). In this case, erythroleukemia cells might have obtained the resistance by means of other than protoporphyrinogen oxidase mutation, because acifluorfen sensitivity of protoporphyrinogen oxidase remained unchanged during the acclimation process (Prasad and Dailey 1995). An *N*-phenylimide peroxidizing herbicide, S-53482 (Fig. 2), showed growth retardation, embryo lethality and teratogenicity in rats, while no developmental toxicity was observed in rabbits due to a lower degree of inhibition of protoporphyrinogen oxidase. Indeed, rabbit embryos did not accumulate protoporphyrin IX even at a high dosage of S-53482. However, the herbicide inhibited rat Protox less than rabbit Protox in in vitro experiments. This inconsistency indicates that the difference in the toxicity of S-53482 may be due to a difference in the metabolism of the herbicide, including placental transfer (Kawamura et al. 1996).

A plasma membrane-associated protoporphyrinogen oxidase was assumed to play a critical role in the cytoplasmic accumulation of protoporphyrin IX caused by the oxidation of protoporphyrinogen IX which is exported from protoporphyrinogen oxidase-inhibited etioplasts in acifluorfen-methyl-treated plant tissues (see Chapter 6). The protoporphyrinogen oxidase on plasma membranes would be different from that of etioplasts because of its resistance to the diphenyl ether herbicide (Lee and Duke 1994). A mechanism of oxyfluorfen resistance for a nonchlorophyllous soybean cell line demonstrated that protoporphyrinogen oxidase of the cells was less susceptible to the herbicide than that of a wild plant (Pornprom et al. 1994). Much evidence supports the view that simple life systems such as *E. coli*, *Neurospora crassa*, *Rhodopseudomonas sphaeroides* and *Anacystis nidulans* are insensitive to S-23142 (Fig. 2) which is the most powerful peroxidizing herbicide. However, biochemical studies on the interaction of the herbicide with the respective protoporphyrinogen oxidases have not been completed yet. Recently, the first isolation of an S-23142-resistant strain was carried out by *Chlamydomonas reinhardtii* which involved protoporphyrinogen oxidase less sensitive to the herbicide (Sato et al. 1994).

In order to obtain the first peroxidizing herbicide resistance in higher plants, an S-23142-resistant tobacco cell line was established using photomixotrophic culture with a stepwise selection. From analysis of the mechanisms of the resistance, the cells increased the activity of protoporphyrinogen oxidase likely due to overproduction of the enzyme (Ichinose et al. 1995). Recently, cloning and characterization of protoporphyrinogen oxidase of *Myxococcus* (Dailey and Dailey 1996a), yeast (Camadro and Labbe 1996), mouse (Dailey et al. 1995) and human (Dailey and Dailey 1996b) were reported, but all of these protoporphyrinogen oxidases were very susceptible to the inhibitory action of acifluorfen. Presently, there is only one acifluorfen-resistant protoporphyrinogen oxidase. This is encoded by the *Bacillus subtilis hem* Y gene (Dailey et al. 1994), and expression of this gene in plants might confer resistance against peroxidizing herbicides.

13.3.2
Selection of Natural Plant Resistance

Rice plants are naturally resistant to oxyfluorfen and do not degrade the herbicide faster than susceptible plant species. Therefore, various factors can be suggested to account for the resistant mechanism, including enzymatic resistance to the herbicide (Lee et al. 1991). Mustard is another example of natural resistance against peroxidizing herbicides. The plant is highly resistant to acifluorfen at the tissue level, but protoporphyrinogen oxidase was inhibited by the herbicides both in in vitro and *in* vivo assays (Sherman et al. 1991). There seem to be various mechanisms of protoporphyrin IX accumulation following the treatment of peroxidizing herbicides, and the mechanisms seem to depend upon plant species. It may be natural for plants or plant enzymes to have evolutionary variation in either biosynthetic or metabolic pathways of porphyrin.

13.3.3 Genetic Engineering

The idea of transformation of crops to obtain norflurazon resistance gave us an idea for donating peroxidizing herbicide resistance to plants, namely by implantation of a foreign gene for herbicide-insensitive protoporphyrinogen oxidase which has been known in *B. subtilis* (Dailey et al. 1996). Recently, the resistance to oxyfluorfen was generated in tobacco plants by expression of *B. subtilis* protoporphyrinogen oxidase (Choi et al. 1998). Under control of cauliflower mosaic virus 35S promoter, *B. subtilis* protoporphyrinogen oxidase was expressed in tobacco plants. The transgenic tobacco plants appeared to be resistant to oxyfluorfen and they showed retardation in photobleaching and electrolyte leakage. Enzymatic tests has been carried out and the mechanism of the resistance is still unclear. However, this is the first report of peroxidizing herbicide-resistant transgenic plants.

13.3.3.1 *Genetic Engineering for Overproduction of the Herbicide Target Enzyme*

While there has been no report of a transgenic plant which overproduces herbicide target enzyme, genetic engineering may be a suitable approach to generating resistant plants. New genes of protoporphyrinogen oxidase have been cloned from tobacco (Lermontova et al. 1997) and *Arabidopsis thaliana* (Narita et al. 1996) through screening for complementation of the *E. coli hemG* mutant which accumulates protoporphyrin IX and these genes may allow the production of peroxidizing herbicide-resistant plants: transfer of a protoporphyrinogen oxidase gene into plant cells under control of a suitable promoter may cause overexpression of the enzyme in the plant, leading to resistance to peroxidizing herbicides.

The relationship between expression of a herbicide target enzyme and concomitant herbicide resistance was discussed for the case of transgenic tobacco plants resistant to chlorsulfuron (Charest et al. 1990; Odell et al. 1990; Saari and Mauvais 1996). When the *A. thaliana* gene of acetolactate synthase was overexpressed in tobacco plants, the transgenic plants were only three-fold more resistant to chlorsulfuron than the wild-type. In contrast, expression of the herbicide-resistant gene *csr1-1*, which was obtained from the *A. thaliana* mutant, conferred a high level of chlorsulfuron resistance on tobacco plants. Results of these pathfinders have given a warning of a transgenic strategy for peroxidizing herbicide-resistant plants that overproduction of the herbicide-susceptible protoporphyrinogen oxidase will probably disappoint us with unexpectedly lower herbicide resistance of resultant plants.

It is well known that soil microflora rapidly decomposes many herbicides into non-herbicidal compounds. Genetic and molecular manipulations of bacteria are easier than those of plants, and biochemical experiments are also readily done in a bacterial system, which affords accurate information on metabolism and allows us to identify enzymes diminishing herbicides. Therefore transfer of genes involved in the degradation of herbicides from bacteria into plants is very attractive for the generation of crops resistant to peroxidizing herbicides (Stalker 1989).

The concept adopted in producing 2,4-D-resistant plants can be useful in producing peroxidizing herbicide-resistant plants. Although genes encoding enzymes that detoxify a protoporphyrinogen oxidase inhibitor have not yet been cloned, breakthroughs towards solving this problem can be seen in some reports. As a source of such genes, the following examples are nominated as candidates: diphenyl ether herbicides are metabolized to be inactive by some plants and microorganisms (Schroeder et al. 1990; Lamoureux et al. 1991, 1993; Schmidt et al. 1992; Tanaka et al. 1996); also fluordifen resistance of peanut (Eastin 1971) and acifluorfen resistance of soybean (Frear et al. 1983) were suggested to be candidates for genes that are involved in metabolic detoxification of peroxidizing herbicides; alternatively, an extensive screening of bacteria is likely to produce the desired gene, because of the promising result in the biodegradation research on peroxidizing herbicides (Tanaka et al. 1996).

13.3.3.2
Degradation of Excess Protoporphyrinogen

The mode of action of peroxidizing herbicides in plants is totally different from other herbicides: inhibition of protoporphyrinogen oxidase itself is not fatal to plants, but the inhibition results in an intracellular accumulation of the photosensitizing protoporphyrin IX followed by light-dependent membrane damage. The cytosolic accumulation of protoporphyrin IX is caused by a flood of protoporphyrinogen IX into herbicide-inhibited chloroplasts. The excess protoporphyrinogen IX flows into the cytosol and is associated with subsequent rapid oxidation due to herbicide-insensitive oxidase(s) at plasma membranes (Duke et al. 1991, 1994; Jacobs et al. 1994a; Lee and Duke 1994). According to this scheme, if there is a suitable mechanism for degradation of the cytosolic protoporphyrinogen IX, this would provide a new method for the production of herbicide resistance in crops. Jacobs et al. reported that plant extracts degraded protoporphyrinogen IX into nonporphyrin compounds which were no longer photosensitizers (Jacobs et al. 1994b). Therefore, gene enhancement of this degradation enzyme may protect crops from the damage of peroxidizing herbicides. Actually, the destruction of protoporphyrinogen IX was least active in young leaves of cucumber, a plant highly susceptible to the herbicide, while higher levels of protoporphyrinogen IX destruction were found in leaves of broadleaf mustard and radish, two plants exhibiting herbicide tolerance (Jacobs et al. 1996).

Today we are on the way to comprehend the determinants of peroxidizing herbicide resistance, which are widely divergent. This chapter offers the conclusion that the widespread development and employment of peroxidizing herbicide-resistant crops will come soon with the use of the sophisticated biotechnology.

References

Abe T, Bae CH, Takahashi H, Kumata S, Yoshida S (1997) Effective plant-mutation method using heavy-ion beams (II). RIKEN Accerelator Progress Report, vol 31:148

Ayotte R, Harney PN, Souza Machado V (1989) The transfer of triazine resistance from *Brassica napus* to *B. oleracea* L. IV. Second and third backcrosses to *B. oleracea* and recovery of an 18-chromosome, triazine-resistant backross. Euphytica 40:15–19

Barsby TL, Kemble RJ, Yarrow SA (1987) Brassica hybrids and their utility in plant breeding. Plant Mol Biol 140:223–228

Bayley C, Trolinder C, Morgan M, Quisenberry JE, Ow DW (1992) Engineering 2,4-D resistance into cotton. Theor Appl Genet 83:645–649

Camadro JM, Labbe P (1996) Cloning and characterization of the yeast HEM14 gene coding for protoporphyrinogen oxidase, the molecular target of diphenyl ether-type herbicides. J Biol Chem 271:9120–9128

Caretto S, Giardina MC, Nicolodi C, Mariotti D (1994) Chlorsulfuron resistance in *Daucus carota* cell lines and plants: Involvement of gene amplification. Theor Appl Genet 88(5):520–524

Chamovitz D, Pecker I, Hirschberg J (1991) The molecular basis of resistance to the herbicide norflurazon. Plant Mol Biol 16:967–974

Chamovitz D, Sandmann G, Hirschberg J (1993) Molecular and biochemical characterization of herbicide-resistant mutants of cyanobacteria reveals that phytoene desaturation is a rate-limiting step in carotenoid biosynthesis. J Biol Chem 268:17348–17353

Charest PJ, Hattori J, DeMoor J, Iyer VN, Miki BL (1990) In vitro study of transgenic tobacco expressing *Arabidopsis* wild type and mutant acetohydroxyacid synthase genes. Plant Cell Rep 8:643–646

Choi KW, Han O, Lee HJ, Yun YC, Moon YH, Kuk YI, Han SU, Guh JO (1998) Generation of resistance to diphenyl ether herbicide oxyfluorfen via expression of *Bacillus subtilis* protoporphyrinogen oxidase gene in transgenic tobacco plants. Biosci Biotech Biochem 62:558–560

Coleman JOD, Blake-Kalff MMA, Emyr Davies TG (1997) Detoxification of xenobiotics by plants: chemical modification and vacuolar compartmentation. Trends Plant Sci 2:144–151

Comai L, Sen LC, Stalker DM (1983) An altered *aroA* gene product confers resistance to the herbicide glyphosate. Science 221:370–371

Dailey HA, Dailey TA (1996a) Protoporphyrinogen oxidase of *Myxococcus xanthus*. J Biol Chem 271:8714–8718

Dailey TA, Dailey HA (1996b) Human protoporphyrinogen oxidase: expression, purification, and characterization of the cloned enzyme. Protein Sci 5:98–105

Dailey TA, Dailey HA, Meissner P, Prasad AR (1995) Cloning, sequence, and expression of mouse protoporphyrinogen oxidase. Arch Biochem Biophys 324:379–384

Dailey TA, Meissner P, Dailey HA (1994) Expression of a cloned protoporphyrinogen oxidase. J Biol Chem 269:813–815

Darmency H, Pernes J (1985) Use of wild *Setaria italica* (L.) Beauv. to improve triazine resistance in cultivated *S. italica* (L.) by hybridization. Weed Res 25:175–180

Donn G, Tischer E, Smith JA, Goodman HM (1984) Herbicide-resistant alfalfa cells; an example of gene amplification in plants. J Mol Appl Genet 2:621–635

Duke SO, Becerril JM, Lydon J, Matsumoto H, Sherman TD (1991) Protopophyrinogen oxidase-inhibiting herbicides. Weed Sci 39:465–473

Duke SO, Lee HJ, Nandihalli UB, Duke MV (1994) Protoporphyrinogen oxidase as the optimal herbicide site in the porphyrin pathway. In: Duke SO, Rebeitz CA (eds) Porphyric pesticides. American Chemical Society, Washington DC, pp 191–205

Dyer WE (1994) Resistance to glyphosate. In: Powles SB, Holtum JAM (eds) Herbicide resistance in plants. Lewis Publishers, Boca Raton, pp 229–242

Eastin EF (1971) Fate of fluorodifen in resistant peanut seedlings. Weed Sci 19:261–266

Frear DS, Swanson HR, Mansager ER (1983) Acifluorfen metabolism in soybean: diphenylether bond cleavage and the formation of homoglutathione, cysteine, and glucose conjugates. Pestic Biochem Physiol 20:299–308

Gupta AS, Heinen JL, Holaday AS, Burke JJ, Allen RD (1993) Increased resistance to oxidative stress in transgenic plants that overexpress chloroplastic copper-zinc superoxide dismutase. Proc Natl Acad Sci USA 90:1629–1633

Haughn GW, Smith J, Mazur B, Somerville C (1988) Transformation with a mutant *Arabidopsis* acetolactate synthase gene renders tobacco resistant to sulfonylurea herbicides. Mol Gen Genet 211:266–271

Holleander-Czytko H, Sommer I, Amrhein N (1992) Glyphosate tolerance of cultured *Corydalis sempervirens* cells is acquired by an increased rate of transcription of 5-enolpyruvylshikimate-3-phosphate synthase as well as by a reduced turnover of the enzyme. Plant Mol Biol 20:1029–1036

Horikoshi M, Mametsuka K, Hirooka T (1998) Molecular breeding of photobleaching herbicide-resistant plant (II): Molecular basis of photobleaching herbicide resistance in tobacco. Abstracts of the 23rd Annual Meeting of Pesticide Science Society of Japan, Matsue, Japan, p 77

Ichinose K, Che FS, Kimura Y, Matsunobu A, Sato F, Yoshida S (1995) Selection and characterization of protoporphyrinogen oxidase inhibiting herbicide (S23142) resistant photomixotrophic cultured cells of *Nicotiana tabaccum*. J Plant Physiol 146:693–698

Jacobs JM, Jacobs NJ, Borotz SE, Guerinot ML (1990) Effects of the photobleaching herbicide, acifluorfen methyl, on protoporphyrinogen oxidation in barley organelles, soybean root mitochondria, soybean root nodules, and bacteria. Arch Biochem Biophys 280:369–375

Jacobs JM, Jacobs NJ, Sherman TD, Duke SO (1994a) Effect of diphenyl ether herbicides on oxidaton of protoporphyrinogen to protoporphyrin in organellar and plasma membrane-enriched fractions of barley. Plant Physiol 97:197–203

Jacobs JM, Wehner JM, Jacobs NJ (1994b) Porphyrin stability in plant supernatant fractions: implications for the action of porphyrinogenic herbicides. Pestic Biochem Physiol 50:23–30

Jacobs JM, Jacobs NJ, Duke SO (1996) Protoporphyrinogen destruction by plant extracts and correlation with tolerance to protoporphyrinogen oxidase-inhibiting herbicides. Pestic Biochem Physiol 55:77–83

Kawamura S, Kato T, Matsuo M, Katsuda Y, Yasuda M (1996) Species difference in protoporphyrin IX accumulation produced by an N-phenylimide herbicide in embryos between rats and rabbits. Toxicol Appl Pharmacol 141:520–525

Kishore GM, Brundage L, Kolk K, Padgette SR, Rochester D, Huynh QK, della-Cioppa G (1986) Isolation, purification and characterization of a glyphosate tolerant mutant. Fed Proc 45:1506

Lamoureux GL, Rusness DG, Schroeder P, Rennenberg H (1991) Diphenyl ether herbicide metabolism in a spruce cell suspension culture: the identificaton of two novel metabolites derived from a glutathione conjugate. Pestic Biochem Physiol 39:291–301

Lamoureux GL, Rusness DG, Schroeder P (1993) Metabolism of a diphenylether herbicide to a volatile thianisole and a polar sulfonic acid metabolite in spruce (*Picea*). Pestic Biochem Physiol 47:8–20

Lee HJ, Duke SO (1994) Protoporphyrinogen IX-oxidizing activities involved in the mode of action of peroxidizing herbicides. J Agric Food Chem 42:2610–2618

Lee JJ, Matsumoto H, Pyon JY, Ishizuka K (1991) Mechanism of selectivity of diphenyl ether herbicides oxyfluorfen and chlomethoxynil in several plants. Weed Res (Tokyo) 36:162–170

Lee JJ, Matsumoto H, Ishizuka K (1992) Light involvement in oxyfluorfen-induced protoporphyrin IX accumulation in several species of intact plants. Pestic Biochem Physiol 44:119–125

Lermontova I, Kruse E, Mock H-P, Grimm B (1997) Cloning and characterization of a plastidal and a mitochondrial isoform of tobacco protoporphyrinogen IX oxidase. Proc Natl Acad Sci USA 94:8895–8900

Linden H, Lucas MM, de Felipe MR, Sandmann (1993) Immunogold localization of phytoene desaturase in higher plant chloroplasts. Plant Physiol 88:229–236

Llewellyn D, Last D (1996) Genetic engineering of crops for tolerance to 2,4-D. In: Duke SO (ed) Herbicide-resistant crops, CRC Press, Boca Raton, pp 159–174

Lyon BR, Llewellyn DJ, Huppatz J, Dennis ES, Peacock WJ (1989) Expression of a bacterial gene in transgenic tobacco confers resistance to the herbicide 2,4-dichlorophenoxyacetic acid. Plant Mol Biol 13:533–540

Matringe M, Camadro JM, Labb P, Scalla R (1989a) Protopophyrinogen oxidase as a molecular target for diphenyl ether herbicides. Biochem J 260:231–235

Matringe M, Camadro JM, Labb P, Scalla R (1989b) Protoporphyrinogen oxidase inhibition by three peroxidizing herbicides: oxadiazon, LS82-556 and M & B 39279. FEBS Lett 245:35–38

Mazur BJ, Chui CF, Smith JK (1987) Isolation and characterization of plant genes coding for acetolactate synthase, the target enzyme for two classes of herbicides. Plant Physiol 85:1110–1117

McHughen A (1989) *Agrobacterium* mediated transfer of chlorsulfuron resistance to commercial flax cultivars. Plant Cell Rep 8:445–449

McKersie BD, Chen Y, De-Beus M, Bowley SR, Inze D, D'Halluin K, Botterman J (1993) Superoxide dismutase enhances tolerance of freezing stress in transgenic alfalfa (*Medicago sativa* L.). Plant Physiol (Rockville) 103:1155–1163

Mets L, Thiel A (1989) Biochemistry and genetic control of the photosystem II herbicide target site. In: Böger P, Sandmann G (eds) Target sites of herbicide action. CRC Press, Boca Raton, pp 2–24

Misawa N, Nakagawa M, Kobayashi K, Yamano S, Izawa Y, Nakamura K, Harashima K (1990) Elucidation of the *Erwinia uredovora* carotenoid biosynthetic pathway by functional analysis of gene products expressed in *Escherichia coli.* J Bacteriol 172:6704–6712

Misawa N, Yamano S, Linden H, de Felipe M.R, Lucas M, Ikenaga H, Sandmann G (1993) Functional expression of the *Erwinia uredovora* carotenoid biosynthesis gene crtI in transgenic plants showing an increase of β-carotene biosynthesis activity and resistance to the bleaching herbicide norflurazon. Plant J 4:833–840

Mouches C, Pasteur N, Berge J, Hyrien O, Raymond M, Vincent B, Silvestri B, Georghiou G (1986) Amplification of an esterase gene is responsible for insecticide resistance in a California *Culex* mosquito. Science 233:778–780

Mullineaux PM (1992) Genetically engineered plants for herbicide resistance. In: Gatehouse AMR, Hilder VA, Boulder D (eds) Plant genetic manipulation for crop protection. Biotechnology in Agriculture Series 7:75–107

Nafziger EM, Widholm JM, Steinrücken HC, Killmer JL (1984) Selection and characterization of a carrot cell line tolerant to glyphosate. Plant Physiol 76:571–574

Nandihalli UB, Duke MV, Duke SO (1992) Quantitative structure-activity relationships of protoporphyrinogen oxidase-inhibiting diphenyl ether herbicides. Pestic Biochem Physiol 43:193–211

Narita S, Tanaka R, Ito T, Okada K, Taketani S, Inokuchi H (1996) Molecular cloning and characterization of a cDNA that encodes protoporphyrinogen oxidase of *Arabidopsis thaliana*. Gene 182:169–175

Odell JT, Caimi PG, Yadav NS, Mauvais CJ (1990) Comparison of increased expression of wild-type and herbicide-resistant acetolactate synthase genes in transgenic plants, and indication of posttranscriptional limitation on enzyme activity. Plant Physiol 94:1647–1654

O'hare K, Benoist C, Breathnach R (1981) Transformation of mouse fibroblasts to methotrexate resistance by a recombinant plasmid expressing a prokaryotic dihydrofolate reductase. Proc Natl Acad Sci USA 78:1527–1531

Padgette SR, Re DB, Barry GF, Eichholtz DE, Delannay X, Fuchs RL, Kishore GM, Fraley RT (1996) New weed control opportunities: development of soybeans with a ROUNDUP READY™ gene. In: Duke SO (ed) Herbicide resistant crops. CRC Press, Boca Raton, pp 54–84

Parker WB, Somers DA, Wyse DL, Keith RA, Burton JD, Gronwald JW, Gengenbach BG (1989) Selection and characteriztion of sethoxydim-tolerant maize tissue cultures. Plant Physiol 92:1220–1225

Pornprom T, Matsumoto H, Usui K, Ishizuka K (1994) Characterization of oxyfluorfen tolerance selected soybean cell line. Pestic Biochem Physiol 50:107–114

Prasad ARK, Dailey HA (1995) Generation of resistance to the diphenyl ether herbicide acifluorfen by MEL cells. Biochem Biophys Res Commun 215:186–191

Reinbothe S, Nelles A, Parthier B (1991) N-(phosphonomethyl)glycine (glyphosate) tolerance in *Euglena gracilis* acquired by either overproduced or resistant 5-enolpyruvylshikimate-3-phosphate synthetase. Eur J Biochem 198:365–374

Reinbothe S, Ortel B, Parthier B (1993) Overproduction by gene amplification of the multifunctional arom protein confers glyphosate tolerance to a plastid-free mutant of *Euglena gracilis*. Mol Gen Gene 239:416–424

Ryan GF (1970) Resistance of common groundsel to simazine and atrazine. Weed Sci 18:614–616

Saari LL, Mauvais CJ (1996) Sulfonylurea herbicide-resistant crops. In: Duke SO (ed) Herbicide-resistant crops. CRC Press, Boca Raton, pp 127–142

Sandmann G, Misawa N, Böger P (1996) Step towards genetic engineering of crops resistant against bleaching herbicides. In: Duke SO (ed) Herbicide-resistant crops. Lewis Publishers, Boca Raton, pp 189–200

Sasarman A, Letowski J, Czaika G, Ramirez V, Nead MA, Jacobs J, Morais R (1993) Nucleotide sequence of the *hem*G gene involved in the protoporphyrinogen oxidase activity of *Escherichia coli* K12. Can J Microbiol 39:1155–1161

Sato R, Yamamoto M, Shibata H, Oshio H, Harris EH, Gillham NW, Boyton JE (1994) Characterization of a protoporphyrinogen oxidase mutant of *Chlamydomonas reinhardtii* resistant to protoporphyrinogen oxidase inhibitors. ACS Symp Ser 559:91–104

Shah D, Horsch R, Klee H, Kishore G, Winter J, Tumer N, Hironaka C, Sanders P, Gasser C, Aykent S, Siegel N, Rogers S, Fraley R (1986) Engineering herbicide tolerance in transgenic plants. Science 233:478–481

Schmidt S, Wittich RM, Fortnagel P, Erdmann D, Francke W (1992) Metabolism of 3-methyldiphenyl ether by *Sphingomonas* sp. SS31. FEMS Microbiol Lett 96:253–258

Schroeder P, Lamoureux GL, Rusness DG, Rennenberg H (1990) Glutathione S-transferase activity in spruce needles. Pestic Biochem Physiol 37:211–218

Sherman TD, Becerril JM, Matsumoto H, Duke MV, Jacobs JM, Jacobs NJ, Duke SO (1991) Physiological basis for differential sensitivities of plant species to protoporphyrinogen oxidase-inhibiting herbicides. Plant Physiol 97:280–287

Sherman TD, Vaughn KC, Duke SO (1996) Mechanisms of action and resistance to herbicides. In: Duke SO (ed) Herbicide-resistant crops. Lewis Publishers, Boca Raton, pp 13–36

Sost D, Amrhein N (1990) Substitution of Gly-96 to Ala in the 5-enolpyruvylshikimate-3-phosphate synthase of *Klebsiella pneumoniae* results in a greatly reduced affinity for the herbicide glyphosate. Arch Biochem Biophys 282:433–436

Stalker DM (1989) Producing herbicide-resistant plants by gene transfer technology. In: Böger P, Sandmann G (eds) Target sites of herbicide action. CRC Press, Boca Raton, pp 147–163
Stark GR, Wahl GM (1984) Gene amplificaton. Annu Rev Biochem 53:447–475
Streber WR, Willmitzer L (1989) Transgenic tobacco expressing a bacterial detoxifying gene are resistant to 2,4-D. Biotechnology 7:811–815
Tanaka Y, Iwasaki H, Kitamori S (1996) Biodegradation of herbicide chlornitrofen (CNP) and mutagenicity of its degradation products. Water Sci Tech 34:15–20
Tourneur C, Jouanin L, Vaucheret H (1993) Over-expression of acetolactate synthase resistant to valine in transgenic tobacco. Plant Sci 88:159–168
Trebst A (1991) The molecular basis of resistance of photosystem II herbicides. In: Caseley JC, Cussans GW, Atkin RK (eds) Herbicide resistance in weeds and crops. Long Ashton Int Symp, Butterworth-Heinemann, Boston, pp 145–164
Von Wettstein D, Chua NH (1987) (eds) Plant molecular biology. Plenum Press, New York
Wang JM, Asami T, Che FS, Murofushi N, Yoshida S (1997) Photobleaching activity of 2-(phenylamino)methylidenecyclohexane-1,3-diones in tobacco (*Nicotiana tabaccum*) cultured cells. J Agric Food Chem 45:2728–2734
Wiersma PA, Schmiemann MG, Condie JA, Crosby WL, Maloney MM (1989) Isolation, expression and phylogenetic inheritance of an acetolactate synthase gene from *Brassica napus*. Mol Gen Genet 219:413–420

CHAPTER 14

Metabolism and Degradation of Porphyrin Biosynthesis Inhibitor Herbicides

HIROYASU AIZAWA[1] and HUGH M. BROWN[2]

Contents

[1] DuPont Agricultural Products, DuPont Kabushiki Kaisha, Tokyo, Japan
[2] DuPont Agricultural Products, Stine-Haskell Research Center, Newark, Delaware, USA

Peter Böger, Ko Wakabayashi
Peroxidizing Herbicides

14.1 Introduction

The metabolism and degradation of herbicides are key factors determining their weed control activity, crop selectivity, environmental fate and safety to non-target organisms. Numerous compounds which inhibit protoporphyrinogen oxidase (i.e. peroxidizing herbicides) have been commercialized or are under development. These compounds span a wide range of use patterns, environmental fate profiles and toxicological properties which result at least in part from their different susceptibilities to plant, animal and microbial metabolism and/or abiotic degradation. In this chapter, we describe the metabolism and degradation of 19 commercialized or candidate peroxidizing herbicides in a number of plant, animal, microbial, soil, water and light systems. These herbicides fall into two subclasses: (1) *p*-nitro diphenyl ethers, and (2) heterocyclic-substituted benzenes. Data from published reports have been augmented by unpublished results from regulatory studies, when available. As for many large herbicide classes (and the peroxidizing herbicides are one of the largest), susceptibility to metabolism and abiotic degradation is quite dependent on chemical structure, and generalizations about such properties are difficult. For example, some peroxidizing herbicides are relatively recalcitrant to metabolism in plants and rely on soil sorption/placement for crop selectivity. Others are readily metabolized in certain plants and postemergence crop selectivity often results from metabolic inactivation. Similarly, there is wide variability among these herbicides in susceptibility to soil microbial degradation. However, one generalization is that peroxidizing herbicides are subject to the same metabolic transformations as are other agrichemicals including aryl and aliphatic hydroxylation, hydrolysis, deesterification, nitro reduction, O-dealkylation, and conjugation with glutathione as well as multiple secondary conjugation reactions.

14.2 Metabolism and Degradation of *p*-Nitro Diphenylether Herbicides

The following section describes the biotic and abiotic transformations of eight *p*-nitro diphenyl herbicides in plants, animals, soils, and water. The compounds included can be clustered into three substructural categories based on the substituents on the 2,4-disubstituted phenyl ring. These subclasses are: 2-chloro-4-trifluoromethyl phenyl derivatives including acifluorfen (and its esters), fluoroglycofen ethyl, fomesafen, and oxyfluorfen; 2,4-dichlorophenyl derivatives including nitrofen, chlomethoxyfen, and bifenox; and fluorodifen, which is a 2-nitro-4-trifluoromethyl phenyl derivative. Although the published literature is incomplete and does not always cover the same metabolic systems for each active ingredient, it is possible to compare some aspects of the metabolism and degradation of these eight analogs.

Nitro Reduction. All commercialized diphenylether herbicides contain a *p*-nitrophenyl substituent, and this *p*-nitro group appears to be generally susceptible to reduction to an amino group. Reports of this reaction in soil, cultures of soil microorganisms, animals, and plants are found for various analogs. Nitro group reduction is also commonly found in sterile aqueous photolytic systems. In living systems, the resulting amino group is often acetylated. Although the literature may be incomplete, the reports of nitro reduction in plants appear limited to diphenylether analogs (nitrofen, chlomethoxyfen, bifenox) which are not readily cleaved and conjugated by reaction with glutathione, possibly indicating that in living systems only the intact diphenylether structure is recognized as substrate for nitro reduction.

Deesterification. Several diphenylethers have carboxyester groups substituted on the *meta* position of the nitrophenyl ring. In each case, the ester is readily cleaved in a variety of systems, producing the carboxylic acid derivative, which is also herbicidally active. Thus, fluoroglycofen, lactofen and acifluorfen methyl are each converted to aciflurofen, and bifenox is deesterifed to its active carboxylic acid form. The ester forms of these herbicides have different activities against weeds, different crop selectivities, and generally tighter soil binding properties (prior to deesterification) than the corresponding carboxylic acids. It should be recognized that the rapid half-lives reported for some of these analogs in soil and other systems do not represent inactivation but rather conversion to a herbicidally active derivative which often degrades or is metabolized more slowly.

Fomesafen is an interesting analog of acifluorfen having an N-methylsulfonyl carboxamide substituent (see Fig. 3). Fomesafen is a weak acid (pKa = 2.7) due to ionization of the sulfonamide moiety, and this substituent mimics the free carboxylic acid group found in the same position of acifluorfen.

Glutathione conjugation. Some diphenylether herbicides are susceptible to inactivation by glutathione conjugation and subsequent cleavage of the ether bridge. This reaction is important to postemergence crop selectivity in cereals and soybeans since it can occur rapidly enough to confer crop tolerance to these exceptionally fast-acting herbicides. Fluorodifen was the first diphenylether herbicide found to be readily conjugated to glutathione in plants (Shimabukuro et al. 1973) and animals (Lamoureux and Davison 1975). It appears that the combination of electron-withdrawing groups in appropriate *ortho* and *para* positions on both phenyl rings makes one of the ether bridge carbons susceptible to nucleophilic attack by glutathione. The reaction is catalyzed by a glutathione-S-transferase (GST) (Frear and Swanson 1973) and in the case of fluorodifen is chemically facilitated by the good leaving group properties of *p*-nitrophenol. Acifluorfen is also susceptible to this reaction, and rapid conjugation to homoglutathione is the basis for soybean tolerance to this herbicide (Frear et al. 1983). In this case, it is interesting that 2-chloro-4-trifluoromethyl phenol is the leaving group (rather than p-nitro phenol), indicating a subtle shift either in chemical reactivity across the ether bridge or in binding to the GST enzyme active site and subsequent reaction. Fluoroglycofen is deesterified in wheat and the resultant acifluorfen metabolite is subsequently conjugated to glutathione, conferring tolerance in this crop (Jacobson 1989). Fomesafen is also cleaved *via* conjugation to glutathione in soybeans (Evans et al. 1987).

However, many diphenylethers are not susceptible to this reaction. Oxyfluorfen is a close analog of acifluorfen (see Fig. 3) which has not been found to conjugate with glutathione in vivo in either plants or animals. Frear and Swanson (1973) found that oxyfluorfen was not susceptible to glutathione conjugation in vitro using a glutathione-s-transferase from peas that could readily catalyze glutathione conjugation with fluorodifen. Although substantially similar in structure to acifluorfen, it is possible that the *m*-ethoxy group of oxyfluorfen contributes electron density to the system, reducing the susceptibility of the ether bridge to nucleophilic attack. Note that both anionic (acifluorfen) and neutral (fluorodifen) diphenylethers are susceptible to this reaction, indicating that the enzyme(s) responsible for this reaction can react with both forms, so the fact that oxyfluorfen is uncharged is not likely to explain its lack of reaction with glutathione.

There are no reports of glutathione conjugation with the 2,4-dichlorophenyl diphenylethers such as nitrofen, chlomethoxyfen, and bifenox. This is consistent with the relatively slower metabolism of these analogs in plants and their general reliance on soil sorption for selectivity. The chlorine substituents of these herbicides appear to provide inadequate activation of the diphenylether bridge to nucleophilic attack, possibly due to weaker electron withdrawing properties *vis-a vis* trifluoromethyl and nitro groups.

14.2.1
Acifluorfen

Acifluorfen {5-[2-chloro-4-(trifluoromethyl)phenoxy]-2-nitrobenzoic acid; CAS No. 62476-59-9} is a diphenylether herbicide used for postemergence weed control in soybeans, peanuts and rice. Application rates of 140–420 g a.i./ha control a range of broadleaf weeds including *Abutilon*, *Amaranthus*, *Chenopodium*, *Datura*, *Ipomea*, *Polygonum*, and *Xanthium* species, among others (Ahrens 1994). Application at early growth stages or higher rates can control or suppress some grass species. Ritter and Coble (1981) found that acifluorfen was more rapidly metabolized in tolerant soybeans than in sensitive *Xanthium pensylvanicum* and *Ambrosia artemisiifolia*. It is not readily absorbed or translocated in these plants, but metabolic inactivation of that

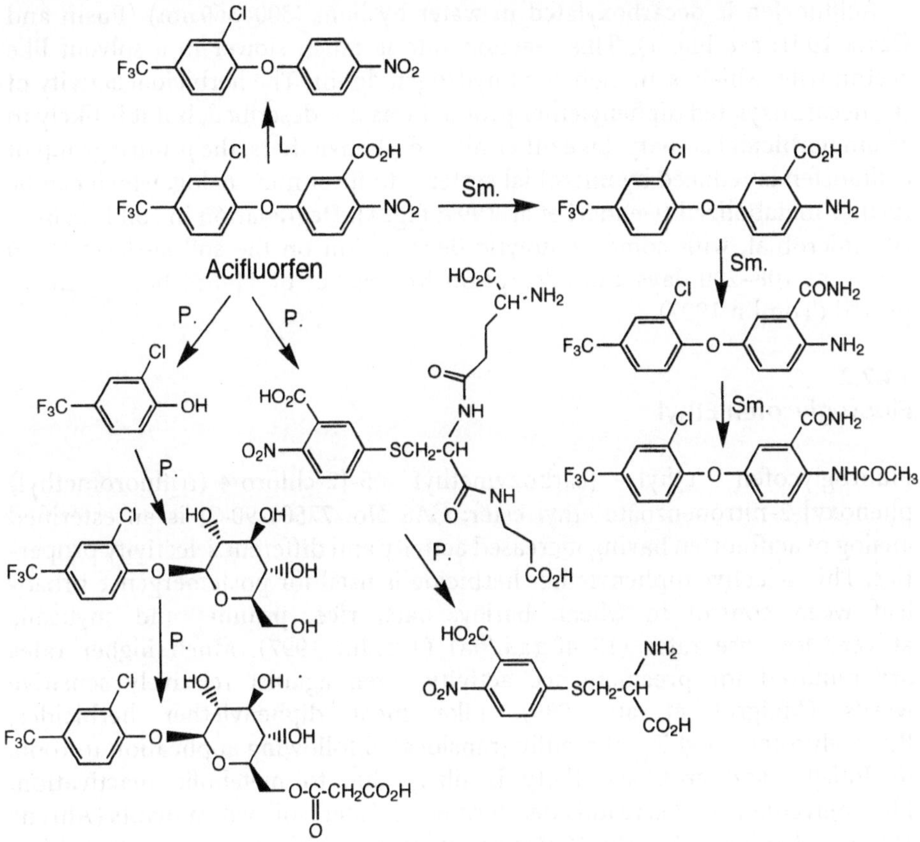

Fig. 1. Metabolism and degradation of acifluorfen in soybeans (*P*), anaerobic cultures of soil microorganisms (*Sm*), and light (*L*)

fraction which is absorbed is important to its crop selectivity. Frear et al. (1983) found that acifluorfen metabolism in soybeans is initiated by nucleophilic attack by homoglutathione [probably catalyzed by a (homo)glutathione-s-transferase] resulting in diphenylether bridge cleavage (Fig. 1). This reaction in soybeans is analogous to that of fluorodifen in peas (Frear and Swanson 1973), peanuts (Shimabukuro et al. 1973), *Picea* spp. (Lamoureux et al. 1991) and rats (Lamoureux and Davison 1975), where glutathione is the predominant thiol. It is interesting to note that with acifluorfen, the 2,4-disubstituted phenol is the leaving group in the reaction whereas in the case of fluorodifen, this phenol receives the sulfhydryl-linked glutathione. As seen in Fig. 1, the homoglutathione conjugate is further metabolized to the cysteine derivative, and the phenol forms a glucoside and a malonyl-D-glucoside. Acifluorofen also appears to be conjugated to glutathione in peanuts; Raub et al. (1996) found the S-(3-carboxy-4-nitrophenyl) cysteine conjugate (CYS) as one of 16 minor metabolites in peanut seeds and hulls.

Acifluorfen is decarboxylated in water by light (300–360 nm) (Pusin and Gessa 1991; see Fig. 1). This reaction rate is much slower in a solvent like acetonitrile, which is an inefficient hydrogen donor. The herbicidal activity of the decarboxylated diphenylether product was not described, but it is likely to retain herbicidal activity. Like other nitro diphenylethers, the *p*-nitro group of acifluorfen is reduced by microbial systems to its amino analog, which can be further metabolized (Gennari et al. 1994; Fig. 1). Degradation in soil is primarily microbial, with some photolytic degradation on the soil surface. DT50 values of 108–200 days and adsorption Koc values of 44–684 have been reported (Tomlin 1997).

14.2.2 Fluoroglycofen Ethyl

Fluoroglycofen ethyl {carboxymethyl 5-[2-chloro-4-(trifluoromethyl) phenoxy]-2-nitrobenzoate ethyl ester; CAS No. 77501-90-7} is an esterified analog of acifluorfen having increased activity and different selectivity properties. This selective diphenylether herbicide is used for postemergence broadleaf weed control in wheat, barley, oats, rice, peanuts and soybeans at very low use rates (15–40 g a.i./ha) (Tomlin 1997). Much higher rates are required for preemergence activity, even against relatively sensitive weeds (Maigrot et al. 1989). Like most diphenylether herbicides, fluoroglycofen ethyl is not readily translocated following application to roots or foliage, and crop selectivity is likely due to metabolic inactivation. Fluoroglycofen ethyl is readily deesterified in water, soil, wheat plants (Ahrens 1994), and animals (Tomlin 1994) to acifluorfen, which is subsequently subject to conjugation to glutathione in cereal plants (Jacobson 1989). It is also subject to the degradative pathways described for acifluorfen in other matrices (Fig. 2).

14.2.3 Fomesafen

Fomesafen {5-[2-chloro-4-(trifluoromethyl)phenoxy]-N-(methylsulfonyl)-2-nitrobenzamide; CAS No. 721178-02-0} is a diphenylether herbicide for early postemergence control of broadleaf weeds at 280–420 g a.i./ha. Soybeans, *Phaseolus* spp. and some leguminous cover crops are tolerant to this herbicide. Fomesafen is a weak acid (pKa = 2.7) (Wauchope et al. 1992) and is also marketed in the sodium salt form. The anionic character of fomesafen leads to higher soil mobility (average Koc = 34–164 ml/g) relative to some other diphenylether herbicides (Ahrens 1994). It is readily photolyzed on the soil surface and degrades more rapidly in soil under anaerobic ($DT_{50} < 1$–2 months) than aerobic ($DT_{50} > 6$ months) conditions (Ahrens 1994; Tomlin 1997). Like fluorodifen and acifluorfen, fomesafen is susceptible to nucleophilic attack and undergoes bridge cleavage through conjugation with

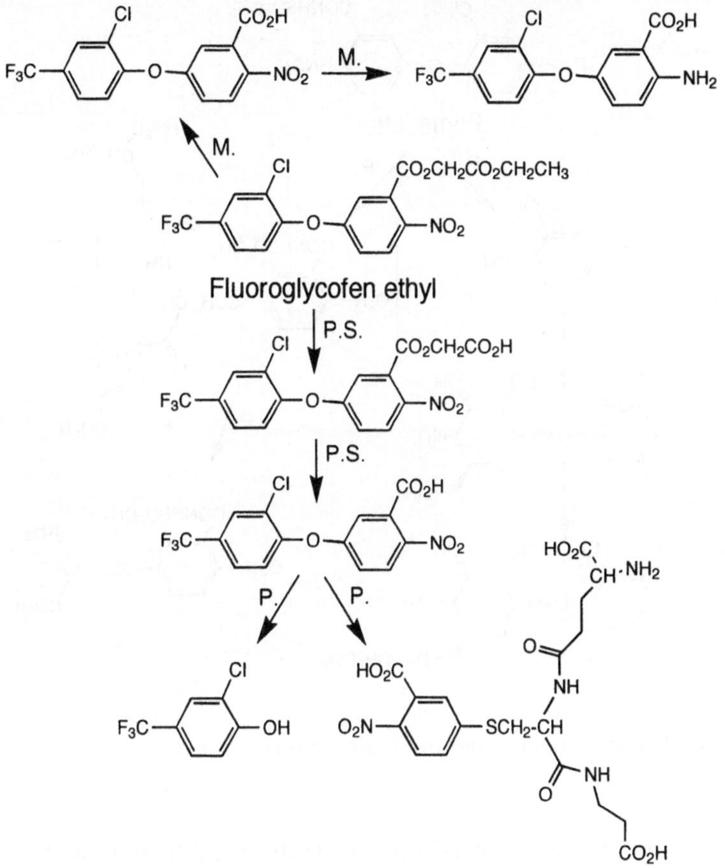

Fig. 2. Metabolism of fluoroglycofen ethyl in wheat (*P*), rats (*M*), and soil (*S*)

(homo)glutathione in soybeans (Evans et al. 1987). Fomesafen was metabolized more rapidly in soybeans than in a susceptible weed (*Xanthium pennsylvanicum*), and metabolic inactivation is the likely basis for soybean tolerance. The metabolic pathway shown in Fig. 3 is partially extrapolated from the work with acifluorofen (Frear et al. 1983) in that Evans et al. (1987) identified the cysteine conjugate of fomesafen, and it is presumed to be derived from the homoglutathione conjugate. Also, acid hydrolysis of polar metabolite fractions produced 2-chloro-4-trifluorophenol, providing some evidence for the glucose conjugates shown (Evans et al. 1987).

14.2.4 Oxyfluorfen

Oxyfluorfen {2-chloro-1-(3-ethoxy-4-nitrophenoxy)-4-(trifluoromethyl)benzene; CAS No. 42874-03-3} is a diphenylether herbicide used in numerous crops for control of broadleaf and grass weeds. Preemergence (weeds) and

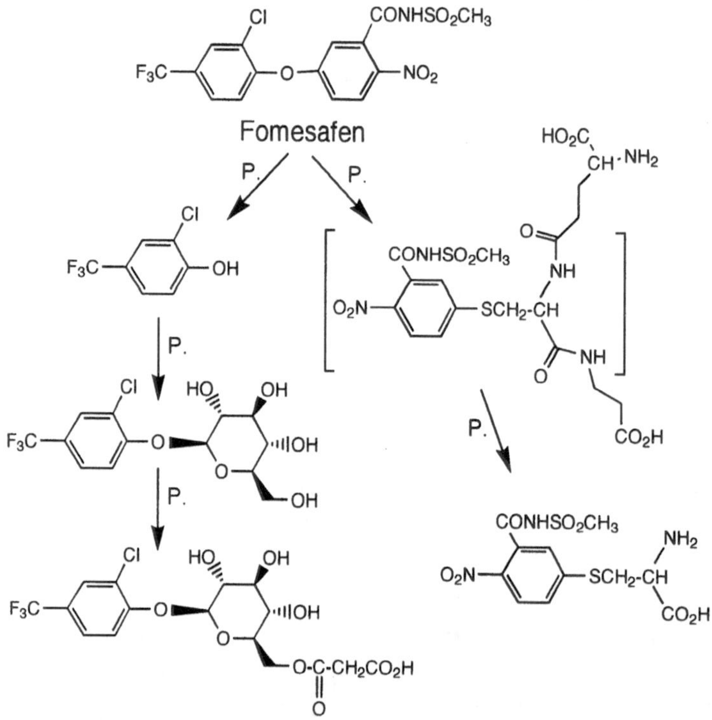

Fig. 3. Metabolism of fomesafen in soybeans (*P*)

postemergence (especially post-directed) application rates range from 0.034 to 2.24 kg a.i./ha and utility is found in conifers, citrus, fruit and nut trees, vines and plantation crops, and some annual field crops including vegetables, cotton, peanuts and maize (Ahrens 1994). Studies in *Vicia faba*, *Setaria viridis* (Vanstone and Stobbe 1978), rice (Guh et al. 1995), and tolerant and sensitive soybean cell cultures (Pornprom et al. 1994) indicate that oxyfluorfen is not readily metabolized in plants. For example, less than 10% of the oxyfluorfen taken up by *Vicia faba* (a moderately tolerant species) and *Setaria viridis* (sensitive) was metabolized within 24 h (Vanstone and Stobbe 1978), and it is unlikely that metabolism plays a significant role in plant tolerance. Even less metabolism was observed in rice and *Echinochloa crus-galli* (Guh et al. 1995). In all plants studied, oxyfluorfen was translocated to a very limited extent following either root or foliage treatment (Fadayomi and Warren 1977). The high Kow (29400) and non-dissociable property led to very tight soil sorption. Oxyfluorfen has measured Koc values ranging from 2891 ml/g (sand) to 32381 ml/g (silty clay loam) (Wauchope et al. 1992; Tomlin 1994) and it is neither readily desorbed nor mobile in soil. The rate of field dissipation ranges from 5 to 55 days (times to reach 50% dissipation; DT_{50}). The preemergence and postdirected crop selectivity of this compound is likely related to soil sorption.

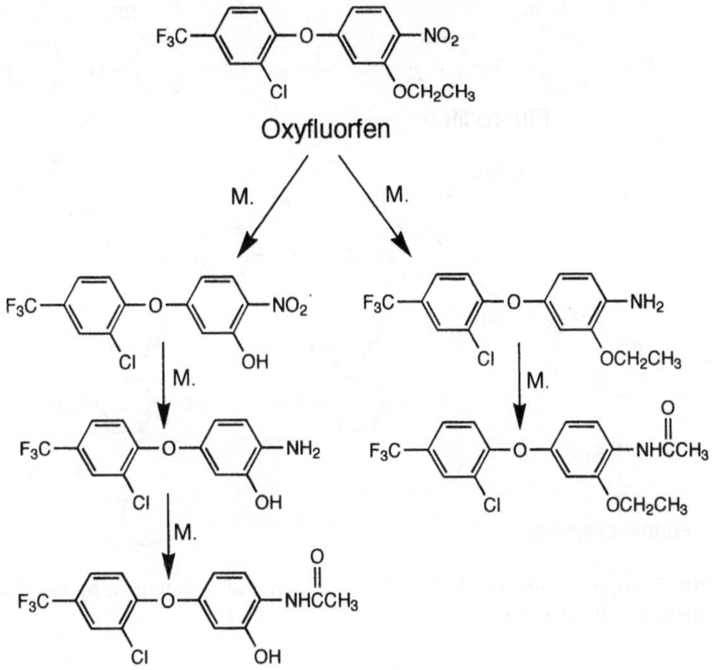

Fig. 4. Metabolism of oxyfluorfen in rats (*M*)

Oxyfluorfen is metabolized in rats through O-dealkylation, nitro reduction and subsequent conjugation and excretion (Adler et al. 1977; Fig. 4).

14.2.5 Fluorodifen

Fluorodifen [2-nitro-1-(4-nitrophenoxy)-4-(trifluoromethyl)benzene; also known as preforan] is a discontinued diphenylether herbicide which was used preemergence at 2–4 kg a.i./ha for control of annual grasses and broadleaf weeds in soybeans, peanuts, cotton, rice and other crops. It is relatively unique among diphenylethers in being the only commercialized example of a 2-nitro-4-trifluoromethyl benzene analog. The metabolism and degradation of fluorodifen has been extensively studied. Like other diphenylethers, fluorodifen is stable in the dark but is susceptible to photolysis. Dry film photolysis of fluorodifen follows the pattern seen with several other diphenylethers and includes ether bridge cleavage, nitro reduction, and (in this case) possible polymerization (Eastin 1972a; Fig. 5). Soil microorganisms in enrichment cultures catalyzed ether bridge cleavage and denitrification (Tewfik and Hamdi 1975). Fluorodifen could serve as the sole carbon and nitrogen source in these enrichment cultures, although its degradation was faster in media supplemented with alternate carbon and nitrogen sources. Fluorodifen metabolism was studied in rats (Lamoureux and Davison 1975)

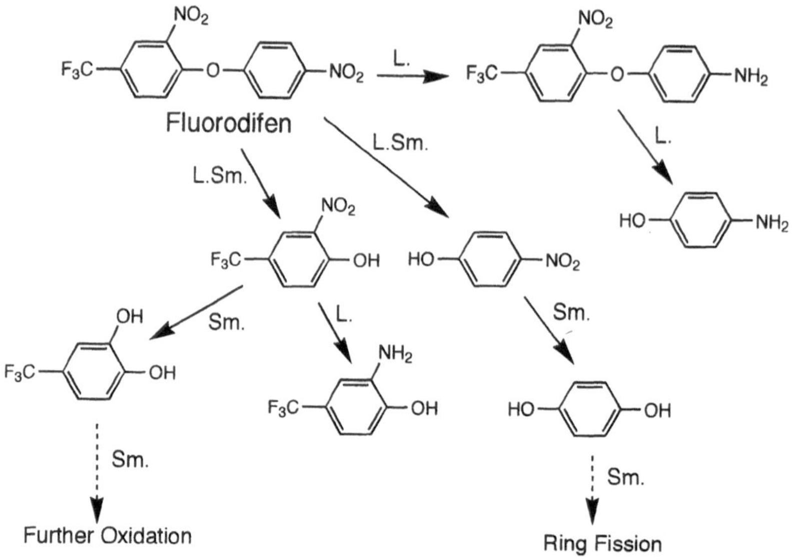

Fig. 5. Aqueous photolysis (*L*) of fluorodifen and metabolism by enrichment cultures of soil microorganisms (*Sm*)

and a wide variety of plants including cucumber (Eastin 1971a, 1972b), peanuts (Eastin 1969, 1971b,c; Shimabukuro et al. 1973), soybeans (Rogers 1971), peas (Frear and Swanson 1973), tobacco (Locke and Baron 1972), and spruce (*Picea*) (Schroder et al. 1990; Lamoureux et al. 1991, 1993). A general correlation was found between rapid metabolism and crop tolerance in soybeans and peanuts while slower metabolism correlated with sensitivity in cucumber. Slow translocation, especially from roots, may also contribute to tolerance in peanut. Fluorodifen was the first analog in this herbicide class found to be susceptible to ether bridge cleavage by glutathione (Fig. 6). The combined effects of the electron withdrawing trifluoromethyl and nitro substituents make the ether bridge susceptible to nucleophilic attack by the sulfhydryl group of glutathione. This reaction is facilitated by the good leaving-group properties of *p*-nitrophenol and is likely catalyzed by glutathione-s-transferases (GST) in these various species. For example, Frear and Swanson (1973) purified and characterized a GST from peas which readily catalyzed this glutathione conjugation with fluorodifen. Neither the amino (nitro-reduced) analog of fluorodifen nor nitrofen served as substrates for this enzyme activity. As seen in Fig. 6, the tripeptide portion of the glutathione conjugate is further metabolized to yield the cysteine derivative, and both it and *p*-nitrophenol are further metabolized to a variety of secondary metabolites and conjugates.

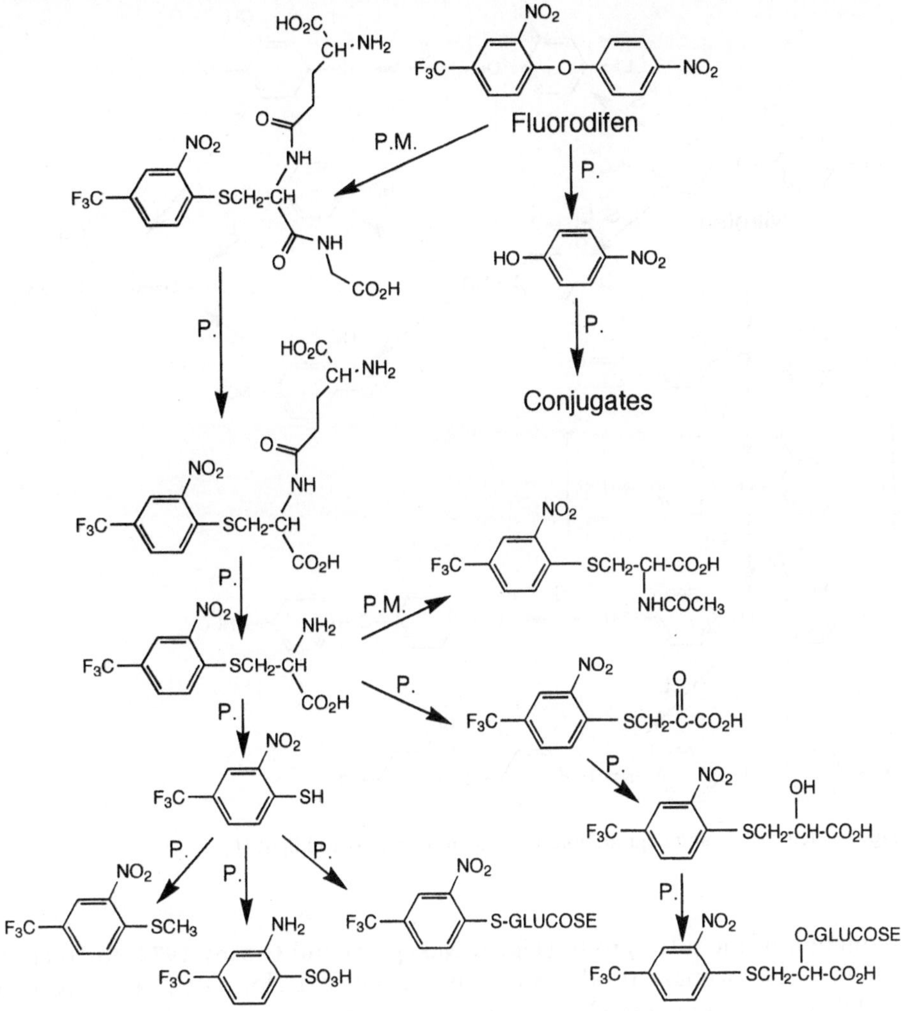

Fig. 6. Metabolism of fluorodifen in plants (*P*) and rats (*M*)

14.2.6 Nitrofen

Nitrofen [2,4-dichloro-1-(4-nitrophenoxy)benzene] is a diphenylether herbicide used for preemergence weed control in rice, cereals and vegetable crops at 2–3 kg a.i./ha. Nitrofen can also be used postemergence in vegetable crops at <1 kg a.i./ha. As an early product from this herbicide class, its metabolism and degradation were relatively extensively studied. Nitrofen is moderately susceptible to metabolism and degradation in a variety of systems. Nitrofen photolyzes in aqueous solution in sunlight through cleavage of the ether bridge and

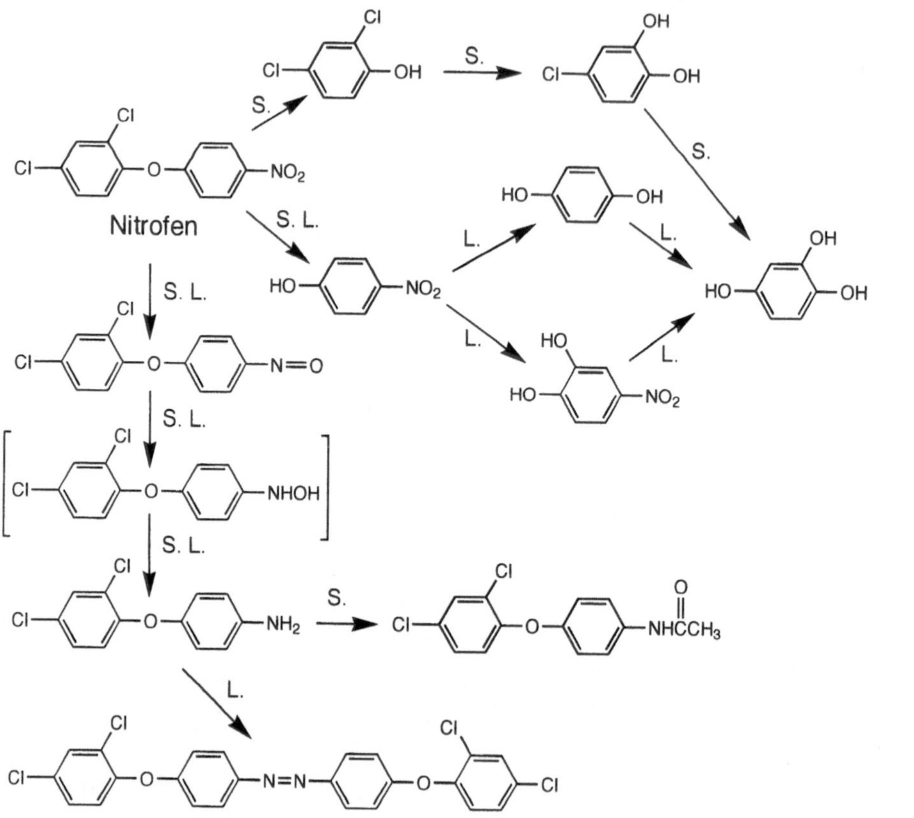

Fig. 7. Metabolism and degradation of nitrofen in soil (*S*) and light (*L*)

reduction of the nitro group (Fig. 7; Nakagawa and Crosby 1974; Ruzo et al. 1980). These primary products are subsequently declorinated, hydroxylated and denitrified. In a model rice paddy ecosystem, nitrofen tended to accumulate in rice, algae, sand and several aquatic organisms and comprised the majority of extractable residues during 3 to 33-day treatment periods (Lee et al. 1976). However, this study used sand and laboratory water, probably minimizing the role of microbial degradation. Later model aquatic ecosystem studies using freshly collected soil and water showed that total nitrofen residues (parent and bound and soluble metabolites) declined to 6% of the applied dose over 120 days (Kale and Raghu 1989). There was negligible accumulation of radiolabeled residues in plants and aquatic microorganisms, possibly due to the sharply increased degradation in the soil and water in this microbially active system. The important role of soil microorganisms in nitrofen degradation was confirmed in sterile and non-sterile soil systems (Kale and Raghu 1994). Additionally, degradation has been repeatedly shown to be faster under flooded soil conditions (Niki and Kuwatsuka 1976; Oh et al. 1981; Qian et al. 1982; Oyamada and Kuwatsuka 1988; Kale and Raghu 1994). Major degrada-

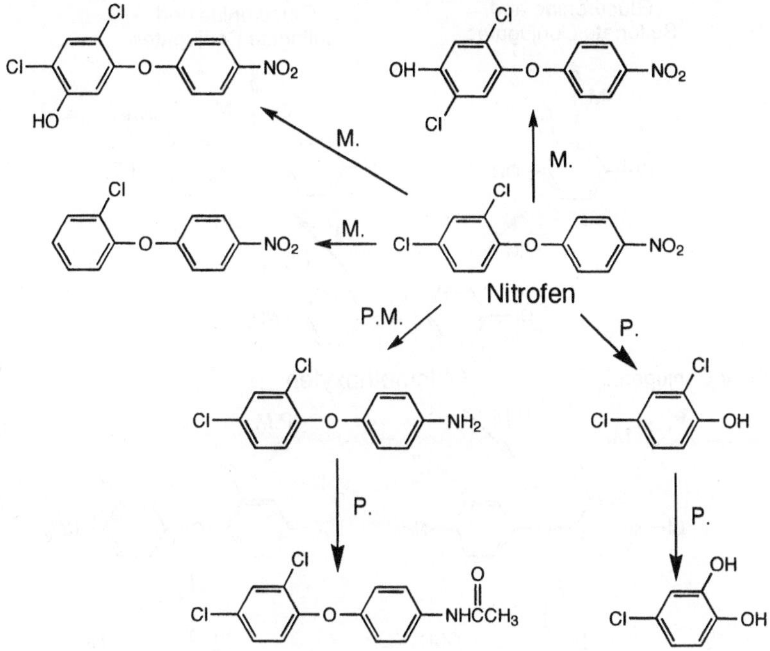

Fig. 8. Metabolism of nitrofen in rice and wheat (*P*) and sheep (*M*)

tion pathways of nitrofen in soil are shown in Fig. 7. The relatively slow metabolism in rice and wheat also involves ether bridge cleavage and nitro reduction (Wargo et al. 1975; Fig. 8). In addition to these reactions, Hunt et al. (1977) found that sheep could dechlorinate and hydroxylate intact nitrofen. In this animal system, 76% of the radiolabeled nitrofen dose was excreted (primarily in feces) during the first 99 h, and the highest levels of nitrofen residues were found in fat tissues, with lesser amounts in organs.

14.2.7 Chlomethoxyfen

Chlomethoxyfen [chlomethoxynil; 4-(2,4-dichlorophenoxy)-2-methoxy-1-nitrobenzene; CAS No. 32861-85-1; formerly X-52] is a diphenylether herbicide used preemergence at 1.5–2.5 kg a.i./ha in transplanted rice for the control of *Echinochloa* spp., *Scirpus* spp. and annual broadleaf weeds. Rice tolerance is based, at least in part, on metabolic inactivation. In two separate studies (Ishizuka et al. 1988; Lee et al. 1991), 20–28% of a dose of foliarly applied chlomethoxyfen remained intact in tolerant rice after 24 h while >85% remained intact in sensitive barnyardgrass. The major initial metabolite in rice was the O-desmethyl metabolite, comprising 33.8% of the dose at 24 h (Fig. 9). Metabolism was rapid in rice in both the light and dark (Ishizuka et al. 1988), and studies with a variety of other sensitive and tolerant crops suggest that

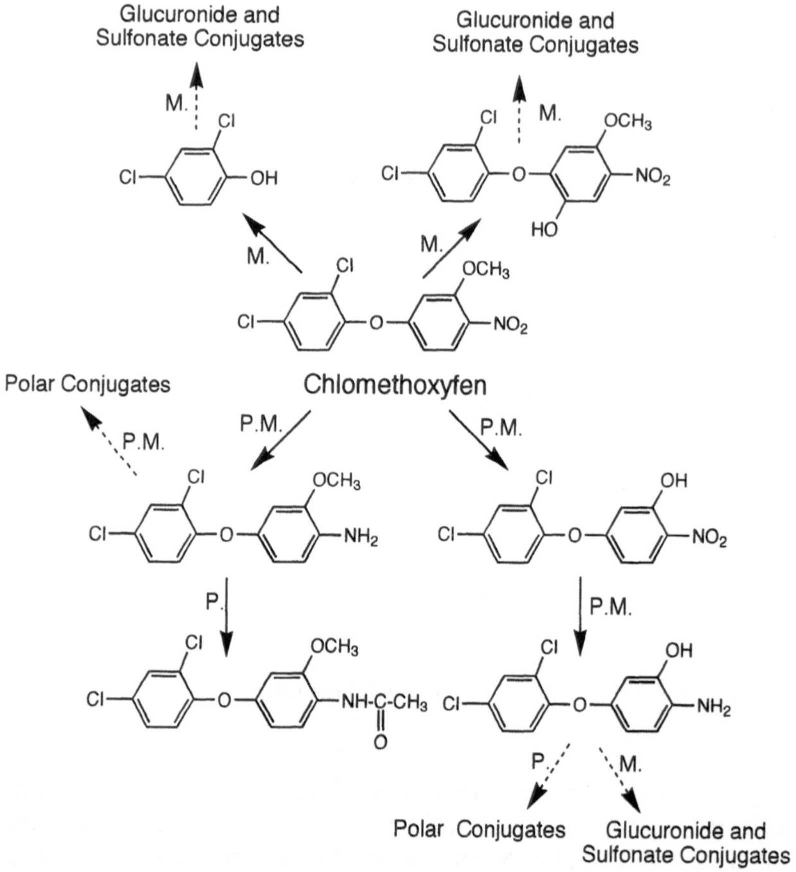

Fig. 9. Metabolism of chlomethoxyfen in rice (*P*) and rats (*M*)

both metabolism and degree of uptake contribute to relative plant sensitivity. Chlomethoxyfen is readily metabolized and excreted from rats (Tsuchiya et al. 1982). After 96h, 95% of the radiolabeled dose had been excreted in urine (78%) and feces (17%). The major metabolite in urine was 2,4-dichlorophenyl sulfate, with lesser amounts of 2,4-dichlorophenyl-B-D-glucuronide. Multiple minor metabolites were also identified (Fig. 9).

14.2.8 Bifenox

Bifenox [methyl-5-(2,4-dichlorophenoxy)-2-nitrobenzoate; CAS No. 42576-02-3] is a diphenylether introduced by Mobil Chemical Co. in 1973. It controls a spectrum of annual broadleaf weeds and some grasses at rates of 0.5–0.9kga.i./ha (postdirected) in cereals. It can also be used preemergence in soybeans, sunflowers, rice and maize at 0.5–2.1kga.i./ha (depending on the crop and utility), and is often used in combination with other products

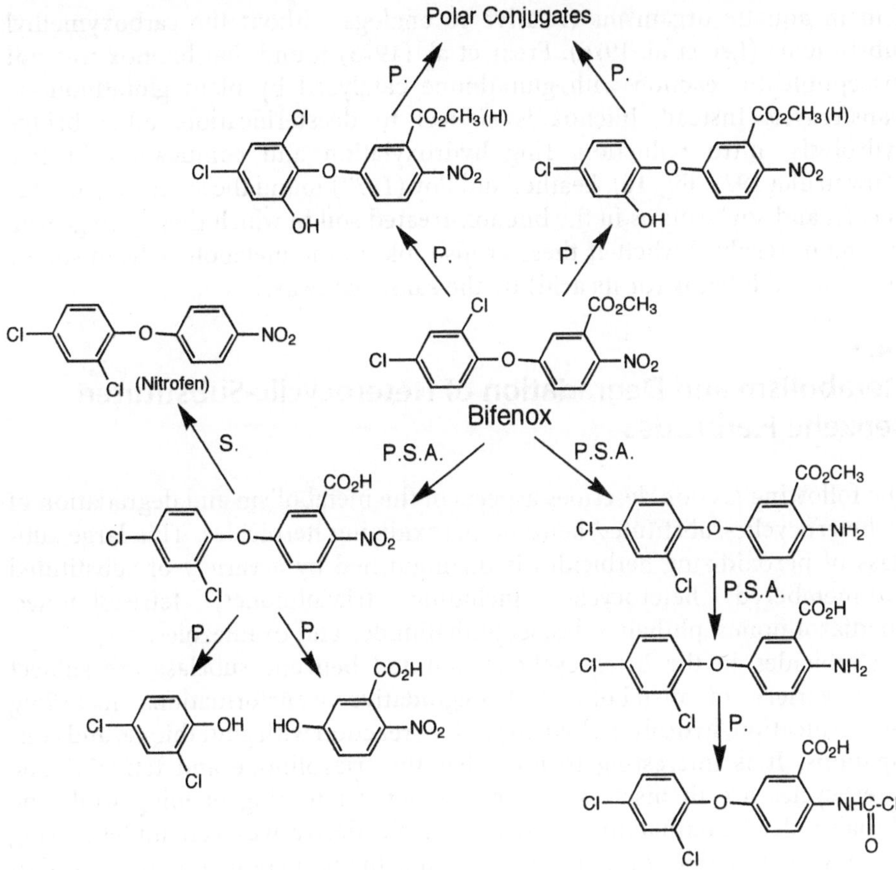

Fig. 10. Metabolism of bifenox in soil (*S*), plants (*P*) and aquatic animals (*A*)

(Tomlin 1997). Bifenox is readily deesterified in plants, animals and soil, forming 5-(2,4-dichlorophenoxy)-2-nitrobenzoic acid which remains herbicidally active (Fig. 10). In soil, bifenox is deesterified with DT_{50} values ranging from 3 to 14 days (Leather and Foy 1977; Lee et al. 1976; Ohyama and Kuwatsuka 1978, 1983). It is strongly adsorbed to soil (Koc = 10 000 ml/g), although the deesterified product was more mobile and was the major component eluted from a soil column treated with ^{14}C-bifenox (Leather and Foy 1977). Bifenox degraded somewhat more rapidly in soil under flooded than aerobic conditions, and sterilization of both flooded and moist soils sharply decreased bifenox degradation (Ohyama and Kuwatsuka 1978). Under flooded conditions, sterilization of the soil did not eliminate nitro reduction of bifenox, indicating that at least some of this activity is abiotic (Ohyama and Kuwatsuka 1983). These authors have suggested that ferrous iron formed from ferric iron by microorganisms under anaerobic conditions is responsible for nitro reduction of this (and possibly other) diphenylethers (Ohyama and Kuwatsuka 1978). In a model aquatic ecosystem, the conversion of bifenox to its corresponding acid led to less accumulation and greater conjugation and metabolism in aquatic organisms (relative to analogs without the carboxymethyl substituent) (Lee et al. 1976). Frear et al. (1973) found that bifenox was not susceptible to reaction with glutathione catalyzed by plant glutathione-s-transferases. Instead, bifenox is subject to deesterification, ether bridge hydrolysis, nitro reduction, ring hydroxylation and conjugation in rice (Kuwatsuka 1977; Fig. 10). Leather and Foy (1977) found the same metabolites in corn and soybeans as in the bifenox-treated soil in which they were grown, leaving unresolved whether these crops took up the metabolites from soil or metabolized bifenox (or its acid) by the same pathways.

14.3 Metabolism and Degradation of Heterocyclic-Substituted Benzene Herbicides

The following section describes aspects of the metabolism and degradation of 11 heterocyclic-substituted benzene peroxidizing herbicides. This large subclass of peroxidizing herbicides is distinguished by a variety of substituted five-membered heterocycles including triazolinones, tetrazolinones, thiadiazolinones, phthalimides, isophthalimides and oxadiazoles.

Herbicides in the heterocyclic-substituted benzene subclass are subject to a variety of metabolic and degradative transformations including deesterification, hydrolysis, hydroxylation, reaction with glutathione, and conjugations. It is interesting to note that the triazolinone and tetrazolinone heterocycles are themselves not readily subject to ring opening while the phthalimide, isophthalimide and oxadiazole derivatives can undergo ring opening degradation. An unusual reaction is the reduction and incorporation of the sulfonic acid group at the double bond of the tetrahydrophthalimide ring of flumipropyn and flumiclorac.

Although glutathione S-transferases (GST) generally catalyze reactions with reduced glutathione that detoxify agrichemicals, the herbicidal activation by the metabolic isomerization of several of these herbicides is a new function of GST and glutathione. This surprising reaction was found in isomerization studies of fluthiacet-methyl, N-(4-bromophenyl)-3,4,5,6-tetrahydroisophthalimide and 5-(4-bromophenylimino)-3,4-tetramethylene-1,3,4-thiadiazolidin-2-one.

14.3.1 Triazolinones

14.3.1.1 *Carfentrazone Ethyl*

Carfentrazone ethyl {ethyl, α,2-dichloro-5-[4-(difluoromethyl)-4,5-dihydro-3-methyl-5-oxo-1H-1,2,4-triazol-1-yl]-4-fluorobenzenepropanoate; CAS No.

128639-02-1} is an aryl triazolinone herbicide for the selective postemergence control of broadleaf weeds in cereals. Application rates range from 9 to 35 g a.i./ha. In Europe, carfentrazone ethyl is especially useful against *Galium*, *Lamium* and *Veronica* spp., while in the USA it is effective against most major broadleaf weeds in wheat including *Salsola kali*, *Chenopodium album*, *Amaranthus*, *Kochia*, and *Ipomoea* spp. and mustards. Optimum application timing is at the 1 to 2-leaf and 2-node crop stages (Van Saun et al. 1993; Tomlin 1997).

Carfentrazone ethyl is not readily translocated. It degrades photolytically in water (half-life of 8.3 days), but does not photodecompose on soil. Carfentrazone ethyl is not volatile following application to soil (Willut 1994). It is metabolized in wheat (ElNagger 1995) and corn (Deng 1995) to a series of acid metabolites including: chloropropionic acid (**I**), hydroxymethyl-chloropropionic acid (**II**) and desmethyl-chloropropionic acid (**III**) carfentraazone derivatives (Fig. 11). A minor sulfonyl-conjugated product is also detected in wheat.

Dayan et al. (1997a) studied the basis for the moderate soybean selectivity of carfentrazone ethyl. They found that while tolerant soybeans and sensitive ivyleaf morningglory (*Ipomoea* spp.) and velvetleaf (*Abutilon theophrasti* Medicus) took up approximately equal amounts of a postemergence application of carfentrazone ethyl, the herbicide was more rapidly metabolized in soybeans, with only 27% remaining after 24 h versus 54% and 61% in morningglory and velvetleaf, respectively. The free acid metabolite (**I**), which is also herbicidally active, was found in all three plant species while polar metabolites were 4–5 times more abundant in soybeans. On the basis of the rate and pathway of metabolism, it was concluded that metabolic inactivation is an important factor in soybean tolerance.

In rats, extensive metabolism and elimination occur within 24 h after dosing. Compounds **I** and **II** are the major metabolites. Hydroxymethyl-propionic acid (**IV**) and cinnamic acid (**V**) metabolites were also detected (Wu 1995; Fig. 11). Similar results were obtained with lactating goats (Robinson 1995).

Carfentrazone ethyl degrades rapidly in aerobic soil, with less than 55% of the parent compound remaining after 24 h. The degradates include benzoic acid (**VI**) and propionic acid (**VII**) carfentrazone ethyl derivatives in addition to **I** and **V** (Fig. 11; Goodyear 1995a). Rapid degradation in soil was also observed under anaerobic conditions, with a DT_{50} of less than 1 day (Goodyear 1995b). Aerobic aquatic (i.e. water/sediment) studies revealed rapid conversion to **I** with increasing amounts of the cinnamic, benzoic and propionic acid (**V**, **VI** and **VII**) degradates forming over time (Purser 1994; Elmarakby 1997). Intact carfentrazone ethyl adsorbs to sterile soil with a Koc value of *ca.* 750, but its deesterification product (**VI**) is much more weakly bound (Koc values 15–35) (Tomlin 1997).

Thus, the primary metabolic pathway of carfentrazone ethyl in mammals, crops and soil involves initial hydrolysis of the ethyl ester resulting in the herbicidally active chloropropionic acid metabolite (**I**). Further metabolism of

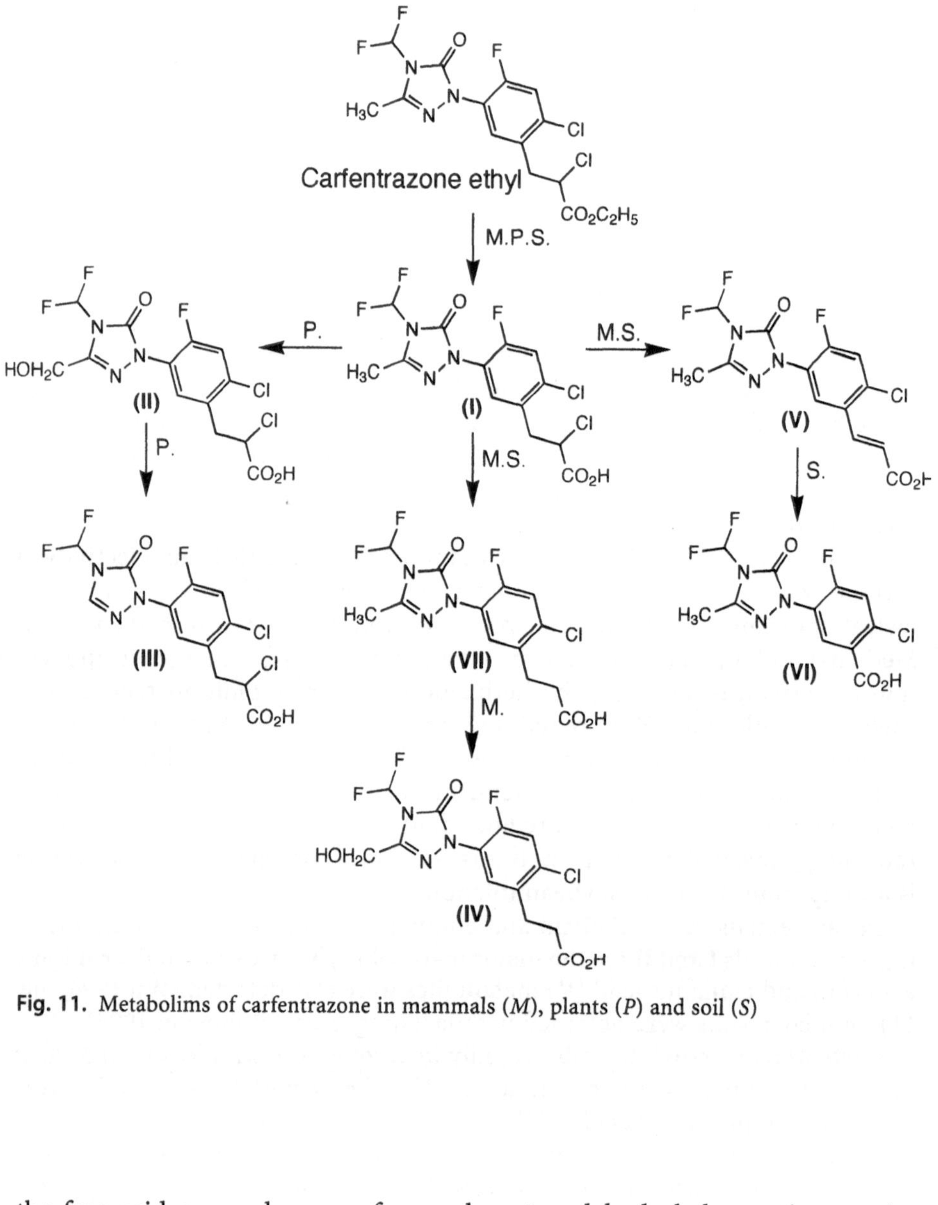

Fig. 11. Metabolims of carfentrazone in mammals (*M*), plants (*P*) and soil (*S*)

the free acid occurs by one of several routes: dehydrohalogenation to give cinnamic acid (**V**) (mammals, soil), dechlorination to give propionic acid (**VII**), or oxidation of the exocyclic, allylic methyl group to form hydroxychloropropionic acid (**II**) (crops). Compound **II** may undergo further oxidation followed by decarboxylation to give the desmethyl product **III**. In mammals, propionic acid (**VII**) may undergo cytochrome P-450 oxidation of the methyl group to form **IV**. A concurrent metabolic pathway (in soil) involving multiple oxidative steps on the aromatic side chain of cinnamic acid (**V**) gives rise to a benzoic acid degradate (**VI**).

14.3.1.2
Sulfentrazone

Sulfentrazone {N-[2,4-dichloro-5-[4-(difluoromethyl)-4,5-dihydro-3-methyl-5-oxo-1H-1,2,4-triazol-1-yl]phenyl]methanesulfonamide; CAS No. 122836-35-5} is an aryl triazolinone herbicide used for selective preemergence or preplant incorporated control of broadleaf and grass weeds in soybeans, tobacco, sugarcane and several species of turfgrass (Ahrens 1994). Application rates of 355–425 g a.i./ha control a range of weeds including *Amaranthus*, *Cyperus*, *Digitaria* and *Solanum* spp. (Van Saun et al. 1991).

When ^{14}C-radiolabeled sulfentrazone was administered orally to rats (10 mg/kg), goats (2 mg/kg, 300 ppm in diet daily, 5 days), or hens (3 mg/kg, 45 ppm in diet daily, 7 days), radioactivity was quantitatively excreted in the urine, feces, or hen excreta. In all of the species studied, unchanged sulfentrazone comprised only a minor portion of the excreted radioactivity (<2% of the excreted radioactivity), and two nonconjugated metabolites were found. These are 3-hydroxymethyl sulfentrazone (I), comprising 88–95% of the excreted radioactivity and its carboxylic acid derivative (III) (0.3–5%) (Fig. 12). The carboxylic acid metabolite was decarboxylated to give 3-desmethyl sulfentrazone as a minor metabolite. In rats, a minor metabolite which is tentatively characterized as the 2,3-dihydro-3-hydroxymethyl derivative (II) (0.5–5%) was detected (Leung et al. 1991).

Similar metabolic transformations of sulfentrazone have also been observed in soybean, barley, wheat, radish, tobacco and lettuce. In each case, the major plant metabolite was 3-hydroxymethyl sulfentrazone (I) which is found in both free and conjugated forms (EINaggar 1992; Ramsey 1994). A minor plant metabolic transformation involves cleavage of sulfentrazone to yield the free triazole ring and its N-glycoside (Fig. 12). The free aromatic ring was not detected and is apparently further degraded and bound in the unextractable fraction (Ramsey and Venables 1997).

Several studies have shown that plant tolerance to sulfentrazone is based on rapid metabolic inactivation. Dayan et al. (1996b) showed that tolerant sicklepod (*Senna obtusifolia*) metabolized 91.6% of the sulfentrazone taken up by roots after 9 h while sensitive coffee senna (*Cassia occidentalis*) metabolized only 17% over the same time period. The relatively small differences in root uptake and translocation of sulfentrazone between these species were inadequate to explain their large difference in sensitivity. The primary detoxification reaction again appears to be oxidation of the methyl group on the triazolinone ring resulting in the formation of the more polar hydroxymethyl derivative (I) (Fig. 12).

Similarly, differences in tolerance between several soybean cultivars are at least partially explained by differences in rates of sulfentrazone metabolism (Dayan et al. 1997b). These authors also showed that intrinsic differences in response to herbicide-induced peroxidative stress may also play a role in tolerance and sensitivity among these cultivars.

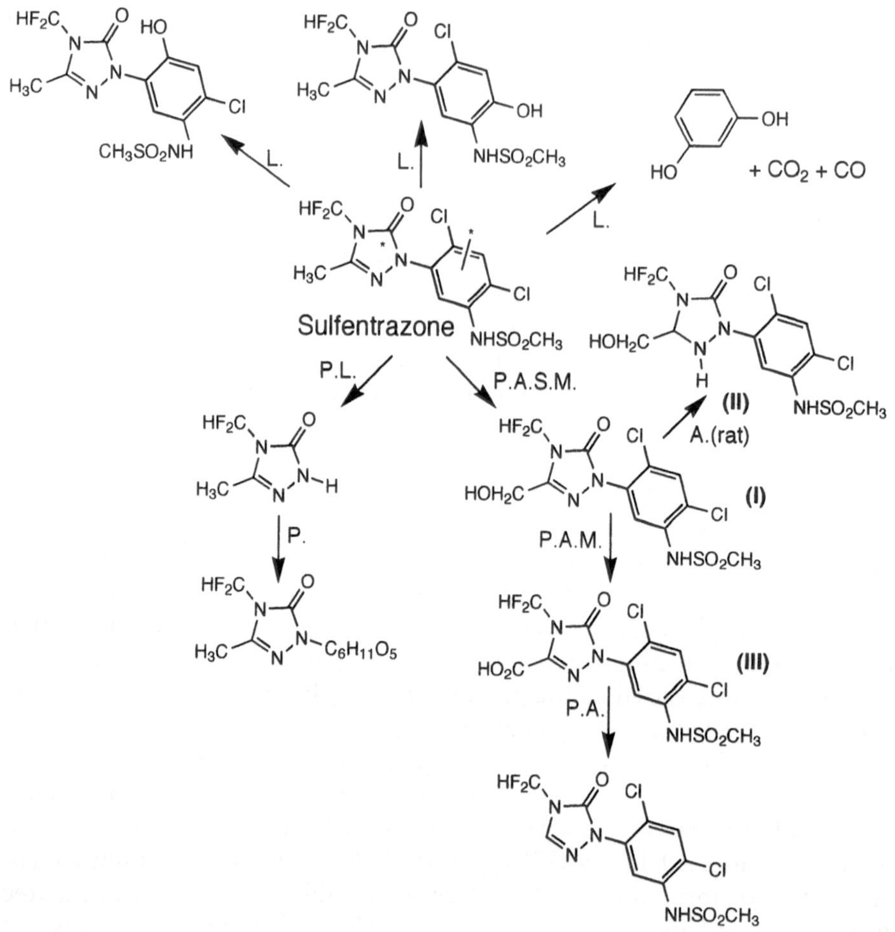

Fig. 12. Metabolism and degradation of sulfentrazone in rats, goats and hens (*A*), plants (*P*), soil (*S*) and light (*L*)

Sulfentrazone is degraded in soil by microbial action to give primarily 3-hydroxymethyl sulfentrazone (**I**) and then further oxidized to sulfentrazone-3-carboxylic acid (**III**) (Singer and Schocken 1991). The half-life in soil ranges from 120 to 300 days under field conditions (Culligan 1994; Becker 1994). It is moderately mobile in soil. Sulfentrazone is not susceptible to photodecomposition following soil application and is resistant to hydrolysis (Kabler and Williamson 1991; Schocken 1994). Photolytic degradation of sulfentrazone in water produces dechlorinated and hydroxylated products. Continued exposure to sunlight causes cleavage of the aromatic and triazole rings to give the triazole fragment and resorcinol, as well as carbon monoxide and carbon dioxide (Fig. 12; Willut et al. 1997).

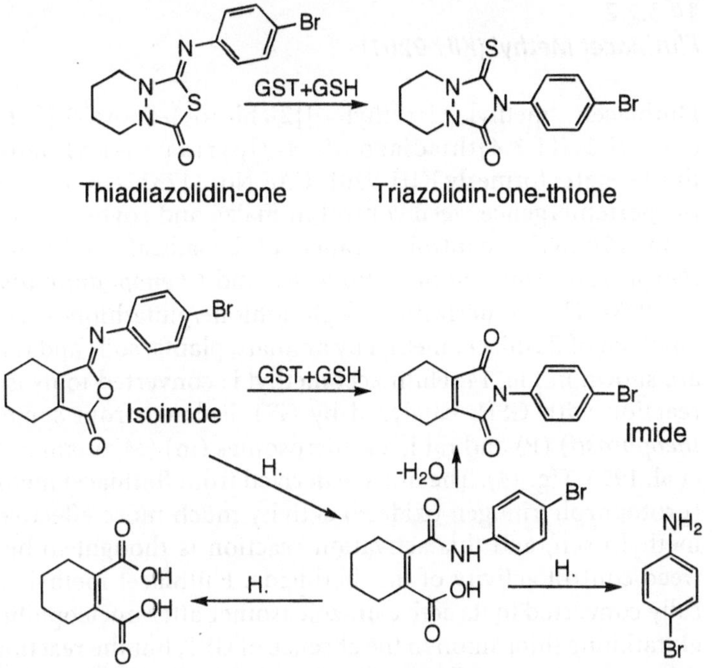

Fig. 13. Glutathione-s-transferase-catalyzed isomerization of two heterocyclic-substituted benzene analogs to active peroxidizing herbicides. Isoimide is also isomerized *via* hydrolysis and dehydration (*H*)

14.3.2
Thiadiazolidinones

14.3.2.1
5-(4-Bromophenylimino)-3,4-tetramethylene-1,3,4-thiadiazolidin-2-one and N-(4-bromophenyl)-3,4,5,6-tetrahydroisophthalimide

Glutathione S-transferase (GST) converts 5-(4-bromophenylimino)-3,4-tetramethylene-1,3,4-thiadiazolidin-2-one (thiadiazolidin-one) into its isomer, 4-bromophenyl-1,2-tetramethylene-1,2,4-triazolidin-3-one-5-thione (triazolidin-one-thione), which strongly inhibits protoporphyrinogen oxidase activity (Sato et al. 1997; Fig. 13). Similarly, N-(4-bromophenyl)-3,4,5,6-tetrahydroisophthalimide (isoimide) is isomerized to N-(4-bromophenyl)-3,4,5,6-tetrahydrophthalimide (imide) rapidly in the presence of reduced glutathione (GSH) catalyzed by GST (Fig. 13). This isoimide is also converted into the imide by hydrolysis and subsequent dehydration. The imide derivative strongly inhibits protoporphyrinogen oxidase activity. Although GST has generally been described as a detoxifying enzyme in pesticide toxicology, herbicidal activation by the metabolic isomerization is an apparently novel function of GST and GSH.

14.3.2.2
Fluthiacet Methyl (KIH-9201)

Fluthiacet methyl {methyl [[2-chloro-4-fluoro-5-[(5,6,7,8-tetrahydro-3-oxo-1H,3H[1,3,4]thiadiazolo[3,4-a]pyridazin-1-ylidene)amino]phenyl] thio]acetate: formerly KIH-9201; CAS No. 117337-16-6} is under development for postemergence weed control in maize and soybeans. Application rates of 5 to 10 g a.i./ha control a range of broadleaf weeds including *Abutilon theophrasti*, *Amaranthus retroflexus*, and *Chenopodium album* (Miyazawa et al. 1993). The isomerization by glutathione/glutathione-s-transferase and metabolism of fluthiacet methyl by animals, plants, soils and rat liver microsomes are shown in Fig. 14. Fluthiacet methyl is converted to its isomer (urazole) by reaction with GSH, catalyzed by GST isolated from some plants (*Abutilon theophrasti*) (P) and rat liver microsomes (m) (Mizutani et al. 1994a; Shimizu et al. 1995; Fig. 14). The urazole derived from fluthiacet methyl inhibits Protox (protoporphyrinogen oxidase) activity much more effectively than fluthiacet methyl itself, and this activation reaction is thought to be important to the weed control activity of this herbicide. Fluthiacet methyl can also be chemically converted to its active urazole isomer after nucleophilic reaction with the glutathione thiol anion in the absence of GST, but the reaction is accelerated by the enzyme, especially at pH values below 7 (Shimizu et al. 1995). It is also suggested that the free acid of urazole, desulfated urazole and the oxidatively ring-cleaved formyl degradation products result from hydrolysis by esterase from soybean seedlings (Mizutani et al. 1994b). When fluthiacet methyl was orally administered to rats (M) at a single dose (100 mg/kg), 63–85% of the dose was excreted in the feces and 10–23% in the urine after 48 h. The main excreted metabolite is the isomer of fluthiacet methyl (urazole) which was also identified as a soil degradation product in upland alluvial and volcanic ash soils (S). Monohydroxylated products on the pyridazine ring were identified as metabolites in rats (Mizutani et al. 1994c).

14.3.3
Tetrazolinones

14.3.3.1
F5231

F5231 {1-[4-chloro-2-fluoro-5-(ethylsulfonylamino)-phenyl]-1,4-dihydro-4-(3-fluoropropyl)-5H-tetrazol-5-one} is a herbicidally active tetrazolinone which was under development by FMC Corp. Figure 15 summarizes the metabolism of F5231 by cultures of the filamentous fungus *Absidia pseudocylindrospora* (ATCC 24169). Most (88–99%) of the initially added radioactivity from ^{14}C-F5231 is ethyl acetate-extractable from the culture filtrates following biotransformation. Six degradation products were identified, none of which include compounds where the aryl or the tetrazolinone rings have

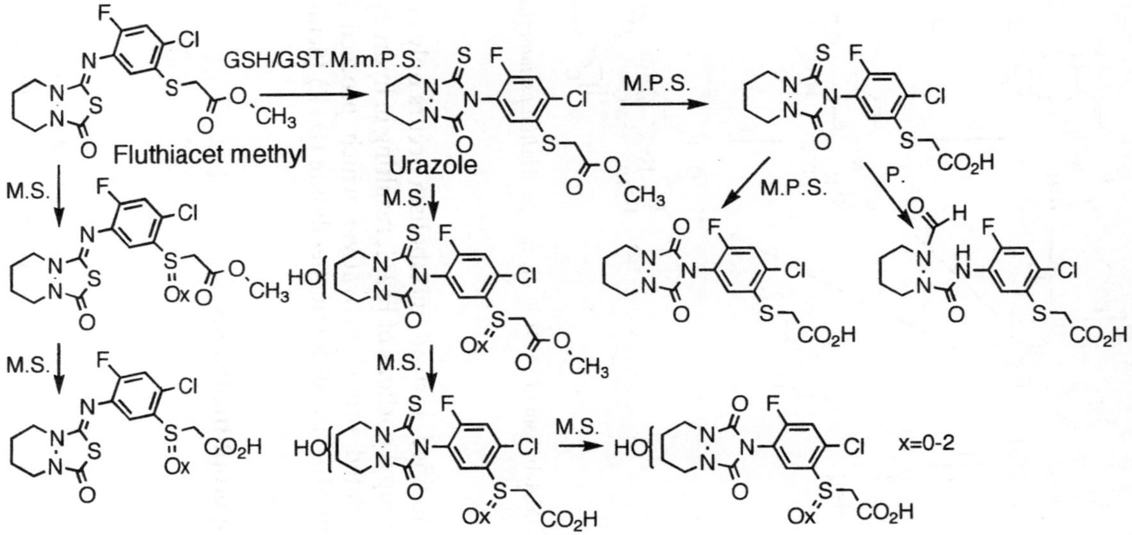

Fig. 14. Isomerization and metabolism of fluthiacet methyl in rat liver microsomes (*m*), rats (*M*), plants (*P*) and soils (*S*)

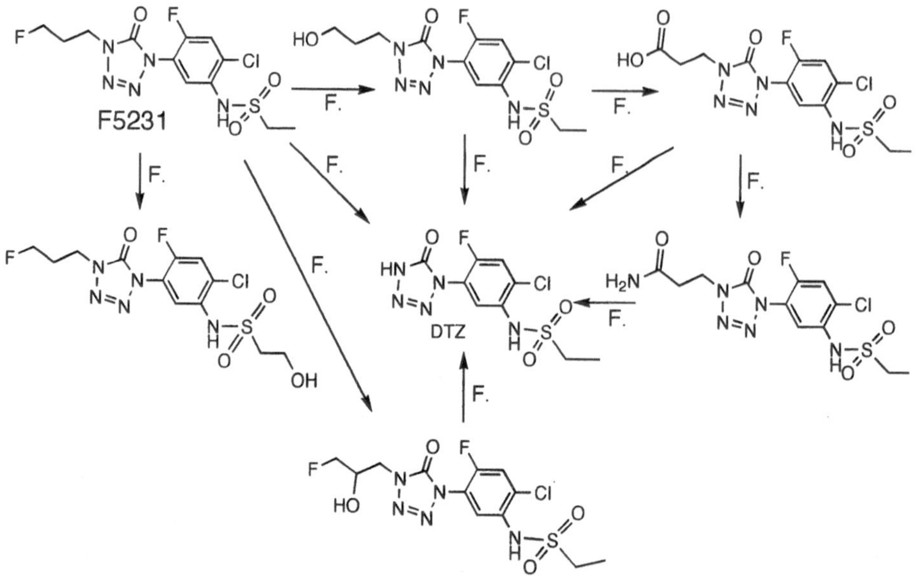

Fig. 15. Metabolism of F5231 in the fungus *Absidia pseudocylindrospora* (*F*)

been modified. Instead, metabolism occurs only at the ethylsulfonamide and fluoropropyl portions of F5231, resulting in hydroxyethylsulfonamide and two hydroxylated propyl derivatives which proceed to the N-dealkylated 1,4-dihydro-5H-tetrazol-5-one metabolite (DTZ) (Schocken et al. 1989).

14.3.4 Tetrahydrophthalimides

14.3.4.1 *MK-129*

MK-129 {N-[4-(4-chlorobenzyloxy) phenyl]-3,4,5,6-tetrahydrophthalimide} (discontinued) was under development for broadleaf and sedge weed control in rice at a proposed use rate of 900 g a.i./ha. The metabolic pathways of MK-129 in animals, plants and soils are shown in Fig. 16. Both rice plants and barnyardgrass (*Echinochloa* spp.; 4th leaf stage) produced twelve metabolites of ^{14}C-MK-129, and of these, 4-chlorobenzoic acid was the main metabolite. Infant seedling rice metabolized MK-129 much more slowly than the 4th leaf stage of barnyardgrass, and the metabolism activity of infant seedlings was almost the same as that of the 4th leaf stage of rice plants (Okada et al. 1980).

When ^{14}C-MK-129 is orally administered to male Wistar rats, about 85% and 2–4% of MK-129 was excreted in the feces and urine, respectively after 1 day, and those amounts were 94% and 3–5%, respectively, after 5 days of

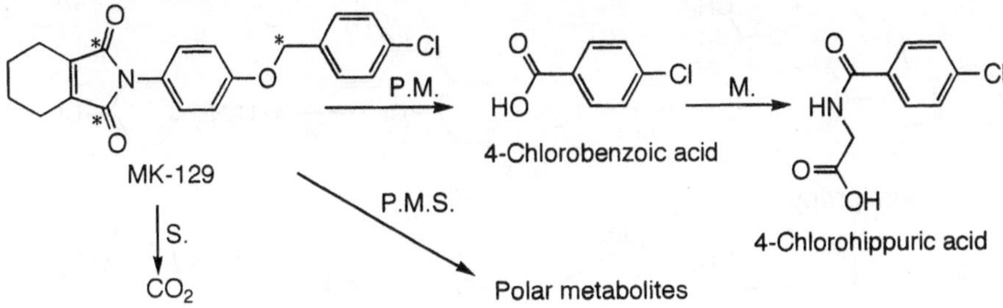

Fig. 16. Metabolism of MK-129 in rats (*M*), plants (*P*) and soils (*S*)

administration. Less than 0.1% of MK-129 was expired as CO_2. Residual amounts of about 0.3–0.4% in the fat tissues resulted from unchanged MK-129. When MK-129 was administered to rats for 7 days consecutively, the residual amount in the fat tissues was three times higher than from 1-day adminstration and decreased to one and a half times 3 days after cessation of its administration. Of the radioactivity excreted in the feces, 80% was unchanged MK-129, but the metabolites which were identified in the urine were 4-chlorobenzoic acid (0.7%) and 4-chlorohippuric acid (2.6%). The other metabolites which were derived from both benzyl and carbonyl ^{14}C-MK-129 were polar metabolites and were not identifed (Ogawa et al. 1980).

The main metabolite of ^{14}C-MK-129 in three different kinds of soils (Tochigi volcanic ash, Saitama mineral and Kanagawa sandy soils) was $^{14}CO_2$ under both oxidative flood and upland conditions (Ohori et al. 1980).

14.3.4.2
Flumipropyn

Flumipropyn {N-[4-chloro-2-fluoro-5-[(1-methyl-2-propynyl)oxy]phenyl]-3,4,5,6-tetrahydro-phthalimide; CAS No. 84478-52-4; formerly S-23121} is under develoment for pre- and postemergence broadleaf weed control in cereals. Application rates of 10–20 g a.i./ha are effective against *Galium aparine*, *Veronica* spp., and *Viola* spp. with good crop safety, and combinations with other herbicides increase the weed control spectrum (Hamada et al. 1989).

The metabolism of flumipropyn in Sprague-Dawley rats is shown in Fig. 17. The major urinary metabolites in male and female rats were 4-chloro-2-fluoro-5-hydroxyaniline and its sulphate and glucuronide conjugates. The major fecal metabolites in addition to the parent compound were six sulphonic acid conjugates having a sulphonic acid group incorporated across the double bond of the 3,4,5,6-tetrahydrophthalimide moiety (Yoshino et al. 1993). The acetylated precursor of 4-chloro-2-fluoro-5-hydroxyaniline (I) was not identified in mammalian cytosol preparations in vitro (Saito et al. 1996).

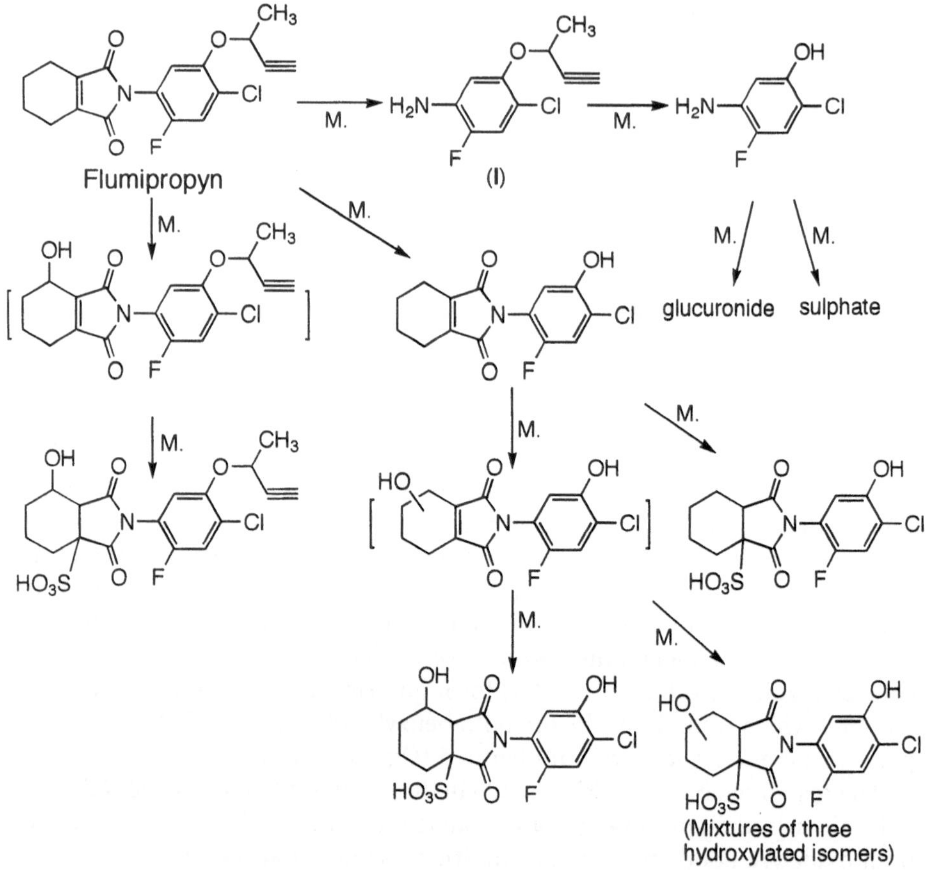

Fig. 17. Metabolism of flumipropyn in rats (*M*)

14.3.4.3
Flumiclorac Pentyl

Flumiclorac pentyl [2-chloro-4-fluoro-5-(3,4,5,6-tetrahydrophthalimido) phenoxyacetate; CAS No. 87546-18-7; formerly S-23031] is a postemergence peroxidizing herbicide for broadleaf weed control in soybeans and maize. Application rates of 30–90 g a.i./ha in soybeans or 30–45 g a.i./ha in maize control important weeds including *Chenopodium album*, *Xanthium strumarium*, *Ambrosia artemisifolia*, *Datura strumarium* and *Abutilon theophrasti*, among others (Kamoshita et al. 1992; Ahrens 1994)

Male and female Sprague-Dawley rats were given single oral doses of [phenyl-^{14}C]- or [tetrahydrophthaloyl-1,2-^{14}C]([THP-^{14}C])-flumiclorac pentyl at 1 or 500 mg/kg. More than 90% of the dosed radioactivity was excreted into feces and urine within 48 h after treatment. While ^{14}C levels in tissues and blood were generally low on the seventh day after dosing, the levels with [THP-

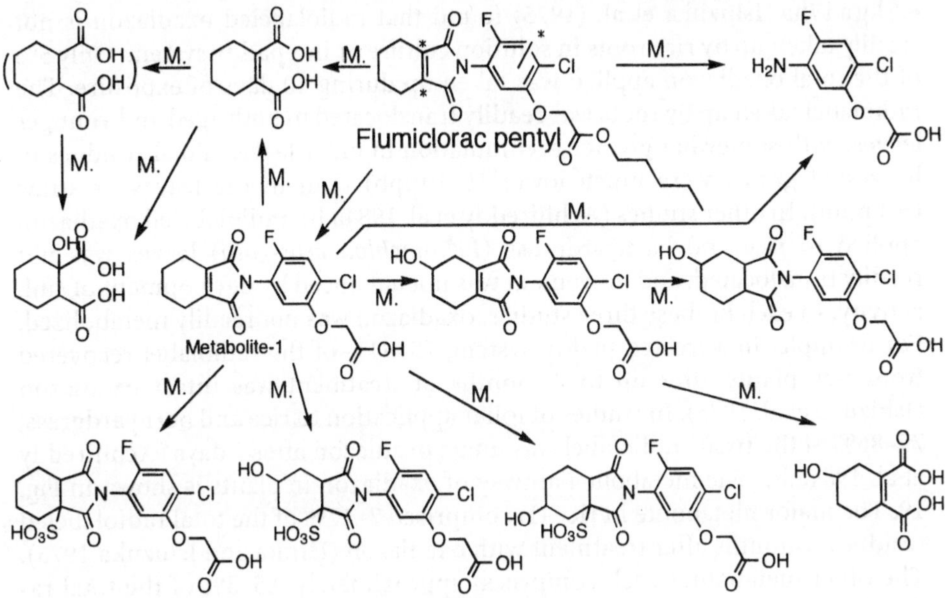

Fig. 18. Metabolism of flumiclorac pentyl in rats (*M*)

[14]C]-flumiclorac pentyl were relatively higher than those with [phenyl-^{14}C]-flumiclorac pentyl is rapidly and extensively metabolized, and sulfonic acid conjugates were identified as the major fecal metabolites while the major urinary metabolite was 5-amino-2-chloro-4-fluorophenoxyacetic acid. The metabolic reactions are shown in Fig. 18 and include cleavage of the ester and imide linkages, hydroxylation and reduction of the cyclohexene ring, the 3,4,5,6-tetrahydrophthalimide moiety and the cyclohexane ring of the cyclohexane-1,2-dicarboxylic acid moiety (Matsunaga et al. 1997a). A sulfonic acid group is also incorporated across the double bond of the 3,4,5,6-tetrahydrophthalimide moiety (Matsunaga et al. 1997b).

Flumiclorac pentyl and its degradates had DT_{50} values of 2–30 days in soil (Tomlin, 1997).

14.3.5
Oxadiazoles

14.3.5.1
Oxadiazon

Oxadiazon {3-[2,4-dichloro-5-(1-methylethoxy)phenyl]-5-(1,1dimethylethyl)-1,3,4-oxadiazol-2-(3H)-one; CAS No. 19666-30-9} is a soil-applied peroxidizing herbicide for preemergence broadleaf and grass control in numerous crops including rice, cotton, ornamentals, fruit trees, soybeans, sunflowers and others (Ahrens 1994; Tomlin 1994). Application rates range up to

4.5 kg a.i./ha. Ishizuka et al. (1975) found that radiolabeled oxadiazon is not readily taken up by rice roots in solution culture or in a paddy system. Only 5% of the total oxadiazon applied was taken up during 10 days of exposure. The radiolabel taken up by roots was readily translocated to both aged and younger leaves, with somewhat greater accumulation in older leaves. Final residues in harvested grains were much lower (4–10 ppb) than in the leaves or straw (~1 ppm). In other studies (Achhireddy et al. 1984a,b), radiolabeled oxadiazon applied to rice and barnyardgrass (*Echinochloa crus-galli*) leaves was not readily translocated, and movement was not enhanced by development of sink activity. In each of these three studies, oxadiazon was not readily metabolized. For example, in a treated paddy system, 75–94% of the radiolabel recovered from rice plants after up to 2 months of treatment was intact oxadiazon (Ishizuka et al. 1975). In studies of foliar application to rice and barnyardgrass, 79–86% of the tissue radiolabel was intact oxadiazon after 7 days (Achhireddy et al. 1984a,b). The metabolic pathway of oxadiazon in plants is shown in Fig. 19. The major metabolite in rice (I) comprised 7–17% of the total radiolabeled residues 2 months after treatment with oxadiazon (Hirata and Ishizuka 1975). The other metabolites each comprised approximately 0.5–3% of the total radiolabeled residues. Chakraborty et al. (1995) isolated a soil fungus, *Fusarium solani* (Mortius) Sacc., using enrichment culture techniques which could metabolize oxadiazon. The major metabolic pathway in this organism also includes metabolite I as well as two additional derivatives of this metabolite not previously identified in plants (Fig. 19). Oxadiazon is strongly adsorbed to soil with an average Koc of 3200 ml/g and has an average field half-life of 60 days (Wauchope et al. 1992). The preemergence selectivity of oxadiazon to many crops is believed to be largely due to soil sorption and consequent protection of the crop roots from this herbicide.

14.3.6 Tetrahydrotriazolopyridinones

14.3.6.1 *Azafenidin*

Azafenidin {2-[2,4-dichloro-5-(2-prop-2-ynyloxy)phenyl]-5,6,7,8-tetrahydro-1,2,4-triazolo[4,3-a]pyridin-3(2H)-one; CAS No. 68049-83-2; formerly DPX-R6447} is an N-phenyl tetrahydrotriazolopyridinone herbicide under development for use in citrus, grapes, olive orchards, sugarcane, and tree fruits and nuts (Amuti et al. 1997). Preemergence application of *ca.* 200–2000 g a.i./ha provides 60–120 days of broadspectrum grass and broadleaf weed control. Postemergence applications to weeds when mixed with a contact herbicide also provide residual control of weeds germinating after the application.

Azafenidin photolyzes readily in aqueous solutions. Using simulated sunlight, it photolyzed in sterile buffer with DT_{50} values of 1.4–1.7 days of natural sunlight equivalents (Massey and Reilly, 1996) at pH 7–8. The major photolytic

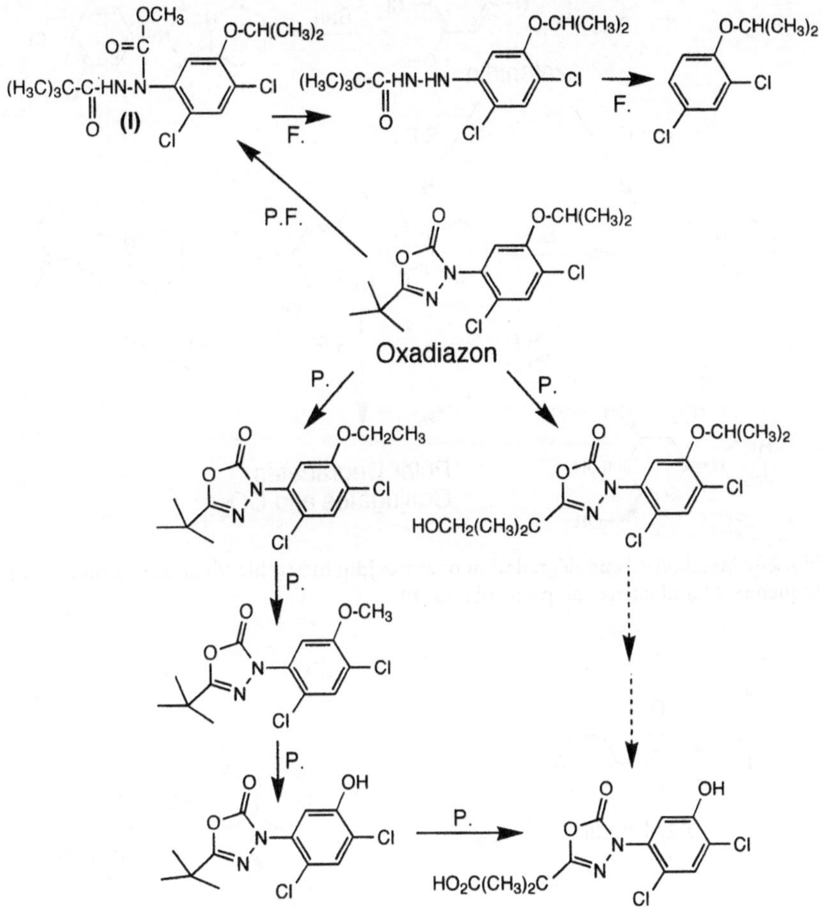

Fig. 19. Metabolism of oxadiazon in rice (*P*) and *Fusarium solani* (*F*)

degradate was the tetrahydrotriazolopyridinone heterocycle (85%) with minor amounts of O-dealkylated azafendin (I) being produced (Fig. 20). Further photolysis produced polar aliphatic acids and CO_2, especially from the radiolabeled phenyl ring.

Azafenidin degrades in soils with DT_{50} values ranging from 50 to 186 days at 20 °C in laboratory studies (Naidu and Fox 1995a; Kim and McEuen 1996) and 4 to 129 days under field conditions at diverse locations in the USA and Europe (Zietz, 1995). The major degradate in soil was the O-dealkylated analog (I) (Fig. 20), which is herbicidally inactive. Interesting minor reactions in soil are the subsequent methylation of the O-dealkylated degradate and the reduction of the propynyl group under anaerobic conditions (Naidu and Fox 1995b). Azafenidin binds to soil with Koc values ranging from 186 to 579 (Armbrust and Reilly, 1992). Laboratory and field leaching studies in sandy soils showed that azafenidin does not move beyond the top few centimeters of soil.

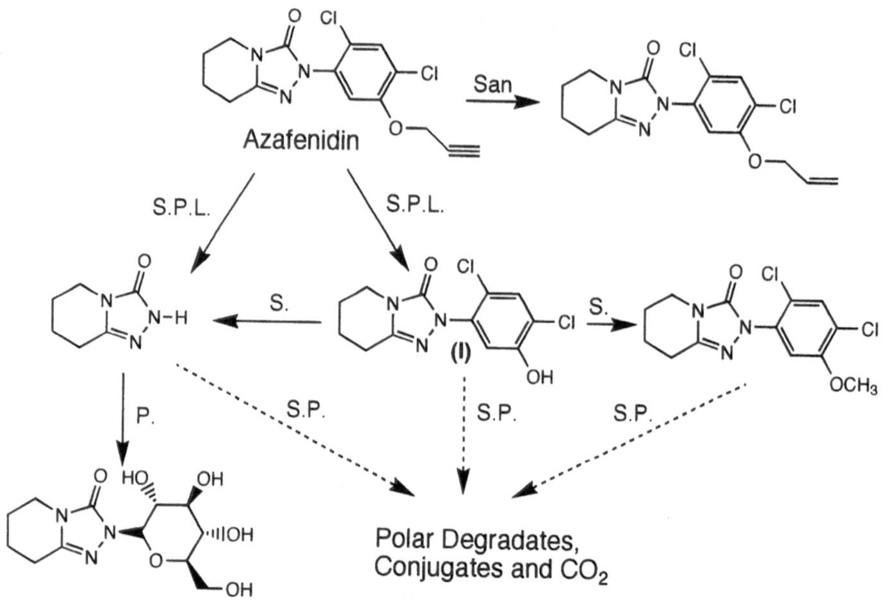

Fig. 20. Metabolism and degradation of azafenidin in aerobic (*S*) and anaerobic (*San*) soil, light (aqueous; *L*) and citrus and grape plants (*P*)

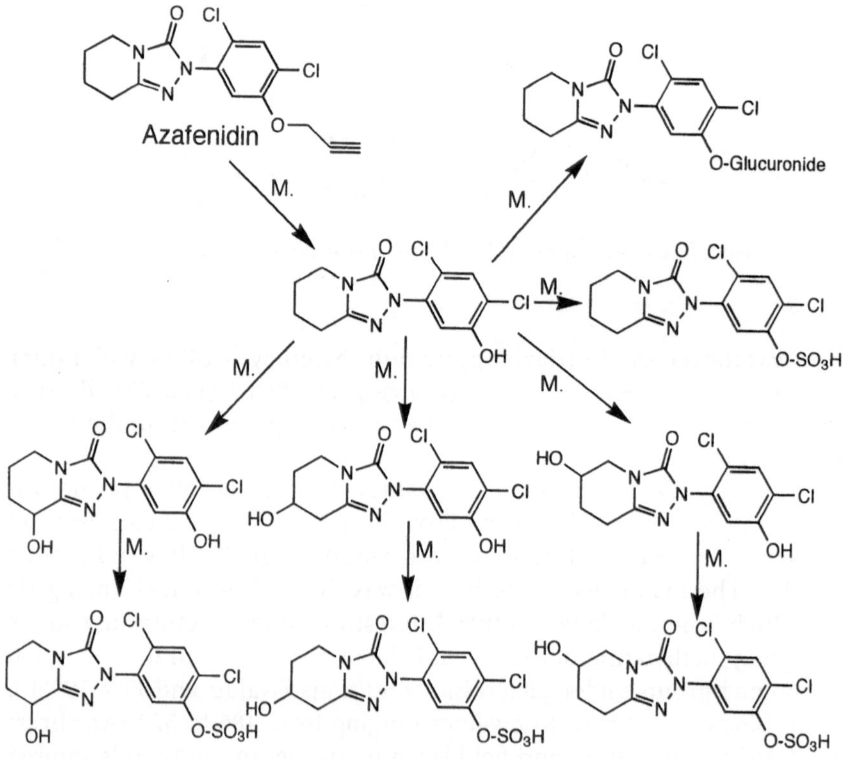

Fig. 21. Metabolism of azafenidin in rats and lactating goats (*M*)

Small amounts of soil degradates of azafenidin may be taken up by citrus (Bollin, 1996) and grape plants (Brisbin, 1996) and conjugated to hexoses and further degraded (Fig. 20). Residues in harvested grapes or citrus fruit were negligible.

Azafenidin is readily metabolized and excreted in rats and lactating goats. During a 5-day dosing period, 80–85% of the radiolabeled dose was excreted in the urine and feces and less than 0.5% of the total dose was found in milk (goat) or any of the tissues in the goat or rat. Azafenidin is metabolized in these animals through O-dealkylation, hydroxylation of the tetrahydrotriazolopyridinone ring in several positions, and glucuronide and sulfonate conjugation (Fig. 21).

Acknowledgments. We are pleased to acknowledge Dr. K. Hirai of the Sagami Chemical Research Center for his valuable assistance in providing a comprehensive list and classification of peroxidizing herbicides, and Dr. William Kenyon and Dr. Angela Klemens of DuPont Agricultural Products for their valuable review of this chapter. We also thank Dr. Ronald Cook and his colleagues at FMC for providing significant additional information on carfentrazone and sulfentrazone.

Abbreviations

The following abbreviations are used in the figures to identify various metabolic and degradative systems: L (light); H (hydrolysis); Sm (soil microorganisms); F (fungi); S (soil); San (anaerobic soil); m (*in vitro* microsomes); P (plant); M (mammals); A (non-mammalian animals).

References

Achhireddy NR, Kirkwood RC, Fletcher WW (1984a) The uptake and distribution of MCPA and oxadiazon in excised leaves. J Pestic Sci 9:611–615

Achhireddy NR, Kirkwood RC, Fletcher WW (1984b) Oxadiazon absorption, translocation and metabolism in rice (*Oryza sativa*) and barnyardgrass (*Echinochloa crus-galli*). Weed Sci 32:727–731

Ahrens WH (ed) (1994) Herbicide handbook. Weed Sci Soc Am, Champaign

Adler IL, Jones BM, Wargo JP Jr (1977) Fate of 2-chloro-1-(3-ethoxy-4-nitrophenoxy)-4-(trifluoromethyl) benzene (oxyfluorfen) in rats. J Agric Food Chem 25:1339–1341

Amuti K, Trombini A, Giammarrusti L, Sbriscia C, Harder H, Gabard J (1997) Azafenidin: A new low use rate herbicide for weed control in perennial crops, industrial weed control and forestry. Proc Brit Crop Prot Conf 1:59–66

Armbrust KL, Reilly D (1992) Batch equilibrium (adsorption/desorption) study of DPX-R6447 in soil. DuPont report AMR 2483-92

Becker JM (1994) Field dissipation of sulfentrazone. FMC internal report

Bollin E Jr (1996) Metabolism of ^{14}C-DPX-R6447 herbicide in grapefruit. DuPont report AMR 2531-92

Brisbin JM (1996) Metabolism of ^{14}C-DPX-R6447 herbicide in grapes at an application rate of 2.0 lbs/ac. DuPont Report AMR 3511-95

Chakraborty SK, Chowdhury A, Bhattacharyya A, Ghosh S, Pan S, Waters R, Adityachaudhury N (1995) Microbial degradation of oxadiazon by soil fungus *Fusarium solani*. J Agric Food Chem 43:2964–2969

Culligan (1994) Field dissipation of sulfentrazone. FMC internal report

Dayan FE, Keener BR, Weete JD, Hancock HG (1996a) Uptake, translocation and metabolism of carfentrazone in soybean and selected weeds. Proc Southern Weed Sci Soc, pp 151–152
Dayan FE, Weete JD, Hancock HG (1996b) Physiological basis for differential sensitivity to sulfentrazone by sicklepod (*Senna obtusifolia*) and coffee senna (*Cassia occidentalis*). Weed Sci 44:12–17
Dayan FE, Duke SO, Weete JD, Hancock HG (1997a) Selectivity and mode of action of carfentrazone, a novel phenyl triazolinone herbicide. Pest Sci 51:65–73
Dayan FE, Weete JD, Duke SO, Hancock HG (1997b) Soybean (*Glycine max*) cultivar differences in response to sulfentrazone. Weed Sci 45:634–641
Deng Y (1995) Field corn metabolism of carfentrazone ethyl. FMC internal report
Draper WM, Casida JE (1983) Diphenyl ether herbicides: mutagenic metabolites and photoproducts of nitrofen. J Agric Food Chem 31:227–231
Eastin EF (1969) Movement and fate of *p*-nitrophenyl trifluoro-2-nitro-*p* tolyl ether-1′ ^{14}C in peanut seedlings. Plant Physiol 44:1397–1401
Eastin EF (1971a) Movement and fate of fluorodifen-1′-^{14}C in cucumber seedlings. Weed Res 11:63–68
Eastin EF (1971b) Degradation of fluorodifen-1′-^{14}C by peanut seedling roots. Weed Res 11:120–123
Eastin EF (1971c) Fate of fluorodifen in resistant peanut seedlings. Weed Sci 19:261–265
Eastin EF (1972a) Photolysis of fluorodifen. Weed Res 12:75–79
Eastin EF (1972b) Fate of fluorodifen in susceptible cucumber seedlings. Weed Sci 20:255–260
ElNaggar SF (1992) Soybean metabolism of sulfentrazone. FMC internal report
ElNaggar SF (1995) Wheat metabolism of carfentrazone ethyl. FMC internal report
Elmarakby SA (1997) Aerobic aquatic metabolism of carfentrazone ethyl. FMC internal report
Evans JDHL, Cavell BD, Hignett RR (1987) Fomesafen – metabolism as a basis for its selectivity in soya. Proc Br Crop Prot Conf 1:345–352
Fadayomi A, Warren GF (1977) Uptake and translocation of nitrofen and oxyfluorfen. Weed Sci 25:111–114
Frear DS, Swanson HR (1973) Metabolism of substituted diphenylether herbicides in plants. I. Enzymatic cleavage of fluorodifen in peas (*Pisum sativum* L.). Pest Biochem Physiol 3:473–482
Frear DS, Swanson HR, Mansager ER (1983) Acifluorfen metabolism in soybean: diphenylether bond cleavage and the formation of homoglutathione, cysteine, and glucose conjugates. Pest Biochem Physiol 20:299–310
Gennari M, Negre M, Asmbrosoli R, Vincenze AV, Vincenti M, Acquati A (1994) Anaerobic degradation of acifluorfen by different enrichment cultures. J Agric Food Chem 42:1232–1236
Goodyear A (1995a) Aerobic soil metabolism of carfentrazone ethyl. FMC internal report
Goodyear A (1995b) Anaerobic soil metabolism of carfentrazone ethyl. FMC internal report
Guh JO, Lee EK, Kuk YI, Park RD (1995) Absorption, translocation and metabolism of oxyfluorfen in rice (*Oryza sativa*) and barnyardgrass (*Echinochloa crus-galli*). Weed Res Jpn 40:245–251
Hamada T, Yoshida R, Nagano E, Oshio H, Kamoshita K (1989) S-23121 – a new cereal herbicide for broad-leaved weed control. Proc Brit Crop Prot Conf 1:41–46
Hirata H, Ishizuka K (1975) Identification of the metabolite (M-1) of 2-*tert*-butyl-4-(2,4-dichloro-5-isopropoxyphenyl)-1,3,4-oxadiazolin-5-one (oxadiazon) in rice plants. Agric Biol Chem 39:1447–1454
Hunt LM, Chamberlain WF, Gilbert BN, Hopkins DE, Gingrich AR (1977) Absorption, excretion, and metabolism of nitrofen by a sheep. J Agric Food Chem 25:1062–1065
Ishizuka K, Hirata H, Fukunaga K (1975) Absorption, translocation and metabolism of 2-*tert*-butyl-4-(2,4-dichloro-5-isopropoxyphenyl)-1,3,4-oxadiazolin-5-one (oxadiazon) in rice plants. Agric Biol Chem 39:1431–1446
Ishizuka K, Matsumoto H, Hyakutake H (1988) Selective inhibitory action of chlomethoxynil on rice and barnyardgrass and its molecular fate in the light and dark. Weed Res Jpn 33:41–48
Jacobson A (1989) Metabolic fate of Compete herbicide in wheat. Abstr Am Chem Soc 198:AGRO 19

Kabler K, Williamson K (1991) Aqueous hydrolysis of sulfentrazone. FMC internal report
Kale SP, Raghu K (1989) Fate of ^{14}C nitrofen in rice paddy ecosystem. Bull Environ Contam Toxicol 42:544–547
Kale SP, Raghu K (1994) Fate of ^{14}C-nitrofen in soils. Bull Environ Contam Toxicol 53:298–302
Kamoshita K, Nagano E, Saito K, Sakaki M, Yoshida R, Sato R, Oshio H (1992) S-23031, a new postemergence herbicide for soybeans. In Copping LG, Green MB, Rees RT (eds) Pest management in soybean. SCI (Elsevier Applied Science), Amsterdam, pp 317–325
Kim EI, McEuen SF (1996) Rate of degradation of ^{14}C-DPX-R6447 on three soils. DuPont report AMR 3190-94
Kuwatsuka (1977) Studies on the fate and behavior of herbicides in soil and plant. J Pestic Sci 2:201–213
Lamoureux GL, Davison KL (1975) Mercapturic acid formation in the metabolism of propachlor, CDAA, and fluorodifen in the rat. Pest Biochem Physiol 5:497–506
Lamoureux GL, Rusness DG, Schroder P, Rennenberg H (1991) Diphenyl ether herbicide metabolism in a spruce cell suspension culture: the identification of two novel metabolites derived from a glutathione conjugate. Pest Biochem Physiol 39:291–301
Lamoureux GL, Rusness DG, Schroder P (1993) Metabolism of a diphenylether herbicide to a volatile thioanisole and a polar sulfonic acid metabolite in spruce (*Picea*). Pest Biochem Physiol 47:8–20
Leather GR, Foy CL (1977) Metabolism of bifenox in soil and plants. Pest Biochem Physiol 7:437–442
Lee AH, Lu PY, Metcalf RL, Hsu EL (1976) The environmental fate of three dichlorophenyl nitrophenyl ether herbicides in a rice paddy model ecosystem. J Environ Qual 5:482–486
Lee JJ, Matsumoto H, Pyon JY, Ishizuka K (1991) Mechanism of selectivity of diphenyl ether herbicides oxyfluorfen and chlomethoxynil in several plants. Weed Res Jpn 36:162–170
Leung LY, Lyga JW, Robinson RA (1991) Metabolism and distribution of the experimental triazole herbicide F6285 [1-[2,4-dichloro-5-[N-(methylsulfonyl)]-1,4-dihydro-3-methyl-4-(difluoromethyl)-5H-triazol-5-one] in the rat, goat, and hen. J Agric Food Chem 39:1509–1514
Locke RK, Baron RL (1972) Preforan metabolism by tobacco cells in suspension culture. J Agric Food Chem 20:861–867
Maigrot PH, Perrot A, Hede-Hauy L, Murray A (1989) Fluoroglycofen-ethyl: a new selective herbicide for broad-leaved weeds in cereals. Proc Br Crop Prot Conf 1:47–51
Massey JH, Reilly D (1996) Photodegradation of DPX-R6447 in water by simulated sunlight. Dupont report AMR 2491-92
Matsunaga H, Isobe N, Kaneko H, Nakatsuka I, Yamane S (1997a) Metabolism of pentyl 2-chloro-4-fluoro-5-(3,4,5,6-tetrahydrophthalimido)phenoxyacetate (flumiclorac pentyl, S-23031) in rats. 2. Absorption, distribution, biotransformation, and excretion. J Agric Food Chem 45:501–506
Matsunaga H, Tomigahara Y, Kaneko H, Nakatsuka I, Yamane S (1997b) Identification of a reduced form metabolite of flumiclorac pentyl (S-23031) in rats. J Pestic Sci 22:133–135
Miyazawa T, Kawano K, Shigematsu S, Yamaguchi M, Matsunari K, Porpiglia P, Gutbrod KG (1993) KIH-9201: a new low-rate post-emergence herbicide for maize (*Zea mays*) and soybeans (*Glycine max*). Proc Br Crop Prot Conf 1:23–28
Mizutani H, Unai T, Shimizu T, Ishikawa K, Yusa Y (1994a) Isomerization of a novel herbicide KIH-9201 and its analogs by glutathione and glutathione-s-transferase. Abstr 19th Meeting Jpn Pest Sci Soc, p 128
Mizutani H, Nakahira Y, Unai T, Ishikawa K, Yusa Y, Yamaguchi M (1994b) Metabolism of a novel herbicide KIH-9201 by soybean and onamomi seedlings. Abstr 19th Meeting Jpn Pest Sci Soc, p 129
Mizutani H, Unai T, Nakahira Y, Ishikawa K, Yusa Y (1994c) Metabolism and degradation of a novel herbicide KIH-9201 in rats and soils. Abstr 19th Meeting Jpn Pest Sci Soc, p 149
Naidu MV, Fox GC (1995a) Aerobic soil metabolism of ^{14}C-DPX-R6447. DuPont report AMR 2690-93

Naidu MV, Fox GC (1995b) Anaerobic soil metabolism of ^{14}C-DPX-R6447. DuPont report AMR 2684-93
Nakagawa M, Crosby DG (1974) Photodecomposition of nitrofen. J Agric Food Chem 22:849–853
Niki Y, Kuwatsuka S (1976) Degradation of diphenyl ether herbicides in soils. Soil Sci Plant Nutr (Tokyo) 22:223–232
Ogawa K, Ohori Y, Aizawa H, Shigeoka T, Yamauchi F (1980) Metabolism and degradation of an imide herbicide, MK-129 (1) in plants. Abstr 5th Meeting Jpn Pest Sci Soc, p 131
Oh BY, Jeang YH, Lee BM (1981) Studies on degradation of butachlor and nitrofen in different soil conditions. Hanguk Nonghwa Hakhoe Chi 24:112–119
Ohori Y, Ogawa K, Aizawa H, Shigeoka T, Yamauchi F (1980) Metabolism and degradation of an imide herbicide, MK-129 (2) in soils. Abstr 5th Meeting Jpn Pest Sci Soc, p 132
Ohyama H, Kuwatsuka S (1978) Degradation of bifenox, a diphenyl ether herbicide, methyl 5-(2,4-dichlorophenoxy)-2-nitrobenzoate, in soils. J Pestic Sci 3:401–410
Ohyama H, Kuwatsuka S (1983) Behavior of bifenox, a diphenyl ether herbicide, methyl 5-(2,4-dichlorophenoxy)-2-nitrobenzoate, in soil. J Pestic Sci 8:17–25
Okada M, Aizawa H, Shigeoka T, Yamauchi F (1980) Metabolism and degradation of an imide herbicide, MK-129 (3) in rats. Abstr 5th Meeting Jpn Pest Sci Soc, p 133
Oyamada M, Kuwatsuka S (1988) Effects of soil properties and conditions on the degradation of three diphenyl ether herbicides in flooded soils. J Pestic Sci 13:99–105
Plowchalk DR (1996) Metabolism of DPX-R6447 in lactating goat. DuPont report AMR 2708-93
Pornprom T, Matsumoto H, Usui K, Ishizuka K (1994) Absorption and metabolism of oxyfluorfen in tolerant soybean cells. Weed Res Jpn 39:180–182
Purser D (1994) Degradation and retention of carfentrazone ethyl in a water/sediment system. FMC internal report
Pusin A, Gessa C (1991) Photolysis of acifluorfen in aqueous solution. Pest Sci 32:1–5
Qian W, Jin W, Li D, Xu R (1982) Persistence of nitrofen in soil. Huanjing Kexue 3:36–39
Ramsey AA (1994) Confined crop rotation study of sulfentrazone. FMC internal report
Ramsey AA, Venables JA (1997) Tobacco metabolism of sulfentrazone. FMC internal report
Reilly D, Armbrust KL (1996) Photodegradation of radiolabeled DPX-R6447 on soil conducted in simulated sunlight. DuPont report AMR 2692-93
Raub MF, Vengurlekar SS, Reiser CA, Panek MG, Veit P, McGown SR, Geiger DS (1996) Metabolism of ^{14}C-sodium acifluorfen in peanut. Abstr Am Chem Soc 211 (1–2):AGRO 51
Ritter RL, Coble HD (1981) Penetration, translocation and metabolism of acifluorfen in soybean (*Glycine max*), common ragweed (*Ambrosia artemisiifolia*) and common cocklebur (*Xanthium pensylvanicum*). Weed Sci 29:474–480
Robinson RA (1995) Goat metabolism of carfentrazone ethyl. FMC internal report
Rogers RL (1971) Absorption, translocation and metabolism of *p*-nitrophenyl trifluoro-2-nitro-*p*-tolyl ether by soybeans. J Agric Food Chem 19:32–35
Ruzo LO, Lee JK, Zabik MJ (1980) Solution-phase photodecomposition of several substituted diphenyl ether herbicides. J Agric Food Chem 28:1289–1292
Saito K, Kaneko H, Sato K, Nakatsuka I, Yamada H (1996) Production of acetylated metabolites of pesticides in mammals: characterization of aniline derivatives in vitro. J Pestic Sci 21:333–336
Sato Y, Boger P, Wakabayashi K (1997) The enzymatic activation of peroxidizing cyclicisoimide: a new function of glutathione S-transferase and glutathione. J Pestic Sci 22:33–36
Schocken MJ (1994) Soil photolysis of sulfentrazone. FMC internal report
Schocken MJ, Creekmore RW, Theodoridis G, Nystrom GJ, Robinson RA (1989) Microbial transformation of the tetrazolinone herbicides F5231. Appl Environ Microbiol 55:1220–1222
Schroder P, Lamoureux GL, Rusness DG, Rennenberg H (1990) Glutathione-S-transferase activity in spruce needles. Pest Biochem Physiol 37:211–218
Shimabukuro RH, Lamoureux GL, Swanson HR, Walsh WC, Stafford LE, Frear DS (1973) Metabolism of substituted diphenylether herbicides in plants. II. Identification of a new fluorodifen metabolite, S-(2-nitro-4-trifluoromethylphenyl)-glutathione in peanut. Pest Biochem Physiol 3:483–494

Shimizu T, Hashimoto N, Nakayama I, Nakao T, Mizutani H, Unai T, Yamaguchi M, Abe H (1995) A novel isourazole herbicide, fluthiacet-methyl, is a potent inhibitor of protoporphyrinogen oxidase after isomerization by glutathione-S-transferase. Plant Cell Physiol 36:625–632
Singer SS, Schocken MJ (1991) Aerobic soil metabolism of sulfentrazone. FMC internal report
Tewfik MS, Hamdi YA (1975) Metabolism of fluorodifen by soil microorganisms. Soil Biol Biochem 7:79–82
Tomlin C (ed) (1994) The pesticide manual 10th edn. Br Crop Prot Council and The Royal Soc Chem, Surrey UK
Tomlin C (ed) (1997) The pesticide manual 11th edn. Br Crop Prot Council and The Royal Soc Chem, Surrey UK
Tsuchiya K, Uchida M, Sugioto T (1982) Metabolism of ^{14}-C-chlomethoxynil in rats. J Pestic Sci 7:187–193
Van Saun WA, Bahr JT, Crosby GA, Fore ZQ, Guscar HL, Harnish WN, Hooten RS, Marquez MS, Parrish DS, Theodoridis G, Tymonko JM, Wilson KR, Wyle MJ (1991) F6285 – a new herbicide for the pre-emergence selective control of broad-leaved and grass weeds in soybeans. Proc Br Crop Prot Conf 1:77–82
Van Saun WA, Bahr JT, Bourdouxhe LJ, Gargantiel FJ, Hotzman FW, Shires SW, Sladen NA, Tutt SF, Wilson KR (1993) F8426 – a new, rapidly acting, low rate herbicide for the postemergence selective control of broad-leaved weeds in cereals. Proc Br Crop Prot Conf 1:19–28
Vanstone DE, Stobbe EH (1978) Root uptake, translocation and metabolism of nitrofluorfen and oxyfluorfen by fababeans (*Vicia faba*) and green foxtail (*Setaria viridis*). Weed Sci 26:389–392
Wargo JP, Honeycutt R, Alder IL (1975) Characterization of bound residues of nitrofen in cereal grains. J Agric Food Chem 23:1095–1097
Wauchope RD, Buttler TM, Hornsby AG, Augustijn-Beckers PWM, Burt JP (1992) The SCS-ARS-CES pesticide properties database for environmental decision-making. Rev Environ Contam Toxicol 123:1
Willut JM (1994) Aqueous photolysis/soil photolysis of carfentrazone ethyl. FMC internal report
Willut JM, McLaughlin TM, Shomo RE, Fang XP, Gravelle WD, Varanyak LA (1997) Formation and decline of major sulfentrazone photoproducts in buffered aqueous solutions by simulated sunlight. Abstr Am Chem Soc AGRO 096
Wu D (1995) Rat metabolism of carfentrazone ethyl. FMC internal report
Yoshino H, Matsunaga Y, Kaneko H, Yoshitake A, Nakatsuka I, Yamada H (1993) Metabolism of N-[4-chloro-2-fluoro-5-[(1-methyl-2-propynyl)oxy]phenyl]-3,4,5,6-tetrahyrophthalimide (S-23121) in rats: I. Identification of a new, sulphonic acid type of conjugate. Xenobiotica 23:609–619
Zietz E (1995) Field soil dissipation of DPX-R6447 herbicide at European sites using an 80% water dispersible granule formulatioin. DuPont report AMR 2922-94

CHAPTER 15

Peroxidizing Herbicides: Toxicology to Mammals and Non-target Organisms

JAN KRIJT[1]

Contents

15.1 Introduction

The peroxidizing herbicides constitute a growing group of industrial chemicals (for a review see Duke et al. 1991; Böger and Wakabayashi 1995). Their exact mode of action was elucidated only quite recently, when it was realised that these compounds block the enzymatic oxidation of protoporphyrinogen to protoporphyrin (Matringe et al. 1989; Witkowski and Halling 1989). In plants, the protoporphyrinogen oxidase-catalysed reaction is a necessary step in the biosynthesis of both chlorophyll and heme, whereas in mammalian cells it is utilised solely for the production of heme. Since heme forms the prosthetic group of mitochondrial respiratory cytochromes and numerous other hemeproteins, it has to be synthesised in every mammalian cell.

[1] Institute of Pathophysiology, First Faculty of Medicine, Charles University, U nemocnice 5, 128 53 Prague, Czech Republic, Fax: 00 420 2 24912834, E-Mail: jkri@lf1.cuni.cz

Peter Böger, Ko Wakabayashi
Peroxidizing Herbicides

15.1.1 Mammalian Heme Biosynthesis

The first enzyme of mammalian heme biosynthesis, 5-aminolevulinic acid synthase (ALAS), produces 5-aminolevulinic acid (ALA) from glycine and succinyl-CoA (for a review see Kappas et al. 1995). In the next step, two molecules of ALA are converted to the monopyrrole porphobilinogen (PBG). Porphobilinogen deaminase condenses four PBG molecules to form a linear tetrapyrrole, which is converted to uroporphyrinogen III. Uroporphyrinogen is decarboxylated to coproporphyrinogen III and the side chains are further modified to form protoporphyrinogen IX. The porphyrinogens do not contain a conjugated double bonds system and are not optically active. Only at the level of protoporphyrinogen oxidase (protox, EC 1.3.3.4), which oxidises protoporphyrinogen to protoporphyrin, does the mammalian cell produce an optically active, fluorescent porphyrin molecule. Finally, insertion of iron in the protoporphyrin IX ring by ferrochelatase results in the formation of non-fluorescent heme. Protoporphyrinogen oxidase was the last enzyme of the heme biosynthetic pathway to be characterised, cloned and expressed (Nishimura et al. 1995; Camadro and Labbe 1996; Dailey and Dailey 1996; Puy et al. 1996; for a review see Chap. 8, 9, this Vol.). The enzyme contains a bound flavoprotein (Proulx and Dailey 1992) and available data indicate that it exists as a single transcript in both erythroid and non-erythroid human cells (Taketani et al. 1995; Dailey and Dailey 1996).

15.1.2 Erythroid Heme Biosynthesis

It is estimated that about 85% of human heme production takes place in the bone marrow as a part of hemoglobin synthesis (Kappas et al. 1995). This erythroid heme synthesis uses specific regulation and an erythroid-specific 5-aminolevulinic synthase gene. In addition, different transcription of a common porphobilinogen deaminase gene in humans also results in two enzyme isoforms, one ubiquitous and one erythroid-specific.

The exact regulation mechanisms of the erythroid heme biosynthesis pathway are not clear, but it is well established that the mRNA for erythroid ALAS contains an iron-responsive element. Thus, the erythroid heme biosynthesis appears to be regulated mainly by the availability of iron (for a review see Ponka 1997).

15.1.3 Hepatic Heme Biosynthesis

A significant part of whole body heme synthesis takes place in the liver. Hepatocytes contain high amounts of diverse, specialised hemeproteins for the metabolism of xenobiotics – the inducible microsomal cytochromes P450.

Cytochromes P450 have a relatively rapid turnover, and the liver heme synthesis has to be efficiently regulated in order to immediately respond to the varying heme demand. It is postulated that the regulation of hepatocyte heme biosynthesis depends on the size of a free heme pool in the cell. A decrease in the free heme pool size results in increased activity of ALA synthase, mediated at the levels of transcription, mRNA stability and translocation of the ALA synthase precursor protein into mitochondria (Kappas et al. 1995; Ponka 1997). An extreme decrease of the heme pool (for example by xenobiotics-mediated destruction of cytochrome P450 heme) increases the hepatic ALA synthase activity by an order of magnitude. Under these conditions, the increased flux of ALA through the pathway exceeds the capacity of porphobilinogen deaminase and excess ALA and PBG are excreted in urine. This increased excretion of ALA and PBG is a typical finding in patients with acute porphyrias.

15.2 Porphyrias

Inherited deficiencies in the heme biosynthetic pathway cause relatively rare diseases known as porphyrias. A deficiency of protoporphyrinogen oxidase results in variegate porphyria. Together with acute intermittent porphyria and hereditary coproporphyria, variegate porphyria is classified as an acute hepatic porphyria, highlighting the fact that the patient may (but also may not) develop an acute porphyric crisis. Most variegate porphyria patients display a reduction of protoporphyrinogen oxidase activity of about 50% in all examined tissues. Clinically, the disease is characterised by cutaneous photosensitivity and/or neurological dysfunction. As in other acute porphyrias, erythroid heme biosynthesis is practically unaffected by the enzymatic defect.

The pathophysiological mechanism of photosensitivity resembles to some extent the mechanism of plant destruction by peroxidizing herbicides. In both cases, decreased protoporphyrinogen oxidase activity causes accumulation of protoporphyrinogen. The colorless, non-fluorescent protoporphyrinogen is then oxidised to fluorescent protoporphyrin – a highly active photosensitizer, mediating photodynamic damage to light-exposed surfaces. In herbicide-treated plants, this secondary oxidation of accumulated protoporphyrinogen is probably caused by herbicide-resistant protoporphyrinogen oxidase activity associated with the plasma membrane and endoplasmic reticulum (Jacobs et al. 1991; Lee et al. 1993; Retzlaff and Böger 1996). In man, the mechanism of the secondary protoporphyrinogen oxidation is unclear – it could be caused by protoporphyrinogen leakage from the sites of overproduction to other sites or organs, or it could be mediated by non-specific peroxidases (Jacobs et al. 1996).

Patients with variegate porphyria can develop life-threatening acute crises, the so-called acute porphyric attacks. These are characterised by severe abdominal pain and peripheral neuropathy, sometimes leading to quadriplegia or respiratory paralysis. A hallmark of an acute attack in all acute porphyrias is a dramatic increase in the urinary excretion of ALA and PBG. The pathophysiology of porphyric attack is at present unclear, but it evidently takes place when the hepatocyte demand for heme is further increased by induction of hemeprotein synthesis or by increased rates of heme degradation.

It is generally accepted that a 50% decrease in enzymatic activity alone is not sufficient to cause an acute attack, and it is even assumed that the majority of patients with protox deficiency can go through their life without ever developing any clinical symptoms of the disease. This latent phase of variegate porphyria is characterised by only a slight increase of urinary ALA and PBG (Meissner et al. 1993), indicating a successful compensation of the inherited enzymatic defect. However, administration of a cytochrome P450-inducing drug (such as phenobarbital) or a cytochrome P450-destroying drug (such as griseofulvin) leads to a further induction of hepatic ALA synthase and can result in an acute episode.

A very rare, homozygous form of the disease (Kordac et al. 1985) is associated with a residual protox activity of less than 10% of normal and a rapid onset of cutaneous photosensitivity. Interestingly, urinary ALA and PBG excretion is not significantly elevated, nor do the children suffering from this disease develop acute porphyric attacks (Hift et al. 1993).

15.3 Inhibition of Mammalian Protoporphyrinogen Oxidase

The ability of peroxidizing herbicides to block protoporphyrinogen oxidation at relatively low concentrations has raised the question of potential interference with the heme biosynthetic pathway in animals and man. Even the very first paper on the mode of action of peroxidizing herbicides (Matringe et al. 1989) demonstrated protox inhibition in mouse liver mitochondria. Since then, a number of studies have demonstrated the susceptibility of mammalian cells to protoporphyrinogen oxidase inhibition. In vitro incubations of rat hepatocytes (Jacobs et al. 1992; Krijt et al. 1993), chick embryo hepatocytes (Sinclair et al. 1994), mouse liver mitochondria (Corrigall et al. 1994; Birchfield and Casida 1996; Birchfield and Casida 1997), pig liver mitochondria (Corrigall et al. 1994), pig placental mitochondria (Corrigall et al. 1994), murine erythroleukemia cells (Prasad and Dailey 1995) and J774.G8 murine macrophages (Krijt et al., unpubl. data) with herbicide concentrations of 10^{-7} to 10^{-4} M resulted in enzyme inhibition and/or porphyrin accumulation.

15.3.1 Inhibition of the Enzyme in Human-Derived Cells

Of special interest with respect to human toxicology are studies with cells or tissues of human origin. A moderate increase in cellular porphyrin content was found in human hepatoma-derived HepG2 cells treated with fomesafen, oxyfluorfen and oxadiazon (Krijt et al. 1993). Acifluorfen-methyl and oxadiazon were reported to elevate intracellular porphyrins in HeLa cells at 10^{-4} M (Halling et al. 1994). Corrigall et al. (1994) demonstrated protox inhibition of mitochondria obtained from human liver and placenta by acifluorfen – in both cases the inhibition was competitive. The same study reported competitive inhibition of human lymphoblast-derived protox. In studies with human erythroblastic progenitors, the herbicide oxyfluorfen inhibited hemoglobin synthesis at concentrations not affecting cellular proliferation and differentiation (Rio et al. 1997). It is therefore clear that human protox is susceptible to inhibition by peroxidizing herbicides, as also demonstrated with the cloned and expressed human enzyme (Nishimura et al. 1995; Dailey and Dailey 1996).

15.3.2
In Vivo Inhibition of the Mammalian Enzyme

The effect of peroxidizing herbicides on mammalian porphyrin metabolism was demonstrated in a number of in vivo experiments, either by changes in the porphyrin excretion patterns and tissue porphyrin accumulation or by tissue protox inhibition (Krijt et al. 1991, 1992; Machemer et al. 1992; Halling et al. 1994; Kawamura et al. 1996a). However, most initial in vivo studies used relatively high doses. Recently, a single dose of oxyfluorfen (4 mg/kg i.p.) was reported to cause a 50% decrease in binding of a specific radioligand probe to the inhibitor/herbicide binding site of the protox component of solubilised mouse liver mitochondria 19 h after treatment (Birchfield and Casida 1996), indicating significant in vivo effect at a relatively low dose.

In our laboratory, we have tested the porphyrogenic effects of oxadiazon (Rhône Poulenc, France), oxyfluorfen (Rohm and Haas, Philadelphia, USA) and fomesafen (ICI, now Zeneca, UK). Dietary administration of these herbicides to male BALB/c mice for a period of 8 days caused dose-related changes in the metabolism of porphyrins and heme. At low doses, the effect of the herbicides was observed mainly in the liver. Oxadiazon elevated bile porphyrin content and fecal porphyrin content when fed to mice at 5 mg/kg diet (5 ppm). Up to 50 ppm, the urinary excretion of the porphyrin precursors ALA and PBG remained in the normal range (Table 1). However, a dramatic increase in porphyrin precursors excretion was evident at dietary oxadiazon concentrations of 200 ppm or more. Oxyfluorfen and fomesafen were less potent in this respect (Table 2).

In parallel with the established patterns of porphyrin precursor excretion during the latent and acute phases of variegate porphyria, the data in Tables 1 and 2 can be mechanistically interpreted as follows: low doses of the peroxidizing herbicides cause a dose-dependent inhibition of hepatic protox, which results in increased protoporphyrin excretion in the bile. This partial block in heme biosynthesis is compensated by physiological regulatory mechanisms and the hepatocyte heme supply is at this stage not seriously compro-

Table 1. Effect of oxadiazon on porphyrin metabolism in male mice

Dietary oxadiazon (mg/kg diet)	Liver porphyrins (nmol/g)	Bile porphyrins (nmol/g)	Fecal porphyrins (nmol/g)	Plasma porphyrins (nmol/ml)
Control	1.0 ± 0.1	5 ± 3	26	ND
5	1.1 ± 0.1	12 ± 6	65	ND
10	1.2 ± 0.1	26 ± 4*	112	<5
25	1.9 ± 0.3*	91 ± 17*	185	20
50	2.5 ± 0.2*	172 ± 29*	435	30

Groups of three male BALB/c mice were fed the indicated dietary concentrations of oxadiazon for 8 days. Liver, bile and fecal porphyrins were determined fluorimetrically by methanol/sulfuric acid extraction; plasma porphyrin content was estimated from direct fluorescence of diluted plasma. Fecal and plasma porphyrins were determined in pooled samples. Protoporphyrin IX was used as a standard with both methods.
ND, not detected.
* Statistically different from control group, $p < 0.05$.

Table 2. In vivo effect of three peroxidizing herbicides on urinary porphyrin and porphobilinogen excretion

Treatment	Urinary porphyrins (nmol/l)		Urinary PBG (μmol/l)	
Control	<1000		<50	
Herbicide in diet:	200 mg/kg	1000 mg/kg	200 mg/kg	1000 mg/kg
Fomesafen	1400	2100	<50	<50
Oxyfluorfen	3100	18600	180	3380
Oxadiazon	8300	13400	1130	2640

Groups of three male BALB/c mice were fed diets containing the indicated herbicide concentrations for 8 days. Porphyrins and porphobilinogen were determined in pooled urine samples.

mised, as indicated by normal urinary PBG and ALA content. At higher herbicide concentrations, the substantial protox inhibition decreases hepatic heme levels and reduces the enzymatic activity of some heme-dependent cytochromes P450 (Fig. 1). A similar decrease in cytochrome P450 activity was reported in variegate porphyria patients (Tokola et al. 1988). In addition, the hepatic heme depletion caused by protox inhibition could be further aggravated by additional effects (De Matteis and Marks 1996) of the herbicides on hepatic cytochrome P450 synthesis and/or degradation. The result is a compensatory increase in 5-aminolevulinic acid synthase activity, coupled with overproduction and excretion of the porphyrin precursors ALA and PBG (Table 2).

According to the proposed mechanism, the in vivo effect of peroxidizing herbicides on the heme biosynthetic pathway is mediated primarily by protox inhibition and further modulated by cytochrome P450 induction or degradation. The different cytochrome P450 induction patterns (Fig. 1) could also

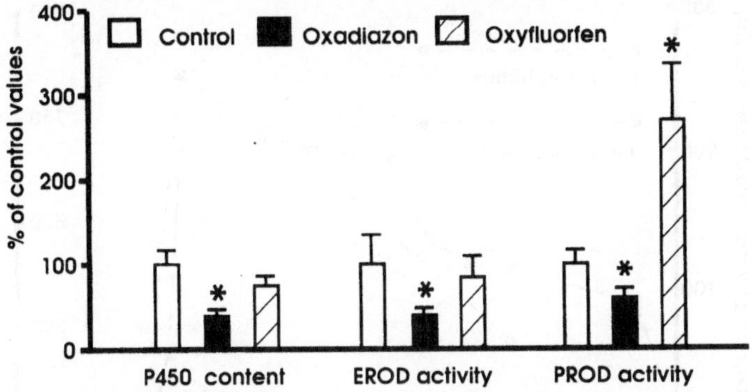

Fig. 1. Effect of oxadiazon and oxyfluorfen on mouse liver cytochrome P450 content and activities. Groups of four male BALB/c mice were fed diets containing 1000 mg oxadiazon or oxyfluorfen/kg for 8 days. *P450* Total cytochrome P450 content; *EROD* ethoxyresorufin (7-ethoxy-3H-phenoxazin-3-one) O-dealkylation activity; *PROD* pentoxyresorufin (7-pentoxy-3H-phenoxazin-3-one) O-dealkylation activity. EROD and PROD are traditional substrates for 3-methylcholanthrene-inducible and phenobarbital-inducible cytochrome P450 isozymes. * Statistically different from controls, $p < 0.05$

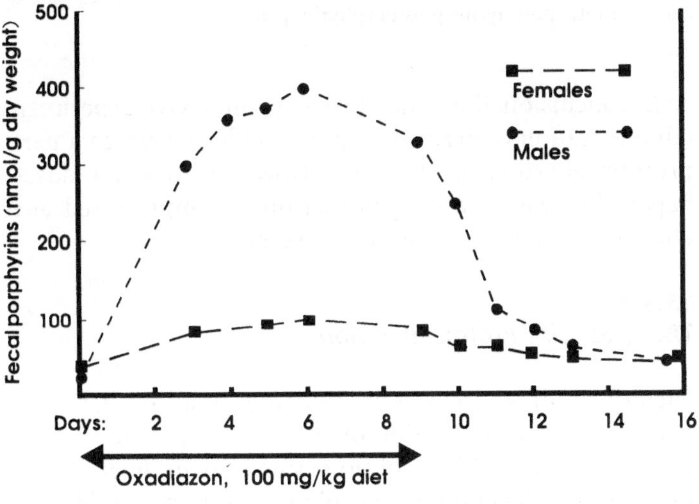

Fig. 2. Effect of oxadiazon (100 mg/kg in the diet) on fecal porphyrin excretion in male and female mice. Groups of three female and three male mice of ICR strain were fed a diet containing 100 mg oxadiazon/kg for 9 days. Porphyrin content was determined by methanol/sulfuric acid extraction of pooled samples

contribute to the observed differences in the in vivo porphyrogenic potency of peroxidizing herbicides (Table 2).

Since protox inhibition is competitive, the excretion of porphyrins should return to normal upon cessation of exposure. This was confirmed in experiments with both oxadiazon and fomesafen (Figs 2, 3).

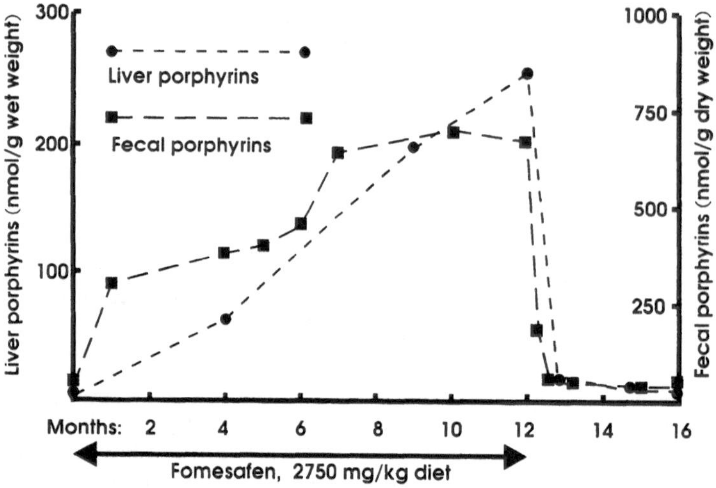

Fig. 3. Effect of long-term, high-dose fomesafen feeding on porphyrin metabolism in male mice. Male ICR strain mice were fed a diet containing 2750 mg fomesafen/kg of for 12 months. Four animals per group were sacrificed at 4, 9, 12, and 16 months; additional groups of two animals were sacrificed after 3 and 9 weeks following cessation of fomesafen treatment. Liver and fecal porphyrins were determined by methanol/sulfuric acid (20 : 1 v/v) extraction (Krijt et al. 1997) as a sum of total porphyrins plus porphyrinogens

In conclusion, the available data from in vivo experiments indicate that oral administration of peroxidizing herbicides results in a partial block in hepatic protoporphyrinogen oxidation. However, at lower doses the availability of hepatic heme is apparently not seriously compromised, as reflected by a lack of effect on urinary ALA and PBG excretion.

15.3.2.1
Tissue-Specific Protox Inhibition

Dietary administration of 200 and 1000 ppm oxadiazon to male BALB/c mice and subsequent incubation of tissue homogenates with protoporphyrinogen demonstrated inhibition of protox activity in the liver and kidney. No inhibition or an elevation of porphyrin content could be observed in brain homogenates (Krijt et al. 1997). In addition, no significant elevation of brain porphyrin content was found in male mice fed high doses (1000 mg/kg diet) of the diphenyl ethers oxadiazon or fomesafen, probably indicating a low inhibitor concentration in brain compared to the liver and kidney.

The insignificant effect of protox inhibitors on brain heme biosynthesis contrasts with the situation in human acute porphyrias, where a decrease of enzymatic activity occurs in all tissues. In fact, heme depletion in the central nervous system is considered to be one of the possible causes of severe neurological dysfunction observed in porphyria patients (Moore et al. 1987).

15.3.2.2
Porphyrinogen Versus Porphyrin Accumulation

Porphyrinogens are reduced, colorless and optically inactive porphyrin precursors. With the exception of protoporphyrin IX, porphyrinogens rather than porphyrins are the physiological intermediates in the heme biosynthetic pathway – once a porphyrinogen is non-enzymatically oxidised to porphyrin, it can no longer be utilised for heme biosynthesis. Under normal conditions, the precursor porphyrinogens are kept in the reduced state and tissue and plasma porphyrin levels are very low. In variegate porphyria, as well as in herbicide-treated plants, the inhibition of protox causes overproduction of protoporphyrinogen. Because the cutaneous symptoms of variegate porphyria must be mediated by protoporphyrin, it was assumed that a secondary oxidation of protoporphyrinogen takes place in the skin and other tissues. However, using a new enzymatic assay for protoporphyrinogen (Jacobs and Jacobs 1993), it was demonstrated that rat hepatocytes contain large amounts of protoporphyrinogen when treated with a diphenyl ether herbicide (Sinclair et al. 1994). The accumulation of porphyrinogens in the hepatocyte could have important toxicological implications, as it was reported that porphyrinogens are inhibitors of porphobilinogen deaminase (Meissner et al. 1993), an earlier enzyme in porphyrin biosynthesis. In experiments with oxadiazon- and oxyfluorfen-treated mice, protoporphyrinogen accounted for more than 75% of the sum of total bile porphyrins plus porphyrinogens (Krijt et al. 1997). The main fraction of liver porphyrins in these animals was formed by uroporphyrin, indicating efficient biliary excretion of protoporphyrinogen. These results confirm that hepatocyte protoporphyrinogen is not to a significant extent oxidised to protoporphyrin (Sinclair et al. 1994), but remains in the reduced state and is partly accumulated and partly excreted in the bile.

15.3.2.3
Species Sensitivity

At present, there is little information on species sensitivity in vivo. In vitro, protox inhibitors are active at 10^{-6} to 10^{-5} M concentration in a variety of mammalian cells, and, with the possible exception of the pig (Corrigal et al. 1994), there are no obvious species differences. However, in in vivo experiments with oral administration of flumioxazin, an N-phenylimide herbicide, it was found that female rabbit is much less sensitive than female rat (Kawamura et al. 1996a). Data from our laboratory indicate that 500 ppm oxadiazon in the diet does not elevate fecal porphyrin content when fed to male Syrian golden hamsters (unpubl. data), whereas 50 ppm causes a more than tenfold elevation of fecal porphyrins in mice (Table 1). The reason for this difference in species sensitivity is at present unclear, but different induction patterns of hepatic cytochrome P450 hemeproteins could be one of the contributing factors.

15.4 Other Toxicologically Relevant Effects

In addition to the well established effect on protoporphyrinogen oxidase inhibition, there are reports describing the carcinogenic potential of some peroxidizing herbicides and their effect on peroxisome proliferation. There are also reports on developmental toxicity.

15.4.1 Hepatocarcinogenesis

The classical DPE-type herbicide, acifluorfen, is classified as a probable human carcinogen (B2) according to the Environmental Protection Agency (USA). In long-term studies in mice and rats, acifluorfen significantly increased the incidence of hepatocellular carcinoma (Quest et al. 1989). Five other diphenyl ether protox inhibitors (bifenox, fomesafen, lactofen, nitrofen and oxyfluorfen) were also reported to cause liver tumors in mice (Quest et al. 1989), and the ability of fomesafen to induce hepatocellular carcinoma was confirmed and discussed in a separate report (Smith and Elcombe 1989). In addition, increased incidence of hepatocellular carcinomas at a relatively low dietary level of 100 ppm was reported in mouse studies with oxadiazon, which lacks the diphenyl ether structure (Von Burg 1994). It thus appears that many peroxidizing herbicides induce liver tumors in rodents, particularly in mice.

Hepatocellular carcinoma is a very common finding in rodents fed higher doses of test chemicals, and it is not clear to what extent the results of rodent studies are predictive for humans (for a review see Grisham 1997). The interpretation of results is further complicated by the fact that the response to a specific chemical may differ between rodent and human hepatocytes. This difference is very well documented in the case of peroxisome proliferation.

15.4.1.1 *Peroxisome Proliferation*

Peroxisomes are subcellular organelles active in lipid and hydrogen peroxide metabolism. Some chemicals, of which the antihyperlipidemic drug clofibrate is the best studied example, cause proliferation of peroxisomes when administered to mice and rats. In long-term rodent studies, nearly all peroxisome proliferators cause hepatocellular carcinomas, although most peroxisome proliferators show negative response in mutagenicity tests. Peroxisome proliferators therefore constitute a discrete group of epigenetic rodent hepatocarcinogens (Ashby et al. 1994).

In contrast to the situation in rodents, guinea pig hepatocytes, monkey hepatocytes and human hepatocytes display no proliferation of peroxisomes when exposed to the same chemicals. Since there is a very strong association between rodent peroxisome proliferation and rodent hepatocellular carcinoma formation, it is argued that results from rodent hepatocarcinogenicity studies are probably not relevant to humans if the tested carcinogenic chemical is a potent peroxisome proliferator (Smith and Elcombe 1989; Ashby et al. 1994). A number of peroxidizing herbicides induce peroxisome proliferation. Of the classical diphenyl ether herbicides, lactofen (Butler et al. 1988), fomesafen (Smith and Elcombe 1989) and acifluorfen (Krijt et al., unpubl. data) are strong inducers. In addition, peroxisome proliferation has recently been reported in experiments with oxadiazon (Richert et al. 1996). It can therefore be postulated that the ability of these herbicides to produce liver tumors in rodents is linked solely to their effect on rodent peroxisome proliferation, and that they present no carcinogenic hazard to human liver (Smith and Elcombe 1989). However, it should be noted that the doses used in most carcinogenicity tests were sufficient to cause a disturbance of heme biosynthesis, coupled with a substantial increase of hepatic porphyrinogen/porphyrin content. Since an increased incidence of hepatocellular carcinoma is reported in patients with porphyria cutanea tarda (Salata et al. 1985), acute intermittent porphyria (Kauppinen and Mustajoki 1988), and variegate porphyria (Kauppinen and Mustajoki 1992), it is possible to speculate that the elevated liver porphyrin content could, by a yet unknown mechanism, contribute to liver neoplasia. In addition, the known hepatocarcinogens hexachlorobenzene and 2,3,7,8-tetrachlorodibenzo-*p*-dioxin (TCDD) induce the same typical pattern of liver porphyrin accumulation (Smith and De Matteis 1990; Smith et al. 1985) as some diphenyl ether herbicides in long-term experiments, suggesting a similar mechanism of hepatotoxicity.

15.4.1.2
In Vivo Liver Uroporphyrin Accumulation

Long-term administration of high doses of the herbicide fomesafen (Krijt et al. 1994) or of the acifluorfen-containing preparation Blazer (unpubl. data) to male mice causes a gradual development of a characteristic pattern of liver porphyrin accumulation. During the first weeks of herbicide administration, fecal porphyrins are moderately elevated due to protox inhibition. Next, there follows a steep increase of both fecal and urinary porphyrins, and the liver contains substantial amounts of uroporphyrin I, uroporphyrin III and heptacarboxylic porphyrin. The same typical pattern of highly carboxylated hepatic porphyrins is encountered in patients with porphyria cutanea tarda, a chronic hepatic porphyria characterised by uroporphyrinogen decarboxylase deficiency (Kappas et al. 1995). At 9 and 12 months of herbicide treatment, the liver is grossly increased in size and macroscopic nodules up to several millimetres in diameter are evident. Upon discontinuation of treatment, liver and fecal porphyrins rapidly return to normal levels (Fig. 3) and normal liver histology is found after a 4-month recovery period. Continued treatment leads to the development of hepatocellular carcinoma in some of the animals at 15 months.

The exact mechanism by which a protox inhibitor causes a secondary, reversible inhibition of uroporphyrinogen decarboxylase is at present unclear. The similarity between the porphyrogenic action of fomesafen and the polychlorinated aromatic compounds such as hexachlorobenzene, TCDD or polychlorinated biphenyls suggests a partially similar mechanism of hepatotoxicity, possibly depending on induction of distinct cytochrome P450 isoforms (Sinclair et al. 1997) and formation of reactive oxygen species. It was proposed that this process could be responsible for both uroporphyrin accumulation and hepatocarcinogenesis (Smith and De Matteis 1990).

15.4.2 Developmental Toxicity of Peroxidizing Herbicides

Since the peroxidizing herbicides are in general use, it must be assumed that they display no significant teratogenic effects. However, the diphenyl ether protox inhibitor nitrofen had to be withdrawn from the market when its teratogenic potential was discovered (Francis 1986). In addition, there are data on the teratogenicity of two new inhibitors. The phenylpyrazole SLA 3992 {1-[2,6-dichloro-4-(trifluoromethyl)phenyl]-4-nitro-1H-pyrazol-5-amine}, developed by Bayer AG, was reported to cause severe embryo malformations and a dramatic increase in fetal liver protoporphyrin when administered to rats at a maternal daily dose of 5 mg/kg body weight (Machemer et al. 1992). The effect on maternal liver porphyrins was only slight, and the dose caused no signs of maternal toxicity. A similar substantial increase of whole embryo protoporphyrin content was reported (Kawamura et al. 1996a) for the N-phenylimide flumioxazin. Again, the administered dose (30 mg/kg per day on days 6 through 15 of gestation) resulted only in a moderate increase of maternal liver protoporphyrin, but produced a significant increase in cardiovascular abnormalities (Kawamura et al. 1995). In contrast to SLA 3992, flumioxazin displayed a marked species sensitivity, with only a minimal liver porphyrin accumulation and no developmental toxicity encountered in rabbits. It was proposed that the malformations caused by flumioxazin result from the effect of the herbicide on erythroid heme biosynthesis, as visualised by increased iron deposits in embryonic blood cells (Kawamura et al. 1996b). These data not only demonstrate a direct effect of peroxidizing herbicides on in vivo erythroid heme biosynthesis, but would also suggest that all protox inhibitors capable of crossing the placenta are potential teratogens. The diphenyl ether nitrofen is indeed a potent teratogen and, like flumioxazin, was reported to induce ventricular septal defects in rats (Costlow and Manson 1981). However, the structurally related herbicide bifenox was not teratogenic at a similar dose (Francis 1986). The developmental toxicity of oxyfluorfen and nitrofen was also compared in a study with American kestrel nestlings (Hoffman et al. 1991). In contrast to nitrofen, oxyfluorfen caused no nestling mortality, despite being a more potent inhibitor in experiments with both plant (Matringe et al. 1990) and animal (Birchfield and Casida 1997) cells. Nevertheless, the data published so far indicate that the effects on fetal porphyrin metabolism should be monitored in developmental toxicity studies with these compounds.

15.5 Discussion of Available Data

Most peroxidizing herbicides display relatively low acute toxicity, with LD_{50} values in the order of 1 g/kg. At low doses, the inhibition of hepatic protox and biliary protoporphyrin excretion is a very sensitive indicator of exposure. For example, the NOEL (no observed effect level) for oxadiazon, derived from a 2-year feeding study in mice, is reported as 10 mg/kg in the diet (Tomlin 1994). As can be seen in Table 1, both fecal and biliary porphyrin content are elevated when the same concentration of oxadiazon is fed to mice in a short-term experiment. This suggests a decrease in target enzyme activity even at an exposure level considered to be without significant biological effects, and highlights the usefulness of porphyrin analyses in future toxicological studies.

The toxicological significance of a possible interference with heme biosynthesis in vivo is at present unclear. In humans, a lifelong reduced activity of protox in all tissues is found in patients with variegate porphyria. Although variegate porphyria patients can display porphyrin-related photosensitivity and acute porphyric crises with primarily neurological symptoms, it is estimated that in the majority of patients the inherited metabolic defect never becomes clinically manifest. It is therefore evident that even a prolonged substantial inhibition of protox in all tissues is still compatible with life.

The inhibition of the liver enzyme causes a block in the conversion of protoporphyrinogen to protoporphyrin. Excess protoporphyrinogen is rapidly excreted in the bile and, at lower doses, does not significantly accumulate in plasma. In mice fed oxadiazon-containing diets, the typical variegate porphyria-specific plasma porphyrin spectrum (Poh-Fitzpatrick 1980) develops only at oxadiazon concentrations above 10 mg/kg diet, and skin fluorescence can be observed at dietary concentrations of 50 mg/kg and more. It thus appears that photosensitivity should not pose a problem in individuals exposed to low herbicide doses. The major part of heme is synthesized in the bone marrow, and protox inhibitors clearly could interfere with erythroid heme biosynthesis (Rio et al. 1997). Long-term feeding of high doses (1000 mg/kg diet) of oxadiazon or fomesafen to mice lowered hematocrit and hemoglobin concentration (Krijt et al., unpubl.) but no data on the effect of lower doses are currently available. Although erythroid heme biosynthesis is usually not affected in variegate porphyria, a possible interference of protox inhibitors with hemoglobin synthesis cannot be ruled out.

The most threatening condition in all acute porphyrias is an acute porphyric attack, which is invariably associated with elevated urinary levels of porphyrin precursors 5-aminolevulinic acid and porphobilinogen. In our experiments, increased urinary PBG was found only in mice fed relatively high dietary levels of oxadiazon (100 mg/kg diet and more) and even higher doses of oxyfluorfen (Table 2). It is therefore highly unlikely that a condition resembling acute porphyric attack would result from professional herbicide exposure.

15.6 Conclusion

Available data indicate that some peroxidizing herbicides are potent in vivo protox inhibitors in rodents, and that a significant inhibition of the liver enzyme takes place at doses less than 5 mg/kg body weight (Birchfield and Casida 1996). However, the enzyme inhibition is reversible and porphyrin excretion returns to normal following cessation of exposure. In addition, protoporphyrinogen oxidase has a relatively high activity when compared to other heme biosynthetic enzymes, since even a continuous 90% reduction of activity, as observed in very rare cases of homozygous variegate porphyria, is still compatible with life. Therefore, it appears possible to speculate that a slight inhibition of protox resulting from temporary exposure to low doses of peroxidizing herbicides would be no significant risk. On the other hand, the effect of peroxidizing herbicides on porphyrin metabolism is a very sensitive indicator of exposure and should be monitored in toxicological evaluations of new compounds of this herbicide class.

Acknowledgments. Supported by Charles University Grants 268/93, 167/94 and 31/96.

References

Ashby J, Brady A, Elcombe CR, Elliott BM, Ishmael J, Odum J, Tugwood JD, Kettle S, Purchase IF (1994) Mechanistically-based human hazard assessment of peroxisome proliferator-induced hepatocarcinogenesis. Hum Exp Toxicol 13[Suppl 2]:S1–S117

Birchfield NB, Casida JE (1996) Protoporphyrinogen oxidases: high affinity tetrahydrophthalimide radioligand for the inhibitor/herbicide-binding site in mouse liver mitochondria. Chem Res Toxicol 9:1135–1139

Birchfield NB, Casida JE (1997) Protoporphyrinogen oxidase of mouse and maize: target site selectivity and thiol effects on peroxidizing herbicide action. Pestic Biochem Physiol 57:36–43

Böger P, Wakabayashi K (1995) Peroxidizing herbicides (I): mechanism of action. Z Naturforsch 50c:159–166

Butler EG, Tanaka T, Ichida T, Maruyama H, Leber AP, Williams GM (1988) Induction of hepatic peroxisome proliferation in mice by lactofen, a diphenyl ether herbicide. Toxicol Appl Pharmacol 93:72–80

Camadro JM, Labbe P (1996) Cloning and characterization of the yeast HEM14 gene coding for protoporphyrinogen oxidase, the molecular target of diphenyl ether-type herbicides. J Biol Chem 271:9120–9128

Corrigal AV, Hift RJ, Adams PA, Kirsch RE (1994) Inhibition of mammalian protoporphyrinogen oxidase by acifluorfen. Biochem Mol Biol Int 34:1283–1289

Costlow RD, Manson JM (1981) The heart and diaphragm: target organs in the neonatal death induced by nitrofen (2,4-dichlorophenyl-p-nitrophenyl ether). Toxicology 20:209–227
Dailey TA, Dailey HA (1996) Human protoporphyrinogen oxidase: expression, purification, and characterization of the cloned enzyme. Protein Sci 5:98–105
De Matteis F, Marks GS (1996) Cytochrome P450 and its interactions with the heme biosynthetic pathway. Can J Physiol Pharmacol 74:1–8
Duke SO, Lydon J, Becerril JM, Sherman TD, Lehnen LP, Matsumoto H (1991) Protoporphyrinogen oxidase-inhibiting herbicides. Weed Sci 39:465–473
Francis BM (1986) Teratogenicity of bifenox and nitrofen in rodents. Environ Sci Health B21:303–317
Grisham JW (1997) Interspecies comparison of liver carcinogenesis: implications for cancer risk assessment. Carcinogenesis 18:59–81
Halling BP, Yuhas DA, Fingar VF, Winkelmann JW (1994) Protoporphyrinogen oxidase inhibitors for tumor therapy. In: Duke SA, Rebeiz CA (eds) Porphyric pesticides. Chemistry, toxicology and pharmaceutical applications. ACS Symp Ser, vol 559. Am Chem Soc, Washington, DC, pp 280–290
Hift RJ, Meissner PN, Todd G, Kirby P, Bilsland D, Collins P, Ferguson J, Moore MR (1993) Homozygous variegate porphyria: an evolving clinical syndrome. Postgrad Med J 69:781–786
Hoffman DJ, Spann JW, LeCaptain LJ, Bunck CM, Rattner BA (1991) Developmental toxicity of diphenyl ether herbicides in nestling American kestrels. J Toxicol Environ Health 34:323–336
Jacobs JM, Jacobs NJ (1993) Porphyrin accumulation and export by isolated barley (*Hordeum vulgare*) plastids. Plant Physiol 101:1181–1187
Jacobs JM, Jacobs NJ, Sherman TD, Duke SO (1991) Effect of diphenyl ether herbicides on oxidation of protoporphyrinogen to protoporphyrin in organellar and plasma membrane enriched fractions of barley. Plant Physiol 97:197–203
Jacobs JM, Sinclair PR, Gorman N, Jacobs NJ, Sinclair JF, Bement WJ, Walton H (1992) Effects of diphenyl ether herbicides on porphyrin accumulation by cultured hepatocytes. J Biochem Toxicol 7:87–95
Jacobs JM, Jacobs NJ, Kuhn CB, Gorman N, Dayan FE, Duke SO, Sinclair JF, Sinclair PR (1996) Oxidation of porphyrinogens by horseradish peroxidase and formation of a green pyrrole pigment. Biochem Biophys Res Commun 227:195–199
Kappas A, Sassa S, Galbraith RA, Nordmann Y (1995) The porphyrias. In: Scriver CR, Beandet AL, Sly WS, Valle D (eds) The metabolic and molecular basis of inherited disease. Mc Graw-Hill, New York, pp 2103–2159
Kauppinen R, Mustajoki P (1988) Acute hepatic porphyria and hepatocellular carcinoma. Br J Cancer 57:117–120
Kauppinen R, Mustajoki P (1992) Prognosis of acute porphyria: occurrence of acute attacks, precipitating factors, and associated diseases. Medicine 71:1–13
Kawamura S, Kato T, Matsuo M, Sasaki M, Katsuda Y, Hoberman AM, Yasuda M (1995) Species difference in developmental toxicity of an N-phenylimide herbicide between rats and rabbits and sensitive period of the toxicity to rat embryos. Cong Anom 35:123–132
Kawamura S, Kato T, Matsuo M, Katsuda Y, Yasuda M (1996a) Species difference in protoporphyrin IX accumulation produced by an N-phenylimide herbicide in embryos between rats and rabbits. Toxicol Appl Pharmacol 141:520–525
Kawamura S, Yoshioka T, Kato T, Matsuo M, Yasuda M (1996b) Histological changes in rat embryonic blood cells as a possible mechanism for ventricular septal defects produced by an N-phenylimide herbicide. Teratology 54:237–244
Kordac V, Martasek P, Zeman J, Rubin A (1985) Increased erythrocyte protoporphyrin in homozygous variegate porphyria. Photodermatology 2:257–259
Krijt J, Sanitrak J, Vokurka M, Janousek V (1991) Liver HPLC profiles in experimental porphyria induced by peroxidizing herbicides. Biomed Chromatogr 5:229–230
Krijt J, Pleskot R, Sanitrak J, Janousek V (1992) Experimental hepatic porphyria induced by oxadiazon in male mice and rats. Pestic Biochem Physiol 42:180–187

Krijt J, Van Holsteijn I, Hassing I, Vokurka M, Blaauboer BJ (1993) Effect of diphenyl ether herbicides and oxadiazon on porphyrin biosynthesis in mouse liver, rat primary hepatocyte culture and HepG2 cells. Arch Toxicol 67:255–261

Krijt J, Vokurka M, Sanitrak J, Janousek V, van Holsteijn I, Blaauboer BJ (1994) Effect of the protoporphyrinogen oxidase-inhibiting herbicide fomesafen on liver uroporphyrin and heptacarboxylic porphyrin in two mouse strains. Food Chem Toxicol 1994 32:641–650

Krijt J, Stranska P, Maruna P, Vokurka M, Sanitrak J (1997) Herbicide-induced experimental variegate porphyria in mice: tissue porphyrinogen accumulation and response to porphyrogenic drugs. Can J Physiol Pharmacol 75:1181–1187

Lee HJ, Duke MV, Duke SO (1993) Cellular localization of protoporphyrinogen-oxidizing activities of etiolated barley (*Hordeum vulgare* L.) leaves. Plant Physiol 102:881–889

Machemer L, Schmidt U, Holzum B (1992) Specific and non-specific developmental effects. In: Neubert D, Kavlock RJ, Merker HJ, Klein J (eds) Risk assessment of prenatally-induced adverse health effects. Springer, Berlin Heidelberg New York, pp 85–100

Matringe M, Camadro JM, Labbe P, Scalla R (1989) Protoporphyrinogen oxidase inhibition by three peroxidizing herbicides: oxadiazon, LS-82-556 and M&B 39279. FEBS Lett 245:35–38

Matringe M, Clair D, Scalla R (1990) Effects of peroxidizing herbicides on protoporphyrin IX levels in non-chlorophyllous soybean cell culture. Pestic Biochem Physiol 36:300–307

Meissner P, Adams P, Kirsch R (1993) Allosteric inhibition of human lymphoblast and purified porphobilinogen deaminase by protoporphyrinogen and coproporphyrinogen. A possible mechanism for the acute attack of variegate porphyria. J Clin Invest 91:1436–1444

Moore MR, McColl KEL, Rimington C, Goldberg A (eds) (1987) Disorders of porphyrin metabolism, Plenum Publishing, New York

Nishimura K, Taketani S, Inokuchi H (1995) Cloning of a human cDNA for protoporphyrinogen oxidase by complementation in vivo of a hemG mutant of *Escherichia coli*. J Biol Chem 270:8076–8080

Poh-Fitzpatrick MB (1980) A plasma porphyrin fluorescence marker for variegate porphyria. Arch Dermatol 116:543–547

Ponka P (1997) Tissue specific regulation of iron metabolism and heme synthesis: distinct control mechanisms in erythroid cells. Blood 89:1–25

Prasad AR, Dailey HA (1995) Generation of resistance to the diphenyl ether herbicide acifluorfen by MEL cells. Biochem Biophys Res Commun 215:186–91

Proulx KL, Dailey HA (1992) Characteristics of murine protoporphyrinogen oxidase. Protein Sci 1992 1:801–809

Puy H, Robreau AM, Rosipal R, Nordmann Y, Deybach JC (1996) Protoporphyrinogen oxidase: complete genomic sequence and polymorphisms in the human gene. Biochem Biophys Res Commun 226:226–230

Quest JA, Phang W, Hamernik KL, VanGemert M, Fisher B, Levy R, Farber TM, Burnam WL, Engler R (1989) Evaluation of the carcinogenic potential of pesticides 1. Acifluorfen. Regul Toxicol Pharmacol 10:149–159

Retzlaff K, Böger P (1996) An endoplasmic reticulum plant enzyme has protoporphyrinogen IX oxidase activity. Pestic Biochem Physiol 54: 105–114

Richert L, Price S, Chesne C, Maita K, Carmichael N (1996) Comparison of the induction of hepatic peroxisome proliferation by the herbicide oxadiazon in vivo in rats, mice and dogs and in vitro in rat and human hepatocytes. Toxicol Appl Pharmacol 141:35–43

Rio B, Parent-Massin D, Lautraite S, Hoellinger H (1997) Effects of a diphenyl-ether herbicide, oxyfluorfen, on human BFU-E/CFU-E development and haemoglobin synthesis. Hum Exp Toxicol 16:115–122

Salata H, Cortes JM, Enriquez de Salamanca R, Oliva H, Castro A, Kusak E, Carreno V, Hernandez Guio C (1985) Porphyria cutanea tarda and hepatocellular carcinoma. Frequency of occurrence and related factors. J Hepatol 1:477–487

Sinclair PR, Gorman N, Walton HS, Sinclair JF, Jacobs JM, Jacobs NJ (1994) Protoporphyrinogen accumulation in cultured hepatocytes treated with the diphenyl ether herbicide, acifluorfen. Cell Mol Biol 40:891–897

Sinclair PR, Gorman N,Walton HS, Sinclair JF, Lee CA, Rifkind AB (1997) Identification of CYP1A5 as the CYP1 enzyme mainly responsible for uroporphyrinogen oxidation induced by AH receptor ligands in chicken liver and kidney. Drug Metab Dispos 25:779–783
Smith AG, De Matteis F (1990) Oxidative injury mediated by the hepatic cytochrome P-450 system in conjunction with cellular iron. Effects on the pathway on haem biosynthesis. Xenobiotica 20:865–877
Smith AG, Francis JE, Dinsdale D, Manson MM, Cabral JRP (1985) Hepatocarcinogenicity of hexachlorobenzene in rats and the sex difference in hepatic iron status and development of porphyria. Carcinogenesis 6:631–636
Smith LL, Elcombe CR (1989) Mechanistic studies: their role in the toxicological evaluation of pesticides. Food Addit Contam 6[Suppl 1]:S57–65
Taketani S, Inazawa J, Abe T, Furukawa T, Kohno H, Tokunaga R, Nishimura K, Inokuchi H (1995) The human protoporphyrinogen oxidase gene (PPOX): organization and location to chromosome 1. Genomics 29:698–703
Tokola O, Mustajoki P, Himberg JJ (1988) Haem arginate improves hepatic oxidative metabolism in variegate porphyria. Br J Clin Pharmacol 26:753–757
Tomlin C (ed) (1994) The pesticide manual, 10th edn. BCPC Publications, Farnham
Von Burg R (1994) Toxicology update - oxadiazon. J Appl Toxicol 14:69–71
Witkowski DA, Halling BP (1989) Inhibition of plant protoporphyrinogen oxidase by the herbicide acifluorfen-methyl. Plant Physiol 90:1239–1242

Subject Index